# INTERNATIONAL HANDBOOK OF ENERGY SECURITY

# International Handbook of Energy Security

*Edited by*

Hugh Dyer
*University of Leeds, UK*

Maria Julia Trombetta
*Delft University of Technology, The Netherlands*

**Edward Elgar**
Cheltenham, UK • Northampton, MA, USA

© Hugh Dyer and Maria Julia Trombetta 2013

All rights reserved. No part of this publication may be reproduced, stored in a retrieval system or transmitted in any form or by any means, electronic, mechanical or photocopying, recording, or otherwise without the prior permission of the publisher.

Published by
Edward Elgar Publishing Limited
The Lypiatts
15 Lansdown Road
Cheltenham
Glos GL50 2JA
UK

Edward Elgar Publishing, Inc.
William Pratt House
9 Dewey Court
Northampton
Massachusetts 01060
USA

A catalogue record for this book
is available from the British Library

Library of Congress Control Number: 2013932959

This book is available electronically in the ElgarOnline.com
Economics Subject Collection, E-ISBN 978 1 78100 790 7

ISBN 978 1 78100 789 1

Typeset by Servis Filmsetting Ltd, Stockport, Cheshire
Printed and bound in Great Britain by T.J. International Ltd, Padstow

# Contents

*List of figures* viii
*List of tables* x
*List of contributors* xi
*Acknowledgements* xiii

### PART I  INTRODUCTION

1 The concept of energy security: broadening, deepening, transforming 3
  *Hugh Dyer and Maria Julia Trombetta*

### PART II  ENERGY SECURITY ISSUES

2 Energy security and liberal democracy: ideals, imperatives and balancing acts 19
  *Steve Wood*
3 Framing new threats: the internal security of gas and electricity networks in the European Union 40
  *Peter Zeniewski, Carlo Brancucci Martínez-Anido and Ivan L.G. Pearson*
4 Resource conflicts: energy worth fighting for? 70
  *Joshua Olaniyi Alabi*
5 Global energy supply: scale, perception and the return to geopolitics 92
  *Susanne Peters and Kirsten Westphal*

### PART III  SECURITY OF ENERGY SUPPLY

6 Securing energy supply: strategic reserves 117
  *Elspeth Thomson and Augustin Boey*
7 Securing energy supply II: diversification of energy sources and carriers 133
  *Kas Hemmes*
8 Energy security assessment framework and three case studies 146
  *Aleh Cherp and Jessica Jewell*

vi  *International handbook of energy security*

| | | |
|---|---|---|
| 9 | National energy strategies of major industrialized countries<br>*Stephan Schott and Graham Campbell* | 174 |
| 10 | Developing world: national energy strategies<br>*Sylvia Gaylord and Kathleen J. Hancock* | 206 |

## PART IV  SECURITY OF ENERGY DEMAND

| | | |
|---|---|---|
| 11 | Energy demand: security for suppliers?<br>*Tatiana Romanova* | 239 |
| 12 | Oil producers' perspectives on energy security<br>*Gawdat Bahgat* | 258 |
| 13 | Energy security governance in light of the Energy Charter process<br>*Andrei V. Belyi* | 273 |

## PART V  ENERGY, THE ENVIRONMENT AND SECURITY

| | | |
|---|---|---|
| 14 | Governance dimensions of climate and energy security<br>*John Vogler and Hannes R. Stephan* | 297 |
| 15 | Energy, climate change and conflict: securitization of migration, mitigation and geoengineering<br>*Jürgen Scheffran* | 319 |
| 16 | Environmental implications of energy production<br>*Yolanda Lechón, Natalia Caldés and Pedro Linares* | 345 |
| 17 | Washing away energy security: the vulnerability of energy infrastructure to environmental change<br>*Cleo Paskal* | 386 |
| 18 | Paradoxes and harmony in the energy-climate governance nexus<br>*Stéphane La Branche* | 402 |

## PART VI  ENERGY AND HUMAN SECURITY

| | | |
|---|---|---|
| 19 | Energy poverty: access, health and welfare<br>*Subhes C. Bhattacharyya* | 423 |
| 20 | Ethical dimensions of renewable energy<br>*Hugh Dyer* | 443 |
| 21 | Low carbon development and energy security in Africa<br>*Chukwumerije Okereke and Tariya Yusuf* | 462 |

| | | |
|---|---|---|
| 22 | The road not taken, round II: centralized vs. distributed energy strategies and human security<br>*Ronnie D. Lipschutz and Dustin Mulvaney* | 483 |
| 23 | Human security and energy security: a sustainable energy system as a public good<br>*Sylvia I. Karlsson-Vinkhuyzen and Nigel Jollands* | 507 |

PART VII   CONCLUSIONS

| | | |
|---|---|---|
| 24 | The political economy of energy security<br>*Hugh Dyer and Maria Julia Trombetta* | 529 |

*Index* 539

# Figures

| | | |
|---|---|---|
| 3.1 | Generation adequacy assessment from ENTSO-E SO-AF 2012–2030 | 50 |
| 3.2 | Projects of pan-European significance: volume breakdown per EU energy policy pillar | 53 |
| 7.1 | Distribution of energy sources for the production of electricity in 2011 for the Netherlands | 134 |
| 7.2 | Supply chain in an energy system | 135 |
| 7.3 | Simple form of multi-input systems combining fossil energy with renewable energy by blending | 139 |
| 7.4 | Classification of energy systems | 139 |
| 7.5 | Schematic illustration of a multisource multiproduct energy system (or energy hub) | 140 |
| 7.6 | Schematic representation of the Superwind concept | 142 |
| 8.1 | Major methodological choices in measuring energy security and the energy systems approach | 148 |
| 8.2 | Vital energy systems covered in the GEA | 153 |
| 8.3 | Aggregate analysis of energy security in future energy scenarios using a two-dimensional plotting | 165 |
| 8.4 | Energy security assessment framework | 168 |
| 9.1 | Russia's primary energy consumption by fuel as a percentage of total consumption | 188 |
| 9.2 | Energy mix outlook for the EU 2030 and 2050 | 196 |
| 9.3 | U.S. primary energy consumption by fuel (quadrillion btu per year) | 197 |
| 16.1 | Airborne pollutant emissions from fossil fuel fired power plants | 348 |
| 16.2 | Lifecycle GHG emissions of electricity generation technologies disaggregated by life cycle stage | 371 |
| 16.3 | Lifecycle NOx and $SO_2$ emissions of electricity generation technologies | 372 |
| 16.4 | Lifecycle particulate and NMVOC emissions of electricity generation technologies | 373 |
| 16.5 | Lifecycle GHG emissions of bioethanol production from different raw materials | 375 |
| 16.6 | Lifecycle GHG emissions of biodiesel production from different raw materials | 376 |

| | | |
|---|---|---|
| 16.7 | External costs of electricity generation technologies disaggregated by impact category | 377 |
| 16.8 | External costs of electricity generation technologies disaggregated by life cycle stage | 378 |
| 18.1 | Climate governance versus energy governance | 414 |
| 18.2 | Climate/energy governance versus social acceptability | 415 |
| 19.1 | Major concentration of population without access to electricity in 2009 | 429 |
| 19.2 | Urban-rural electricity access disparity in major concentrations in 2009 | 430 |
| 19.3 | Ten least electrified countries in the world in 2009 | 430 |
| 19.4 | Distribution of lack of cooking energy access in the world in 2009 | 431 |
| 19.5 | Share of different cooking fuels in developing countries in 2007 | 432 |
| 19.6 | Regional distribution of countries with targets for electricity and clean cooking energy access | 434 |
| 19.7 | Energy access improves with per capita income | 435 |
| 19.8 | HDI and electricity access | 436 |
| 19.9 | HDI and cooking energy access | 436 |
| 19.10 | Life expectancy at birth and cooking energy access | 437 |
| 19.11 | Mean schooling years against electricity access | 437 |
| 19.12 | Intervention options for reducing health impacts of solid fuel use | 440 |
| 21.1 | Number of people without electricity (actual and projected) by region under current policies | 465 |

# Tables

| | | |
|---|---|---|
| 4.1 | Civil wars, rebellion and militancy in selected oil and gas dependent states | 74 |
| 7.1 | Main security of supply issues on different length scales and time scales | 137 |
| 8.1 | Three perspectives on energy security | 156 |
| 8.2 | Indicators of crude oil supply security used in MOSES | 159 |
| 8.3 | Indicators used in the GEA energy security assessment | 160 |
| 8.4 | Vulnerabilities of future energy systems and related indicators | 161 |
| 8.5 | Ranges of indicators for crude oil supply in MOSES | 162 |
| 8.6 | Aggregating indicators for external resilience of crude oil supply in MOSES | 164 |
| 8.7 | Results of the crude oil analysis for MOSES | 166 |
| 8.8 | Cases of energy security assessments presented in this chapter | 170 |
| 9.1 | Export-import dependence of selected countries | 177 |
| 9.2 | Characteristics of national energy strategies in countries with high export and balanced export-import dependence | 178 |
| 9.3 | Characteristics of national energy strategies in countries with high import dependence | 192 |
| 9.4 | Dramatic supply shifts in Japan by 2030 | 200 |
| 11.1 | Key oil and gas exporting countries | 240 |
| 16.1 | Impact categories, pollutants and effects considered in the ExternE methodology | 369 |
| 19.1 | Factor goalposts for 2010 EDI | 425 |
| 19.2 | Example of EDI for India | 426 |
| 19.3 | Commonly used national and international indicators of energy poverty | 426 |
| 19.4 | Level of electrification in various regions in 2009 | 428 |
| 19.5 | Reliance on biomass for cooking energy needs in 2009 | 431 |
| 19.6 | Expected number of people without electricity access in 2030 | 433 |
| 19.7 | Outlook for biomass use for cooking in 2030 (million) | 433 |

# Contributors

**Joshua Olaniyi Alabi**, University of Leeds, UK

**Gawdat Bahgat**, National Defense University, Washington, USA

**Andrei V. Belyi**, University of Tartu, Estonia

**Subhes C. Bhattacharyya**, De Montfort University, UK

**Augustin Boey**, National University of Singapore, Singapore

**Carlo Brancucci Martínez-Anido**, European Commission Joint Research Centre, Belgium

**Natalia Caldés**, CIEMAT, Madrid, Spain

**Graham Campbell**, Carleton University, Canada

**Aleh Cherp**, Central European University, Hungary, and Lund University, Sweden

**Hugh Dyer**, University of Leeds, UK

**Sylvia Gaylord**, Colorado School of Mines, USA

**Kathleen J. Hancock**, Colorado School of Mines, USA

**Kas Hemmes**, Delft University of Technology, The Netherlands

**Jessica Jewell**, International Institute for Applied Systems Analysis, Austria, and Central European University, Hungary

**Nigel Jollands**, European Bank for Reconstruction and Development, London, UK

**Sylvia I. Karlsson-Vinkhuyzen**, Wageningen University, The Netherlands

**Stéphane La Branche**, PACTE, Institute of Political Studies, Grenoble, France

**Yolanda Lechón**, CIEMAT, Madrid, Spain

**Pedro Linares**, Universidad Pontificia Comillas, Spain

**Ronnie D. Lipschutz**, University of California, Santa Cruz, USA

**Dustin Mulvaney**, San Jose State University, USA

**Chukwumerije Okereke**, University of Reading, UK

**Cleo Paskal**, Royal Institute of International Affairs, UK

**Ivan L.G. Pearson**, European Commission Joint Research Centre, Belgium

**Susanne Peters**, Kent State University, Geneva, Switzerland

**Tatiana Romanova**, St. Petersburg State University, Russia

**Jürgen Scheffran**, University of Hamburg, Germany

**Stephan Schott**, Carleton University, Canada

**Hannes R. Stephan**, University of Stirling, Scotland, UK

**Elspeth Thomson**, National University of Singapore, Singapore

**Maria Julia Trombetta**, Delft University of Technology, The Netherlands

**John Vogler**, Keele University, UK

**Kirsten Westphal**, German Institute for International and Security Affairs (SWP), Berlin, Germany

**Steve Wood**, Macquarie University, Australia

**Tariya Yusuf**, University of Reading, UK

**Peter Zeniewski**, European Commission Joint Research Centre, Belgium

# Acknowledgements

The editors would like to thank Alexandra O'Connell and all the Edward Elgar Publishing team, in particular Rebecca Wise, Jane Bayliss and Jennifer Wilcox for their efficiency and devotion to this publication.

Many thanks are owed to the contributors, who agreed readily to the request for contributions and dealt with editorial requests in a professional and timely manner.

Julia Trombetta would like to thank the Energy Delta Gas Research (EDGaR) Programme for a research grant that allowed her to work on energy security in Europe, and the section "Economics of Infrastructures" (Faculty of Technology Policy and Management) of the Delft University of Technology for constant support.

# PART I

# INTRODUCTION

# 1. The concept of energy security: broadening, deepening, transforming
*Hugh Dyer and Maria Julia Trombetta*

Over the last decade the term energy security has gained relevance in political and academic debates. Framing energy as a security issue is not a new phenomenon. The last great debate on energy security dated back to the 1970s (Schultz 1973; Nye 1982) and it was prompted by the crises that followed the cut of oil supply by OPEC countries in 1973. As the price of oil quadrupled, triggering an economic crisis, the vulnerability of the energy system was fully exposed (Cherp and Jewell 2011, 203). Over the subsequent two decades, however, energy was considered more an economic than a security issue – at least in the main political and academic debates. A global liquid market (for oil) and relatively low fossil fuel prices have prioritized economic aspects, often ignoring the premises on which the energy market has developed, and in particular how its working is shaped by security considerations whose appropriateness is now being questioned.

Several issues are behind the questioning and the renewed concerns and quests for energy security. Tight oil markets and volatile prices have created concern for an affordable and secure supply of energy. Several disruptions in gas supply determined by disputes between Russia and various Eastern European countries have evoked the spectre of an energy weapon and questioned the reliability of Russia. Similarly, the reliance on potentially, or historically 'unstable' areas of the world for oil supplies raises concern for energy independence in Western countries. Besides, the transformation of energy into an issue of high politics was speeded up by the realization that increases in oil prices since 2002 have not only been the result of passing crises – like the war in Iraq or strikes in Venezuela – structural changes in the global energy market were largely to blame (Westphal 2006, 48). Growing demands from emerging economies have changed the energy landscape and prompted traditional geopolitical arguments and questioned the limited mechanisms of global energy governance. Environmental concern and raising consumption have questioned the sustainability of the existing energy system. The global security landscape has changed as well, with growing concerns about threats from non-state actors. In the case of energy systems they include piracy and terrorist attacks, including cyber ones.

Energy is fundamental for the working of modern societies and that poses political questions about access to energy services, and yet those energy services are provided by complex technical systems and global markets. The contemporary energy system is largely based on dwindling fossil fuels in distant places; it involves different actors and requires long term planning, as well as short term action; it requires different kinds of expertise often with competing logics and perspectives. Not surprisingly energy security is understood in different ways in different contexts (Chester 2010). While the term security is largely evocative of state's intervention, energy security is also about the smooth function of liquid markets and safe and reliable energy infrastructures (Cherp and Jewell 2011). Energy security is both a strategic issue and public policy issue (Goldthau 2012), even if the two perspectives are based on different assumptions and different instruments to provide security. While a better understanding of the different perspectives and a broader and integrated approach is called for, an all-encompassing concept of energy security would lose relevance (Ciuta 2010). Yet, the existing energy system is based on several assumptions about who is supposed to be secured, by whom and by what means. These are the issues that are addressed by (critical) security studies through conceptual and critical analyses.

However, despite the burgeoning literature and growing interest in energy security, there is little from a (critical) security studies perspective (Ciuta 2010: Cherp and Jewell 2011). On the one hand, energy security is left to economists and engineers; on the other hand, there are a few warnings against the transformation of energy into a security issue. The first has led to various attempts to apply public policy tools to deal with market failures or to measure energy insecurity, quantifying risks without questioning who or what is supposed to be secured and how. The second is related to a specific understanding of security as allowing exceptional measures, including the use of force and military intervention, which is associated with the state and *raison d'état*. This resonates with a zero sum, antagonistic understanding of security, which characterizes realism and creates concern for NATO's involvement in energy disputes, or for the militarization of energy (Moran and Russell 2009).

This division is reflected in the identification of two storylines for the development of the global energy landscape by Correljé and van der Linde (2006). The first one, based on 'regions and empires', describes a world divided in rival blocs competing for resources. The second one, 'markets and institutions', identifies a multilateral world with effective institutions and markets (Correljé and van der Linde 2006, 532). This reflects, as Keating et al. (2012) have pointed out, the divide between realist/geopolitical approaches and neoliberal ones, where the geopolitical ones focus on the state and on access to resources (often ignoring the security concerns of producer countries, emphasizing security of supply over security of demand) while neoliberal approaches tend to focus on market mechanisms and identify energy security with free markets (often ignoring that liberalization can be problematic for energy security as well). In this way the debate tends to perpetuate a rather simplistic division between politics and economics (Keating et al. 2012, 3).

The problem is further complicated when environmental considerations are brought into the discussion. Climate change has been considered as a post-normal scientific issue, where the term post-normal science has been introduced to identify a situation in which 'facts are uncertain, values in dispute, stakes high and decisions urgent' (Ravetz, quoted in Friedrichs 2011). This resonates with the debate about energy security, and debates about conceptualizations of security aspire to make these issues explicit. However, as Friedrichs notes, while environmental debate has engaged with post-normal science, much of the discussion about future energy is shaped by an economic approach which implies that future energy demand will be met by market mechanisms, and technological fixes will always be available (Friedrichs 2011, 469). Against this challenging conceptual background, this book is an attempt to bring together energy security experts to explore the implications of framing the energy debate in security terms, both in respect of the governance of energy systems and the practices associated with (energy) security. It reviews and analyses the key aspects and research issues in the emerging field of energy security. It tests the current state of knowledge and provides suggestions for reflection and further analysis. This involves providing an account of the multiplicity of discourses and meanings of energy security and contextualizing them. It further requires using the insights from security studies debates to understand the implications of framing an issue as a security issue. Finally, it means outlining the challenges and peculiarities of framing energy as a security issue. This might suggest a rewriting of the evolution of security discourses and their representation, especially by outlining the links with economics and their implications.

As regards the first point, several commentators have noticed the polysemic nature of energy security, and how the term means different things in different contexts. In different ways, commentators have identified how different disciplinary backgrounds shaped energy security discourses identifying geopolitical, economic and technical perspectives (Keppler 2007; Cherp and Jewell 2011). The renewed interest in energy security challenges that intellectual division of labour, and calls for integration and dialogue; however, different disciplines have different epistemological perspectives, different understandings of what counts as security, and how to provide it. In providing a review of the relevant aspects of the energy security debate, it is important to explore how different logics of security interact and shape different energy security discourses, challenging disciplinary boundaries and security practices.

The tools for understanding these transformations can be provided by critical security studies, and this points to the second aspect characterizing this handbook. Largely grounded in the discipline of International Relations which considers security as a key concept to explain interaction among states and global politics, the debate about rethinking security to make it relevant to understand contemporary dynamics has raised fundamental theoretical and practical issues whose relevance goes beyond the debate within the discipline. Since the 1980s the field of security studies has increasingly challenged a narrow understanding of security as dealing with the threat and use of military force in inter-states relations (Walt 1991), which has led to a broadening and a deepening in the understanding of security (Rothschild 1995; Haftendorn 1991; Krause and Williams 1996). Vaughan-Williams and Peoples (2012), for example, organize their four-volume collection, which provides an overview of the field of critical security studies, around interrelated themes of defining, deepening, broadening and extending security. Broadening refers to the expansion of the threats and issues to be considered. Deepening invokes other entities than the state as the referents for security, like individuals or political communities. The two dimensions are correlated since new threats calls for new tools and new actors to counteract them. In the energy security debate this reflects a shift from considering energy as a cause or an instrument of war, and dealing with threats posed by external actors, mainly states, being unreliable exporters or competitors for scarce resources, to a broader approach that considers different kinds of threats and security provisions. Suggesting that the state is no longer the focal point of security analysis opens up the space for a variety of threats at different levels, from local to global. The term energy security is itself a broadening of the security agenda. However, once a narrow definition of security is challenged the number of issues that can be included in the agenda becomes infinite, as well as the entities that can call for protection. The issue then becomes how to discern legitimate threats and, in turn, how and to what extent states are supposed to provide security in a globalized environment. In the critical security studies debate, these issues are related with other two aspects: the constructions of threats (vs. their allegedly 'objective' discovery) and the different logics of security (vs. the statist, exceptionalist one of national security).

Arguing for a broader security agenda, constructivist approaches have suggested that identifying threats is not a straightforward, objective process. Threats are not objective entities waiting to be discovered and counteracted, to a large extent they are the result of a process of social construction that depicts an issue as dangerous and calls for action. Considerations of whom or what is valuable and deserves to be protected play a role in framing an issue as a threat and priority. So for instance, in the case of immigration, foreigners can be represented either as a resource or as a threat. The identification of threats is more straightforward in the energy sector, however, a lot of social construction is at work in representing Russia as a threat to European energy security or in considering a cut in supply as more threatening than a blackout due to lack of investment in infrastructures. Securitization theory (Wæver 1995; Buzan et al. 1998), which focuses on the process of social construction of a security issue (through a successful speech act that transforms the way of dealing with an issue), has been applied to show how appeals to energy security have transformed energy policy.

Considering new threats and new referents is not the only challenge, as different threats and contexts are characterized by different security practices and logics. Securitization theory, for instance, is based on a specific logic of security which inscribes enemies in a context, and allows exceptional measures and the breaking of rules. It is a logic borrowed from realism, which is evident in some of the geopolitical and resource-war arguments that have re-emerged in the energy security debate (Correljé and van der Linde 2006; Klare 2008). This logic is at odds with the risk-management, cost-benefit analysis that characterizes energy policy. So for instance the mechanisms to ensure energy security after the oil crises in the 1970s have been identified with differentiation of supply, liberalization of the energy system or mechanism to deal with crises like the strategic oil reserves. Other approaches, however, have considered different logics of security like those based on risk (Rasmussen 2001; Aradau et al 2008), insurance mechanisms and other kinds of security apparatus (Dillon and Lobo-Guerrero 2008). Some of them suggest an approach to security which is based on maintaining circulation of goods and services around normal parameters (Dillon 2008). Emergencies can occur, and to some extent they are necessary to maintain the incentives that allow the system to work, as far as they do not challenge the survival of the system. Part of the challenge is to consider how the different logics interact, gain relevance and acquire legitimacy.

Much concern about broadening the security agenda has been about transferring a specific logic of security to sectors other than the military (Buzan and Wæver 1998; Huysmans 2006); little has been said about transferring other logics to the security field, such as the precautionary approach that characterized the environmental debate, or the emphasis on resilience (Trombetta 2010). In this way exploring the peculiarities of the energy sector can provide useful insights into the transformation of security practices and provisions and shed light, for instance, on the changing role of the state as security provider.

This volume emphasizes the broadening and transforming of energy security perspectives. The energy security agenda is broadening, and the first extension of this agenda is to consider not only security of supply, but also security of demand. A second extension involves considering the environmental dimension (as illustrated in Parts III, IV and V). This environmental discourse in particular reflects the limitations of the conventional economic approach to ensuring energy security, where the economic logic is concerned with matching supply and demand, and this is challenged by the issues posed by environmental degradation and climate change. The conventional response to these issues is to incorporate them within the economic logic, where they are inevitably seen as externalities (if perhaps giving rise to processes of ecological modernization (Hajer 1995)). The challenge of energy security (and environmental security) is deeper, and poses questions about whose security and by what means it may be achieved. In this respect, the insights from security studies are relevant, as are the debates about energy and human security, as they question and challenge existing practices. This points to the issues raised in Part VI, where Karlsson and Jollands' concept of 'deep energy security' reflects the aspiration of the volume to show how the concept of energy security has broadened, deepened and transformed.

## STRUCTURE

In order to review the existing debates, while dealing with the range of issues mentioned above, the handbook is structured in seven parts including this introduction, a conclusion and five substantive sections dealing with different perspectives on energy security. This introduction analyses the implications of conceptualizing energy as a security issue and frames energy security within the contemporary debate within security studies. It considers the theoretical issues as well as the historical evolution of what is understood as energy security and the ways of providing it, and extends to considerations of geopolitics, economics and technology perspectives and governance. Thus it provides a framework for analysis based on broadening, deepening and transforming the meaning of energy security and the practices associated with it, the competing conceptual and temporal framework that different disciplines bring to the debate, and the origin of the existing energy security order and assumptions underpinning it.

The remaining parts explore how existing discourses are broadening, deepening and transforming what is understood as energy security.

Parts II to VI deal with emerging energy security issues, security of supply, security of demand and environmental security and human security respectively. For each the key aspects and research issues are reviewed. This is an attempt to make sense of the expansion of the energy security debate while taking into account the multiplicity of dimensions involved. This is certainly the result of a process of interpretation of existing trends and literature, nevertheless it aims to explore existing discourses rather than suggesting what energy security necessarily entails, with a view to normative as well as practical implications.

Part II sets energy security in the context of wider challenges, such as democratic representation, complex interdependence, potential for conflict, and the geopolitical context. Wood suggests that presumed energy imperatives test the quality and integrity of liberal democracy in its external and internal dimensions, where there is a tension between principle and pragmatism. Brancucci, Pearson and Zeniewski present the EU as a case of challenges to network stability arising from developments in both economic and physical infrastructures, where the criteria selected for defining energy security often inform policy. Alabi uses the case of Nigeria to indicate the significant historical role oil has had both in fuelling modern industrial economic development and in domestic and international conflicts, as powerful actors adopt various strategies to influence the direction of supply and price of oil in their own interest, with implications for the modern energy security agenda. Peters and Westphal examine the geopolitical and theoretical bases for a renewed emphasis on energy security, with a changing landscape of energy and the central role of prices presenting challenges for both states and markets.

Part III provides a review and assessment of the security of supply debate. Security of supply has been the dominant concern in the energy security debate, which has often considered the two as being synonymous. To a large extent concerns over security of supply are embedded in the realist understanding of security and the role of the state in providing it; however, what counts as security of supply is changing. Not only have new threats gained relevance (and with them new actors and security practices), but also security of supply is transforming into security of energy services (Faas et al 2011). This part analyses security mechanisms of diversification and strategic reserves, reviews approaches to measuring energy security, and explores different national energy strategies. Thomson and Boey indicate the buffering role of energy reserves, and explore the tension between genuine security and mere stability, given that supply or price shocks are harmful to both developed and developing economies. Cherp and Jewel outline a systematic framework for measuring energy security, with definitional issues presented in historical context, in order to test the prospects for a common measure to help deal with the complex interconnectedness of energy security. Schott and Campbell examine the structure of national energy strategies among the G8 countries and Norway, and evaluate the compatibility of different strategies with an emerging world energy order. Gaylord and Hancock note that energy strategies are particularly relevant to the developing world, with limited finance and other capacities informing the primary concerns of increasing electricity access and long-term economic development.

Part IV deals with security of demand. This is an issue that has been raised by producer countries, concerned with price volatility and investments. This suggests that expanding the energy security agenda is not only about broadening the list of potential threats but also about including different dynamics and referents. Concerns for security of demand suggests interest in a less antagonist logic of security, implying a shift from a zero sum approach in which 'my security is your insecurity' to a more cooperative one in which one's security depends on the security of others. This, in turn, calls for different security practices and provisions. This part covers issues such as the perspectives of different producer countries, and institutional arrangements producing countries' political objectives, where interests and concerns are expressed by both individual producing countries and by organizations that bring them together, given the interconnectedness of supply security and demand security. Bahgat gives the International Energy Forum as an example of cooperation between producers and consumers, which is necessitated by a global mismatch between consumption and production which has made energy products the world's largest traded commodities. Belyi assesses opportunities for energy security governance in the context of the Energy Charter Treaty, given that in a state-centered view energy security is a factor in the unpredictability of international relations.

Part V deals with the environmental dimension of the energy security debate. The growing demand for energy and reliance on fossil fuels make the existing energy system unsustainable. The awareness of global warming and its implications have gained relevance and are a challenge to energy policies. In the EU, for instance, concerns for security of supply and for global warming have jointly contributed to the development of a common energy policy. However, environmental security and energy security – while strictly related and often invoked jointly to promote action on either the environmental or the energy side – can be at odds. Also, while global warming has become the main environmental concern in the energy debate, contemporary energy systems are responsible for a variety of environmental impacts, such as acid rain, and soil and water contamination. A secure and sustainable energy system cannot ignore these dimensions and several issues have begun to emerge. This part includes contributions on the cross-pollination and compatibility of environmental and energy security discourses, and on the environmental impacts of the energy sector and their security implications. Vogler and Stephan test the governance challenges arising from an expansive re-definition of security in which energy and climate are not only materially intertwined, but also interdependent politically. Scheffran notes that energy security depends on its geographical and geopolitical context, as the components of the energy mix shape the global conflict landscape, and climate change issues shape securitization discourses. Lechon, Caldés and Linares examine the environmental implications of energy technology investment and operational decisions, and discuss methods to quantify environmental implications and externalities of energy technologies. Paskal discusses the vulnerability of energy infrastructure to environmental change, given that much energy infrastructure lies in areas that are becoming increasingly physically unstable. La Branche explores the paradoxes and harmonies in the energy-climate nexus, given that climate change is seen as an accelerator and amplifier of natural risks and the single most important obstacle to development efforts in the third world.

Part VI on human security shifts the focus from the security of the state (if still instrumental) to that of individuals and the protection of their basic rights. It suggests a broader, non-zero-sum understanding of security which includes different dimensions, such as income security, food security, health security, environmental security, community/identity security and security of political freedom as specified by the UNDP's report first using the term (UNDP 1994) (Hampson et al. 2002). Energy security is not mentioned in the UNDP's list, however energy has a pervasive role in contemporary society: energy is 'not just another commodity, but the precondition of all commodities, a basic factor equal with air, water, and earth' (Schumacher, quoted in Sovacool and Mukherjee 2011, 5343). Ensuring access to energy services is fundamental to human security. And yet, 'two billion human beings –almost one-third of the planet population – experience evening light by candle, oil lamp, or open fire, reminding us that energy modernization has left intact – and sometimes exacerbated – social inequalities that its architects promised would be banished' (Byrne and Toly 2007: 1; Smil 2003: 370–73). If security is emancipation (Booth 1991) ensuring access to energy services can certainly contribute to the fulfilment of human potential, but decarbonizing the energy system while assuring universal access to modern forms of energy represents a paramount challenge with implications for energy security (AGECC 2010). Securing access to energy services, however, is not only an issue for developing countries: the challenge of sustainable energy, and transition to a low carbon economy and decentralized energy systems, poses questions about priorities and reliability of services. Are energy services a public good? Who is supposed to ensure supply in a decentralized system where consumers are also producers? This part provides contributions on energy as human security, on the policy and ethical implications of shifting security considerations from the state to individuals (e.g. is EU energy policy shifting from national to human security?), on the debate about energy services as public goods, and on the role of the UN in promoting access to energy (and energy governance) to increase human security. Bhattacharyya presents the energy poverty debate and discusses the link between energy poverty and health and welfare, mainly in the developing countries where the implications of lack of access to energy for health and social welfare link energy poverty with human development and economic development. Dyer assumes that justice and equity must underwrite the feasibility of any energy strategies, requiring an ethical framework for energy which addresses the lack of human security in allocations of limited resources, in which an ecological ethic more directly reflects the limits we face and renewable energy offers some promise. Okereke and Yusuf observe that energy poverty is a critical development challenge for Africa, where most do not have access to electricity but depend on biomass for their basic energy needs, and suggest that a low carbon development path offers the prospect of achieving energy security while contributing to the global effort to fight climate change. Lipschutz and Mulvaney note that socio-technical systems, such as those that make possible electrification, have social tendencies independent of human intention with differing technological paths having different kinds of social impacts and, by extension, effects on human well-being and security. In the context of rising risks and impacts of continued reliance on fossil fuels and nuclear energy they provide a comparative assessment of the impacts of centralized and decentralized energy sources and strategies. Karlsson and Jollands note the entrapment of energy policymaking within the paradigm of national security and suggest that linking the sustainability of the energy system with the human security paradigm can address this constraint, and offer the idea of 'deep energy security' to introduce an ethical basis for sustainability.

The editors' conclusion then seeks to capture the range of contributions made in this volume to understanding the emerging political economy of energy security.

# CONCLUSION

The various dimensions of a larger security agenda are taken as a framework to explore the problems and implications involved in the contemporary energy security debate. This book aims to capture the main dimensions of the debate, while providing an analytical framework to explore how different security practices and logics emerge, are legitimized, or are challenged and resisted. Winzer (2011) noted the obvious point that the conceptual framing of research has considerable impact on results in attempting to quantify energy security; in order to distinguish between security, sustainability and economic efficiency (and avoid securitization), he proposes the term 'energy supply continuity' in respect of commodities, services and economic activity, while acknowledging that other energy security concerns will remain. This volume will show that energy security extends far beyond security of supply, and cannot easily be segregated from broader policy concerns. A broadening security agenda is related to its deepening, and can impact on security practices and logics. The broadening of the energy security discourse along the lines of security of supply, security of demand and environmental and human dimensions, indicates the significance of framing energy as an economic issue in current political debates. The interest in security of supply and security of demand suggests an attempt to provide equilibrium in the market, emphasizing the relevance of interdependence and of neoliberal approaches to security provisions (see Dillon 2005). Indeed the larger challenge which energy security concerns reflect may itself be subject to conceptual development: Kanninen's (2012) historical examination of the 'limits to growth' suggests a paradigm shift from 'sustainability' to 'survivability' as global spatial constraints are joined by temporal ones, such as climate change. From the neoliberal perspective, for example, the inclusion of environmental aspects suggests a concern for eliminating externalities. However, the challenge of energy security in both theory and practice is far greater than could be met by a simple adjustment to current theory and practice.

# BIBLIOGRAPHY

AGECC (2010). 'Energy for a Sustainable Future. Summary Report and Recommendation.' New York, UN Secretary-General's Advisory Group on Energy and Climate Change.

Aradau, C., Luis Lobo-Guerrero and Rens Van Munster (2008). 'Security, Technologies of Risk, and the Political: Guest Editors' Introduction.' *Security Dialogue* **39**(2–3): 147–154.

Booth, K. (1991). 'Security and Emancipation.' *Review of International Studies* **17**(4).

Buzan, B. and O. Wæver (1998). *Liberalism and Security: The Contradiction of the Liberal Leviathan*, Copenhagen, COPRI.

Buzan, B., O. Wæver and J. de Wilde (1998). *Security: A New Framework for Analysis*, Boulder, Colorado, Lynne Rienner.
Byrne, J. and N. Toly (2007). 'Energy as a Social Project: Recovering the Discourse.' In J. Byrne, N. Toly and L. Glover, *Transforming Power. Energy, Environment, and Society in Conflict*, New Brunswick (USA) and London, Transaction Publishers.
Cherp, A. and J. Jewell (2011). 'The three perspectives on energy security: intellectual history, disciplinary roots and the potential for integration.' *Current Opinion in Environmental Sustainability* **3**(4): 202–212.
Chester, L. (2010). 'Conceptualising energy security and making explicit its polysemic nature.' *Energy Policy* **38**(2): 887–895.
Ciuta, F. (2010). 'Conceptual Notes on Energy Security: Total or Banal Security?' *Security Dialogue* **41**(2): 123–144.
Correljé, A. and C. van der Linde (2006). 'Energy supply security and geopolitics: A European perspective.' *Energy Policy* **34**(5): 532–543.
Dillon, M.G. (2005). 'Global security in the 21st century: circulation, complexity and contingency.' In 'The globalization of security', ISP/NSC Briefing Paper 05/02, Chatham House, pp. 2–3.
Dillon, M. (2008). 'Underwriting security.' *Security Dialogue* **39**(2–3): 309–332.
Dillon, M. and L. Lobo-Guerrero (2008). 'Biopolitics of security in the 21st century: an introduction.' *Review of International Studies* **34**(02): 265–292.
Faas, H., F. Gracceva, G. Fulli and M. Masera (2011). 'European Security: A European Perspective.' In A. Gheorghe and L. Muresan, *Energy Security*, Springer, Netherlands, pp. 9–21.
Friedrichs, J. (2011). 'Peak energy and climate change: the double bind of post-normal science.' *Futures* **43**(4): 469–477.
Goldthau, A. (2012). 'A public policy perspective on Global Energy Security.' *International Studies Perspectives* **13**(1): 65–84.
Haftendorn, H. (1991). 'The Security Puzzle: Theory-Building and Discipline-Building in International Security.' *International Studies Quarterly* **35**(1): 3–17.
Hampson, Fen Osler, Jean Daudelin, John B. Hay, Todd Martin and Holly Reid (2002) *Madness in the Multitude: Human Security and World Disorder*, Don Mills, Ontario, Oxford University Press.
Hajer, M.A. (1995). *The Politics of Environmental Discourse: Ecological Modernization and the Policy Process*, Oxford, Oxford University Press.
Huysmans, J. (2006). *The politics of insecurity: fear, migration and asylum in the EU*, London, Routledge.
Keating, M., C. Kuzemko, A. Belyi and A. Goldthau (2012). 'Introduction: Bringing Energy into International Political Economy.' In Caroline Kuzemko, Andrei V. Belyi, Andreas Goldthau and Michael F. Keating (eds) *Dynamics of Energy Governance in Europe and Russia*, Houndmills Basingstoke, Palgrave MacMillan.
Keppler, J.H. (2007). *International Relations and Security of Energy Supply: Risk to Continuity and Geopolitical Risk*, Brussels, Directorate General External Policies of the Union, European Parliament.
Kanninen, T. (2012). *Crisis of Sustainability*, London, Routledge.
Klare, M.T. (2008). *Rising powers, shrinking planet: the new geopolitics of energy*, New York, Metropolitan Books.
Krause, K. and M.C. Williams (1996). 'Broadening the Agenda of Security Studies: Politics and Methods.' *Mershon International Studies Review* **40**(2): 229–254.
Moran, D. and J.A. Russell (2009). *Energy security and global politics: the militarization of resource management*, London, New York, Routledge.
Nye, J.S., Jr. (1982). 'Review: Energy and Security in the 1980s.' *World Politics* **35**(1): 121–134.
Rasmussen, M.V. (2001). 'Reflexive Security: NATO and International Risk Society.' *Millennium – Journal of International Studies* **30**(2): 285–309.
Rothschild, E. (1995). 'What Is Security?' *Daedalus* **124**(3): 53–98.

Schultz, C.L. (1973). 'The Economic Content of Security Policy.' *Foreign Affairs* **51**(3): 522–540.
Smil, V. (2003). *Energy at the crossroads: global perspectives and uncertainties*, Cambridge, Mass., MIT Press.
Sovacool, B.K. and I. Mukherjee (2011). 'Conceptualizing and measuring energy security: A synthesized approach.' *Energy* **36**(8): 5343–5355.
Trombetta, M.J. (2010). 'Rethinking the Securitization of the Environment: Old believes, New insights.' In T. Balzacq (ed.), *Securitization theory: how security problems emerge and dissolve*, London, Routledge.
UNDP (1994), *Human Development Report 1994*, New York and Oxford: Oxford University Press.
Vaughan-Williams, N. and C. Peoples (eds) (2012). *Critical Security Studies*, London, Routledge.
Wæver, O. (1995). 'Securitization and Desecuritization.' In R. D. Lipschutz, *On Security*, New York, Columbia University Press, pp. 46–88.
Walt, S. (1991) 'The Renaissance of Security Studies.' *International Studies Quarterly* **35**(2): 211–239.
Westphal, Kirsten (2006). 'Energy Policy between multilateral governance and geopolitics: Whiter Europe?' *Internationale Politik und Gesellschaft* **4**(2006): 44–61.
Winzer, C. (2011). 'Conceptualizing Energy Security.' EPRG Working Paper 1123, Cambridge Working Paper in Economics 1151, Electricity Policy Research Group, Judge Business School, University of Cambridge, available online at http://www.eprg.group.cam.ac.uk/wp-content/uploads/2011/08/EPRG1123_complete.pdf (accessed 8 November 2012).

# PART II

# ENERGY SECURITY ISSUES

## 2. Energy security and liberal democracy: ideals, imperatives and balancing acts
*Steve Wood*

### 1   INTRODUCTION

Contemporary juxtapositions of energy security and liberal democracy are, subtly or dramatically, changing understandings of both. Presumed energy imperatives test the quality and integrity of the liberal democracy that defines most post-industrial societies. This pertains not so much to their domestic functioning – though this is also under stress – as their capacity and will to uphold the values of that philosophy globally. A growing body of inquiry into political and normative dimensions accompanies technical and econometric analyses of resource demand and extraction. Uncomfortable circumstances envelop multifaceted questions: how much and what forms of energy do populations, business and industry, and state sectors need; what should be done to ensure this; how are environmental, job, and societal security affected; what is the order of priority, and how is it determined? Media attention accentuates the existential, economic and political sensitivities that permeate this commotion of factors and possibilities.

While this chapter recognises the frequent intersecting of domestic and foreign affairs, for organisational purposes it designates external and internal dimensions. The external concerns relations between liberal democracies and illiberal regimes where energy is a consideration: an illiberal regime possesses resources that a liberal democracy wants, it controls access to or transit of such, or otherwise influences their supply in ways that have impact, including through pressure on a liberal democracy's partners and allies. The internal pertains to how electoral ambitions, public attitudes, and economic constraints, affect energy policy in individual polities. It is a field of tension between 'principled' and 'pragmatic' responses, from parties and publics (cf. Tavits 2007). Occasionally this tension becomes acute. Governments may be confronted with the convergence and intensification of external and internal pressures. Such situations could involve dependency on illiberal regimes for oil or gas, and on indigenous coal or nuclear power. Sections of electorates may oppose one or other, or both options, without proposing viable alternatives. Conversely, majorities may accept them as reasonable, or undesired but requisite.

The challenge to resolve these political, strategic, economic, security, technical, and normative problems is not about to recede. One foreword in a major volume noted, 'since the Industrial Revolution, energy and the need to secure its supply have been fundamental to any position of power in the world', 'wind, solar, and nuclear power remain limited alternatives', and 'a hydrogen-based economy' is 'decades away' (Schlesinger 2005: xiii). 'Energy', 'security' and 'power' form a triangular bond exerting pervasive influence in world politics. 'Values' further complicate and intensify the predicaments facing liberal democratic states and societies.

## 2 THE EXTERNAL DIMENSION

### 2.1 Liberal Democratic and Illiberal Regimes

This section aims to clarify terms and assumptions, not to provide an extensive examination and critique. Democracy is presented here as a 'liberal' version of that concept, meaning that besides popular election, certain features of civil and political life are guaranteed, including freedom of association, speech and opinion, a free media, an independent judiciary, rights to private property, and that elections are regular, free and fair. 'Liberal democracy' is understood as an ideal type, because nothing so complex operates perfectly and entirely predictably in practice. Easton (1965), Dahl (1971), Bollen (1993), Bollen and Paxton (2000) and Munck and Verkuilen (2002), offer definitions and analyses, discussion of components, quantitative measurements and taxonomies. Downs' (1957) influential 'economic theory' argued that liberal democracy was characterised by the attempts of parties to maximise votes, win elections and enjoy the 'intrinsic rewards of holding office', while citizens make decisions based on perceived self-interest. Employing a survey of 36 states, Kotzian (2011: 24) informed that 'support for liberal democracy is contingent on economic growth being sustained ... economic performance matters more than institutional and political developments'. Energy requirements, and the opportunities for profit or influence they enable, have sometimes, if not in the magnitudes typical of illiberal regimes, conspired to circumvent or degrade institutional probity. This includes climate change strategies, where enormous funds are at stake (cf. Hopkin 1997; Transparency International 2007; Transparency International 2011).

The incidence of the liberal democratic form of polity has varied. So-called 'waves' of democratisation (Huntingdon 1991; Kurzman 1998; Przeworski et al. 2000) reached many states, without excluding relapse or regression. Academics, think tanks and international and non-governmental organisations monitor the functioning, quality or absence of liberal democracy. Its alternates, and sometimes antagonists, are illiberal regimes. These do not hold free and fair elections, have no genuine separation of powers and suppress real or fabricated opponents. Peaceful replacement of rulers rarely occurs. Whether individual, dynastic or juntas, these are coerced to leave, are forcefully eliminated or remain in office until terminated by nature. Most raw materials needed for energy, as currently created, are located in the territories of such regimes. Some dispute land and marine regions where sovereignty is not conclusively established (Daly 2011; Fogarty 2011; Dosch 2011; Mail Online 2007; Catley and Keliat 1997), attempting to expand the volume and spatial scope of their resource base. Liberal and illiberal states are also active in poorer countries (Taylor 2006; McSherry 2006; Koike et al. 2008), where energy competition has an accompanying secondary contest of political systems and political culture.

The nominal obligation of liberal democracies to uphold universal values, prominently as advocates and defenders of human rights, is one element of a dilemma. A less important effect is accusations of 'neo-imperialism' directed at democracy promotion endeavours (Youngs 2011a). More serious is the leverage exerted by authoritarian oil or gas producers. Dealings of liberal democratic governments with autocracies and theocracies suggest that the clash between values and material-strategic interests has been decided. These arrangements demonstrate that selectivity is a more crucial feature of politics than consistency, especially in foreign affairs. The US sustains the House of Saud, a regime rating near the bottom of political rights and civil liberties indices, while overthrowing neighbouring tyrants. European Union (EU) states consorted with Gaddafi, another of the worst human rights transgressors, for years before becoming active or complicit partners in his removal. Others abhor similar violations and sell energy resources to China, a 'hardline autocracy' (Bertelsmann Stiftung 2012; Freedom House 2012; Human Rights Watch 2011).

## 2.2 Energy crises, Peak Oil, Scarcity

Many scholarly studies, government reports, industry assessments and other works have analysed energy crises, or, 'the energy crisis' (Hussey 1974; Williams and Alhaji 2003; Campbell 2005; Newman 2008; Heinberg 2009). Key elements recur: the estimated remaining quantities of a resource viewed as essential: low or rapidly reducing; the availability of that resource and its assured supply: under threat; the price of the resource: already high, rising, fluctuating. Data influencing the declaration of a 'crisis' might not be reliable and complete, leaving best guess scenarios. Politics is sometimes a better guide than geological or economic indicators.

Energy crises are often linked with 'peak oil', for which a plausible consensus forms around 2000 to 2010 (Hubbert 1949; Hubbert 1956; Aleklett and Campbell 2003). Global production of conventional oil was steady over that decade: 64 million barrels per day (Mb/d) in 2000 and 63 Mb/d in 2010. The two biggest producers, Russia and Saudi Arabia, reached outputs between 8.3 and 8.8 Mb/D. These volumes will gradually reduce. Iraq, Kazakhstan and the United Arab Emirates are likely to increase production until 2020. The US, UK, Canada and Norway have passed dates of peak exploration, discovery and production for conventional oil (Campbell 2011). The story is similar in conventional gas, with some divergences. In 2010, Russia and the US each produced 23 trillion cubic feet (tcf). By 2020, Russia will be dominant, producing 30 tcf. US volume will drop to 8.5 tcf. Qatar, Saudi Arabia, China, Turkmenistan and Kazakhstan, which will remain illiberal indefinitely, will increase production. Australian production will also increase, from 1.8 tcf in 2010 to 4.8 tcf in 2020. Production in Canada, the Netherlands and the UK will decrease. Some states, whose political direction, economic orientation and stability are very unclear, including Nigeria and Egypt, are predicted to increase production (Campbell 2011; Energy Information Agency 2011a; Energy Information Agency 2011b; BP 2011).

The inexorable energy thirst of China, India and other new industrialisers is another vector of international pressure. China's Communist Party (CCP) will try to incorporate these needs in ways that assist it to maintain authoritarian rule. India is challenged to prevent the impetus for economic growth, and the aligned requirement for enormous amounts of energy, from threatening its liberal democratic credentials. Both will compete for scarcer raw materials.

### 2.3. The Resource Curse

One version of the resource curse phenomenon is that conceptualised as a problem of 'crowding', whereby energy and mining sectors dominate an economy to the extent that other industries, and especially services, are excluded or suffocated (Sachs and Warner 2001; Auty 2001). Another version equates richness in resources with a virtual absence of political rights and widespread poverty (Collier and Hoeffler 2005). A third interpretation emphasises the quality of institutions, economic and political, as decisive (Mehlum et al. 2006). Ross (2001) investigated the proposition 'does oil hinder democracy', testing three possible explanations or 'effects': 'rentier', 'repression' and 'modernization'. He concluded that oil does tend to hinder (liberal) democracy, but with some qualifications. Ross's modelling suggests that 'oil harms democracy more in poorer countries than in rich ones'. Windfall profits through oil or mineral discoveries would also damage democracy, or its prospects, in poor countries more than rich ones, for which there are 'no discernible effects'.

Many resource-rich states are illiberal and some extremely so. But this is not universal. Canada, Australia and the US have substantial resources, even if the US has depleted much of its conventional oil. Resource curses occur where atavistic political structures coincide with valuable geological deposits. To counter the efforts of a few liberal democratic governments and NGOs to transform them, the rulers of illiberal states exploit resources to retain power. Though it always entails costs, a liberal democratic direction reduces the prospect of a resource curse afflicting a developing or transitional country. Political culture is more important than income levels or the configuration of economies.

## 2.4 Security

Energy is a component in diversification of the security concept, beyond a military-state focus, into security concepts. An early alternative perspective was 'human security', conceived by Blatz (1966) as a condition related to the control or surmounting of anxiety. International relations scholars, sociologists, development specialists and others expanded on this breakthrough, adding environmental, job and societal security (Wæver et al. 1993; Lipschutz 1995; McSweeney 1999; COT et al. 2007). Increases in resource prices and reduced reliability of supply coincided with rising interest in 'energy security'. Most other security notions are connected, given energy's central importance to modern or industrialising societies. Anxiety is also a common if latent ingredient.

The emergence of new ideas on security does not necessitate the abandonment of rationalist or statist interpretations; for government analysts and policymakers it will not. The confluence of location and type of political system is critical. 'Traditional' geopolitical aspects remain relevant because of the global distribution of raw materials and transit routes. This makes Goldwyn and Billig's (2005: 514) observation that, 'For the United States, price shocks are the modern threat to energy and economic security', interesting. Perhaps reflecting presumptions that the US has other options, analyses focusing on prices generally do not posit illiberal regimes as a security problem. Elevating financial and economic costs in determinations of whether energy is considered secure or not downplays the geopolitics that US policymakers and analysts are criticised for overly concentrating on. A Strategic Protection Reserve (SPR) can, for a period, mitigate financial-economic and political-geostrategic concerns, but it cannot be a lasting solution.

European academics and policy intellectuals tend to have more elaborate and sometimes utopian conceptions of security than those in the US. The EU has discovered that such cogitations do not have great resonance in practice. Youngs (2010) explains that 'The EU has constructed an impressively comprehensive edifice predicated on a singular line of reasoning: politics matter to security . . . In theory, democracy is diplomacy's multi-purpose tool . . . the remedy to a host of soft security challenges, including . . . energy supplies.' Rethinking security is an admirable intellectual endeavour that has also fed disingenuousness in some critiques of US policy, contemporaneous to other societies relying on it to ensure a supply of oil. The US, the EU and its member states are all ostensibly defenders and extenders of liberal democracy, and concomitantly seek to maintain vital interests. Downs (1957) emphasised an enigmatic influence, very applicable to security and energy affairs: uncertainty (cf. Trombetta 2007). Uncertainty precipitates and prolongs crises. It makes government, business and academic attention to energy security *de facto* an interest in or concern about *insecurity* (Wæver 1998). If most deposits and transit routes were located in and between liberal democracies, one problem would be resolved. There would be less disquiet about 'energy security'.

## 2.5 The US

In 1950 the US had zero net energy imports. It produced about 5Mb/d of oil, most for domestic consumption. In 1970 it reached a production peak of just over 10Mb/d. By 2008 it had dropped to 7Mb/d. Imports rose to 60 per cent of consumption, then fell to 53 per cent in 2010. Depletion of US oil coincided with rising dependence on Middle Eastern regimes, manipulation by them and military involvement in the region. Removing Saddam Hussein had its justifications but the risks of doing so would not have been accepted without the belief that indigenous US oil was lacking, continued dependency on that commodity, and external producer intransigence. Similar embroilments are possible as long as a variant on that nexus exists (cf. Klare 2004; Heinberg 2004; Peters 2004). Some statistics (cf. Energy Information Agency 2011a), however, raise questions on whether *Iraqi* oil was so necessary. Before and during the 1973–74 shocks, Canada was the US's largest supplier of petroleum products. It has been the largest since 2000. Mexico became the second largest, surpassing Saudi Arabia. In 1990, when US oil imports from Iraq were at their highest, the quantity was smaller than that from Canada and Mexico. They declined again after 2003, while those from Nigeria, Venezuela and Russia, hardly model liberal democracies, increased.

Kalicki and Goldwyn's (2005) compilation, written mainly from US perspectives, contains little direct treatment on how energy concerns affect liberal democracy. In places this is implicit, in the sense that regimes with influence over prices or supply could cause material difficulties for the US, rather than challenge its proclaimed political values. Other scenarios suggest that the US could again become the largest oil producer. This may involve a change in the definition of oil (Oil Drum 2011). If shale oil becomes viable, or discoveries lead to rises in conventional production, or energy efficiency is dramatically improved, it can be anticipated that US pressure for reform on illiberal regimes, which have relied on energy resources to avoid it, will intensify.

Until these as yet hypothetical scenarios transpire, the US will proceed with a diversification strategy, both in forms of energy generation and the geographic regions from which resources are obtained. Conventional oil remains a major part of the agenda. In the Marxist-realist analyses of Stokes and Raphael (2010; 2011), a counter to Kalicki and Goldwyn's volume of largely liberal-realist viewpoints, Africa and Latin America are among the prominent targets in a global search-and-secure-oil strategy. A few details might lend additional perspective to both of these orientations. Africa provides about 10 per cent of US oil consumption and Latin America less than 20 per cent. Almost 40 per cent is produced in the US, where in 2011 total petroleum products exceeded total imports for the first time since 1949 (Energy Information Agency 2012a).

Among the Latin American suppliers of crude oil to the US, Mexico is the largest. Diplomatically and rhetorically, there is a marked divergence in relations with the US between antagonistic Venezuela, its second largest crude oil supplier, and Venezuela's neighbour, the more US-friendly Colombia. Yet 43 per cent of Venezuelan oil exports go to the US. For the US, imports from Venezuela have halved from about 16 per cent in the mid-1990s to 8 per cent in 2010 (Energy Information Agency 2011c). Colombia has slipped under the radar somewhat as a rising energy producer. Oil and coal are its main exports, with the US the biggest importer of both. The gas industry is also growing (Energy Information Agency 2012b). While the government of the late President Hugo Chavez nationalised or otherwise controlled much of the oil industry within Venezuela's borders, Colombia's energy sector has experienced privatisation and liberalisation. Contrasting developments in the two states have been influenced by relations with the US. Others, including China and Iran, are also interested.

## 2.6 The EU

The most difficult balancing act of maintaining energy security and actively defending liberal democratic values is that experienced by the institutions, member states and populations of the EU. The EU is more exposed than most political entities in material and ideational-value terms. Its members have relatively little oil or gas, it is a champion of clean-green energy and environmental protection and in 2011 the governing coalition of its largest state, Germany, declared an apparently definitive phase-out from nuclear power. Its populations want jobs, high quality services and comfortable lifestyles, dependent on economic growth and reliable energy, of which the EU is the world's third largest consumer. Concurrently, the EU is presented as the world's chief promoter of normative political change, towards a social-liberal-humanitarian expression of justice and order.

Are all these demands and preferences simultaneously possible? According to Duke (2010: 329) 'the mixed response from all parties [to the Russia-Georgia war] can be explained by a number of reasons, such as domestic factors, energy supplies and geopolitics'. And:

> Normative power is less likely to be asserted when there is a strategic interest to be advanced. The Nigeria and Russia cases illustrated that the presence of clear strategic interests, such as secure and dependable energy supplies, can push normative considerations to the back seat. (Duke 2010: 333)

This exemplifies disjunction of normative preferences and material-strategic imperatives: when adhesive binding values and policy begins to dissolve and publics are left with rhetoric, part of the stock-in-trade of governmental or institutional pragmatism. In such scenarios, 'policy' means inaction or divergent action (cf. Zimmermann 2008; Wood 2009b, 2011; Mattlin 2010). A European Development Co-operation project (FRIDE 2011) summarised: 'Democratic governance does not appear as a main driver for EU developing policy in many producer states that remain authoritarian and marred by deep internal poverty . . . not only in Russia and the Middle East but also in Central Asia and Africa'. Energy security suffers what Youngs (2010) terms 'single issue syndrome', which results in a 'parcelled up approach', whereby 'energy questions and deliberation on democracy support are simply not linked up within EU policy-making':

> In several countries, such as Russia and Iran, it is clear that the conjoining of economic and political power causes profound problems for energy security . . . EU policy limits itself to improving governance standards in other countries' energy sectors . . . ringfenced from systemic-level political problems and considerations.

An EU official (author interview 2008) explained that there are 'two essential problems':

> we don't have a common energy policy – there is one in practice but not in name – nor the means in law to enforce one. We don't have a common democracy policy; that doesn't mean we don't do democracy support, it is done through other means, *à la carte* and *ad hoc* ... 27 member states have 27 traditions of democracy, with different understandings of what it means and what it means to support it.

According to Giuli (2011: 2), altering EU-Russia interdependence 'would reduce the incentives for the Russian leadership to move towards a more open and rule-of-law-based society'. He opposes a phase-out of oil or 'diversification options'. The latter would 'make the EU dependent on countries even less democratic than Russia ... reduce the rent of someone to increase the rent of someone else has clearly little to do with the aim of making energy security and democracy consistent'. Guili optimistically proposes that the EU-Russia energy dialogue be 'de-politicised' and 'restructured':

> the high political sensitiveness of concepts such as 'energy security' and 'democratisation' can only be detrimental to this process, but the unavoidable spill-over of preferential cooperation will necessarily translate into both energy security [and] more political openness. (Giuli 2011: 5)

EU membership is not prioritised when energy imperatives are at issue. European states tend to pursue a traditional course: their own 'national interests', sometimes in cooperation with others.

## 2.7 Russia and Central Asia

Contemporary Russia is characterised by authoritarian-nationalist rule: a 'sovereign democracy' without real political choice, and potentially without other liberties if the wrong people are opposed. Vladimir Putin, alternately occupying the offices of president or prime minister, aimed at resurgence based on the nationalisation of energy resources, production, transportation and exports. He took over management of this sphere. The clique around him disavows liberal democracy as the system of interfering external powers, among which only the US is capable and marginally willing to challenge the Russian establishment over internal governance or in its 'near abroad'. George Bush II did not risk war over Georgia in 2008, where energy resources and transit were again involved. The EU and its member states, with a partial exception for the UK, will do little of substance. Putin's posturing has appealed to others, such as Chavez in

Venezuela. While it is unsurprising that they viewed each other as allies, relations with their 'enemies' again betrayed contradictions. The US purchases oil from Venezuela, as does the EU from Russia. Each pairing are simultaneously adversaries and business partners.

Building engagement with Central Asian states could relativise Russia's position and alleviate some of the coercion it is able to impose on the EU. This also has its price. Boonstra (2011) restates persisting EU dilemmas and offers recommendations. The relevant interlocutors are aware of the restrictions and imperatives determining how the EU operates. Turkmenistan and Uzbekistan are rated 164 and 165 out of 167 states in one liberal democracy index (Economist Intelligence Unit 2010) and similarly evaluated by others. Some EU states interpret Central Asia's energy wealth as compelling tolerance of regional rulers' illiberalism, and this perspective prevails.

**2.8 China**

Contemporary China presents the liberal democratic world with apparently intractable quandaries. While the US, EU states, Japan and others have no desire for China to become omnipotent, its near double-figure annual growth is now depended on for their own economic welfare. They also want to restrain China from emitting ever-increasing quantities of $CO^2$. Appeals have little effect unless the CCP perceives benefits. It manipulates China's international status: UNSC permanent member (with attendant rights and privileges); 'developing country' (with associated concessions and assistance); and leading exporter, investment seeker and provider and stockpiler of foreign currency reserves. China's national ambition is contingent on reliable, high volume, energy supplies. Western governments and commercial entities usually have formal legal and political conditions, reportable in parliaments and in national and global media, to contend with. China has practical advantages in that it is not constrained by internal governance, civil rights and political liberties when energy opportunities are identified. African freedom movements, now in political power, exploit China as a counterweight to former European colonialists and exchange resource extraction rights for Chinese investment (Frankfurter Allgemeine Zeitung 2012a).

The urgency with which its rulers attempt to secure long-term access to raw materials signals that energy may still become China's Achilles heel. Li (2008) argues that China is 'likely to face an insurmountable energy crisis beyond 2020'. Shortages could even compel a democratic turn if the 8–10 per cent growth, upon which the authoritarian political class relies, is restrained. The civil disobedience which this would incite would build a momentum unstoppable even for China's rulers.

### 2.9 The Middle East and the 'Arab Spring'

Contradictory and tense are two descriptors that could be applied to relations between liberal democracies and Middle Eastern oil producers. In the wake of the first big oil shocks, Paust and Blaustein (1974) proclaimed the 'Arab oil weapon' as a 'threat to international peace'. West (2005) outlined that the 'legitimacy' of the Saudi Arabian regime is based on the accrual and (piecemeal) domestic distribution of oil revenues, and the provision of security, including at Mecca and Medina. Despite an 'interdependence' with the US, the Saudi government assisted al Qaeda members to avoid capture in Afghanistan, even offering them 'jobs in the public sector' (West 2005: 199). A reform process begun in 2003 had stalled by 2005. The 'Arab Spring' of 2011 threatened to restart it in a less controllable manner. Common to most individual cases is a confluence of political upheaval and the presence of oil and/or gas reserves. Western states assisted in, or did not prevent, the removal of several autocracies, but what the revolutions of 2011–2013 will lead to – a quasi-liberal-democratic politics and governance; the replacement of old-style despots and tyrants by other brands of illiberal rule; or something else – is obscure. Revised energy production and supply arrangements involving cooperation with western states and companies are also unclear. Were Israel to become a 'major energy producer' (King 2011) it would recast more than regional affairs. Some distance to the south, China has experimented with mediation in intra-Sudan conflicts (Economist 2012a). Against this background, Youngs (2011b) argues that in its regional policy, the EU must become more 'realist', and less concerned about 'political incorrectness'.

## 3 INTERNAL DIMENSION

Energy also has extensive political effects within liberal democracies. Agency is a feature of this system, as are positions and outcomes that many voters disagree with. Electoral majorities may opt to continue using coal, or to support military interventions to sustain oil supplies. Despite the proliferation of transnational movements and networks, liberal democracy functions in state jurisdictions, rather than a common globalism. Far left and far right oppose the intrusion of foreign forces into national polities. They assert that this degrades democracy, implicitly extending to state borders rather than crossing them. Concurrently, some of the same people advocate global action on climate change, or oppose wars fought partly to secure oil or gas sources. What then – if they do not concur – is the greater necessity: political freedoms and the implementation of majority preferences; or planetary conservation, pacifism, and other 'global goods'?

### 3.1 Climate Change/Global Warming, $CO^2$ Emissions, RES and All That

Uncertainty is not only relevant to politics and security. Who, for example, predicted the Tohoku-Oki earthquake and tsunami, or can say precisely how much of a resource is left? Uncertainty also applies to climate change/global warming, the modulations, variables and measurements of which are beyond the understanding of many involved politically, on any side (Knorr 2009; Scafetta and West 2008). If energy use has a major impact on higher temperatures, what are the most pertinent and exact figures? In liberal democracies, energy efficiency has improved over the past century, as $CO^2$ emissions have increased. Whatever facts are or may be proven, political responses to developments in the physical world are largely national, reflected in taxes, subsidies and regulations. For many citizens, foreign attempts to influence directions are unwelcome, even denounced. Global perspectives and decisions are restricted.

In Australia, energy and environmental policy are intensely politicised. The country's fortunes are influenced, though not overwhelmingly, by its export of coal and other resources and their domestic consumption. After the election victory of the Labor Party (ALP) in 2007, it was expected that a raft of green measures would be implemented. New prime minister, Kevin Rudd, announced climate change as 'the great moral challenge of our generation'. As the 2010 election approached, the government detected a groundswell of aversion to the Emissions Trading Scheme (ETS) that Rudd had proposed. A leader of the opposition Liberal Party (LPA) had been discarded for adopting a similar approach. Rudd's response to popular and business discontent, and the failure at Copenhagen, was to 'delay' the ETS. But many supported it. Rudd's party duly deposed him. The election returned a minority ALP government, dependent on Greens and independents. A 'super profits resource tax' floated under Rudd metamorphosed into a 'mining tax' under his successor, Julia Gillard. This targeted coal, iron ore and 'big' miners (Economist 2012b). It preceded a more controversial 'carbon tax', introduced in July 2012. In the same month Gillard declared that Australia would be dependent on fossil fuels 'for a long time to come' (Ludlow 2012). The opposition says that if elected it will repeal the tax and require definitive international action before embarking on major $CO^2$ reduction schemes that 'damage the economy'.

Canada is well positioned with energy options. It is in the world's top three exporters of natural gas and the top 10 of crude oil, is second for hydro-generated and among the top ten in nuclear-generated electricity. It also has copious tar sand reserves. Its main customer, the US, is one of the world's two largest energy consumers, is contiguous and is a liberal democracy. Despite the further wealth that could be created, public attitudes are ambivalent. Much of the electorate is concerned about ecological damage through new pipeline and other infrastructure and extraction projects. The government's 'Responsible Resource Development' strategy aims to overcome opposition and promote a sector that in 2012 employed more than 750,000 people. It would 'tap into the tremendous appetite for resources ... we have in abundance' and enhance already 'high environmental standards' (Natural Resources Canada 2012).

Although Germany was a leader in transition to RES (Renewable Energy Sources), its Fukushima-inspired *Energiewende* (Bundesministerium für Wirtschaft 2010) ran into difficulties in expanding and accelerating that agenda. One consequence was the sacking of the responsible minister. The change to green/er energy is mainly borne by the state: as pioneer investor, infrastructure provider, lender or underwriter, now when energy and environment are overwhelmed by concerns about how much Germany will have to pay to bail out debt-ridden Eurozone states. Turkey, meanwhile, has become sceptical about the value of joining the EU. This ambitious nation-state is unlikely to compromise its economic and technological progress in order to fulfil others' environmental preferences and the modifications to energy use that might entail. Turkey impresses itself as an 'energy hub', which its political and business sectors perceive as providing vast opportunities. It is politically easier to sell related benefits, including lucrative, redistributable receipts, than a RES agenda. Turkey's historic rival, Greece, is even more desperate for hydrocarbon-transit rents (Wood and Grant 2010).

### 3.2 Nuclear Power

The Euratom treaty of 1957 foresaw an EEC-wide nuclear energy industry. Two decades later, anti-nuclear protest reached its zenith, prominently in what are now EU states. Green movements gradually realised, however, that viable energy options were not as absolute as they preferred. Multiple problems – scientific-technical, financial, political – associated with implementing a sustainable energy future influenced a turn to compromise. Some activists conceded that nuclear power was a lesser evil than the maintenance or expansion of electricity that emitted large amounts of $CO^2$. In 2011, 30 states, including disputed Taiwan, had nuclear industries.

All who lived in them relied upon it, directly or indirectly, and it had gained support as a partial solution.

The March 2011 earthquake and tsunami, and ensuing calamity at the Fukushima plant, reversed that sentiment. The reaction in Germany, 9000 kilometres away, was pandemonium, marches, and a government proclamation to close all atomic energy plants within a decade. An anti-nuclear undercurrent and the political class's capacity for opportunism were reinvigorated. Popular and party attitudes may change again, as energy-related developments intensify and others emerge, along with unimagined 'events' that overwhelm extant circumstances and possibilities. In Japan, operations at some nuclear plants were restarted. Protests regularly occur and nuclear power will be a 'wild card' in the next elections (Frankfurter Allgemeine Zeitung 2012b; Sieg and Yoshikawa 2012). Taiwan also has a precarious energy situation, exacerbated by non-recognition as a sovereign state, being claimed by the PRC and competing with that powerful rival for partners and resources. Controversy has accompanied the building of a fourth nuclear plant. A sectoral manager claimed few people opposed nuclear power, which generated about 20 per cent of Taiwan's electricity, but that the 'noise of these few is very loud' (Wood 2012). Fukushima provoked another burst of protest and the nuclear issue was central in the 2012 presidential and legislative elections. The pro-nuclear incumbents were re-elected (Engbarth 2012; De Changy 2012).

Even if no more nuclear energy was generated, the disposal of radioactive waste is an unavoidable question, with transnational implications. Nearly 40 years ago, one scientist claimed that the 'problem would no longer be millions of years, but would be reduced to the order of a century'. Single countries are unlikely to solve it alone, however: 'problems raised by the relationship between nuclear energy and democracy in Switzerland are going to be, more and more, problems connected with security and international solidarity' (Zangger 1974).

**3.3 Money**

Nuclear plants require high expenditure for construction, waste disposal and decommissioning. Per watt energy is comparatively cheap, more so if charges on carbon emissions are imposed across entire sectors. Nuclear is an option. Whether it is one that EU states can do without, while moving towards a clean energy future, is unanswered. The macro-economic costs of higher energy prices resulting from environmental regulation were estimated by one pre-financial crisis and pre-Fukushima study to be 'fairly limited ... usually below 1% of GDP. Eco-taxes even engender macroeconomic gains by way of double-dividend effects. The results for Germany are fully in line with the results for the EU as a whole' (Dannenberg et al. 2008: 1319). Another study (Technische Universität Berlin 2011: 19–25, 60–61) calculated the direct and indirect costs of Germany's *Energiewende* from 2010 to 2030 at €335 billion. A systemic shift to RES was difficult but manageable before the current crisis. Germany is more than ever relied on to keep the EU afloat. Research and development, infrastructure and implementation are expensive. Money is in shorter supply. As the world's premier promoter of democratisation and green energy, the EU sets higher expectations of and for itself than apply to other political entities. Concurrently, it is among those most dependent on external sources and least able to mobilise the political will to fulfil its rhetorical declarations. Contradictory processes occur. Biofuels, for example, are not as straightforward or equitable as some proponents believe (Williams and Kerr 2011; De Santi et al. 2008).

The Japanese government introduced a feed-in tariff scheme to promote RES and partly compensate for a fall in capacity after operations were halted at many nuclear plants (Ministry of Economy, Trade and Industry 2012). In France, the magnitude of a shift from nuclear power to more RES would be greater than for other countries. About 78 per cent of France's electricity is generated by its 58 nuclear reactors, which have contributed to a relatively low level of $CO^2$ emissions. The state is the principal owner of this infrastructure and France is a big electricity exporter. The nuclear energy and defence sectors have close links. This makes for intriguing politics. In Australia, an array of government investments, incentives and disincentives are foreseen as stimulating a 'Clean Energy Future'. It involves a conservative increase from a 5 per cent RES share of total primary energy in 2009 to 9 per cent in 2034. A carbon price, or tax, of $A23 per tonne was set in July 2012 and projected to become part of an internationalised, and then floating, 'cap and trade' ETS from 2015 (Syed and Penney 2011). Within a few weeks the internal scheme, which included a floor price of $15, was scrapped in favour of a two-stage linkage with the EU's ETS, the first from 2015 and a second from 2018 (ABC 2012). Further vagaries of domestic politics will determine if, when and which of these measures will be implemented, repealed or altered.

The costs for richer societies to convert to RES are huge. But how will developing countries pay? A UN study (2009) estimated the global RES market at $1.6 trillion in 2007–8. It calculated that 'additional investment' of $100 billion p.a. for 15 years would be needed to reduce the cost of the main technologies, solar and wind, to affordable levels. Two-thirds of this 'additional capacity would be deployed in developing countries'. It is hydropower that actually dominates the sector, because it is the cheapest per watt form. The sun is free but solar energy is expensive. $257 billion was invested in all RES in 2011, indicating that a global shift is possible if private enterprise envisages profits. To date this has required large state subsidies and other incentives, provided or guaranteed by taxpayers. This means more debt. If citizens and governments are prepared to accept that, including through funding of projects in China, India and Africa, RES industries will expand. Associated dumping and trade disputes will increase. If public subsidies end, the interest in RES start-ups will decline. Meanwhile, over 80 per cent of primary energy is still generated by fossil fuels, industries with substantial capital stock, investment and employees.

## 4 CONCLUSION

Easton (1965) distinguished between support for an incumbent leadership and support for the liberal democratic system. The system enables popular dissatisfaction with parties, leaders and policies to translate into their removal. What many voters see as an ethically superior and requisite option is regularly countered by an alternative view for which the support of one constituency or one parliamentarian more has been assembled. Polarisation in electorates induces radicalisation of policies as parties try to gain by moving away from the centre, or by moving the centre itself (cf. Downs 1957; Münckler 2012). This trend is apparent in the US, Australia and some European states. Energy policy is one field with potential for a more adversarial form of politics. In a recent US poll, 14 per cent of respondents thought that the country was 'headed in the right direction on energy'. 84 per cent were worried about dependence on foreign oil and 76 per cent about a lack of progress in energy efficiency and RES; 37 per cent ranked economic growth more important than 'preventing harm to the environment', which was prioritised by 33 per cent. Energy prices were the 'most compelling issue' and 71 per cent were dissatisfied with the contribution of Congress to 'tackling energy issues' (Koch 2011).

'Linearization' is commonly understood as continuity without unpredictable change. Hubbard (1969: 238), with whom the term is associated, observed that 'the period of rapid population and industrial growth that has prevailed during the last few centuries, instead of being the normal order of things and capable of continuance into the indefinite future, is actually one of the most abnormal phases of human history'. That period introduced powerful economic, social and technological dynamics in national and international contexts, and made energy political. Economic growth became regarded as indispensable: because it increases wealth or because it contributes to creating and sustaining employment. This condition matured into a dilemma. Even when energy is readily available, growth is not guaranteed. For contemporary economies, energy shortages, or unaffordable expense, would ultimately mean a growthless society.

The future will not just be a continuation of the past. The relative decline of the 'West', and for much of it a deep economic-financial malaise, coincide with pressure in its electorates for conversion from fossil fuels. Many voters also demand phase-outs of nuclear power. Others oppose both of these choices. It cannot be expected, however, that the 'non-West' will lead a transformation in consumption patterns. An end to authoritarianism, which is not on the horizon, does not mean support for a green agenda. It is not guaranteed that newly emancipated electorates would demand what environmental lobbies in established liberal democracies call for. Neither would the emergence of a 'global demos' necessarily result in universal green attitudes and policy. National and global energy challenges have multiple tangents. Trade-offs will always be required. Political machinations in the international arena confront national self-understandings or self-images. Some liberal democracies have reached a *modus vivendi* with illiberal energy suppliers. A few have with illiberal customers. Youngs (2010: 6) suggests that 'European diplomats are entirely candid in arguing that pressure on human rights and peace-building is and will be foregone in order to get the maximum number of countries signed up to internationally-agreed climate change commitments'. Some of them 'even argue that democratisation':

> ... would make it harder for regimes to make the concessions necessary to cut carbon emissions. The growing feeling that liberal democracy is irrelevant to these kind of existential problems may herald a return to elitist visions of technocratic experts running the international system. (Youngs 2010)

## REFERENCES

ABC, Australian Broadcasting Corporation (2012), 'Combet ditches carbon floor price in deal with Europe', 29 August, at http://www.abc.net.au/news/2012-08-28/combet-announcement-on-carbon-floor-price/4228260 (accessed 28 September 2012).
Aleklett, Kjell and Colin Campbell (2003), 'The Peak and Decline of World Oil and Gas Production', *Minerals and Energy* 1(1) 5–20.
Author Interview (2008), Policy Unit, General Affairs Council, Brussels.
Auty, Richard (2001), *Resource Abundance and Economic Development*, Oxford: Oxford University Press.
Bertelsmann Stiftung (2012), *Bertelsmann Transformation Index*, Gütersloh: Bertelsmann.
Blatz, William (1966), *Human security: Some reflections*, Toronto: University of Toronto Press.
Bollen, Kenneth (1993), 'Liberal Democracy: Validity and method factors in cross-national measures', *American Journal of Political Science* 37(4) 1207–1230.

Bollen, Kenneth and Pamela Paxton (2000), 'Subjective measures of liberal democracy', *Comparative Political Studies* **33**(1) 58–86.
Boonstra, Jos (2011), 'The EU's Interests in Central Asia: Integrating Energy, Security and Values into Coherent Policy' EDC2020 Working Paper 9, Madrid: FRIDE.
BP, British Petroleum (2011), *Statistical Review of World Energy*, London: BP.
BP, British Petroleum (2012), *Energy Outlook 2030*, London: BP.
Bundesministerium für Wirtschaft (BMWi) (2010), *Energiekonzept: für eine umweltschonende, zuverlassige und bezahlbare Energieversorgung*, Berlin: BMWi.
Campbell, Colin (2005), *Oil Crisis*, Multi-Science: Brentwood.
Campbell, Colin (2011), 'Energy Politics and Peak Oil Update', *Energy Politics* **25** (Winter) 17–28.
Catley, Bob and Makmur Keliat (1997), *Spratlys: The Dispute in the South China Sea*, Aldershot: Ashgate.
Collier, Paul and Anke Hoeffler (2005), 'Resource Rents, Governance and Conflict', *Journal of Conflict Resolution* **49**(4) 635–633.
COT Institute for Safety, Security and Crisis Management et al. (2007), *Notions of Security: Shifting Concepts and Perspectives*, COT/European Commission.
Dahl, Robert (1971), *Polyarchy: Participation and Opposition*, New Haven: Yale University Press.
Daly, John (2011), 'Russia claims new Arctic Hydrocarbon Finds Effectively Double Nation's Reserves', at http://oilprice.com/Energy/Energy-General/Russia-Claims-New-Arctic-Hydrocarbon-Finds-Effectively-Double-Nations-Reserves.html (accessed 23 March 2012).
Dannenberg, Astrid, Tim Mennel and Ulf Moslener (2008), 'What does Europe pay for clean energy? – Review of Macroeconomic Simulation Studies' *Energy Policy* **36**(4) 1318–1330.
De Changy, Florence (2012), 'Taiwan presses ahead with nuclear power plant despite safety fears', *Guardian Weekly*, 14 February.
De Santi, Giovanni (ed.), Robert Edwards, Szabolcs Szekeres, Frederik Neuwahl and Vincent Mahieu (2008), *Biofuels in the European Context: Facts and Uncertainties*, Petten, Netherlands: European Commission/JRC Institute for Energy.
Dosch, Jörn (2011), 'The Spratly Island Dispute: Order-Building on China's Terms?' *Harvard International Review* 18 August.
Downs, Anthony (1957), *An Economic Theory of Democracy*, New York: Harper.
Duke, Simon (2010), 'Misplaced "other" and normative pretence in transatlantic relations', *Journal of Transatlantic Studies* **8**(4) 315–336.
Easton, David (1965), *A Framework for Political Analysis*, Englewood Cliffs: Prentice-Hall.
Economist Intelligence Unit (2010), *Democracy Index 2010*, London: Economist Intelligence Unit.
Economist (2012a), 'Africa's next big war?' 28 April, p. 14.
Economist (2012b), 'Your tax or mine? "Lucky" Julia chalks up another political victory', 24 March.
Energy Information Agency (2011a), *Annual Energy Review 2010*, Washington: US Department of Energy.
Energy Information Agency (2011b), (Gruenspecht, Howard) *International Energy Outlook 2011*, Washington: US Department of Energy.
Energy Information Agency (2011c), *Country Analysis Briefs: Venezuela*, Washington: US Department of Energy.
Energy Information Agency (2012a), 'U.S. petroleum product exports exceeded imports in 2011 for the first time in over six decades', at http://www.eia.gov/todayinenergy/detail.cfm?id=5290# (accessed 26 September 2012).
Energy Information Agency (2012b), *Country Analysis Briefs: Colombia*, Washington: US Department of Energy.
Engbarth, Dennis (2012), 'Taiwan: Poll Will Decide Nuclear Plant's Fate', 7 January, at http://www.ipsnews.net/2012/01/taiwan-poll-will-decide-nuclear-plantrsquos-fate/ (accessed 3 August 2012).

Frankfurter Allgemeine Zeitung (2012a), 'China sagt Afrika Milliardenhilfen zu', 20 July, p. 13.
Frankfurter Allgemeine Zeitung (2012b), 'Japan fährt Atomkraftwerke wieder hoch', 9 June, p. 7.
Fogarty, Ellie (2011), *Antarctica: Assessing and Protecting Australia's National Interests* Policy Brief, Sydney: Lowy Institute for International Policy.
Freedom House (2012), *Freedom in the World*, New York: Freedom House.
FRIDE (2011), 'Energy Security, Democracy and Development' European Development Co-Operation to 2020 Project, Seventh Framework Program, Madrid: FRIDE et al.
Giuli, Marco (2011), 'Energy Security and the Prospect for Democracy: EU-Russia Relations as a Case Study', Madariaga Speech, Madariaga: College of Europe (June).
Goldwyn, David and Michelle Billig (2005), 'Building Strategic Reserves' in Kalicki and Goldwyn (eds), *Energy Security* . . . pp. 509–530.
Heinberg, Richard (2004), *The Party's Over: Oil, War and the Fate of Industrial Societies*, Gabriola Island: New Society.
Heinberg, Richard (2009), *Blackout: coal, climate and the last energy crisis*, Gabriola Island: New Society.
Hopkin, Jonathan (1997), 'Political parties, political corruption, and the economic theory of Democracy', *Crime, Law and Social Change* **27**(3–4) 255–274.
Hubbert, M. King, (1949), 'Energy from Fossil Fuels' *Science* 109, 103–109.
Hubbert, M. King (1956), *Nuclear Energy and the Fossil Fuels*, Publication 95, Shell Development Company, Houston (June), at http://www.energybulletin.net/node/13630 (accessed 26 March 2012).
Hubbert, M. King (1969), 'Energy Resources' in *Resources and Man*, San Francisco: Freeman, pp. 157–242.
Human Rights Watch (2011), *World Report: Events of 2010*, New York: Human Rights Watch.
Huntington, Samuel (1991), *The Third Wave: Democratization in the Late Twentieth Century*, Norman: University of Oklahoma Press.
Hussey, Hugh (1974), 'Energy Crisis', *JAMA* **228**(8) 1031.
Kalicki, Jan and David Goldwyn (eds) (2005), *Energy Security; Toward a New Foreign Policy Strategy*, Washington: Woodrow Wilson Center Press.
King, Ian (2011), 'Oil shale reserves can turn Israel into major world producer', *The Times*, 21 March.
Klare, Michael (2004), *Blood and Oil: The Dangers and Consequences of America's Growing Dependency on Imported Petroleum*, New York: Metropolitan Books.
Knorr, Wolfgang (2009), 'Is the airborne fraction of anthropogenic $CO^2$ emissions increasing?' *Geophysical Research Letters* **36**(21) L21710.
Koch, Wendy (2011), 'Energy Poll: Americans fret about Congress, foreign oil' *USA Today*, 19 October, at http://content.usatoday.com/communities/greenhouse/post/2011/10/energy-poll-americans-worried-congress/1#.UCjDCNTMoRY (accessed 20 June 2012).
Koike, Masanari, Gento Mogi and Waleed Albedaiwi (2008), 'Overseas oil-development policy of resource poor countries: A case study from Japan', *Energy Policy* **36**(5) 1764–1775.
Kotzian, Peter (2011), 'Public support for liberal democracy', *International Political Science Review* **32**(1) 23–41.
Kurzman, Charles (1998), 'Waves of Democratization', *Studies in Comparative International Development* **33**(1) 42–64.
Li, Minqi (2008), 'Peak Energy and the Limits to China's Economic Growth: Prospect of Energy Supply and Economic Growth from Now to 2050', WP189, Political Economy Research Institute, Amherst: University of Massachusetts.
Lipschutz, Ronnie (ed.) (1995), *On Security*, New York: Columbia University Press.
Ludlow, Mark (2012), 'Gillard says fossil fuel here for a "long time"', *Australian Financial Review*, 11 July.
Mail Online (2007), 'Putin's Arctic Invasion: Russia lays claim to the North Pole – with all its

gas, oil, and diamonds', at http://www.dailymail.co.uk/news/article-464921/Putins-Arctic-invasion-Russia-lays-claim-North-Pole--gas-oil-diamonds.html# (accessed 23 March 2012).

Mattlin, Mikael (2010), 'A Normative EU Policy Towards China: Mission Impossible?', WP67, Helsinki: Finnish Institute of International Affairs.

McSherry, Brendan (2006), 'The Political Economy of Oil in Equatorial Guinea', *African Studies Quarterly* **8**(3) 23–45.

McSweeney, Bill (1999), *Security, Identity and Interests: A Sociology of International Relations*, Cambridge: Cambridge University Press.

Mehlum, Halvor, Karl Moene and Ragnar Torvik (2006), 'Institutions and the Resource Curse', *Economic Journal* **116** (January) 1–20.

Ministry of Economy, Trade and Industry (METI) (2012) at http://www.meti.go.jp/english/press/2012/0618_01.html (accessed 18 July 2012).

Münckler, Herfried (2012), *Mitte und Maß: Der Kampf um die richtige Ordnung*, Reinbek: Rowohlt.

Munck, Gerardo, and Jay Verkuilen (2002), 'Conceptualizing and Measuring Democracy: Evaluating Alterntive Indices', *Comparative Political Studies* **35**(1) 5–34.

Natural Resources Canada (2012), 'The Honourable Joe Oliver, Minister of Natural Resources, Responds to Environmental Groups', 7 May, at http://www.nrcan.gc.ca/media-room/news-release/2012/53/6184 (accessed 7 July 2012).

Newman, Sheila (2008), *The Final Energy Crisis*, 2nd ed., London: Pluto.

Oil Drum (2011), 'Sunday Times Predicts US As Top Oil Producer in 2017', 15 September, at http://www.theoildrum.com/node/8367 (accessed 23 February 2012).

Paust, Jordan and Albert Blaustein (1974), 'The Arab Oil Weapon – A threat to international peace', *The American Journal of International Law* **68**(3) 410–439.

Peters, Susanne (2004), 'Coercive Western Energy Security Strategies: "Resource Wars" as a New Threat to Global Security', *Geopolitics* **9**(1) 187–212.

Przeworski, Adam, Michael Alvarez, José Cheibub and Fernando Limongi (2000), *Democracy and Development: Political Institutions and Well-Being in the World, 1950–1990*, New York: Cambridge University Press.

Ross, Michael (2001), 'Does Oil Hinder Democracy?', *World Politics*, **53**(3) 325–361.

Sachs, Jeffrey and Andrew Warner (2001), 'Natural Resources and Economic Development: The Curse of Natural Resources', *European Economic Review* **45**(4–6) 827–838.

Scafetta, Nicola, and Bruce West (2008), 'Is climate sensitive to solar variability?', *Physics Today* **61**(3) 50–51.

Schlesinger, James (2005), in Kalicki and Goldwyn (eds) *Energy Security* . . . pp. xiii–xvi.

Sieg, Linda and Yuko Yoshikawa (2012), 'Nuclear energy wild card in Japan poll which Democrats likely to lose', 6 August, Reuters, at http://www.reuters.com/article/2012/08/06/us-japan-politics-idUSBRE8750A620120806 (accessed 10 August 2012).

Stokes, Doug, and Sam Raphael (2010), *Global Energy Security and American Hegemony*, Baltimore: Johns Hopkins University Press.

Stokes, Doug, and Sam Raphael (2011), 'Globalizing West African Oil: US "energy security" and the global economy', *International Affairs* 87(4) 903–921.

Syed, Arif and Kate Penney (2011), *Australian Energy Projections to 2034–35*, Canberra: Bureau of Resources, Energy and Economics.

Tavits, Margit (2007), 'Principle vs. Pragmatism: Policy Shifts and Political Competition', *American Journal of Political Science* **51**(1) 151–165.

Taylor, Ian (2006), 'China's Oil Diplomacy in Africa', *International Affairs* **82**(5) 937–959.

Technische Universität Berlin (TUB) (2011), *Kosten des Ausbaus der erneuerbaren Energie*, Berlin: TUB/VBW.

Transparency International (2007), *Global Corruption Report: Corruption in Judicial Systems*, Cambridge: Cambridge University Press.

Transparency International (2011), *Global Corruption Report: Climate Change* London: Earthscan.

Trombetta, Julia (2007), 'Prevention or securitization? Contending perspectives on security', *Network Industries Quarterly* **9**(3) 9–10.

United Nations (UN) (2009), *A Global Green New Deal for Climate, Energy, and Development*, New York: UN Department of Economic and Social Affairs.

Wæver, Ole (1998), 'Insecurity, security and asecurity in the West European non-war community', in Emmanuel Adler and Michael Barnett (eds) *Security Communities*, Cambridge: Cambridge University Press, pp. 69–118.

Wæver, Ole, Barry Buzan, Morten Kelstrup and Pierre Lemaitre (eds) (1993), *Identity, Migration and the New Security Agenda in Europe*, London: Pinter.

West, J. Robinson (2005), 'Saudi Arabia, Iraq and the Gulf' in Kalicki and Goldwyn (eds) *Energy Security* . . . pp. 197–218.

Williams, Alphanso and William Kerr (2011), 'Wishful Thinking in Energy Policy: Biofuels in the US and EU', *Energy Politics* **25** (Winter) 34–46.

Williams, James and A. Alhaji (2003), 'The Coming Energy Crisis?', at http://www.wtrg.com/EnergyCrisis/index.html (accessed 10 April 2012).

Wood, Steve (2009a), 'The European Union: A Normative or Normal Power?', *European Foreign Affairs Review* **14**(1) 113–128.

Wood, Steve (2009b), 'Energy Security, Normative Dilemmas, and Institutional Camouflage: Europe's Pragmatism', *Politics and Policy* **37**(3) 611–635.

Wood, Steve (2011), 'Pragmatic Power EUrope?', *Cooperation and Conflict* **46**(2) 242–261.

Wood, Steve (2012), 'Energy Issues in the EU and Taiwan' in Roland Vogt (ed.) *Europe and China: Strategic Partners or Rivals?* Hong Kong: Hong Kong University Press, pp. 175–199.

Wood, Steve and Richard Grant (2010), 'Energy and Borders in the EU and its Neighbourhood' in Anathasios Kalogeris (ed.) *The Multifaceted Economic and Political Geographies of Internal and External EU Borders*, Association for Borderlands Studies: Thessaloniki, pp. 139–154.

Youngs, Richard (2010), 'Security through democracy: between aspiration and pretence', Working Paper 103, Madrid: FRIDE.

Youngs, Richard (2011a), 'Misunderstanding the maladies of liberal democracy promotion', Working Paper 106, Madrid: FRIDE.

Youngs, Richard (2011b), 'The EU and the Arab Spring: from munificence to geo-strategy', Policy Brief 100, Madrid: FRIDE.

Zangger, Claude (1974), 'Nuclear Energy and Democracy in Switzerland', *IAEA Bulletin* **16**(6) 8–18.

Zimmermann, Hubert (2008), 'How the EU negotiates trade and democracy: The Cases of China's Accession to the WTO and the Doha Round', *European Foreign Affairs Review* **13**(2) 255–280.

# 3. Framing new threats: the internal security of gas and electricity networks in the European Union[1]
*Peter Zeniewski, Carlo Brancucci Martínez-Anido and Ivan L.G. Pearson*

## 1 INTRODUCTION

Most studies investigating the concept of energy security begin by drawing attention to the different interpretations thereof (Chester 2010; Winzer 2012). The large number of conceptual approaches arise from differences between academic disciplines (economics, engineering, political science), historical contexts (the prevailing market and political conditions at the time), levels of development (advanced/developing/underdeveloped countries), timeframes (short-term/medium-/long-term outlooks), market dimensions (demand or supply, liberalised or regulated), value chain (upstream/transport/downstream/end use), levels of analysis (individual, sectoral, national, regional, or global) and the primary or transformed fuel in question (oil, coal, gas, nuclear and renewables on the one hand and electricity, heat or refined products on the other).

The variable interaction of these factors lends an inordinate level of complexity to the issue of energy security, which is further compounded by the varying perceptions by different actors depending on their role and position in the energy supply chain. Indeed, for an energy-exporting firm, security is predominantly about maintaining the ability to serve foreign markets and to keep up a sufficient level of demand and revenue from energy production. For net energy-importing states, security is often a political question of ensuring adequate, reliable and diversified sources of supply; but even within these categories a great deal of variation exists depending on the institutional structure and politics of a given state, societal or market actor. Moreover, there is also differentiation at the transnational and sub-state level; for example, multinational energy firms may view security mainly in terms of a hospitable investment climate in the territories in which they operate, whereas household and industrial consumers assess security in terms of the price and efficiency of energy.

The way in which these factors shape the definition of energy security is not simply a semantic squabble – one's criteria in defining energy security often informs policy outlook. For example, market-based analyses that restrict the essence of energy security to quantifiable phenomena such as physical availability and price tend to posit minimal government intervention as a desired state for energy markets (Noël 2009). Studies that incorporate wider qualitative conditions such as affordability, sustainability or the impact of a loss of supply on social welfare tend to favour greater regulatory oversight of energy markets (Chester 2010). On a practical level, the definition of energy security and its scope conditions will crucially affect how both policy-makers and academics identify, order and manage risks and vulnerabilities affecting the energy system, in whatever form it is analysed.

Thus, in order to meaningfully contribute to the issue of energy security, a great deal of domain-setting and analytical precision is required. This chapter deals with the internal security of the European Union's gas and electricity grids, laying particular emphasis on future challenges to network stability arising from the liberalisation of the internal market for natural gas, the integration of renewable energy sources in the electricity network and the cyber-security challenges to interconnected energy infrastructures.

## 2 CHALLENGES ARISING FROM LIBERALISED GAS MARKETS

The transport of natural gas – particularly across borders – is one of the most commonly analysed phenomena by political scientists when they address the issue of energy security (Hoogeveen and Perlot 2005; Heinrich 2007; Orbán 2008; Pascual and Elkind 2010: Le Coq and Paltseva 2011; Ratner et al. 2012). It is here that geopolitics is a salient parameter; this is particularly true for the European Union, which depends to a large extent on an extensive network of high-volume transmission pipelines carrying gas imported from Russia, Norway and Algeria. That the placement of and flow from these pipelines carries much geopolitical significance is particularly apparent for countries of Central and Eastern Europe as well as the former Soviet Union, for reasons which have been discussed elsewhere in this book. For present purposes it suffices to note that most studies investigating the security of natural gas supply focus predominately on the external context, such as the risks of relying on Russian gas or Ukrainian transit pipelines (Hadfield 2008; Luft and Korin 2009; Mott MacDonald Consultants 2010).

However, there is an internal side to the EU's security of gas supply which is rarely acknowledged in political science literature. In fact, many such analyses effectively halt their discussion of supply risks and vulnerabilities at the EU's borders, assuming that gas produced and transmitted inside the Union is free of the political and commercial risks prevalent in the external sphere (Finon and Locatelli 2008; Bahgat 2011; Le Coq and Paltseva 2011). Yet venturing inwards prompts consideration of a different kind of security, one based less on geopolitically-motivated cuts to supply and more on the challenges arising from gas market liberalisation. Here, security of supply considerations are present in the wider push to build an internal market for gas, manifesting themselves through issues such as third party access and capacity allocation, spot market trading and price volatility. These issues, in fact, warrant close attention in terms of their impact on security of gas supply (understood in this case as the ability of the gas system to satisfy energy services that are demanded over a long-term investment horizon). But doing so requires an awareness of the context in which these security of supply challenges arise.

The European Commission has been attempting to liberalise the gas market for several years. The first meaningful Directive (98/30/EC) was aimed at opening gas networks, which were dominated by vertically-integrated monopolies, to competition from third parties. Further structural measures to achieve full market opening were deemed necessary under the Lisbon agenda, where the second gas directive (2003/55/EC) strengthened provisions to ensure third party access and legal unbundling of transmission system operators. However, because the impact of these directives was not practically felt by customers (where little switching was occurring) nor the wider European network (a continued lack of interconnections precluded new supplies), the European Commission launched the 'third package' of energy legislation (European Commission 2007b). This package constitutes a concerted push for liberalisation of both gas and electricity networks on the basis of structural changes to both industries. Greater authority has been vested in pan-European institutions such as the Agency for the Cooperation of Energy Regulators (ACER) and the European Network of Transmission System Operators, for both gas and electricity (ENTSO-G; ENSTO-E). These bodies play important coordinating roles and are responsible for promoting common rules and the harmonisation of different national operating procedures. In this context, EU-wide guidelines and network codes to facilitate cross-border gas transactions have been developed by ACER and ENTSO-G, addressing issues such as interoperability, capacity allocation, network balancing and congestion management across borders. However, it is easy to lose sight of the broader objectives behind restructuring the gas market; though not always recognised by policy-makers in Brussels, market liberalisation is not an end in itself. Rather, it should be viewed as the means to better quality services, greater customer choice and lower end-user prices.

## 2.1 The Role of Security in Europe's Liberal Gas Market

The EU's energy policy rests on three 'pillars': competitiveness, sustainability and security. Many reports have been devoted to unpacking these concepts and explaining their relationship to one another (e.g. IEA 2008). Broadly, it appears that the guiding philosophy in Brussels is that the market is the most effective tool for ensuring security of supply and should therefore take precedence over state management of the gas sector (whether in the form of active intervention during supply crises, state-led investments in security projects or robust public service obligations relating to strategic storage, back-up fuel capabilities, etc).[2] This reasoning is based on the idea that greater competition and third party access to infrastructure will increase the number of market players and facilitate cross-border gas trade. An open, liberalised gas market, moreover, shifts the responsibility for stable gas supplies from governments and state-owned monopoly energy firms to a broader spectrum of actors – traders, shippers, network operators and end-users. This lends a certain level of supply flexibility via the wholesale gas market which is difficult to achieve in a model dominated by a handful of vertically-integrated firms.

However, shared responsibility for security of supply also carries with it several new challenges and risks. Indeed, the unbundling of vertically-integrated energy firms means that no single entity is responsible for gas supply across the whole gas market; rather, each company is responsible for its own customers and those customers that are eligible, in turn, can opt to pay a premium for additional security (in the form of firm contracts or ensuring their own means to substitute gas supplies). But the main actors responsible for gas flows – traders, suppliers and operators – principally respond to market signals when they consider whether to invest in new infrastructure, redundant capacity or a diversified portfolio of suppliers. In most cases, these actors will not provide extra investments for purposes of reliability if they are not compensated by a competitive return on investment. Moreover, uncertainty is created by the separation of gas supply and transportation activities, in that transporters will not have complete information about future sources of supply.

Liberalisation will also challenge the traditional model whereby vertically integrated, national gas monopolies secure supply by negotiating long-term, oil-indexed contracts (LTCs) with external suppliers. Historically, the long-term nature of these contracts guaranteed a stable source of supply over several decades and hence justified the heavy investment costs of building long-distance gas pipelines. However, this model is being challenged by the EU's internal market reforms; the recent 'Gas Target Model' endorsed by the Madrid Forum (a high-level meeting that brings together all European stakeholders in the gas sector) is effectively a fundamental challenge to the bilateral relationships between buyers of gas and their (external) producers. Indeed, the GTM puts forward the expectation that all trading will be carried out on the spot-market, implying that once long-term contract gas sold by external suppliers enters the theoretical single entry/exit zone of the EU, it can be traded from hub to hub until it is delivered to a distributor or end-user. The problem with this hub-based model is not only that major investments in transmission infrastructures may be discouraged, but also that flows within the EU become less predictable as they will respond to market signals rather than pre-arranged deliveries of gas. This, in turn, will increase the complexity of balancing procedures (ENTSO-G 2012). This is a contrast to an alternative approach more amenable to security of supply, in which base-load gas demand is satisfied by LTCs while peak load periods will become the responsibility of the internal spot market.

The changing nature of gas supply and transport is not the only area where liberalisation can impact on security. There is also the challenge stemming from price volatility. Spot-based trading on virtual and physical gas 'hubs' (where buyers and sellers meet to determine the market price based on supply and demand) increase the frequency and intensity of price fluctuations. Whereas power generators and large industrial consumers of gas can take advantage of this volatility by varying their demand with the price level and using risk management instruments, small customers such as households and small businesses do not benefit from volatility; for these end-users the resulting uncertainty of the gas bill is not linked to the physical but rather the traded market for gas. Thus, in a liberal market, the demand profile has an important bearing on security of supply. For customers with inelastic demand, that is, where their demand for services provided by gas supplies does not change substantially due to price movements,[3] security of supply is a very important dimension. This is true for households and essential social services. For customers with elastic demand, for example industrial companies or power generators, where consumption varies with the price of gas, security of supply is a consumer choice.

With this in mind, there are two additional risks to security of supply that manifest themselves in cases where a liberal market services a demand profile that is predominately made up of gas-fired power plants (which are increasingly built as combined-cycle gas turbines, or CCGTs). The first is a price risk, whereby price volatility in natural gas markets is transmitted to electricity prices; this is amplified by the use of gas as the marginal source of fuel for 'peak' demand periods. The second risk is supply-related; disruptions to gas supplies threaten the reliability of electricity generation. In a liberalised market, this threat is amplified by the lack of economic incentives for CCGT operators to store alternative back-up fuels, given the high costs involved in maintaining on-site lighter distillates that can technically replace gas volumes (e.g. fuel oil). Although it has been argued that volatility may provide the incentive to gas market players to invest in system flexibility (e.g. storage or LNG terminals that enable one to take advantage of price fluctuations), the question remains as to whether volatile prices spur speculative behaviour instead of investment.

Finally, during a supply interruption the incentives to allocate gas are fundamentally different between the two market models; a de-centralised, trading-based market with little public service obligations may allocate gas to the highest bidder, rather than conserve it for essential social services and customers that are unable to switch fuels.

## 2.2 The Role of Network Development in Overcoming Security Challenges in a Liberalised Gas Market

Despite the challenges noted above, it should be acknowledged that liberalisation can also benefit security of supply in a number of important ways. Indeed, a well-functioning liberal market will improve security of supply by providing transparency and information to a wide range of market players involved in the transmission, trade and sale of natural gas. The benefits accrued from this model, among which are lower prices, better services and greater accountability, would otherwise be lacking under the traditional model of monopoly state-owned gas firms importing gas through opaque and confidential long-term contracts. Moreover, as gas import dependence continues to rise a larger number of players in the downstream market may compensate for the increasing level of supply concentration – and the associated risks and vulnerabilities – that many Member States, particularly in import-dependent regions such as Central and Eastern Europe, will have to contend with in the future.

However, no market design is perfect. Liberal gas markets will not by themselves invest in additional flexibility to cope with high impact/low probability events, such as extreme weather conditions (witnessed most recently during the February 2012 cold snap) or external supply shocks (e.g. the Russia/Ukraine gas crises in 2006 and 2009). Thus, the onus is on regulators and governments to provide an institutional and regulatory framework that facilitates security of supply objectives. Although the instruments available are far-reaching, this section focuses in particular on the importance of encouraging cross-border interconnections, where the twin EU energy policy objectives of security of supply and liberalised markets meet. After all, the greater the number of LNG terminals and pipeline interconnections between markets, the less gas transport becomes a captive market that is bound by the need to physically link producer A to consumer B. These infrastructures also bring security of supply benefits by diversifying sources and increasing the resilience of the network to supply shocks from any one source.

The European Commission has recognised the need for infrastructure investments to prop up the internal market and, taking into account the principle of subsidiarity, has put forward a number of initiatives to strengthen natural gas links between Member States. In October 2011, the EC adopted the proposal for a Regulation on 'Guidelines for trans-European energy infrastructure', with the objective of completing strategic energy networks and storage facilities by 2020. To this end, the EC has identified 12 priority corridors, among which gas networks play a prominent role, and has put forward a financing mechanism, the 'Connect Europe Facility' to expedite projects of common interest that would otherwise be considered too risky by the market. The CEF will be one of several tools to fulfil the estimated investment target of €200 billion for energy transmission networks by 2020, as forecast by the EC's in-house system model, PRIMES (European Commission 2011a).

For gas, the priority areas include North/South corridors for Western Europe and Central/South-Eastern Europe, the Baltic Energy Market Interconnection Plan (BEMIP) as well as the Southern Corridor. The purpose of the two North/South corridors is two-fold. The first goal is to better interconnect the Mediterranean area (supplied by sources in northern Africa) with supplies from Norway and Russia. The second goal is to diversify the energy supplies of Central and Eastern Europe by linking sources from the Baltic, Adriatic, Aegean and Black Seas and to deliver gas infrastructure projects that improve the interconnectivity of the region in a North-South direction. Both of these goals will serve to increase short-term gas delivery and enable the full use of possible existing infrastructure, notably LNG terminals. The corridors will also remove internal bottlenecks which currently prevent free gas flows in the regions, and enhance the responsiveness of the system to potential supply disruptions.

Besides these broader initiatives, which are further explored and refined in ENTSO-G's ten-year network development plans as well as the Gas Regional Investment Plans (GRIPS), the EU has also put in place some important regulatory provisions to encourage member states to interconnect with one another's gas infrastructures. For example, the latest regulation on security of gas supply EC/994/2010 is an enabler for investing in cross-border gas infrastructure, obliging member states to enact bi-directional capacities on all cross-border gas pipelines so that gas flows are more flexible during supply crises.

Yet despite all the forceful rhetoric about achieving a competitive internal gas market, most EU legislation recognises the continued importance of long-term bilateral contractual arrangements between non-EU gas producers and EU buyers.[4] Moreover, the emphasis on using public funding to build large-scale infrastructure projects linking Member States constitutes a tacit admission that the liberalised market may fall short of investing in sufficient capacity in the future. Thus, what appears as a broader liberalisation push is actually an effort to build a hybrid gas market structure, one in which the 'anti-competitive' long-term gas supply contracts sit alongside a liberalised, wholesale market for transmission and storage, built around third party access and hub-based trading. This market design is partly the product of attempts by policy-makers in Brussels to reconcile the tension between two of the three pillars of European energy policy: competitiveness and security.

All the while, the wholesale gas market will benefit from physical interconnections, driven by EU policy and sponsored by public funding, that it would not otherwise have the incentive to create. In this sense, ironically, the wholesale gas market becomes a free rider capitalising on the provision of public goods. This must be borne in mind when end-users assess the benefits of liberalisation in terms of secure and affordable supplies of natural gas.

## 3  NEW CHALLENGES TO THE SECURITY OF ELECTRICITY SUPPLY

On February 2011 the European Council (EUCO) concluded that in order to reach the EU objective of reducing greenhouse gas emissions by 80–95 per cent by 2050 compared to 1990 levels, a revolution in energy systems is needed and it must start now (European Council 2011). In order to reach such an ambitious goal, electricity will have to play a key role. The European Commission (EC) called for a European power sector which 'can almost totally eliminate $CO_2$ emissions by 2050' (European Commission 2011c). Several studies have concluded that full decarbonisation of the power sector is technically feasible but will require substantial investments in renewable energy sources (RES), network development, carbon capture and storage (CCS) deployment, energy storage and increased energy efficiency (Haller et al. 2010; Delucchi and Jacobsen 2011; European Commission 2011b).

48   *International handbook of energy security*

The European Council called for an Energy Policy for Europe with three main pillars: security of supply, competitiveness and environmental sustainability (European Council 2006). The changes and developments which are expected to take place in the transition of the European energy system will have an impact on security of supply. Accordingly, the EC's Energy Roadmap 2050 (European Commission 2011b) explores the challenges arising from the need to achieve the EU's decarbonisation objective while at the same time ensuring security of supply and competitiveness. It states that the RES share rises significantly in all the scenarios considered, achieving at least 55 per cent of gross final energy consumption (today's level is around 10 per cent). In other words, the EC concludes, that in order to meet the desired objectives, RES will have to represent the biggest share of energy supply technologies by 2050.

Such a large-scale integration of renewables in an electricity system that is currently dominated by fossil fuels poses a number of challenges. To address these challenges in an adequate institutional framework, the European Commission's Third Energy Package formalised co-operation between Transmission System Operators (TSOs) by establishing the European Network of Transmission System Operators for Electricity (ENTSO-E). Its main objective is 'to promote the reliable operation, optimal management and sound technical evolution of the European electricity transmission system in order to ensure security of supply and to meet the needs of the Internal Energy Market' (ENTSO-E 2010).

The objective of this section is to discuss the potential challenges that an extensive RES deployment would bring to the electricity network in terms of security of supply. In addition, the role and the need for electricity transmission development are discussed, especially in terms of cross-border interconnections.

### 3.1   RES Challenges to Security of Supply

The EC defines security of electricity supply as the 'ability of an electricity system to supply customers with electricity' (European Commission 2005). However, the simplicity of this definition belies the complexity of electricity generation and transport. Indeed, unlike primary fossil fuels such as gas, oil and coal, electricity is a transformed form of energy that to a large extent cannot be stored. Because of this unique attribute, most electricity must be consumed whenever it is produced. This intrinsic characteristic challenges the integration of renewables into the electricity generation portfolio due to their variability and their partial predictability. Indeed, intermittent RES produce electricity only when the wind is blowing or when the sun is shining, thus not necessarily when it is needed. According to most analyses, wind and solar electricity retain the greatest potential to contribute and increase the shares of renewable electricity production in the short to medium term. For this reason, whenever RES is mentioned throughout this section, reference is particularly made to wind and solar energy sources.

RES integration brings several challenges to electricity networks. First of all, the structural characteristics of the electricity system must undergo fundamental changes in order to accommodate the large-scale deployment of RES. The traditional structure of the power system is evolving by introducing electricity generators at lower voltage levels and at more widely distributed locations. In this evolution electricity consumers can become producers as well (also known as 'prosumers'), depending on their RES electricity output at given moment in time. In addition, RES integration affects generation adequacy and back-up needs. Traditional electricity generation technologies are characterised by specific availability factors which may depend on several aspects, such as maintenance time, overhaul, reserves and potential unplanned interruptions. For RES, however, availability factors depend on the wind and solar radiation resources in the geographical location of the wind turbines or the photovoltaic (PV) panels. The overall annual availability of RES is therefore much lower than for traditional fossil-fired or nuclear power plants. In the case of hydro energy sources, their availability depends on the characteristics of the power plant, if it is driven by run of river, its power output is partly controllable and it can be quite accurately predicted; in the case of storage and/or pumping facilities, the power output can be fully controlled and its management depends on seasonal hydro inflows.

Operators and electricity market players must address these challenges by planning their investments, inter alia, around the expected deployment of RES. ENTSO-E plays a supporting role to this end, publishing an annual Scenario Outlook & Adequacy Forecast (SO&AF). This document 'presents the scenarios included in the Ten-Year Network Development Plan (TYNDP) in compliance with Regulation (EC) n. 714/2009 and the assessment of the adequacy between generation and demand in the ENTSO-E interconnected power system on mid- and long-term time horizons' (ENTSO-E 2010). The SO&AF 2012–2025 (ENTSO-E 2012a) describes the generation adequacy assessment of the countries served by ENTSO-E's Transmission System Operator (TSO) members for the period 2012–2025. In addition, the assessment is also presented for six regions and for the whole ENTSO-E region. Figure 3.1 shows a schematic of the generation adequacy assessment by ENSTO-E (ENTSO-E 2012a).

An important measure for the analysis of the generation adequacy for a specified country or region is the estimation of the Reliable Available

50  *International handbook of energy security*

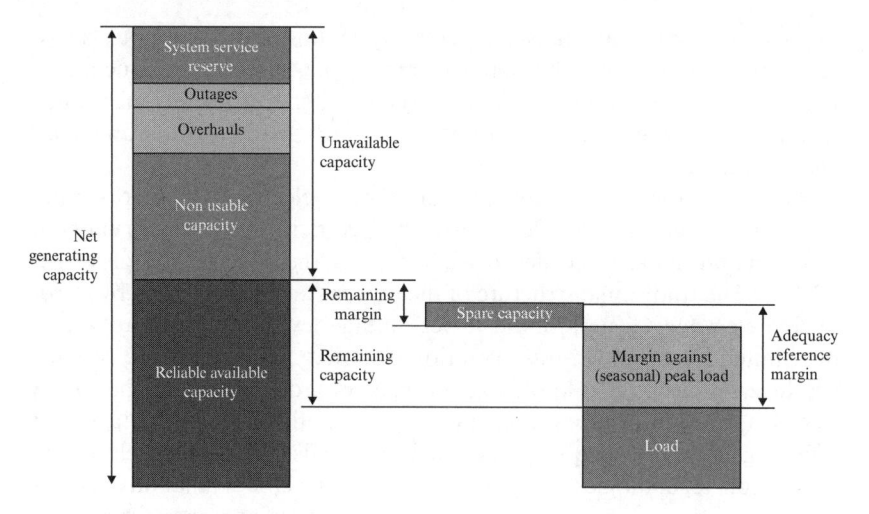

*Source:* ENTSO-E 2012a.

*Figure 3.1  Generation adequacy assessment from ENTSO-E SO-AF 2012–2030*

Capacity (RAC), which is equal to Net Generating Capacity (NGC) minus the Unavailable Capacity. The latter consists of Non Usable Capacity, maintenance, overhauls, outages and system reserves. For the scope of this discussion, we will only focus on the contribution of RES to the Unavailable Capacity, more precisely to the way in which the Non Usable Capacity is calculated for RES. The Unavailable Capacity is an important tool for estimating the additional investments necessary for ensuring security of supply, particularly in cases where a large amount of electricity is generated by RES. However, ENTSO-E (ENTSO-E 2012a) shows how European TSOs account for the unavailability or the Non Usable Capacity of wind and solar energy sources in very different and conflicting ways when assessing the generation adequacy of the countries that they serve. Some TSOs consider that wind generation capacity must be considered totally (100 per cent) or almost totally (94–96 per cent) as Non Usable generation when assessing generation adequacy due to the variable and uncertain characteristics of wind power generation. On the other hand, other TSOs only consider the average unavailability factor (70–75 per cent) of wind power generation.

The assessment of generation adequacy is of crucial importance when evaluating the security of supply of a country or a region. In order to efficiently plan the future of European electricity infrastructure, it is necessary that common definitions are used by all stakeholders. As the Agency for the Cooperation of Energy Regulators (ACER) stated, 'it is important that ENTSO-E promotes new methodological approaches to estimate reliable capacity of wind and solar power plants' (ACER 2012).

The RAC of RES depends on the size of the region that is considered and on the location and distribution of the wind turbines and the solar panels. Several studies (Yi et al. 2009: Hoicka and Rowlands 2011; Widen 2011) have analysed the correlation and potential complementarity of wind and solar energy sources. In some cases, these correlations increase the RES contribution in the RAC estimates. As mentioned above, it is important to analyse carefully and understand the definition of Reliable Available Capacity and its implications in terms of generation adequacy and security of supply, which in turn have an impact on the needs for electricity transmission investments.

A recent study (Grave et al. 2012) defines the concept of 'secure capacity', which results from a combination of several probabilistic distributions on the availability of each type of generation capacity. The increase in the secured capacity of the whole generation fleet provided by wind or solar energy generation is defined as their 'capacity credit'. The study focuses on Germany and assumes that the secured capacity for solar energy generation is 0 per cent, since the annual peak demand happens during hours of relative darkness (between 6 and 7 p.m. on a winter evening). However, this result will probably not be applicable to Southern European countries which observe annual peak load demand in the middle of a summer day due to the high air conditioning demand. In this case the 'secure capacity' of solar energy generation would be higher and significant in the case of a high PV penetration. The study (Grave et al. 2012) calculates that the capacity credit for wind lies between 5.2 per cent and 6.2 per cent of total installed wind generation capacity. The capacity credit is dependent on the distances between wind parks. When the latter increase, the correlation between the wind resources decreases and therefore the distribution function of wind energy generation flattens and the capacity credit grows.

The calculation of the capacity credit for wind or solar energy generators in a region or a country is based on meteorological statistical databases. The longer the period under analysis, the more accurate the capacity credit values. Intuitively, and due to the previously mentioned geographical correlation of natural wind resources, considering an interconnected larger area provides a higher RES capacity credit, which means that secured capacity needs from other energy sources (such as gas-fired power plants) will be lower.

The challenges that RES pose to security of supply, mainly in terms of 'secure capacity', can be balanced by other technological means such as electricity storage and the development of electricity transmission networks.

## 3.2 Impact of Network Development on RES Integration and on Security of Supply

Electricity transmission is an enabling technology that can be used to alleviate, to a certain extent, the challenges that variable RES pose to security of electricity supply in Europe. Several studies state that electricity transmission development, including cross-border interconnections, is essential in order to cope with the variability of RES and to reach the almost total decarbonisation of the European power system by 2050 (European Commission 2011b; Jaureguy-Naudin 2012). For instance, ENTSO-E's latest TYNDP (ENTSO-E 2012b) claims that 80 per cent of the planned electricity transmission projects for the next decade will bring high benefits for the expected RES integration; half will directly connect RES and the other half will accommodate inter-area imbalances triggered by RES.

Development of the European electricity transmission network in order to manage variations from a high penetration of wind and solar electricity generation reduces, partially, the need for energy storage and back-up generation capacity (European Commission 2011b). These three technologies of the electricity system can be seen, in some measure, as complementary. This does not mean that any of them can fully substitute one of the other two. Some of their characteristics and their contributions to the functioning of the electricity system are similar. Nonetheless, the present and especially the future European electricity transmission network will not be able to pursue the three pillars of EU energy policy without any of these technologies. They are all necessary. However, the fundamental question is: to what extent and with which proportion does the European electricity system need transmission capacity development, storage and back-up generation capacity?

In order to answer this question and to adequately plan the future European electricity system, several aspects must be considered. First of all, investment costs for each of these technologies must be taken into account. Another very important and urgent concern is the design of the markets in which each of these technologies will play a role. The rules established by these markets will foster investments towards a certain direction and they will shape the future electricity system. For instance, who will invest in cross-border transmission capacity? And who will benefit from it? Another crucial question is, how can back-up generation capacity be incentivised? With an increasingly large RES share in the electricity generation mix, back-up capacity requirements (mainly gas-

*Internal security of gas and electricity networks in the EU* 53

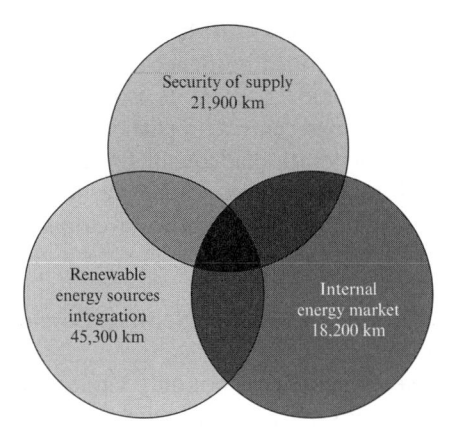

*Source:* ENTSO-E 2012b.

*Figure 3.2 Projects of pan-European significance: volume breakdown per EU energy policy pillar*

driven) will increase, but their load factor will significantly decrease. How do we ensure that market players sufficiently invest in these technologies? An additional interesting question addresses the role of storage: how will investments in storage be remunerated? And more than this: who will manage/operate the storage devices? Are the generation sources to be owned by private or balancing tools to be used by the system operators?

European electricity transmission development plays an important role in the European Union's 'sustainability' pillar. Indeed, the other two pillars – security and competitiveness – definitely benefit from additional transmission capacity (both national and across borders). The projects identified by ENTSO-E for the development of the European electricity network in the next decade amount to a total of 52300 km of either upgraded or new assets. Figure 3.2 shows the volume breakdown per EU energy policy pillar of the projects with pan-European significance (ENTSO-E 2012b).

ENTSO-E's latest TYNDP (ENTSO-E 2012b) has identified 100 potential bottlenecks in the European network by the end of the decade. Half of them are directly related to market integration; transmission development would facilitate grid access to all market participants and it would contribute to social welfare by internal market integration and harmonisation. 20 per cent of the identified bottlenecks directly relate to security of supply issues. Investments in transmission development would ensure safe operation of the electricity network providing a high level of security of supply.

ENTSO-E's TYNDP (ENTSO-E 2012b) has identified over 100 projects to solve the previously mentioned bottlenecks over the next decade: '33% of these projects are required to integrate isolated systems such as the Baltic States, secure large load centres (in particular capital cities), or even countries with negative generation adequacy forecast in the coming years (ex: Belgium)'. In addition, the challenges that variable RES bring to security of supply would also be alleviated by the development of the electricity transmission network.

### 3.3 Challenges to European Electricity Network Development

The extension of a power network is not only hindered by technical barriers. Most of the limitations are political, economic and physical. First of all, it is important to mention that 'the time required to get the authorisations to build new transmission assets is generally much longer than the time needed to build new power plants' (REALISEGRID 2009). The coordination of both types of investments are challenged by the long realisation periods and by the different lifetimes of generation and network infrastructures.

Another important challenge that electricity transmission development must face is the uncertainty about future investments in electricity generation. For example the EC estimated that in order to reach a decarbonised European power sector, the 'cumulative grid investment costs alone could be 1.5 to 2.2 trillion Euros between 2011 and 2050'. The higher value corresponds to a greater investment in RES. The wide range of expected investment costs shows the uncertainty level in terms of future electricity generation investments in Europe.

EU Member States (MS) are solely responsible for their national energy mix. The 27 EU MS together with the other European interconnected countries have their own strategies for shaping the future national electricity networks. In order to fully achieve the three EU energy policy goals, a more European vision must overrule the current national strategies. As Jaureguy-Naudin (2012) argues, EU MS should discuss and agree on a common EU electricity generation mix. In such a way, variable RES would be located where the natural resources are more favourable, a more optimised electricity network could be planned, and clearer signals for investments would arise.

Building a sustainable European electricity network is not without its challenges. This section highlighted some of the tensions apparent in simultaneously achieving a sustainable and secure electricity grid, particularly in the context of the integration of renewable sources of energy. However, it has been demonstrated that the development of the electricity transmission network will not only increase its resilience but also enhance its ability to accommodate large-scale deployment of RES, while at the same time facilitating the creation of a competitive market. In this way, future investments in electricity transmission infrastructure will satisfy all three pillars of the EU's energy policy goals, including security of electricity supply. The essential preconditions for reaching the aforementioned goals should be the synchronisation and harmonisation of the European countries' management and strategies for defining energy security as well as market designs that would incentivise investments in RES and network development.

## 4 CYBER SECURITY AND OUR INCREASINGLY INTERCONNECTED ENERGY INFRASTRUCTURES

Thus far, this chapter has focused on the physical system in place to transmit two different forms of energy: natural gas and electricity. However, the smooth operation of this 'hardware' needs to be underpinned by reliable 'software' that can securely manage the flows and information characterising this system. Thus, we now turn our attention to the threats arising in a different setting, namely cyberspace.

Since 2007, several serious cyber attacks on EU Member States have come to light (Myrli 2009). The most notable attack was that experienced by Estonia in April and May 2007. Because Estonia had embraced ICT as driver for economic growth during the 1990s, the societal impact of the three-week-long series of cyber attacks was substantial (Estonian Ministry of Defense 2008). Although authorities across the globe had recognised the theoretical threat posed by cyber attacks for over a decade, the scale and politically-motivated nature of the attacks on Estonia significantly raised the profile of the problem. The Estonia attacks also spurred an analytic shift. Whereas cyber security was previously examined predominantly through a law-enforcement lens, the pressing international security implications of cyber attacks had now been highlighted.

The year 2010 marks a dubious milestone being the first year that Chatham House devoted a significant section of its annually compiled Military Balance to cyber warfare – chief among a list of asymmetric techniques the think-tank believes are changing the face of modern conflict (International Institute of Strategic Studies 2010). This move is joined by major reports by the Center for Strategic and International Studies in Washington DC (CSIS Commission on Cybersecurity for the 44th Presidency 2008) and the United Kingdom's House of Lords (EU Home

Affairs Sub-Committee 2010), as well as numerous notable pieces on cyber security in leading European and international journals (Ruus 2008; Clark and Levin 2009; Geers 2009; Hughes 2010).

The debate indicates a consensus that potentially all digital systems are vulnerable, whether via their networked interconnections or engineered hardware defects. Networks are especially vulnerable because they consist of a vast number of interdependent devices and components, a great proportion of which may use off-the-shelf hardware or software that has been brought to market as quickly as possible. Because this commercially available product also tends to be the most affordable, economic considerations mean that it becomes widespread. Cyber attackers therefore often only need to discover just a single vulnerability in order to bring disruption at a systemic level. The end result is the current 'patch and pray' computer security paradigm – a defensive posture whereby potential exploits in deployed systems are continuously sought and fixed in light of the pervasive vulnerability of networks.

Most disturbingly of all, the discourse clearly suggests that our networks will increasingly become the battleground for conflicts involving both state and non-state actors such as terrorist organisations and criminal groups – a trend that policy-makers in all fields of government cannot afford to ignore. With specific regards to the energy sector, the effects of cyber attacks have the potential to affect a greater number of stakeholders than ever before because both our gas and electricity networks are becoming increasingly interconnected.

## 4.1 Cyber Attacks on Gas Transmission SCADA Systems

Especially of interest for existing energy infrastructures are attacks on Supervisory Control and Data Acquisition (SCADA) systems – real-time industrial control systems that are used to monitor and control dispersed physical assets or field devices (such as switches, valves, pumps, relays, etc.) from a central location while collecting and logging field data. SCADA systems are widely deployed in national infrastructures such as electricity, telecommunications, water and waste control, as well as oil and gas refining and transportation.

Generically, a SCADA system consists of central and remote site hardware, applications software and communications services. These systems encompass the transfer of data between a SCADA central host computer and a number of Remote Terminal Units (RTUs) and/or Programmable Logic Controllers (PLCs), and the central host and the operator terminals (National Communication System 2004). A SCADA control centre typically has advanced computation and communication facilities. Modern control centres have data servers, Human-Machine Interface (HMI) stations and other servers to aid the operators in the overall management of the utility's processes. The control centre is usually connected to the outside corporate network and/or the Internet through specialised gateways (Igure et al. 2006).

SCADA plays an essential role in natural gas transmission, allowing operators to analyse pipeline conditions and make timely responses to operational problems. Industry surveys reveal that 85 per cent or more of the world's operating oil and gas pipelines of more than 25 kilometres are controlled by a computer-based SCADA system (Treat and Bartle 2009).

Pipeline systems are almost always remotely managed from control centres where key variables, such as moisture, quantity, pressure and temperature are monitored and mechanical components, such as switches, valves, compressors, odorising stations, and pumps are operated via the SCADA system. Essentially, SCADA systems in the gas transmission system also monitor pipelines for total volumetric flow rate in order to provide yield data (Ryu et al. 2009). SCADA systems used for supervising pipeline transportation almost always have full (or as near as is possible) redundancy. For very critical pipelines there might even be a back-up operating site with another complete SCADA system sitting ready. Essential leak detection systems are usually incorporated into the SCADA system, using RTUs and PLCs for collecting field data and automatically executing certain necessary commands (Williams 2003; Shaw 2009).

Because of their sensitive and important nature, SCADA systems pose a target to groups and individuals seeking to attack national infrastructures. For example, computers seized from al-Qaida show details about a range of SCADA systems in the United States (Aylon 2009). An attack on the SCADA systems of Europe's gas transmission network could also have the added effect of crippling downstream facilities. For instance, a cyber attack on compressor stations that fuel natural gas power plants could result in blackouts that would affect many EU Member States in the increasingly interconnected European electricity system. In such a case, a cyber attack on a pipeline's SCADA system would probably not be intended to harm the SCADA system itself, but rather use the SCADA system to trigger actions harmful to the pipeline and associated facilities. It would be possible for an attacker to send forged messages into the SCADA system, causing the operating room to see what the attacker wants it to see (open valves looking closed, running pumps apparently stopped, pressure and flow measurements apparently stable, etc.). Less sophisticated attacks could aim to overload communication networks by means of a so-called 'data storm' in which too much data in the control network blocks its ability to monitor and control the physical process

(Averill and Luiijf 2010). This would blind operators to a dangerous situation for a moderate period of time, or deprive them of communications with field equipment. Many nodes on SCADA networks are embedded computing devices that run real-time operating systems (RTOS) and other real-time control software, making them especially susceptible to even minor disruptions of this sort.

It's important to differentiate here between a loss of control and a deliberate attempt to infiltrate and take over the system. When control is lost, as above, the system will generally fail safe, whereas if the system is hijacked, the consequences are potentially far graver (Munro 2008).

The end-effect of such attacks could be physical damage to a pipeline, release, pressure drop, product loss, environmental contamination, explosions, fire, death, injury or other serious consequences. The degree of damage attacks of this nature could generate depends on the intelligence level of the RTU (how much local, semi-autonomous regulatory control and sequence-safety logic it performs), the types of equipment it directly operates (valves, pumps, motors, etc.), and the presence (or lack) of hard-wired safety logic that would override the RTU's dangerous actions (Shaw 2009).

If such attacks are detected, completely reformatting hard drives and reloading the affected computers (and other infected network components) from 'clean' back-up media, usually puts things back as they were before the attack. But this takes time (once it is actually realised an attack is in progress) and requires thoroughly tested, well-documented and well-rehearsed procedures for performing a system restoration. Because the hardware and processes SCADA usually control are very time sensitive, significant commercial losses can be expected in even minor cyber attacks.

The SCADA systems of the gas and oil sectors have already been the target of successful cyber attacks. In the winter of 2002–3, attackers were able to penetrate the SCADA system responsible for tanker-loading at a marine terminal in eastern Venezuela. Once inside, the hackers erased the programs in the system's PLCs that operated the facility, preventing tanker loading for eight hours (Byres 2009). In a recently published book, a senior US national security official reported how the USSR was allowed to steal compromised pipeline control software from a Canadian company including malicious code that caused a major explosion of the Trans-Siberian gas pipeline in June 1982. The code ran during a pressure test on the pipeline and massively increased the usual pressure, causing the explosion (Reed 2005). This attack highlights the fact that any pipeline system is susceptible to catastrophic failure as a result of loss of control over the hardware measuring or managing pipeline pressure. The highly flammable nature of natural gas exacerbates this problem.

Pipelines and related field sites are typically located in remote, unmanned areas with relatively infrequent maintenance visits, making physical sabotage comparatively uncomplicated. Cyber attacks would therefore be more likely in situations where the attacker's goals are more substantial – for example, the systematic exploitation of vulnerabilities by criminal or even terrorist organisations for financial or political gain. This has led some analysts to suggest that control centres would be the target of the most sophisticated attacks (Averill and Luiijf 2010). Alternatively, motivated attackers may pursue cyber-attacks as a low-risk alternative to physical attacks.

## 4.2 Cyber Attacks on the Electricity Grid

Notwithstanding the focus on natural gas, electricity networks also play a pivotal, albeit understated, role in European energy security. A sizeable 37 per cent of Europe's total primary energy consumption – regardless of fuel source or origin – is destined for the generation of electricity.[5] Given the grid's essential role in delivering energy to the citizen, the security-of-supply implications of this statistic are clear: weaknesses in the electricity network have the potential to circumvent the benefits of a diversified, affordable and readily accessible primary energy mix. Moreover, reliance on electricity will almost certainly continue to grow as Europe's society and its economy become increasingly digitised.

Major blackouts in Italy in 2003, and across Europe in 2006, demonstrated the devastating societal effects a cascading failure of the increasingly interconnected electricity grid can have. During these blackouts, millions of households were deprived of electricity, high-speed trains were halted, members of the public were trapped in lifts and international airports were forced to switch to emergency generation. The blackouts also demonstrated the high level of transnational interdependency that exists in the European electricity transmission system. For example, although the 2006 blackout originated in Germany, it affected an estimated 5 million households in France, as well as 10 million households across Belgium, Germany, Italy, Portugal, Spain and Eastern Europe. Austria, Croatia and the Netherlands were also reportedly affected, and the disturbance even reached as far as North Africa via the Spain-Morocco submarine cable (Stefanini and Masera 2008; van der Vleuten and Lagendijk 2010a, 2010b).

In Europe, a number of factors are driving a revolution in electricity transmission (between the power plant and substation) and distribution (between the substation and the consumer). As mentioned in Section 3.2, the electricity sector will become increasingly reliant on intermittent and distributed wind and solar power. Upgrading ICT is necessary to increase the overall efficiency of the grid, and to increase the grid's reliability in this new challenging context.

### 4.2.1 Growing ICT dependence in the European grid

Three fundamental physical realities already make the electricity grid more dependent on ICT than any other energy infrastructure. First, electricity flows at close to the speed of light, demanding high-speed decision-making and response times for its management. Secondly, it must be used or stored the instant it is produced or else it goes to waste, making maximising efficiency a constant challenge. Thirdly, the flow of alternating current electricity through the grid cannot be straightforwardly controlled by opening or closing a valve in a pipe, or switched like a call on a telephone network. Electricity flows freely along all available paths from the generator to the load in accordance with the laws of physics, and although this flow can be manipulated to a degree using devices such as phase shifting transformers, accurately determining flow along the grid's complex architecture is a complex task (US-Canada Power System Outage Task Force 2004).

The net result is that ICT has become an integral part of the modern electricity transmission system, gathering network information and automatically issuing control actions in near real-time to keep the grid stable as demand and generation fluctuates. The interdependence between the electricity and communication infrastructures is now so profound that some researchers even suggest that they need to be conceived of within an 'energy-and-information', or E + I, paradigm: disturbances in the telecommunication infrastructure can cripple the electricity infrastructure as much as is true visa versa (Gheorghe et al. 2006).

As we look to do more with less fossil fuel, this interdependence is only set to grow. ICT will incentivise consumer energy efficiency by providing real-time feedback on household energy consumption. Studies suggest that direct feedback on electricity consumption, such as that provided by smart meters, leads to domestic energy savings in the region of 5–15 per cent (Darby 2006; Fischer 2008). The 'smart' grids of the future will be more efficient, allowing operators to reduce system losses by more quickly and accurately balancing loads, managing feeder voltages, and analysing power flows. They will be significantly more robust thanks to their ability to 'island', whereby self-sustaining portions of the grid isolate themselves from disrupted sections and help to 'black start' the system after a widespread blackout (L'Abbate et al. 2007).

Decentralised management, diverse network routes and a variety of interlinked utilities and operators have the potential to make the

European electricity system highly reliable and even self-healing (European Commission 2006; United States Department of Energy 2008). Most importantly, however, smart grids will mitigate the effects of climate change and hydrocarbon import-dependence by enabling Europe to incorporate more electricity from indigenous renewable sources that are currently too widely distributed and intermittent for the feed-forward networks in service (Ackermann et al. 2012). This is an especially urgent consideration for Europe given the sharp rise of renewable generation expected (International Energy Agency 2009) and reports that installed wind and solar power is already testing the infrastructure of some Member States (EurActiv with Reuters 2010).

**4.2.2 Network reliability and security challenges in the transition to smart grids**
Whilst this shift towards 'smart', or at least 'smarter', distribution networks could play a key role in enabling Europe to address many of its most pressing energy priorities, it is widely accepted that a number of challenges will need to be overcome.

The next generation of Europe's electricity grids will face an even greater variety of cyber vulnerabilities than those of the gas and electricity networks of today. The remainder of this section reviews developments in the field of cyber security, analysing five of the most pressing challenges the electricity network will face in coming years: 1) the large amount of sensitive customer information the grid will transmit; 2) the greater number of control devices in the smart grid; 3) the poor physical security of a great proportion of these devices; 4) the move away from industry-specific communication standards and hardware; and 5) the greater number of stakeholders the grid will rely on for its smooth operation.

*4.2.2.1 Customer security*   Perhaps the most widely-acknowledged challenge is that smart grids will create the theoretical possibility of a type of attack hitherto unknown in the energy sector: cyber attacks through the electricity network, as opposed to on the network itself. Smart grids will transmit a larger amount of sensitive customer information than the electricity network has at any time before. Additionally, smart metering technology may eventually allow system operators to attenuate customer demand in order to balance the grid during an incident (European Commission 2007a). Given this enhanced functionality, security failures in the grids of tomorrow may not only mean reduced overall grid reliability, but also the possibility of the exposure of highly sensitive information on household activity, and selective or coordinated blackouts. Put simply, the system operators of tomorrow will be more closely charged with the well-being of European citizens than ever before. This will require nothing short of a paradigm shift in industry from the current hardware-centric focus on system adequacy and reliability[6] (Endrenyi and Welsow 2001) towards the inclusion of a more directly consumer-oriented view of security that addresses issues such as the integrity of data communicated: the authentication of communications; the prevention of unauthorised modifications to smart grid networks; the physical protection of smart grid networks and devices; and the potential impact of their unauthorised use on the bulk-power system.

*4.2.2.2 A greater number of intelligent devices* Smart Grids will be reliant on an exponentially greater number of digital devices than today's grids – devices that will be decisively involved in managing both electricity supply and demand in the network. Remote sensors dispersed throughout the transmission system will increasingly help system operators identify bottlenecks, and keep the grid operating reliably in spite of the greater transnational electricity flows anticipated. The 'active' distribution networks of the near future will rely on intelligent substations to quantify and manage the ever increasing amount of electricity put back into the grid from distributed renewable generation. And smart meters will enable greater consumer demand response than ever before.

In absolute terms, this proliferation of intelligent control devices may not significantly increase the grid's reliance on ICT when compared to the present situation: as described earlier, the supply of electricity to vast numbers of consumers is already acutely dependent on the reliable operation of a relatively small number of centralised and well-protected Supervisory Control and Data Acquisition (SCADA) systems. However, if we accept that each node in the network also represents a theoretical point of access, the transition to more intelligent grids will unavoidably increase the grid's permeability to those with malicious intent. Moreover, bigger networks are harder to monitor and police – a stark fact given some estimates that the smart electricity network of the future will be between 100 and 1,000 times larger than even the Internet (LaMonica 2009).

*4.2.2.3 The problem of physical security* Most experts agree that physical security – controlling the physical access to machines and network attach points – is perhaps more critical than any other aspect of computer network security. But building the smart grid not only entails increasing the number of critical control devices, but also their diffusion to a greater number of insecure physical locations than ever before. Because sensitive SCADA systems are already housed in secured sites in the transmission system, most attention will be needed at the level of the distribution network where the passive, radial architecture of the past will give way to a new, meshed structural design that requires the introduction of a plethora of intelligent control devices where once there were few.

Most pressing of all, systemic exposure to faults and malicious activity originating from smart meters will need to be minimised in light of their location within customers' homes, where physical security will be impossible to guarantee. This will have a profound impact on the way network security for the smart grid must be conceived of as it will make the conventional distinction between 'safer' internal networks (such as local area networks) and 'more dangerous' external networks (such as the Internet) a false crutch for security design. Even though there are no plans to connect smart grids to the Internet, direct and indirect connectivity between smart meters and every other node in the electricity network – including power plants – will demand an extremely disciplined approach to internal network security.

*4.2.2.4 The use of internet protocol (IP) and commercial off-the-shelf hardware and software* The use of IP as a communication standard will compound the problems of large network size and physical insecurity. IP offers a practical advantage over other standards in that it facilitates interoperability between a broad range of components, and is flexible enough to accommodate the evolution of the smart grid into as-yet-unforeseen application areas. Depending on the data transmission medium, IP may be more reliable because of its dynamic routing capabilities. And IP is also more affordable to use, being both widespread and well-supported by industry.

One significant drawback of IP is that it is a common network standard with numerous, widely-known vulnerabilities. For example, messages to be transmitted over an IP network are broken into manageable chunks called packets, but attackers have learned to use a number of techniques to mask the origin of a packet as it 'hops' from router to router along an IP network's potentially complex infrastructure, making attacks difficult to trace (Lipson 2002).[7] Hiding the source of messages is just one of many IP vulnerabilities that malicious users have learned to exploit, and researchers have already cautioned against a recent industry trend to replace specialised software and communications systems with IP for reasons of cost and convenience (Markulec 2008; Munro 2008).

Also of concern is the likelihood that the substantial costs associated with the size of the infrastructure upgrade will mean commercial off-the-shelf hardware and software will become widespread in smart grids. This would subject them to the same forces that have contributed to the systemic insecurity of the Internet and corporate IT networks as many smart grid components and applications will be manufactured and distributed on a mass scale, making any vulnerabilities they carry also exploitable on a mass scale. Additionally, smart grid components will be built with the minimum computational capabilities thought necessary at the time of their design to keep production costs low, possibly limiting the security measures they can support in the future.

Because interoperability and affordability will be key challenges in the transition to smart grids, it will be difficult to resist the broad use of IP and commercial off-the-shelf hardware and software in the networks of the future. However, their use also means that potential attackers will not need to be SCADA experts, or have detailed information on industry-specific hardware vulnerabilities, in order to launch an effective and widespread attack on a smart grid. Techniques and vulnerabilities discoverable on the Internet and commercial networks will be adaptable to smart grid attacks, and such expertise will also become more widely known.

*4.2.2.5 More stakeholders* The liberalisation of the European electricity market in recent years has led to the unbundling of the vertical public monopolies that were once dominant. Electricity generation, the operation of the grid and the provision of wholesale and retail electricity services to customers are all now carried out by a greater number of increasingly separated stakeholders that continuously need to share data with each other in a marketplace that looks set to become ever more diversified. While this offers the customer the opportunity to benefit from new services and efficiencies, networks are especially fragile to internal threats and negligence (Cisco Systems 2007).[8] Minimising the impact of the 'human element' will become even more difficult with the smart grid as the traditional mechanisms co-ordinating grid reliability and security fade away, and the number of utility companies, contractors and smart meters increases (De Bruijne and van Eeten 2007; van der Vleuten and Lagendijk 2010a).

The heart of the problem is how to deal with reduced accountability – an unwanted by-product of the new electricity market. As Marcelo Masera writes:

> The current decentralised nature of liberalised electricity infrastructure has as a consequence that individual operators cannot be held responsible for the way the system as a whole functions. . . . Nobody owns, designs, or operates the infrastructure. The state of the infrastructure is the result of many independent decisions taking by all the participant actors, at the technical level, but also at the market level. (Masera 2010)

In particular, how can sufficient investment in security be incentivised where many actors have a collective, but fuzzy, stake in grid reliability?

Meeting this challenge without undue bureaucratic burden will only become more vexing as an increasing number of small scale generation units – and even customers – become integral to supply in the market in the coming years (de Vries et al. 2010).

## 5 CONCLUSION

Interconnections imply both solutions and challenges to security of energy supply. This is particularly true for the European Union, where all three pillars of the EU energy policy converge around a common goal of cross-border infrastructure development. Although additional interconnections are often seen as beneficial to security, the accompanying increase in interdependence should be taken into account. Indeed, this chapter has introduced some of the internal challenges and vulnerabilities facing the European Union as it attempts to modernise and inter-connect some of the most complex energy systems on the globe. That this task carries with it a number of uncertainties is clear; special attention has been paid to the problems for security of supply engendered by gas market liberalisation, renewable energy deployment and the expansion of ICT systems controlling complex energy networks. These three issues, moreover, impact on the external dimension of EU energy security in important ways. For example, just as the liberalisation of the internal EU gas market will challenge long-standing relationships with external suppliers, so too will the problems arising from large-scale deployment of renewables require external solutions, such as the import of electricity from solar panels in the Sahara desert. Thus, the interdependencies between these external and internal dimensions will crucially affect the EU's overall energy security, something that should be further explored in subsequent research.

## NOTES

1. The views expressed are purely those of the authors, and may not in any circumstances be regarded as stating an official position of the European Commission.
2. This market-based reasoning is reflected in the recent EC Regulation 994/2010, which emphasises the use of market-based measures to deal with supply disruptions.
3. Bear in mind that demand *can* change for these customers due to other factors, such as weather conditions.
4. Paragraph 42 of Directive 2009/73/EC.
5. Of the 1,806,336 thousand tonnes of oil equivalent (TOE) of energy consumed by the EU27 countries in 2007, 683,744 were used to generate electricity (source: Eurostat 2012). All figures are for gross consumption, and the figure for transformation comprises the combined input to conventional thermal power stations and nuclear power stations.

6. Adequacy and reliability can be respectively defined as 'the ability of the system to supply the aggregate electric power and energy requirements within current ratings and voltage limits, taking into account planned and unplanned component outages' and 'the ability of the system to respond to disturbances arising within that system'.
7. Source addresses may be forged, logs and other audit data may be destroyed and the so-called IP tunnelling technology used in virtual private networks allows IP packets to be encrypted and hidden within other IP packets.
8. A 2007 *CSO Magazine*/Cisco poll reveals that 59 per cent of chief security officers think that employee error is the number one threat, while 37 per cent reported employee or partner sabotage as the number one threat.

# REFERENCES

ACER (2012). ACER opinion on the ENTSO-E Ten-Year Network Development Plan 2012.
Ackermann, T., E. Tröster, R. Short and S. Teske (2012). 'Renewables 24/7', Greenpeace International.
Averill, B. and E.A. M. Luiijf (2010). 'Canvassing the Cyber Security Landscape: Why Energy Companies Need to Pay Attention.' *Journal of Energy Security*, May.
Aylon, J.P. (2009). 'Soundings: "The Cyber-Threat Grows".' *City Journal*, Fall 2009.
Bahgat, G. (2011). *Energy Security: An Interdisciplinary Approach*, Wiley.
Byres, E.J. (2009). 'Cyber Security And The Pipeline Control System.' *Pipeline and Gas Journal* 236(2).
Chester, L. (2010). 'Conceptualising energy security and making explicit its polysemic nature.' *Energy Policy* 38(2): 887–895.
Cisco Systems, I. (2007). 'Measuring and Evaluating an Effective Security Culture', at http://www.cisco.com/web/about/security/cspo/docs/measuring_effective.pdf (last accessed March 2013).
Clark, W.K. and P.L. Levin (2009). 'Securing the Information Highway: How to Enhance the United States' Electronic Defenses.' *Foreign Affairs* 88(6): 2–10.
CSIS Commission on Cybersecurity for the 44th Presidency (2008). 'Securing Cyberspace for the 44th Presidency.' Washington, DC, Center for Strategic and International Studies.
Darby, S. (2006). 'The effectiveness of feedback on energy consumption: A review for DEFRA of the literature on metering, billing, and direct displays.' Oxford, Environmental Change Institute, University of Oxford.
De Bruijne, M. and M. van Eeten (2007). 'Systems that Should Have Failed Critical Infrastructure Protection in an Institutionally Fragmented Environment.' *Journal of Contingencies and Crisis Management* 15(1): 18–29.
de Vries, L.J., M. Masera and H. Faas (2010). 'The Way Forward.' In S. Lukszo, G. Deconinck and M.P.C. Weijnen, *Securing Electricity Supply in the Cyber Age: Exploring the Risks of Information and Communication Technology in Tomorrow's Electricity Infrastructure*, New York, Springer: 171–180.
Delucchi, M.A. and M.Z. Jacobson (2011). 'Providing all global energy with wind, water, and solar power, Part II: Reliability, system and transmission costs, and policies.' *Energy Policy* 39(3): 1170–1190.
Endrenyi, J. and W.H. Welsow (2001). 'Power system reliability in terms of the system's operating states.' IEEE Porto Power Tech Conference, Porto.
ENTSO-E (2010). System Adequacy Forecast 2010–2025, European Network of Transmission System Operators for Electricity (ENTSO-E).
ENTSO-E (2012a). Scenario Outlook and Adequacy Forecast 2012–2030, European Network of Transmission System Operators for Electricity (ENTSO-E).

ENTSO-E (2012b). Ten-Year Network Development Plan 2012, European Network of Transmission System Operators for Electricity (ENTSO-E).
ENTSO-G (2012). Gas Balancing Launch Documentation BAL0125-11, European Network of Transmission System Operators for Gas (ENTSO-G).
Estonian Ministry of Defence (2008). 'Cyber Security Strategy', at http://www.kmin.ee/files/kmin/img/files/Kuberjulgeoleku_strateegia_2008-2013_ENG.pdf (last accessed March 2013).
EU Home Affairs Sub-Committee (2010). 'Protecting Europe against large-scale cyber-attacks.' Paper 68. London, House of Lords.
EurActiv with Reuters (2010). 'Czech Renewable Energy Boom Tests Grid Safety Limits' at http://www.euractiv.com/en/energy/czech-renewable-energy-boom-tests-grid-safety-limits-news-329169 (last accessed June 2012).
European Commission (2005). Directive 2005/89/EC of the European Parliament and of the Council of 18 January 2006 concerning measures to safeguard security of electricity supply and infrastructure investment. Brussels, Directorate General for Energy.
European Commission (2006). Vision and Strategy for Europe's Electricity Networks of the Future. EUR 22040. Luxembourg, Office for Official Publications of the European Communities.
European Commission (2007a). European Technology Platform Smartgrids: Strategic Research Agenda for Europe's Electricity Networks of the Future. EUR 22580. Luxembourg, Office for Official Publications of the European Communities.
European Commission (2007b). 'Third Energy Package' at http://ec.europa.eu/energy/gas_electricity/legislation/third_legislative_package_en.htm (last accessed June 2012).
European Commission (2011a). 'Energy infrastructure priorities for 2020 and beyond – A Blueprint for an integrated European energy network.' Brussels.
European Commission (2011b). 'Energy Roadmap 2050.' Brussels.
European Commission (2011c). 'A roadmap for moving to a competitive low carbon economy in 2050.' Brussels.
European Council (2006). 'Presidency Conclusions.' Brussels.
European Council (2011). EUCO 2/1/11 REV 1. Brussels.
Eurostat (2012). http://epp.eurostat.ec.europa.eu/portal/page/portal/eurostat/home/ (last accessed August 2012).
Finon, D. and C. Locatelli (2008). 'Russian and European gas interdependence: Could contractual trade channel geopolitics?' *Energy Policy* 36(1): 423–442.
Fischer, C. (2008). 'Feedback on household electricity consumption: A tool for saving energy?' *Energy Efficiency* 1(1): 79–104.
Geers, K. (2009). 'The Cyber Threat to National Critical Infrastructures: Beyond Theory.' *Information Security Journal* 18(1): 1–7.
Gheorghe, A.V., M. Masera, M. Weijnen and L.J. de Vries (2006). *Critical Infrastructures at Risk. Securing the European Electric Power System*, New York, Springer.
Grave, K., M. Paulus and D. Lindenberger (2012). 'A method for estimating security of electricity supply from intermittent sources: Scenarios for Germany until 2030.' *Energy Policy* 46(0): 193–202.
Hadfield, A. (2008). 'Energy and foreign policy: EU-Russia energy dynamics.' In S. Smith, A. Hadfield and T. Dunne, *Foreign Policy: Theories, Actors, Cases*, Oxford, Oxford University Press.
Haller, M., S. Ludig and N. Bauer (2010). 'Fluctuating renewable energy sources and long-term decarbonization of the power sector: insights from a conceptual model.' International Energy Workshop, Stockholm.
Heinrich, A. (2007). *Poland as a Transit Country for Russian Natural Gas: Potential for Conflict*, Koszalin Inst. of Comparative European Studies, KICES.
Hoicka, C.E. and I.H. Rowlands (2011). 'Solar and wind resource complementarity: Advancing options for renewable electricity integration in Ontario, Canada.' *Renewable Energy* 36(1): 97–107.
Hoogeveen, F. and W. Perlot (2005). 'Tomorrow's Mores: The International System,

Geopolitical Changes and Energy.' Clingendael International Energy Programme, CIEP 02/2005, the Hague, NL, available at http://www.clingendael.nl/publications/2006/20060117_ciep_study_hoogeveen_perlot.pdf (last accessed March 2013).
Hughes, R. (2010). 'A Treaty for Cyberspace. International Affairs.' *International Affairs* 86(2): 523–541.
IEA (2008). 'Development of Competitive Gas Trading in Continental Europe.' IEA Info Paper, at http://www.iea.org/publications/freepublications/publication/gas_trading.pdf (last accessed March 2013).
IEA (2009). 'World Energy Outlook' Paris, at http://www.worldenergyoutlook.org/media/weowebsite/2009/weo2009_es_english.pdf (last accessed March 2013).
Igure, V.M., S.A. Laughter and R.D. Williams (2006). 'Security issues in SCADA networks.' *Computers & Security* 25: 498–506.
International Institute of Strategic Studies (2010). 'The Military Balance.' London.
Jaureguy-Naudin, M. (2012). 'Decarbonization and Cost Reduction: Lost in Transmissions?' Institut Français des Relations Internationales.
L'Abbate, A., G. Fulli, F. Starr and Stathis D. Peteves (2007). 'Distributed Power Generation in Europe: Technical Issues for Further Integration.' EUR 23234 EN-2007. Brussels, European Commission.
LaMonica, M. (2009). 'Cisco: Smart grid will eclipse size of Internet. CNET News' at http://news.cnet.com/8301-11128_3-10241102-54.html (last accessed June 2012).
Le Coq, C. and E. Paltseva (2011). 'Assessing Gas Transit Risks: Russia vs. the EU.' *Energy Policy*, 42 (March): 642–650.
Lipson, H.F. (2002). 'Tracking and Tracing Cyber-Attacks: Technical Challenges and Global Policy Issues.' Pittsburg, Carnegie Mellon University/Software Engineering Institute.
Luft, G. and A. Korin (2009). *Energy Security Challenges for the Twenty-first Century*, Praeger Security International.
Mott Macdonald Consultants (2010). 'Supplying the EU Natural Gas Market.' Report for the European Commission, November 2010, Croydon UK, at ec.europa.eu/energy/international/studies/doc/2010_11_supplying_eu_gas_market.pdf (last accessed June 2012).
Markulec, M. (2008). 'SCADA systems: unknown connections could spell trouble.' *Power Engineering* 112(11): 188–191.
Masera, M. (2010). 'Governance: How to Deal with ICT Security in the Power Infrastructure?' In S. Lukszo, G. Deconinck and M.P.C. Weijnen, *Securing Electricity Supply in the Cyber Age: Exploring the Risks of Information and Communication Technology in Tomorrow's Electricity Infrastructure*, New York, Springer: 111–128.
Munro, K. (2008). 'SCADA – A critical situation.' *Network Security* 2008(1): 4–6.
Myrli, S. (2009). 'NATO and Cyber Defence.' 027 DsCFC 09 E. Brussels, NATO Parliamentary Assembly.
National Communications System (2004). 'Technical Information Bulletin 04-1.' Supervisory Control and Data Acquisition (SCADA) Systems. Arlington, Office of the Manager National Communications System.
Noël, P. (2009). 'A Market Between us: Reducing the Political Cost of the Europe's Dependence on Russian Gas.' University of Cambridge EPRG Working Paper.
Orbán, A. (2008). *Power, energy, and the new Russian imperialism*, Praeger Security International.
Pascual, C. and J. Elkind (2010). *Energy Security: Economics, Politics, Strategies, and Implications*, Brookings Institution Press.
Ratner, M., P. Belkin, J. Nichol and S. Woehrel (2012). 'Europe's Energy Security: Options and Challenges to Natural Gas Supply Diversification'. Congressional Research Service, at http://www.fas.org/sgp/crs/row/R42405.pdf (last accessed March 2013).
REALISEGRID (2009). 'Review of existing methods for transmission planning and for grid connection of wind power plants.' D3.1.1, at http://realisegrid.rse-web.it/content/files/File/Publications%20and%20results/Deliverable_REALISEGRID_3.1.1.pdf (last accessed March 2013).

Reed, T. (2005). *At the Abyss: An Insider's History of the Cold War*, Random House Publishing Group.
Ruus, K. (2008). 'Cyber War I: Estonia Attacked from Russia.' *European Affairs* Winter/Spring 2008.
Ryu, D.H., H. Kim and K. Um (2009). 'Reducing security vulnerabilities for critical infrastructure.' *Journal of Loss Prevention in the Process Industries* 22(6): 1020–1024.
Shaw, T. (2009). 'SCADA system protection requires independent barriers.' *Oil & Gas Journal*, 107(37: 68–71.
Stefanini, A. and M. Masera (2008). 'The Security of Power Systems and the Role of Information and Communication Technologies: Lessons from the Recent Blackouts.' *International Journal of Critical Infrastructures* 4(1/2): 32–45.
Treat, R. and A. Bartle (2009). 'New regulations drive expanded SCADA curriculum.' *Pipeline & Gas Journal*, 236(9), September.
United States Department of Energy (2008). 'The Smart Grid: An Introduction.' Washington, DC.
US-Canada Power System Outage Task Force (2004). 'Final Report on the August 14, 2003 Blackout in the United States and Canada: Causes and Recommendations.' Washington, DC.
van der Vleuten, E. and V. Lagendijk (2010a). 'Interpreting transnational infrastructure vulnerability: European blackout and the historic dynamics of transnational infrastructure governance.' *Energy Policy* 38(4): 2053–2062.
van der Vleuten, E. and V. Lagendijk (2010b). 'Transnational infrastructure vulnerability: The historical shaping of the 2006 European "Blackout".' *Energy Policy* 38(4): 2042–2052.
Widen, J. (2011). 'Correlations Between Large-Scale Solar and Wind Power in a Future Scenario for Sweden.' *IEEE Transactions on Sustainable Energy* 2(2): 177–184.
Williams, R.I. (2003). 'New threats prompt renewed security scrutiny for product storage sites.' *Oil & Gas Journal*, March, at http://www.ogj.com/articles/print/volume-101/issue-9/transportation/new-threats-prompt-renewed-security-scrutiny-for-product-storage-sites.html (last accessed March 2013).
Winzer, C. (2012). 'Conceptualizing energy security.' *Energy Policy* 46(0): 36–48.
Yi, L., V.G. Agelidis and Y. Shrivastava (2009). 'Wind-solar resource complementarity and its combined correlation with electricity load demand.' 4th IEEE Conference on Industrial Electronics and Applications, 2009. ICIEA 2009.

# 4. Resource conflicts: energy worth fighting for?
*Joshua Olaniyi Alabi*

## INTRODUCTION

Because of its significant role in fuelling modern industrial economies and military forces, oil has been the subject of domestic and international politics and conflicts. The history of the international oil market and energy security has been one of prolonged conflicting interests, where both the major oil exporting and the industrialised consuming countries have resorted to various strategies to ensure security of supply and of demand or to influence prices, in such a way to meet individual interest. 'From the nineteenth-century battles over the Caspian Sea to the two Gulf Wars, oil has been the prize in numerous military conflicts' (Spero and Hart, 2003:229–30). In the same way the conflicts about offshore tracts in the Gulf of Guinea between Nigeria and Cameroon at Bakassi Peninsula, and between Nigeria and Equatorial Guinea at the Zafiro undersea oil field, were on oil (Klare, 2001:230).

Conflicts over oil resources in many producing countries are generally at two levels: the first, at the macro level, involves international oil companies (IOCs) and governments in the global North; the second is at local level in the oil producing countries. I argue that the main aim is to secure energy security by constant access to cheap oil and gas at reasonable prices, through engaging in whatever means possible and necessarily, including military interventions. An example is the USA-led gulf wars in Iraq in 1991 and 2003. This macro level also includes conflict of interest between national governments of oil producing states and international oil companies over fiscal regimes, corrupt practices, environmental degradation etc. The second analysis of oil conflicts takes place at the micro level within the national boundaries of oil producing countries, where conflicts emerge on how to share the huge revenues from oil between the national governments and sub national governments and other various stakeholders. The various dimensions of these crises will be critically examined to determine why conflicts emerge over energy resources, and why the economic benefits from both the buying and selling of oil and gas are worth fighting for.

This chapter will specifically focus on resource conflicts as they relate to energy security and conflicts generated by oil and gas resources. In doing this I will look at the historical overview of global oil conflicts issues from the Caspian and Middle East in Kuwait and Iraq to Africa in the gulf of Guinea Nigeria and Cameroon. I will analyse how resource conflicts are changing through the role of institutions, piracy, private security companies and the international oil companies (IOCs), using the traditional realist approach to resource conflicts. The realist analysis of the politics of international energy draws on the extent of state power to access and control over energy resources, and it also argues that such resources are becoming insecure and scarce and assumes that the contestation around resource capture will continue (Dannreuther, 2010:2–3). This will be explained in detail with the case of oil conflict in the Niger Delta, Nigeria.

## AN OVERVIEW OF OIL GEOPOLITICS

The history of modern oil and gas industry began in the late 19th century when the first major oil company, Standard Oil, was founded by John D. Rockefeller in 1870 in the United States of America (Business Reference Services, 2012). In the early 20th century the use of oil eventually replaced coal as the world's primary source of industrial power. During the same century the oil industry also expanded rapidly at about 6.5 per cent; from 1913 to 1948 and from 1948 to 1973 it was 7.5 per cent (Bromley, 2005:235 and Business Reference Services, 2012). The availability and control of oil and gas played a foremost role in both the First and Second World Wars and still remains the critical fuel source that powers modern industrial societies (ibid).

Consequently, the emerging industrialisation and the growing dependence on imported materials at the advent of the 20th century necessitated the need to access resources from abroad for an expanding oil industry (Le Billion, 2004:3). For example Standard Oil expanded its business to the Caspian and bought 100,000 tonnes of oil at $33 per tonne from the Azerbaijani government in June 1919 (O'Hara, 2004:142). The significant role of oil in the First World War led to the scrabble for more oil after the war and the partitioning of the Middle East among the various Western Nations for secure supply of oil (Le Billion, 2004:3). This was further reinforced during the Second World War, Le Billion concludes thus:

> ... in their search for resource security and strategic advantage, industrialised countries continued to take a diversity of initiatives including military deployment near exploitation sites and along shipping lanes, stockpiling of strategic

resources, diplomatic support, 'gunboat' policies, proxy wars or coup d'état to maintain allied regimes in producing countries, as well as support to transnational corporations and favourable international trade agreements. (2004:3)

A good example was the military coup that overthrew the Mossadeq's government because of the nationalisation of the British oil corporations in Iran and replaced it with a pro-western Shah in 1953 (Peters, 2004:202).

There is now a 'return' of geopolitics due to structural changes after 2000 which followed two decades of low energy prices under a market based regime. The September 11, 2001 attack on the World Trade Centre twin towers in the US, the US-led invasion of Iraq in 2003, political struggles in Russia and the Niger Delta crisis in Nigeria resulted in rapid increases in oil prices and structural changes in the international oil market (Trombetta, 2012:15). Also, due to the rapid economic growth and development in developing countries particularly in the BRICS countries (Brazil, Russia, India, China, and South Africa), there has been a global contestation on how to secure energy security and also its attendant potential conflicts (Smith, 2010:121).

The competition for access to oil resources will intensify as the global energy consumption of oil and gas mostly from non-OECD countries is set to increase by 40 per cent in 2030 (Smith, 2010:122). Over the same period of time the production of oil and gas will decrease both in Europe and America, this puts more pressures on the international oil supply market leading to the 'securitization'[1] of energy, making energy security issues a national priority for all oil importing countries (Trombetta, 2012:15 and Smith, 2010:120).

Most of the substantial proven oil and gas reserves are now located in the Middle East, Caspian and Africa region, where security uncertainties due to conflicts induced by oil and gas are major challenges to trade (Smith, 2010:121). There is also competition among the oil importers with each one trying to scheme out the other: while China is close to Iran and Russia, the Western oil consumers are also in alliance over Iraq and other countries in the Caspian (ibid). Some of the Western countries are also present in the Gulf of Guinea oil region in Africa, and China has also gained a foothold for oil operation in Angola and Nigeria.

# ANALYSIS OF VARIOUS ARGUMENTS ON RESOURCE CONFLICTS AND ENERGY SECURITY

Studies by various scholars, economists and political scientists and others have produced series of explanations and arguments on conflicts generated by natural resources, in this case oil and gas and energy security (Klare, 2001, 2008; Watts, 2004; Ross, 1999, 2001; Le Billion and El Khatib, 2004; Spero and Hart, 2003; Raphael and Stokes, 2011 and Collier, 2007, 2009 among others). Collier (2009) asserts that the explanations of natural resources conflicts are more deep rooted political economy issues, 'an interplay between politics and valuable natural assets'.

Resource conflicts and war in resource-dependent countries, particularly oil, are not limited only to the boundaries of their respective countries, but go far beyond it. The global consequences of the invasion of Iraq by the US-led allied forces in 1990 to 1992, and since 2003, have been far reaching. The intermittent skyrocketing increases in oil prices is the main direct cost to the world economy, and with its macroeconomic repercussions is 'approximately $1.1 trillion' which 'dwarfs all other economic cost' (Stiglitz and Blimes, 2008:132–160).

Part of the consequences is the soaring cost of transportation in most countries, including the oil exporters, and increasingly resulting in 'higher food prices' (ibid). Beyond the threat to global peace and security through the growth of resistance and extremism throughout the world, the humanitarian and economic costs are also alarming not only in Iraq but also in the neighbouring countries of Syria, Lebanon, Jordan and Egypt; civilian death estimated at over 100,000 people and '4.6 million people – one of every seven Iraqis uprooted from their homes' by December and September 2007 respectively; an estimate of over $1 billion for Jordan alone and $123 million budgeted by the UNHCR in 2007 for the up-keep of Iraqi refugees (Stiglitz and Blimes, 2008:132–160).

The struggles amongst various factions and political groups to control political power at the centre in resource-dependent countries goes beyond corruption, it actually leads to conflicts and may degenerate in many cases to civil war if not properly managed. This is because those who control political power at the centre are the custodians of the revenues that accrue from the resources. Studies show that oil and mineral resource-exporting countries are more vulnerable to violent conflicts that are particularly secessionist in nature; they also last longer with higher casualties than conflicts with no resources to capture, and are mostly located in the regions where the natural resources are located (Collier and Hoeffler, 2004, 2006; Lujala et al 2005).

Table 4.1 below shows past and present civil wars, rebellion and militancy in 8 selected oil and mineral-dependent countries. The wars have lasted from three to 36 years, the free Aceh movement (GAM) rebelled against the Indonesian authorities and fought the national army (TNI) to break away the natural gas rich Aceh region. The conflicts in some of the countries have occurred intermittently and still persist, for example the Sudan and South

74   *International handbook of energy security*

*Table 4.1   Civil wars, rebellion and militancy in selected oil and gas dependent states*

| Country | Duration | Resources |
|---|---|---|
| 1  Algeria | 1991–2001 | Oil |
| 2  Angola (Cabinda) | 1992–2001 | Oil |
| 3  Congo, Republic | 1997–1999 | Oil |
| 4  Indonesia (Aceh) | 1975–2005 | Natural Gas |
| 5  Iraq | 1974/75, 1985–92, 2003–Present | Oil |
| 6  Nigeria (Niger Delta region) | 1967–1970, 1998–Present | Oil |
| 7  Sudan | 1983–Present | Oil |
| 8  Yemen | 1986–87, 1990–94 | Oil |

*Source:*   Ross (2001:15) and Alabi (2010:58).

Sudan, Nigeria's Niger Delta region and particularly in Iraq from 1974 to 1975, then 1985 to 1992 and from 2003 up to October 2012.

'The discovery of resource wealth in a discontented region may add fuel to separatist sentiments' (Ross, 2001:15). For example the Biafra rebellion in Nigeria, the Cabinda enclave in Angola, and the Aceh rebellion in Indonesia. Karl argues that 'Oil may be the catalyst to start a war; petrodollars and pipelines may serve to finance either side and prolong conflict and this, of course, is the biggest resource curse of all' (Karl, 2005:26). The Azerbaijan government's additional US$35 million in military spending in the 2004 budget did not provide any direct benefit to the citizens (Shultz, 2005:35).

Collier (2009) posited that an increase in the prices of natural resources in the international market also increases not just the propensity of conflict in resource-dependent countries, but also the period of war once it has started. In his analysis he identified three reasons for this: the first is that the atmosphere of conflicts creates the possibility for the rebel groups to have illegal access to the natural resource, be it oil or gas, as is the case in Angola and Nigeria, and use part of the proceeds from the sales to broaden and sustain the conflict. Secondly, most recruits to rebellion are motivated by loot-seeking rather than fighting for any political cause: even when the conflict began with a political motivation it will in the long run result in loot-seeking. Thirdly, rebellion might be a consequence of inadequate or lack of accountability to their citizens by governments of most natural resource-dependent countries (Collier, 2009:5–6).

Conflicts create an atmosphere of fear, and may allow rebel or militant groups to illegally tap pipelines to steal oil, or seize gas fields, or carry out other nefarious acts like kidnapping and extortion to finance and elongate conflict periods. The main motivation behind insurrections, rebellions and wars in resource-dependent developing countries is not just about greed or the desire to loot; in most cases it is resistance to and reaction against long-term neglect and socio-economic deprivation in the mineral resource rich regions, the capture and corrupt control of the revenues accruing from such resources by the central governments and their lack of accountability to the citizens. This explains why these type of conflicts over resources are not happening in Norway or other developing resource rich countries such as Qatar and United Arab Emirates. If the causes of wars are merely because of the presence of natural resources or greed to loot, and not resistance against socio-economic deprivation, marginalisation and lack of accountability, the same conflicts should be happening in other developing resource-rich countries. Thus Ake (2000) concludes that most conflicts labelled as ethnic conflicts are nothing but democratic conflicts where groups or societies who have been denied benefits from the national resources for so long struggle to regain their rights and access to the commonwealth as citizens (Ake, 2000:10).

## THE INTERRELATIONSHIP BETWEEN ENERGY SECURITY AND RESOURCE CONFLICTS

This section will help to explain the interrelationships between the energy security agenda of the global North and the nature of conflicts and underdevelopment in the oil and gas producing countries of the global South. From the end of the cold war to the early 1990s there has been an increase in internal and regional conflicts in many resource developing nations, which have led to massive humanitarian interventions by international NGOs and the United Nations, and also a new form of global governance by governments in the global North (Duffield, 2001:1). In essence, it is clear that by 'the mid-1990s the need to address the issue of conflict became a central concern within mainstream development policy', which was previously in the purview of international and security studies (ibid).

The major explanation for this could be traced to the unequal integration of the South into the global capitalist system and the nature of this continued exploitative relationship between the global North and South in terms of trade, raw materials and cheap energy (oil and gas by all means and at all cost) to power the industrial development in the North. The conflicts were further aggravated by the concentration of formal trade, technology and international finance in the North and East Asia since the 1970s, to the severe exclusion of the global South (Duffield, 2001:2). This exclusion and declining investment particularly in Sub-Saharan Africa has now led to escalating contestation around resource capture – including the state, criminalisation, breakdown of law and order and a global criminal economy network (ibid:6–7). The crisis in the oil producing Niger Delta of Nigeria is a classical example, where the militants in the early 1990s began to attack oil installations, kidnapping IOCs personnel and demanding ransoms for their release. This soon graduated into a large network of gangs in collaboration with high ranking military officers and top government officials stealing oil from the pipelines and selling to foreign vessels at a cheaper price on the high seas. Some of the proceeds are in turn used to buy arms and ammunitions to perpetuate their hold on the oil resources. The adverse social and economic effects of this are not only felt locally within Nigeria but also globally, as any major disruptions to oil production in the Delta or in the Gulf of Guinea sends the price of oil higher in the international oil market.

Therefore the current global security challenges go beyond interstate wars: 'the threat of an excluded South fomenting international instability through conflict, criminal activity and terrorism is now part of a new security framework' in which 'underdevelopment has become dangerous' (Duffield, 2001:2). This has led to the radicalisation of development policy, which is merging conflict resolution with a neoliberal global governance agenda of social transformation of societies in the global South (ibid). Although it is exclusion and contestation for the few resources that has led to the conflicts, viewing underdevelopment as dangerous and as an agent of destabilisation is enough reason for policing and engagement with the governments of the South (ibid:7). For example in the Gulf of Guinea the establishment of the United States African Command (AFRICOM) Africa Partnership Station to police the oil facilities in the region is unsolicited assistance to protect the energy security interests of the North (Raphael and Stokes, 2011:909–911).

Internationalisation of public policy is also part of the new security and governance framework; governments in the South are controlled by means of 'conditional selective inclusion' (ibid:7–8). Thus, they are required to execute international laws such as anti-terrorism, anti-money laundering and the Kimberley process, for the sake of global peace and security. The processes of inclusion are sometimes stratified, depending on which side is to profit most (ibid).

## ENERGY SECURITY AND OIL CONFLICTS: THE CASE OF THE NIGER DELTA, NIGERIA

This section focuses on the persistency of the Niger Delta crisis and competition for resource control by those in the Delta, and argues that the acceleration of struggle and contestation around resource capture has escalated dramatically and shaped policy regarding fiscal federalism. It then examines the resultant socio-economic and environmental impacts of the crisis and how that challenges the sustainability of oil production in Nigeria. The conflict in the Niger Delta has a fairly long trajectory: the structures for the present conflicts were laid by the colonialists and the first Nigerian political leaders, which successive regimes have built on and nurtured. After independence in 1960, the post-colonial Nigerian state continued unabated the totality of power it inherited from the colonialists over land and all mineral resources (Omoweh, 2006:40). Consequently, the first post-colonial Nigerian leaders jettisoned the Willink Commission's report, abandoned its implementation and the development of the Niger Delta Development Board, NDDB and thereby laid the foundation for the current crisis.

The first post-colonial oil conflict in the Niger Delta region was sparked by Isaac Adaka[2] Boro in February 1966 (Tebekaemi, 1982:6 and International Crisis Group, 2006:4).

The contemporary crisis threatens not just the oil exploration and exploitation in the Niger Delta region but 'everything' about Nigeria: unity, politics and socio-economic development to its foundation. Echoing this fact Bush (2007:613–4) posits that:

> Resistance is at fever pitch in the Niger Delta. It is opposition to international oil companies and Nigerian state collusion with exploitation in the region. It is opposition to corrupt local politicians and leaders and it is driven significantly by the youth.

By the late 1990s, the situation in the Niger Delta had transformed into aggressive agitations and violent protest of unprecedented dimensions that shook the very foundations of the Nigerian nation (NDDC, 2002:5). Violence and youth restiveness increased greatly as Nigeria approached the return to democracy in May 1999. Youths in the region began to lay siege to oil installations and kidnapped oil workers from various IOCs, making it unsafe for any socio-economic activity to thrive (Djebah, 2003:3). The armed resistance started by Isaac Adaka Boro's 12-day revolution in February 1966 transformed to the commando-style operations of the Movement for the Emancipation of the Niger Delta (MEND).[3] Consequently, militant groups have increased deadly attacks on oil facilities and IOCs personnel from 2000 onwards and have taken a higher dimension since 2005 with activities of MEND, attacking offshore oil facilities 200 kilometres below sea such as the Bonga FPSO platform and shooting at IOCs servicing helicopters.

The conflicting claims to oil resources in the Niger Delta between the federal government and the oil producing states provide major challenges to the Nigerian state, due to the intensity with which several oil producing states are clamouring for control of oil resources in their region. As asserted by the International Crisis Group (ICG) 'a potent cocktail of poverty, crime and corruption is fuelling a militant threat to Nigeria's reliability as a major oil producer' (ICG, 2006:1). Corroborating the ICG, Soyinka (2008) noted:

> What is happening in the Delta today, points a finger of guilt in so many directions and of course, the primary is within Nigeria itself and what Nigerian leaders have made of this incredible opportunity and however, but at the expense of some of the people within the Delta region. Then of course the oil exploration companies with their contempt for basic minimum standard of conduct towards the areas, the indigenes from whom this wealth is been dredged. So very often it is a matter of embarrassment, a guilt feeling and so they pretend it is just a few people restless, violence and kidnapping and so on whereas this problems goes back decades.

**The Niger Delta Crisis: From MOSOP to MEND 1990–2012**

From 1990, the late Ken Saro Wiwa,[4] the founding leader of the Movement for the Survival of the Ogoni People (MOSOP), led the Ogoni struggles against the Shell Petroleum Development Company (SPDC) and the federal government to put an end to the environmental destruction orchestrated by SPDC against the Ogoni people in the name of oil exploration (Rowell et al, 2005:2; Watts, 2007:652). MOSOP produced the Ogoni Bill of Rights, which was signed in August 1990 and presented to the federal government. It demanded 'political control of Ogoni affairs by Ogoni people and the right to protect the Ogoni environment and ecology from further degradation' (Rowell et al, 2005:3). Instead of responding to the Bill of Rights, the IOCs and the federal government deployed the security forces to all the oil installations and the towns of Ogoniland, banned all public protest, and decreed any demand for self-determination and obstruction of oil production in any form as an act of treason which is punishable by death (Osaghae, 1995:336). The threats by the government only propelled the full scale mobilisation and mass attendance of thousands of protesters at the Bori rally of 3 January 1993 (ibid).

This Bill of Rights and the tenacious campaign both at local and international level to popularise and enforce its aims drew the attention of the international community to the Ogonis' plight. The publicity and campaigning forced SPDC to call off its production throughout the Ogoni land in 1993 (International Crisis Group, 2008:2 and Osaghae, 1995:336). The ensuing circumstances brought MOSOP, led by Ken Saro Wiwa and the Ogonis, into direct confrontation with the Nigerian government and SPDC. On 21st May 1994 a crisis led to violence between the pro-SPDC Ogoni elders, who were accused by the radical Ogoni youths of receiving largesse from SPDC to scuttle the Ogoni struggle process; four Ogoni elders were killed. It was the moment the government had been waiting for to finally nail MOSOP and its leaders (International Crisis Group, 2008:2–4).

Ken Saro Wiwa and other MOSOP leaders (including Mr Ledum Mitee, the chairman of the Federal government 2008 Technical Committee on the Niger Delta who was arrested but later released), were tried by a Special Military Tribunal that was constituted by the Abacha regime to try their case in 1995. They were sentenced to death by hanging with the collaboration of Shell Petroleum Development Company (SPDC) and other IOCs. Although the SPDC strongly denied their involvement in Saro Wiwa and other MOSOP leaders' trial and death, they had a legal representative, a hired Queens Counsel (QC) flown into Nigeria from London on the account of SPDC, who monitored the whole trial process (Rowell et al, 2005: 4–5).

On 10th November 1995, Saro Wiwa and eight others were executed by hanging after 18 months incarceration. To make matters worse the government created more psychological torture for the families of the dead MOSOP leaders: rather than releasing the bodies to the families, acid was poured on to the bodies to be sure that they were destroyed, even after they were certified dead by medical personnel at the hanging (Rowell et al, 2005:4–5). Apart from protests against SPDC in Europe and America where it had operational bases, and in a couple of Nigerian foreign missions, the Commonwealth reacted to the killings at its summit in New Zealand by suspending Nigeria, and its delegation led by Mr Tom Ikimi, the Nigerian former foreign minister, walked out. The Nigerian government and Shell were also condemned by the United Nations Human Rights Commission, the European Union and the US States Department.

However, all of them failed to act with any real effect, because the issue at hand was about security of supply;[5] an oil embargo on Nigeria 'was deemed unacceptable to the United States' (Rowell et al, 2005:12). With the inability of the international community to reprimand Nigeria's military regime, it was 'mission accomplished' and the resumption of business as usual by both SPDC and the federal government in the Niger Delta. As the security forces continued their crackdown on the residents of oil producing communities Soyinka (2008) noted:

> The international community to some extent felt that this crime committed against the representatives of the Nigerian people, the defenders of the ecology,

the protectors of the right of indigenous people, Ken Saro Wiwa and his companions, when this crime was committed under such a public global glare that it would compel the oil companies to tidy up their acts, and to some extent it did for a short while. But they become complacent once again (Soyinka, 2008).

The Rivers State Internal Security Task Force (a military special squad put together by the military regime) in conjunction with Lt Colonel Musa Komo, the state military governor, started negotiating with SPDC to resume operations in Ogoni. At the same time the chiefs and villagers were also intimidated into signing documents for SPDC to return to Ogoni. At the final sitting of the Special Military Tribunal in 1995 Saro Wiwa made this closing statement:

> I repeat that we all stand before history. I and my colleagues are not the only ones on trial. Shell is here on trial . . . but its day will surely come and the lessons learnt here may prove useful to it. There is no doubt in my mind that the ecological war that the company has waged in the Niger Delta will be called to question sooner than later and the crimes of that war be duly punished. (Rowel et al, 2005:211)

This statement by Saro Wiwa found fulfilment on 4th June 2008 when the federal government announced the final withdrawal of SPDC's operating licence which has been suspended since 1993, and ordered the closure of all its operations in Ogoni land by December 2008. According to late President Yar'Adua it was evident that 'there was a total loss of confidence between the Ogoni people and Shell, government decided that another operator acceptable to the Ogonis will take over all oil operations. Nobody is gaining from the conflict and stalemate, so this is the best solution' (Nigerian Tribune, 2008). On 6 June 2008 Corroborating President Yar'Adua, Ledum Mitee said 'we agree that the relationship between Shell and the Ogoni people has been damaged irreparably and therefore, the president's decision is timely and appropriate' (Nigerian Tribune, 2008).

After the Ogoni crisis, from April 1997 onwards protest against all the IOCs in the Niger Delta increased considerably and this time it was not just against SPDC. The military government and the oil technocrats had 'feared all along; that other communities, inspired by Ken Saro Wiwa and the Ogoni struggle, would rise up against the devastation of their environment' (Rowell et al, 2005:18). Thus, the various youth groups began to form resistance movements across the Delta region. The first of such groups was the 'Chicoco movement' formed in August 1997 by Oronto Douglas[6] in Ijaw town of Aleibiri, Bayelsa State with over 1,000 youths (Rowell et al, 2005:18).

The Chicoco movement was a forerunner to the formation of the Ijaw National Congress (INC) which managed the implementation of the

Kaiama declaration. In Bayelsa State, the Ijaw youths, under the umbrella 'Egbesu Supreme Assembly', drawn from over 500 communities and 40 clans that made up the Ijaw nation and 25 representative organisations, met in Kaiama on 11th December 1998, 'to deliberate on the best way to ensure the continuous survival of the indigenous peoples of the Ijaw ethnic nationality of the Niger Delta within the Nigerian state' and put forward the 'Kaiama Declaration'. The declaration states that:

> As a step towards reclaiming the control of our lives, we, therefore, demand that all oil companies stop all exploration and exploitation activities in the Ijaw areas. We are tired of gas flaring; oil spillages, blowouts and being labelled saboteurs and terrorists. It is a case of preparing the noose for our hanging. We reject this labelling. Hence, we advise all oil companies' staff and contractors to withdraw from Ijaw territories by 30th December, 1998 pending the resolution of the issue of resource ownership and control in the Ijaw area of the Niger Delta. (Ijaw Youths Centre, 1998)

This declaration was synonymous to declaring an 'Ijaw Nation' and war on almost all the oil companies in their area (Essien, 1999:5). They also set up the Ijaw Youth Council (IYC) to coordinate their struggles for self-determination and justice (ibid). Installations owned by oil companies were not only seized, but oil production ground to a halt which also led to a substantial revenue loss to the Nigerian government and increased oil prices in the international market.

**Federal Government and IOCs Security Responses to the Oil Conflicts**

The state continued with the militarisation of the oil producing communities as a response strategy at the instance of community protest. The position of the government and the IOCs was that the community protests, agitations and losses which result from these disasters amounted to acts of sabotage, and this prompted their actions in an attempt to stem the rate of increasing communal clashes and violence. Whenever there are crises, the government often reacts aggressively by drafting the security forces for reprisal attacks to quell the militants, who are branded criminals, looters and saboteurs[7] of the economy. Generally when the militants could not be found, the security forces turned their grievances on the communities resulting in many casualties and loss of innocent lives, and in the process exterminating some villages and communities in the Delta. This led to the emergence of more youth groups: from 1998 to 2012, apart from MOSOP and the Chicoco movement, over 30 ethnic-based minority rights groups have emerged across the Delta region (Adejumobi, 2002:6 and Ojakorotu, 2005:3). The strategies and tactics of these movements have changed over the years from non-violence to increasingly militaristic and commando-style armed groups; this was to 'counter what local groups viewed as military's repressive tactics' by the Nigerian government's security forces (International Crisis Group, 2006:5). Thus there are constant conflicts between state and various movements in the Delta over resource extraction.

The IOCs are too confident of protection from their own security forces and that of the federal government, and are quick to call for the State's security forces to quell community protest. That normally leads to heavy casualties on the side of the host communities due to the excesses of the security forces. For example on 29th October 1990 the SPDC Divisional Manager in Port Harcourt ordered state security forces through the Rivers State Commissioner of Police to quell the 'impending militant attack' on Umuechem flow station (Frynas, 2001:50). The events of 30th and 31st October 1990 were a serious crisis as 'Mobile Police moved in with teargas and gunfire, killing about 80 people and destroying almost 500 houses' (ibid and International Crisis Group, 2006:6). The judicial commission of inquiry later found there was no planned attack, as alleged by SPDC's manager, because the youth protests were non-violent and the youths were not armed (Rivers State Government, 1991).

The oil producing communities are therefore at the mercy of armed security personnel, police and military, who constitute the Joint Task Forces. Some of these use the cover of providing security and maintaining law and order to torture and terrorise innocent citizens, openly demanding bribes from motorists and cyclists. Thus, 'the ugly relationship between oil and violence has played out along very clear lines in Nigeria' (Klare, 2004:127). From 1966 to 1999 the country was ruled at various times by the military, which having forcefully taken power had squandered massive oil revenues, leaving the Niger Delta region to wallow in poverty: this led to the increase in ethnically based militias in the Niger Delta (Klare, 2004:127). Consequently, the few key actors in Nigerian domestic policy facilitate and encourage a mutually dependent process of systematic looting of the country's oil wealth (White and Taylor, 2001:324), and therefore have no incentive to change the fraudulent mutual dependency between the IOCs and the various regimes that have ruled the country. The responses of the IOCs and the government through militarisation have been counterproductive as they generate more crises. The next section will examine the impacts of such crises.

**The Socio-Economic Impacts of the Conflict on Nigeria and the International Oil Market**

The conflicts in the Niger Delta have impacted negatively on Nigeria and is difficult to quantify. The international community feels the effects particularly in terms of volatility in the price of oil in the international market. This section will examine the impacts of the conflicts in terms of economic and other social losses including kidnapping, sabotages, oil bunkering,[8] gangs and the resultant effects.

Events in the Niger Delta took another dimension in January 2006 when four expatriate oil personnel were kidnapped. All were employees of the Shell contractors Tidex and Ecidrill and were kidnapped from a support vessel in the Exploration Area (EA) shallow offshore field operated by Shell Petroleum Exploration and Production Company (SNEPCO) (Oduniyi and James, 2006). The Movement for the Emancipation of the Niger Delta (MEND) claimed responsibility for the kidnapping; MEND demanded the release of Mujahid Asari Dokubo, the leader of another militant group, Niger Delta Peoples Volunteer Force (NDPVF), who had been incarcerated but was later released by the federal government as part of the efforts to calm the militants' activities. They also demanded US$1.5 billion compensation for oil pollution to the Ijaws, an ethnic group in the Niger Delta (Daniel and Olaniyi, 2006).

As a result of the attacks on oil installations across the Niger Delta, IOCs including SPDC, Italian energy giant Eni SpA, ChevronTexaco and ExxonMobil among others, declared force majeure intermittently from the emergence of MEND in December 2005, culminating in up to 1.3 million barrels per day (bpd) shut-in of crude oil by the end of 2006 (Nigerian Guardian, 2008). MEND sent a clear signal to the IOCs and federal government that no oil installation within the Delta was unreachable when they launched a major attack on the 225,000 barrels per day floating production storage facility in SPDC's Bonga oil FPSO, about 200 kilometres below sea on 19th June 2008. It was automatically shut down and resumed production after a few weeks (Amaize, 2008) and according to General Boyloaf[9]:

> We really wanted to prove that nowhere is untouchable that is why we visited there. We wanted to make this point because Shell and Chevron, all of them are moving offshore. So we visited them to prove that there is nowhere to hide.
> (Lloyd-Roberts, 2009)

The Commission for Africa (CFA), in its report on the British All Party Parliamentary Group (APPG) delegation's evaluation of the Niger Delta, urged the Nigerian government to stop the yearly theft of an estimated N625 billion (£2.5 billion) of the country's oil resources (Obayuwana, 2006:1). Mutiu Sunmonu, the Managing Director of SPDC, also told the delegates at the Nigerian Oil and Gas conference and exhibition in Abuja on 24th February 2009 that 'even with low oil prices, the Nigerian government loses between $1 billion and $1.5 billion every year to crude theft and pose a threat to oil loadings from export terminals' which was then reduced to six hours daily because of the security situation in the Niger Delta (Igbikiowubo, 2009). The restive Niger Delta youths exchanged the smuggled oil for firearms from their buyers, and the youths in return used firearms to attack and vandalise oil installations for the accumulation of more crude oil to be smuggled out for sale (Laba, 2004:3). In his overview of the Shell Petroleum Development Company (SPDC) of Nigeria to the British All Party Parliamentary Group (APPG) on the Niger Delta in August 2005, Basil Omiyi, the former Nigeria Chair, declared that 'it is clear for this level of oil theft to continue at its current rate, it is highly organised' (Obayuwana, 2006:1).

Other costs included huge annual expenditure on security for oil workers and facilities by the IOCs. According to the country security manager of Addax Petroleum Nigeria, at the 2008 Offshore West Africa (OWA) conference in Abuja, IOCs operating in Nigeria jointly spent an estimated $3.7 billion on security in the Delta in 2008 alone, while the figures for 2007 stood at $3.5 billion (Alike, 2009). It is disheartening to note that in 2008 alone security against the militants cost the IOCs $3.7 billion (N432.9 billion) (ibid), almost five times as much as the government's budget for the region in 2009 (N77 billion).

## Intra-Community Crisis, Inter-Community Conflicts, Inter-Ethnic Clashes and Conflicts Between Communities and IOCs

Since the early 1990s, intra and inter-community, inter-ethnic conflicts and community protest and violence against the IOCs in the oil producing areas have been increasingly rampant, resulting in serious socio-economic loss to the Delta region and the country. The increasing unrest has been mostly due to the high level of underdevelopment in the Delta.

### The intra-community crisis
The intra-community crisis is usually between members of a clan, village or ethnic group with communal identity (UNDP, 2006:113). A good example was in Nembe, Bayelsa state, where the IOCs reached an agreement with the local chiefs on the mode of payment of compensation through cash, job opportunities and contracts for the locals, but through greed the chiefs allocated more of the benefits to themselves and their associates. The youths and other excluded members of the community began a protest: they occupied flow stations, harassed oil workers and took hostages to force the IOCs to renegotiate and pay them their dues (ibid). Because such payments are huge, at least in local terms, the youth group soon disintegrated into several gangs and continued to extort money from the IOCs.

The result was a reign of terror in Nembe, with violent clashes between the various rival youth groups resulting in mass loss of life and property (ibid, 115).

The manner in which the pursuit for compensation was carried out among the communities has over time helped to exacerbate conflict. Onosode (1998) posits that:

> There is a considerable mistrust between elders and youths, between traditional rulers and their subjects. The older groups are perceived by the youths to have 'sold out'; others are branded 'Oil company chiefs'. (Onosode, 1998:39–41)

According to a World Bank report, 'compensation may not be paid to the affected community or individuals – because other communities, disbursement agents, or powerful individuals may keep the compensation funds' (World Bank, 1995: 72).

In an article on this subject, Ibeanu echoes the World Bank's view by asserting that 'The leaders of communities in the Niger Delta are very rich, and at the same time their people are extremely poor' (Ibeanu, 2002:163). By the end of February 1998, about 14,000 claims for compensation for oil related damages totalling an estimated US$100 million had been submitted to Nigerian courts by individuals, groups and communities in the region (ibid).

**Inter-community conflicts**
Inter-community conflicts occur when communities are not satisfied with the outcome of a settlement by the government, traditional authorities or court. In an on-going agitation, usually over ownership of land and other natural resources, mostly the male youths of two or more warring communities begin fighting, setting houses and business premises ablaze (UNDP, 2006:115). For example, the conflict that emerged as a result of the location of SPDC's oil flow station in the Olomoro and Oleh communities of Delta State, occupying 20 per cent and 80 per cent of land in the two communities respectively. The government responded swiftly and drafted the security forces to maintain law and order, but the security forces instead burnt down the Oleh community (ibid, 118).

**Ethnic clashes**
Ethnic clashes have also been on the increase in various parts of the Delta region; these include Urhobo versus Itsekiri, Ijaw versus Itsekiri in Delta state, Ogoni versus Okrika, Andoni versus Ogoni in River state and the Ilaje versus the Ijaw in Ondo state. While some of these conflicts are historical and predate the discovery of oil in the Niger Delta, others were

provoked by the actions of the government and IOCs (UNDP, 2006:118). For example the violent conflicts which involved the Ogonis and Andonis from July to September 2003, and the Ogonis and the Okrikas in December of the same year, seemed to have some support from the government and IOCs in order to thwart the Ogoni agitations, as the leaders of both the Ogonis and Andonis asserted publicly through the media that they had no disagreements before the attacks (Osaghae, 1995:37–38). At the end of the conflict, over 1,000 Ogonis had been killed and 30,000 people displaced from their communities (ibid).

**Conflicts between communities and IOCs**
Conflicts between communities and IOCs emerge as a result of general dissatisfaction with the benefits from oil. They bear the impact of the consequences of oil exploration –gas flaring, oil spillages, blowouts and so on – and government policies failed over the years to translate oil wealth into meaningful socio-economic development for the communities. The reactions to their frustrations over the years sometimes resulted in protest against the IOCs, epitomised by hostage-taking, sabotage to oil installations, damage to pipelines, flow stations and oil export terminals. This series of reactions from the communities in the Delta propelled the IOCs led by SPDC to establish a 'supernumerary police' in the early 1990s in conjunction with the Nigerian Police Force in the Delta to protect their facilities. On 1st December 1993, Phillip Watts, then SPDC's Nigeria Managing Director, requested the police authorities to increase the 'supernumerary police guards' (or the 'spy police') from 1,200 in 1993 to 1,400 in 1995; this was in addition to other private security personnel engaged by the company (International Crisis Group, 2006, HRW, 1999:116; Frynas, 2001:51). The Joint Task Force (JTF) – the Army, Navy and the Police – was also deployed by the federal government to provide security in the Delta region and the waterways, which come under the federal government's authority (SPDC, 2006:8).

# CONCLUSIONS

This chapter started by examining the significant role of oil in fuelling modern industrial development, and then examined the role of oil in both domestic and international politics and conflicts. It noted that the history of the international oil market has been one with prolonged conflicting interest where both producers and consumers strive to influence the direction of the market in their favour, and argued that the main aim on the part of the consumer nations is to secure energy security. Various factors are responsible for the instability of oil prices in the international oil market and conflicts in different oil and gas geopolitical zones of the world. Some of the major factors include the two gulf wars of the 1990s, the September 11 terrorist attack by Al-Qaida on United States in 2001, and the US-led invasion in Iraq in 2003. After these events oil prices began to rise and there was a related realignment in global geopolitics of oil and gas.

This chapter also posits that there is an interrelationship between the energy security agenda of the Western world and the nature of the conflicts and underdevelopment in the oil and gas producing countries in the global South. The unequal integration of the South into the global capitalist system and the nature of this continued exploitative relationship between the global North and South is driven by uneven terms of trade in relation to raw materials and cheap energy (oil and gas by all means and at all cost) to power the industrial development in the North.

Two levels of resource conflicts were also analysed: at the macro level which concerns conflicts of interest between the IOCs, the Western nations and producing nation governments over fiscal regimes and contracts, negotiations and unwillingness to grant access to particular IOCs; and at the micro or local level between national governments and other stakeholders in the producing nations. A detailed illustration was provided by the case study of the Niger Delta in Nigeria where various militia groups turned the oil producing region into a battle ground as a result of lack of access to socio-economic development and opportunities from the oil industry in their region. The main motivation behind insurrections, rebellions and wars in resource-dependent developing countries is not just about greed or the desire to loot; in most cases it is resistance to and reaction against long-term neglect and socio-economic deprivation in the mineral resource rich regions, the capture and corrupt control of the revenues accruing from such resources by the central governments and their lack of accountability to the citizens. Therefore, I conclude that most conflicts labelled 'resource conflicts' occur where groups or societies who have been denied benefits from their national resources struggle to regain their rights and access to these benefits. The Western nations, IOCs and oil producing governments need an alliance to ensure the benefits of the resources reach the citizens of the producing nations. Otherwise, the conflicts will continue as it will always be worth fighting for denied benefits.

# NOTES

1. Securitisation is 'the discursive process through which an inter-subjective understanding is constructed within a political community to threat something as an existential threat to

88  *International handbook of energy security*

   a valued referent object and to enable a call for urgent and exceptional measures' (Buzan and Wæver 2003, 491). This process 'legitimizes actions outside the normal political process' (Buzan et al, 1998, 24, quoted in Trombetta, 2012:2).
2. Isaac Boro an Ijaw, former school teacher and police officer in the Nigerian police force, was born on the 10th September 1938 in Oloibiri where the first oil was struck by Shell Petroleum Development Company (SPDC). He landed in Tontoubau in Bayelsa State on the 23rd February 1966 accompanied by 159 men, and launched a guerrilla war against the Nigerian government. He led an armed showdown against the Nigerian Police Force in a bloody battle and defeated them with a ragtag force of native comrades. They then grounded all oil exploration activities in the region, declared the State of the Niger Delta People's Republic and instructed all IOCs to deal directly with Boro. They engaged in battle for 12 days before they were rounded up by the federal government forces with the help of the SPDC and were tortured, tried and condemned to death (Tebekaemi, 1982:6).
3. Movement for the Emancipation of the Niger Delta is an armed and formidable militant group based in the cities and creeks of the Niger Delta, particularly in the western region of the Delta in and around Warri (the so-called 'Warri axis'). MEND is responsible for 'shutting-in' 40 per cent (November, 2008: approximately 900,000 barrels per day) of Nigeria's oil industry through making direct attacks on facilities, taking hostages, and generally creating an inhospitable and unsafe environment for the oil industry (Kashi and Watts, 2008:3).
4. Ken Saro Wiwa is an author, environmental activist and businessman.
5. The energy security (security of supply) issue is also about geopolitics: consider the geopolitical position that Nigeria seems to have in relations to the ECOWAS region, and the US Africa Command (AFRICOM). Nigeria is seen as a steady regional policeman for the whole Gulf of Guinea up to Angola. Michael Watts posits that a geo-strategic driver was involved (Watts, 2004).
6. An environmental activist and lawyer and one of the youngest attorneys on Saro Wiwa's defence team against the Nigerian government tribunal in 1995. Senior Special Assistant to the Nigerian President on Research and Strategy 2011 to the present.
7. The accusations by the governments and the IOCs against the people of the Delta of sabotage and belonging to criminal gangs are similar to the way oppressive regimes treat resitance to law and order issue. This started in the colonial period when the Colonial government reacted in a similar way to the attacks on the Royal Niger Company headquarters by the youths who protested against RNC's monopoly in the palm oil trade in Brass in 1894.
8. Oil bunkering is the illegal tapping of both crude and refined oil into barges or other containers and loading them on to tankers or cargo ships for sale locally or to neighbouring countries and other buyers in the international oil market.
9. One of MEND's senior commanders.

# BIBLIOGRAPHY

Adejumobi, Said (2002), 'Ethnic Militia Groups and the National Question in Nigeria', paper presented at the Conference on Urban Violence, Ethnic Militia and the Challenges of Democratic Consolidation in Nigeria, at: http://www.programs.ssrc.org/gsc/gsc_quarterly/newsletter8/content/ade, accessed 7 March 2009.

Ake, Claude (2000), *Feasibility of Democracy in Africa*, Dakar: CODESRIA Book series, pp. 10–25.

Alabi, Joshua (2010), 'The Dynamics of Oil and Fiscal Federalism: Challenges to Governance and Development in Nigeria', PhD Thesis, University of Leeds, p. 58.

Alike, Ejiofor (2009), 'Oil firms spent $3.7bn on security', N'Delta *Thisday Online* at http://www.Thisdayonline.com/nview.php?id=134544, accessed 8 February 2009.

Bromley, Simon (2005), 'The United States and the Control of World Oil', *Government and Opposition*, 40(2).
Bush, Ray (2007), 'Class, Resistance & Social Transformation', *Review of African Political Economy No. 114*, London: Francis and Taylor, pp. 613–617.
Businessday (2008), 'Jomo Gbomo MEND spokesman in an email sent to the Press', *Business Day*, at https://www.businessdayonline.com/index.php?view=articles&catid, accessed 12 November 2008.
Business Reference Services (2012), 'History of Oil and Gas Industry', at http://www.loc.gov/rr/business/BERA/issue5/history.html, accessed 31 October 2012.
Buzan, Barry and Wæver, Ole (2003), *Regions and powers: The structure of international security*, Cambridge University Press, Cambridge; New York, p. 491.
Buzan Barry, Ole Wæver and Jaap De Wilde (1998), *Security: A New Framework for Analysis*, London: Lynne Rienner Publishers, p. 24.
Collier, Paul (2007), *The Bottom Billion: Why The Poorest are Failing and What Can be Done*, Oxford: Oxford University Press.
Collier, Paul (2009), 'The Political Economy of Natural Resources: Interdependence and its Implications', paper presented at the 10th Anniversary Conference of the *Global Development Network (GDN)* Kuwait, 3–6 February, pp. 1–18.
Collier, P and A. Hoeffler (2004), 'Greed and Grievance in Civil Wars', Oxford Economic Papers 56, pp. 663–695.
Collier, P., A. Hoeffler and D. Rohner (2009), 'Beyond Greed and Grievance: Feasibility and Civil War', Oxford Economic Papers 61, pp. 1–27.
Daniel, Soni and Bisi Olaniyi (2006), 'Hostages: FG strikes deals with militants', *Punch*, at http://www.punchng.com/news, accessed 21 February 2006.
Dannreuther, Roland (2010), 'International Relations Theories: Energy, Minerals and Conflict', POLINARES working paper no. 8, September, pp. 1–3,
Djebah, Oma (2003), 'Tackling The Niger Delta Conundrum . . .', *Thisday*, at http://www.thisdayonline.com, accessed 29 July 2007, p. 3.
Duffield, Mark (2001), *Global Governance and the New Wars: The Merging of Development and Security*, London: Zed Books, pp. 1–2, 5–9, 16.
Essien, Udo (1999), 'Stuck in an Oil Mess', *Thisday*, Issue 14, pp 3–8.
Frynas, J. George (2001), 'Corporate and State Responses to Anti-Oil Protest in the Niger Delta', *African Affairs*, 100, 46–47.
Guardian Newspaper (2008), 'Nigeria's crude oil shut-in hits 1.3million bpd', *The Guardian* at https://www.ngrguardiannews.com/news/article01/indexn2_html?pdate=290408, accessed 18 October 2008.
Hoogvelt, Ankie (2001), *Globalisation and The Postcolonial World: The New Political Economy of Development*, London: Palgrave, pp. 8–9.
HRW (1999), *The Price of Oil: Corporate Responsibility and Human Violations in Nigeria's Oil Producing Communities*, New York: Human Rights Watch, pp. 20–22.
Ibeanu, Okechukwu (2002), 'Janus Unbound: Petrol-business & Petropolitics in the Niger Delta', *Review of African Political Economy*, 91, 163.
Igbikiowubo, Hector (2009), 'Nigeria loses $1.5billion yearly to oil theft – Shell', *Nigerian Vanguard Newspaper* at https://www.vanguardngr.com/content/view/30034/49/, accessed 3 March 2009.
Ijaw Youths Centre (1998), 'Kaiama Declaration' by the Ijaw Youths December 1998 at http://www.ijawcenter.com/kaiama_declaration.html, accessed 6 March 2009.
International Crisis Group (ICG) (2006), 'Swamps of Insurgency', International Crisis Group, Working Paper, Brussels/Dakar: ICG, pp. 1–29.
International Crisis Group (ICG) (2008), 'Nigeria: Ogoni Land after Shell', International Crisis Group, Working Paper, Brussels/Abuja/Dakar: ICG, pp. 1–9.
Karl, Terry Lynn (2005), 'Understanding the Resource Curse', in Svetlana Tsalik and Anya Schiffrin (eds), *Covering Oil: A Reporters Guide to Energy and Development*, New York: Open Society Institute, pp. 21–28.

Kashi, Ed (2008), 'An interview with Ritz Khan on the Niger Delta Crisis', *Al-Jazeera Television*, 22 July 2008.
Kashi, Ed and Michael Watts (2008), *Curse of the Black Gold: 50 Years of Oil in the Niger Delta*, New Jersey: PowerHouse Books.
Klare, Michael T. (2001), *Resource Wars: The New Landscape of Global Conflict*, New York: Metropolitan Books, p. 230.
Klare, Michael T. (2004), *Blood and Oil: The Dangers and Consequences of America's Growing Petroleum Dependency*, Hamish Hamilton: Penguin, pp. 124–127.
Laba, Oghenekevwe (2004), 'Code Nigerian Oil to Check Bunkering', *Thisday* at http://www.thisdayonline.com/news/20040509news13, accessed 9 May 2004.
Le Billion, Philippe (2004), 'The Geopolitical economy of "resource wars"', *Geopolitics*, 9(1), 3.
Le Billion, P. and F. El Khatib (2004), 'From free oil to "freedom oil": terrorism, war and US Geopolitics in the Persian Gulf', *Geopolitics*, 9(1), 121–123.
Lloyd-Roberts, Sue (2009), 'Fighting for Nigeria's Oil Wealth', BBC documentary at http://www.news.bbc.co.uk/1/hi/ programmes/newsnight/7816654.stm, accessed 24 February 2009.
Lujala, P., N.P. Gleditsch and E. Gilmore (2005), 'A diamond curse? Civil war and a lootable resource', *Journal of Conflict Resolution*, 49, 538–562.
NDDC (2002), 'Policies and Programme of NDDC', Port Harcourt, Nigeria.
Nigerian Tribune (2008), 'Ya'radua Sends Shell out of Ogoniland', *Ibadan: Nigerian Tribune*, Monday 7 July.
Nigerian Tribune (2010), 'MEND threatens to resume hostilities before Jan 20', *Ibadan: Nigerian Tribune*, Monday 11 January.
Obayuwana, Oghogho (2006), 'African body says Nigeria losses N625 billion yearly to oil theft', *The Guardian* at http:// www.guardiannewsngr.com/ News/article01, accessed 25 January 2006.
Oduniyi, Mike and James, Segun (2006), 'Militants Attack: Shell Shuts Down 226,000 Crude Production Facility', *Thisday* at http:// www.thisdayonline.com/news/2006, accessed 12 April 2007.
O'Hara, Sarah (2004), 'Great game or Grubby game? the struggle for control of the Caspian', *Geopolitics*, 9(1), 142.
Ojakorotu, Victor (2005), 'Social Movements versus Shell: The Impact of Globalization on the Environmental Violence in the Niger Delta Nigeria', Conference proceeding on contemporary governance and the question of the social, Humanities Centre, University of Alberta Canada, 11–13 June 2004, pp. 1–13.
Omoweh, Daniel (2006), 'Natural Resource Struggles in the Niger Delta and Democratization in Nigeria', *International Journal of African Studies*, 5(1), Spring, 29–68.
Onosode, G. (1998) 'Curbing flaring tempers, Pollution in the Niger Delta', *The Guardian*, 14 December.
Osaghae, Eghosa E. (1995), 'The Ogoni Uprising: Oil Politics, Minority Agitation and the Future of the Nigerian State', *African Affairs*, 94, 325–344.
Peters, Susanne (2004), 'Coercive western energy security strategies: "resource wars as a new threat to global security"', *Geopolitics*, 9(1), 202.
Raphael, Sam and Doug Stokes (2011), 'Globalizing West African Oil: US "energy security" and the global economy', *International Affairs*, 87(4), 909–911.
Rivers State Government (1991), 'Report of the Judicial Commission of Inquiry into Umuechem Disturbances', Port Harcourt, Nigeria.
Ross, Michael (1999), 'The Political Economy of Resource Curse', *World Politics*, 51, 297–322.
Ross, Michael (2001), 'Extractive Sectors and the Poor', Oxfam America Report, Washington DC: Oxfam America, p. 7.
Rowell, A., J. Marriott and L. Stockman (2005), *The Next Gulf: London, Washington and Oil Conflict in Nigeria*, London: Constable and Robinson.
Shultz, Jim (2005), 'Lifting the Resource Curse', in *Follow the Money: A Guide to*

*Monitoring Budgets and Oil and Gas Revenues*, New York: Open Society Institute, pp. 13, 14, 37–39.

Smith, Ben (2010), 'International Energy Security: Energy security will play an ever growing role in foreign policy', in Mellows Facer (ed.) *Key Issues for the New Parliament 2010*, London: House of Common Library Research Publication.

Soyinka, Wole (2008), 'An interview with Ritz Khan on the Niger Delta Crisis', *Al-Jazeera Television*, 22 July 2008.

SPDC (2006), 'Shell Nigeria Annual Report', Port Harcourt: Shell Petroleum Development Company Ltd, pp. 6–9.

Spero, Joan and Jeffrey Hart (2003), *The Politics of International Economic Relations*, Belmont CA: Wadsworth/Thomson, pp. 174–179, 229–300.

Stiglitz, Joseph E. and Linda J. Blimes (2008), *The Three Trillion Dollar War: The True Cost of the Iraq Conflict*, New York: Norton & Co, pp. 132–160.

Tebekaemi, Tony (ed.) (1982), *The Twelve Day Revolution*, Benin City, Idodo Umeh Publishers.

Trombetta, Julia (2012), 'European energy security discourses and the development of a common energy policy', Working Papers of the Energy Delta Gas Research, no 2, pp. 1–3.

UNDP (2006), 'Niger Delta Human Development Report', Abuja: UNDP, pp. 9–48, 111–129.

Watts, Michael (2004), 'Resource Curse? Governmentality, Oil and Power in the Niger Delta, Nigeria', *Geopolitics*, 9(1).

Watts, M J., P.M. Lubeck and R.D. Lipschutz (2007), 'Convergent interests: US Energy Security and the "Securing" of Nigerian Democracy', International Policy Report, Washington DC: Centre for International Policy.

White, Gregory and Scott Taylor (2001), 'Well-Oiled Regimes: Oil & Uncertain Transitions in Algeria & Nigeria', *Review of African Political Economy*, 89, 324–325.

William, Graf D. (1988), *The Nigerian state: political economy. State, Class and political system, in the post-colonial era*, London: Currey, pp. 133–141.

World Bank (1995), *Nigeria: Poverty in the midst of plenty, the challenges of growth with inclusion*, Washington DC: Oxford University Press, p. 10.

# 5. Global energy supply: scale, perception and the return to geopolitics
*Susanne Peters and Kirsten Westphal*

## 1 INTRODUCTION

Energy security is back on the agenda. No longer is it an issue of the past like it was during the 1973/74 and 1979 crises.

During the heyday of globalization in the 1990s, energy was increasingly framed as a commodity among others with markets taking care of trade flows between producers and consumers. Since the early 2000s, the growing consumption of China and India has resulted in tightening of the hydrocarbon markets and consequently in rising prices. Moreover, with the 2003 Iraq War it became evident that a geopolitical approach to energy has revived again as a policy paradigm. This large-scale war between a third world country and a Western coalition, led by the United States, which was also fought to gain access to Iraq's vast untapped potential brought home the message that "territory", "geography" and "state interest" are major categories for constructing energy relations again.[1]

Whereas in the United States the geopolitical narrative has been closely related to oil, the storyline in Europe is somewhat different: in the EU, natural gas deliveries from Russia have been framed in geopolitical terms. United States criticism on the Soviet-German gas-pipe deals accompanied the trade relationship from the very beginning. More recently, the new Eastern and Central European member states raised the issue of energy security vis-à-vis Russia in a more pronounced way and the Russian-Ukrainian gas crises in 2006 and 2009 were framed as a case in point. We will focus on oil and natural gas, because indeed, these are the two strategic resources around which geopolitical strategies center, and which are seen as potential sources of conflicts over access, use and distribution.

In a multipolar age in which power is diversified among many international actors across nations, and with multilateralism in crisis, these noteworthy incidents raise the question whether disputes over territory and natural resources will again turn into important factors in international politics and whether interstate war is indeed back on the agenda. The concern about "energy security" and its geopolitical framing and hence "geopolitics of energy" and the securitization of energy is back and now a priority issue for governments.

In this brief chapter, we examine the geopolitical and theoretical bases for a renewed emphasis on energy security. Starting from a discussion of theoretical concepts we then look at how "sector fundamentals" have affected scholarly discussion of energy supply. It has to be emphasized, though, that energy is about geography. Thus, we will start with a mapping of hydrocarbon resources as a given underlying reality of producer – consumer and transit relations. We will also make the point that the landscape of energy is changing profoundly when you take unconventional energy resources and most prominently the shale gas revolution in the US into the picture. We emphasize that the supply situation is characterized by "unprecedented uncertainty". Third, we analyze different phenomena of the past decade – high and volatile prices, increasing resource concentration, perceived scarcity and state interventions – that have fuelled the fear of resource conflict. This will also shed light on the fact that the dominating mechanism for access and distribution of resources is pricing, not state intervention. Having said this, it is important to preclude that globalization is in retreat. Price structures start to differ between the various regions. Fourth, we focus on prominent case studies that have been analyzed under the geopolitical lens. These case studies illustrate that it makes a difference how analysts describe and frame the actual development in the markets. "Energy security" is indeed back on the agenda, but how governments and/ or markets pursue such an elusive goal is very much – following these analysts – a question of their integration into the global economy, their pluralistic structures, "good governance" in the sector, as well as the perception of the tightness of the future supply market and the availability of power projections forces.

## 2 RETURN OF TRADITIONAL GEOPOLITICS – WHAT IS "GEOPOLITICS"?

The concept of geopolitics which dates back to the late 19th century currently is enjoying a revival.[2] While used in the post war era with caution – because of its connection to Nazi Germany – the term was still a useful tool to analyze the relationship between the superpowers and its effects on different regions of the world. Concepts like "geo-strategies" and theories of "falling dominos" were used to explain "influence and control over other states and strategic resources" (Profant 2010, 41). These approaches were deeply rooted with theories of realism and neo-realism, focusing on the state as the main actor in the international system. According to these approaches, dominance and hegemony over territories, population and endowment of natural resources shape international relations.

Under a geopolitical lens, energy relations are analyzed as state relations driven by state and foreign policy interests. Energy resources are seen as a source of conflict but also as a power tool in international relations. This "securitization" of energy relations has been lately (again) promoted first by US scientists and political circles and then spread to Europe (Götz 2012, 436). Whereas in the US the major focus is on oil, in the EU the gas deliveries from Russia and the EU's import dependency have fuelled geopolitical approaches.

Michael T. Klare explicitly adheres to a geopolitical approach using traditional and realist concepts. He is a prominent and prolific scholar in the field of the geopolitics of energy security. For more than a decade he has warned that the international oil system is soon to break down, because the growing "energy gap" is going to lead to a geopolitical rivalry and an implosion of cooperation among the main players. He sees a struggle among the major energy-consuming nations in the "new international energy system" absorbed in a competitive struggle over some indispensable resources that are becoming more scarce. The focus is on the United States as the most important player in a new geopolitical game, challenged by new emerging players like China, Russia and India.

These power relations which are "(a)lready edgy and competitive (. . .) hint at future scenarios of conflict among the so-called Great Powers of a far more dangerous sort." (Klare 2008, 21) In this "new international energy order" governments – and not private corporations – would be the main players and the subjects of geopolitics since governments have increasingly lost confidence in private firms to acquire new oil and gas properties in the last decade. Therefore, governments have to take on a "commanding role" again by taking key strategic decisions leading to new forms of "resource nationalism" (Klare 2008, 22–23). In a world which is divided into energy surplus and energy deficit countries the latter are striving to develop good relationships with the former. It is a logical consequence that the fiercer the competition among the energy-deficit countries over "what is left", the more the competing states will feel compelled to take recourse to means of military involvement with the key resource supplying areas in particular in Africa and Central Asia:

> As the scramble for vital resources intensifies, however, governments will also become increasingly likely to employ more forceful means. In all probability, countries with major resource deposits will receive more weapons, military training, technical assistance, and intelligence support from states that wish to curry favor or establish closer ties. At the same time, combat forces will be deployed abroad to defend friendly regimes and protect key

ports, pipelines, refineries, and other critical installations. (Klare 2012, 220–221)

Here Klare sees the cause for possible conflicts which might spin out of control. As the key *rival* for the United States Klare points to China where the competition includes all corners of the world where oil is being produced in big quantities: from the Gulf region over South America to the Caspian to Africa, but in particular in Africa.

## 2.1 Critical Geopolitics

But like realism, the concept of "geopolitics" has also experienced cycles of recognition and dismissal as a valuable analytical tool. With the emergence of globalization in the 1980s and 1990s, theories with a focus on the role of economics in international relations seemed to be more apt to explain the new reality. Apart from mainstream theories which clustered around the school of neo-liberalism, an interesting school called "critical geopolitics" developed as an alternative method of analyzing geopolitics and global change. Scholars of this approach recognized that "along with the changing ways in which the international political economy operates (new patterns of flows, transfers and interactions) come new representations of the division and patterning of global space." (Agnew and Corbridge 1995, 7) In this transforming world of the 1990s Agnew and Corbridge identified a process of "deterritorialization" (Agnew and Corbridge 1995, 100) in which "geography" and "territory" cannot be attributed any meaning and in which "relative economic power has begun to displace military force and conquest as an important feature of international relations." (Agnew and Corbridge 1995, 3)

For the purposes of our discussion, we use Gearóid Ó Tuathail's and John Agnew's definition of "geopolitics" as a "discursive practice by which intellectuals of statecraft 'spatialize' international politics in such a way as to represent it as a 'world' characterized by particular types of places, peoples and dramas."[3] For pursuing geopolitics these intellectuals have to take recourse to a "geo-strategization" which Tuathail defines as "the making of a discursive claim that a particular foreign policy crisis or challenge has the locational and transcendent material national interest qualities that make it 'strategic'." (Tuathail 2004, 96) This might apply to conflict over a particular material resource or might have an impact on consensual national security interests, but in any event an inventory of these dimensions is constructed and represented as part of the material self interest of the state" (Tuathail 2004, 96). This is at the core of our interest in this analysis: to determine whether after a long phase of globalization states are again refocusing their attention to spatialize international politics according to their national interests and "securitize" energy relations. A geo-strategization is understood in this context as not excluding military measures as part of a comprehensive concept.[4]

## 2.2 Energy Cycles and Corresponding Theoretical Views

Like "geopolitics", "energy security" has experienced certain cyclical ups and downs following oil price developments. In very broad terms, "energy security" is defined as stable and secure supplies at affordable prices. The attention of governments to energy security follows the fundamental situation in the respective oil and gas markets: In the 1980s and 1990s, thanks to a vigorous diversification strategy from the Middle East on part of the OECD countries and later the dissolution of the Eastern Bloc the precious resource oil was abundant again and available for very low prices. Energy was starting to be seen as a "commodity" and no longer as a trophy in a geopolitical great game. Consequently, energy was defined as a commodity or a service (electricity).

Ups and downs in the supply situation are part of the raw materials' market features: physical availability and consequently price developments are closely intertwined with the issue of appropriate and sufficient investments. The raw materials investment cycle is characterized by the fact that projects in the energy sector have remarkable lead times until they are fully developed and on stream, e.g. the respective pipeline is being filled. A cycle can unfold: price volatility and uncertainty discourage investment, as a consequence producing country governments under-invest in productive capacity, a behavior that is constraining the capacity over time. With the increase in demand, e.g. oil prices rise thereby resulting in a tightening of the supply/demand balance. Both sides take actions: in the face of high prices, consuming country governments take action to curb oil-demand growth. Oil demand slows with a lag. At the same time the investment rebounds in producing countries, boosting capacity with a lag. Together this is leading to over-capacity and causing prices to fall back again (IEA 2010, 141) The (vicious) cycle starts again. These cyclic investment swings are known as the "pig-cycle" and are reinforced by intransparent markets and very limited price elasticity of both demand and supply.

After having being treated like a commodity for two decades with the beginning of the 2000s energy seemed to be back on the agenda of politicians (Yergin 2006). We saw prices skyrocketing and the ever alarming term "peak oil" made it for the first time into mainstream analyses of oil supply.[5] The first articles appeared indicating that energy security would be a very dominant subject again in the future, and back were articles which interpreted "energy security" not as a matter which could be solved by the "markets and institutions", but rather by pursuing "geopolitics" and "geo-strategy". For those arguing in this line the Iraq War was telling evidence of this new trend.

But, first we start with an analysis of the geopolitical implications of today's supply issues which are characterized by unprecedented risks and uncertainties.

## 3  THE UNPRECEDENTED UNCERTAINTY AND THE INTERNATIONAL LANDSCAPE OF SUPPLY

"Unprecedented uncertainty" (IEA 2010) is indeed the main feature of today's energy system, as put forward by the International Energy Agency (IEA). This is equally true for the demand as well as for the supply side on which we focus.[6] Unprecedented uncertainty characterizes the present energy system as the size of ultimately recoverable conventional and unconventional reserves is a major source of uncertainty for the long-term outlook (IEA 2010, 48). Equally unclear are demand developments in light of the economic crises but also of possible fuel switching and substitution effects.

A key issue for determining the quantity of reserves left is how progress in technology is evaluated against geological fundamentals. The "Peak Oil" discussion has raised public consciousness because hydrocarbons are exhaustive and non-recoverable. Estimates differ widely about when oil production has reached or will reach its plateau. The Association of Peak Oil (ASPO) argues that the peak of oil production was reached somewhere between 2005 and 2010; multinational companies are much more optimistic. BP estimates, that the known reserves are able to cover the demand for the next 40 years. Maugeri has predicted a new revolution in oil supply capacities and even an "oil glut" based on "deconventionalization" (Maugeri 2012). Other papers discuss the estimates on ultimately recoverable resources and price elasticities of supply and demand by bringing together the opposing approaches of geology and technology (Benes et al. 2012). The new dreams of resource endowment are fuelled by thoughts of unconventional resources. "Unexplored and untapped" areas such as deep waters, the Arctic, East Siberia as well as oil sands, tight oil and shale oil have contributed to increase the global reserve basis. However, a clear definition for *unconventional* is lacking, but is meant for resources that can only be exploited by new technologies and is unclear with respect to economical recovery. Exploring (un)conventional oil seems more a matter of costs and prices and of course is related to new environmental risks. The above gives evidence that the time of cheap oil is over. And in our geopolitical context, the most important message is that the use of the unconventional hydrocarbon resources may change the landscape of energy relations.

The conventional picture is one of growing concentration in few world regions and in countries that are characterized by unstable and undemocratic regimes. One of the main particularities of conventional fossil fuels is their uneven global distribution: In the future, the world will have to rely to an ever larger extent on the energy-abundant countries of the "strategic ellipsis": the geographical area stretching from Siberia to the Caspian Basin, the Persian Gulf to the Arabian Peninsula. The region contains 63.5 per cent of global oil reserves, compared to a 47 per cent share in overall production in 2009 (BP 2012, 6, 8). The future role of the Organization of Petroleum Exporting Countries (OPEC) and the implications for security of supply is among the geopolitical uncertainties. OPEC countries control more than 76 per cent of global reserves and 63 per cent are located in the Gulf region. The OECD disposes over only 6 per cent of global oil reserves (BGR 2009, 37) and conventional oil production in the OECD area has most likely achieved the plateau.

On a global scale the resource base of natural gas is abundant in comparison to those of oil. For gas, the share of conventional gas reserves in the strategic ellipsis is even higher with more than 70 per cent, while the bulk of almost 60 per cent is located in four countries: Russia, Qatar, Iran and Turkmenistan. The countries of the strategic ellipsis also dominate gas production with around 37.5 per cent (BP 2012, 20, 22). In a global perspective, most of the reserves are conventional gas. The OECD share is only about 10 per cent of the world total.

The Arab Spring is a case in point for the (geo)political impacts and has caused widespread fears about the prospect of oil (and gas) supply disruptions. Very closely related is the issue of internal reforms that may go hand in hand with new depletion strategies, a revisiting of existing Production Sharing Agreements, changes in the managements of National Oil Companies etc. This in turn, may affect the business conditions for IOCs and the access regime. Moreover, political and socio-economic reforms may affect the volumes and direction of exports and there are strong indications that necessary investment into new sites is on hold, funds are redirected and domestic price reforms are reversed.

The uprisings in the Arab world that unfolded in 2010 and 2011 question existing alliances. This is important from a geopolitical perspective and indeed, the movements mark a watershed (Darbouche and Fattouh 2011, 1). The OECD world had long depended on autocratic regimes because they seemed to provide stability. This proved to be an error: rather, it is "stable societies(,) that hold the key to future reliance on MENA hydrocarbons." (Roberts 2011, 3) Regionally, there is an immediate risk of contagion and changing balance of power. Iran and the Shiite minorities in the Gulf countries are a source of concern. This is particularly sensitive when it comes to the three countries of the Gulf Cooperation Council (GCC): Saudi Arabia, United Arab Emirates and Kuwait. Saudi Arabia's strategic importance for world oil markets cannot be overstated. It is not just the sheer size of its production and exports, but its spare capacities to increase oil production at short notice. Saudi Arabia produces around 10 million barrels daily, and has another estimated 2.5 million barrel capacity,[7] which gives the Kingdom the opportunity to act as a balancing swing producer. Even though the real potential has been questioned (Simmons 2005), there is no doubt about the relative market position: there is no significant spare capacity outside the Gulf region. Moreover, the region hosts some of the world's famous chokepoints and transport arteries with the Strait of Hormuz, through which 20 per cent of all globally traded oil passes, the Bab-el Mandeb Strait, the Suez Canal and the Sumed pipeline.

What Saudi Arabia is for oil, Russia is for (conventional) gas. Russia has been the dominant gas player in the world and has built its regional (read: in the EU and the CIS) energy power on its pipeline network. Change is in the air or is even taking place already. The dominating geopolitical theme has been Europe's (inter)dependency from Russian gas supplies. With respect to diversification, the Caspian Basin with its vast, partly unexplored, and untapped resources is one of the regions that could change the future pattern of resource development and exports (IEA 2010, 523). Moreover, the natural gas reserves in the MENA region represent some 45 per cent of the world's total reserves, its marketed production amounts to 20 per cent of the world's total output (Darbouche and Fattouh 2011, 21). Qatar is the major player on the world's LNG market. Another trend observable in the landscape of supply or more precisely in supply *and* demand is the steeply rising demand in energy producing countries. The gas consumption in the MENA countries is a case in point with the growing indigenous demand. Most countries of the region, with the exception of Qatar, have faced supply shortages. Gas will be extensively used in the power sector: the North African countries will see a quadrupling of their electricity consumption till 2030, given annual growth rates of between 4 and 8 per cent. It will also serve as a feed stock for energy-intensive industries, whilst at the same time upstream production costs will increase as "easy gas resources" are depleting. Gas consumption in the Middle East is expected to grow by 3.9 per cent per annum between 2010 and 2030 (BP 2011, 51). The rising internal demand will most likely affect the disposable export volumes and constrain the region's export capabilities.

Rising gas demand and antagonistic relations between neighboring countries are a threat to regional stability. After the Egyptian revolution in February 2011, the gas export contracts of the country with Israel and its other neighbors have been constantly questioned, and the pipeline through the Sinai has been blown up several times. This is a source of geopolitical instability in the region, as Israel gets 40 per cent of its imports from Egypt. The country will be forced to exploit disputed off-shore resources in the Eastern Mediterranean Sea. The Levant Basin has witnessed the world's largest deepwater gas discoveries in 2009 and 2010. This may lead to conflict between Israel, the Palestinian Authorities and the Lebanon. Moreover, these vast reserves have increasingly been a source of dispute between Cyprus and Turkey in 2011 (Popvici 2011).

The dominance of few exporters and the concentration of the conventional gas reserves have raised concerns over the creation of the Gas Exporting Countries Forum (GECF). Pipeline-dominated gas exports and the existence of oil price indexed contracts have since its creation been seen as hindering the formation of a cartel. However, since spot market transactions in Europe and global Liquefied Natural Gas (LNG) trade has increased, the room for maneuver for gas exporting countries to steer export volumes and directions has enlarged. As in OPEC, an effective cartelization is constrained by diverging (geo)political and commercial interests of the members. Nevertheless, the developments are a source of concern which consumer states should keep an eye on.

The shale gas revolution in the US has fundamentally changed the gas markets. Whether this boom can be reproduced in other parts of the world, namely in (Eastern) Europe and China is uncertain, though. The steep increase in shale gas production in the US resulted in a remarkable fall of LNG imports to the US. This LNG was then exported to Europe, helping to increase the liquidity of its gas hubs and spot markets significantly. The gas glut also unfolded because the economic crisis in 2008/2009 resulted in decreasing demand. Gas prices eroded in Continental Europe and consumers were able to buy gas at a much lower price than under the Long Term Contracts with its major Norwegian and Russian suppliers. As a consequence of these latest developments, the gas markets are in flux with respect to the marketing areas, the different price regimes and the relevant actors (Westphal 2012, 2013). At the time of writing three (or even four) regional gas markets exist with very different price levels. LNG trade has not led – as expected less than a decade ago – to a globalized gas market and converging prices. In the North American market, gas prices are very low due to the shale gas revolution. In the European market the price is more than three times higher than in North America, but still much lower than in the Asian-Pacific market.

Also the pricing mechanisms differ. South America is the fourth region but on a nascent stage.

The above described geopolitics around hydrocarbons will certainly change with a transformation toward a more sustainable energy system. Yet, a more renewable energy based system will not be free of geopolitics,[8] because it shapes future regions through interconnecting infrastructure and new (inter)dependencies. Moreover, inter- and transnational cooperation is a key condition for the expansion of renewable energy and developing green electricity markets as reflected in the Seatec concept of large off-shore wind parks in the North Sea or envisioned in the Desert power concept linking the EU with the MENA region. Moreover, the case of biofuels or hydropower is closely related to the water and food usage. Energy is part of the water-energy-food nexus which itself displays the linkages between territory, access to resources, usage and distribution.

The unknown level and the future structure of energy demand is another uncertainty. However, the preconditions are very different around the world: the industrialized countries have almost reached their peak in energy consumption, but the emerging countries and the developing countries have to cope with a (sharp) increase in demand. This can be described as the global energy dilemmas (Bradshaw 2010) that complicate the search for new global energy solutions at a time when we observe a general retreat of globalization.

## 4 RETREAT OF GLOBALIZATION – EFFECTS ON ENERGY SECTOR

The following section will try to shed light on certain trends that provide evidence for a "retreat of globalization".

The energy world will soon become much more differentiated in terms of energy production, consumption paths and the landscape of actors. Major uncertainty relates to the different development paths and their effective implementation. Depending on the locational prerequisites needed for renewable energies, storage facilities etc., the energy production and consumption map will become much more diversified being tailored to local circumstances. Moreover, oil, gas and coal will likely each share a third of the hydrocarbon energy mix. This brings about many uncertainties with respect to the development of costs and winning technologies.

A retreat of globalization also occurred as oil (and even indirectly gas and coal) markets became global because of the price mechanism. However, this has always been less true for the physical flow of oil. The recent price developments as seen in the remarkable gap between West

Texas Intermediate Oil and Brent Oil prices are illustrative. The US has managed to become increasingly independent from the MENA region. Europe has proved to be much more vulnerable, in particular from trade disruptions in North Africa. Yet, the bulk of GCC crude oil exports go to the Asia-Pacific region.[9] What do the physical oil trade flows tell us? In the future, the North American oil markets may become less interconnected depending on the amount of deep water off shore exploration in the Mexican Gulf, unconventional (shale) oil production in the US and Canada, and the prospected off-shore production in Brazil. The US and the North American market have taken significant steps to become largely self-suppliers. The US has become a self supplier, most likely also with (shale and offshore) oil. This may have geopolitical implications as it results in less immediate interest in the Middle East. This may result geopolitically in a more isolationist attitude towards issues of supply security, in particular with regard to securing oil and LNG flows through the major chokepoints. More than 40 per cent of global oil exports originate in the Gulf Region. One-third of all sea-based oil trade transits through the Strait of Hormuz. The revolution in Egypt in February 2011 has raised concerns over the SUMED pipeline and the Suez Channel through which the LNG exports from Qatar to Europe cross. Will Europe and China step in? Also as yet unknown is how China as a major consumer of oil and gas flows from the Middle East will behave as "Ordnungsmacht". So far, consumer countries' cooperation is limited by the institutional landscape of international governance. Neither China nor India are members of the IEA. However, it is common sense that it will be the emerging consuming countries that will make the difference.

Another question is whether the US will become a gas exporter and challenge Russia in that respect. It has already overtaken the Cold War antagonist in production. If the two clash on market shares in the energy trade that could mean a qualitatively new situation with regard to already sensitive international energy relations. Whether Europe will still see an increased availability of LNG (redirected from the US) depends on the level of demand increase in the Pacific region. For Europe, the major challenges to be faced are the strategic partnership with Russia to be developed further and the energy relationship with China. Both regions are looking to the same resources in the Caspian region and in Central Asia.

Second, the cyclical development of oil prices has implications for the national governance of access, use and distribution of hydrocarbons, and subsequently for the redistribution of rents. Oil prices are the lead currency for most of the other raw materials. Oil prices are the major reference point for investment into oil exploration, production and infrastructure, and they are the major incentive to reduce oil consumption. Moreover, they determine the income to the state budget of oil producing states. Last but not least, oil prices are set in a complex interplay of the market fundamentals, market expectations and financial transactions and speculation.

Third, with respect to globalization in retreat is "resource nationalism" and the renationalization and creation of national oil companies (NOCs). At the beginning of the 2000s, when a steep increase in demand drove oil prices to record levels, many oil (and gas) producing states took a more assertive stance towards foreign investments in these strategic sectors. In particular, the policies of Russia and Kazakhstan to reconsolidate the state's (read: the elites') grab and control over resources has deeply shaken up the market. It was the dissolution of the Soviet Union, the new volumes of Russian (privatized) oil companies and the prospect for new super/giant fields that fuelled the dreams of the International Oil Companies (IOCs) but also resulted in plummeting oil prices. By that time, also under the neo-liberal market paradigm Western consuming countries and their IOCs achieved significant inroads into energy-abundant countries (Mommer 2002).

In the early 2000s, Russia and Kazakhstan alarmed energy investors: renationalization and wide scale corruption challenged fundamental rights guaranteed under production-sharing-agreements. The empirical bases for *Friedman*'s first law of petropolitics stating that the price of oil and the pace of freedom move in the opposite direction (Friedman 2006) unfolded. Whereas the 1990s witnessed a phase of significant inroads into producing countries, in the early 2000s the pendulum swung back. Re-nationalization of the oil industry or at least of its core parts (as exemplified by the Yukos oil company in Russia that was taken over by state company Rosneft after the arrest of the Yukos' CEO Khodorkovsky) significantly changed the business environment for the multinational oil companies, who were the major instruments for OECD countries to secure timely, stable and affordable supply. Nowadays, it is the National Oil and Gas companies (NOC) that control over 80 per cent of the reserves (BGR 2009, 37). NOCs are subjected to political considerations and serve as a major instrument for the ruling elites to stay in power. "Resource nationalism" results in limited access for IOCs.

However, the motivation behind the waves of renationalization in the former Soviet Union and Latin America are different. But high oil prices have enabled these steps. When the oil price increased money was soaring into state budgets and enlarged the room for maneuver. Latin America, Ecuador and Bolivia for instance took the opportunity to revise and change unfavorable deals that had been agreed upon under different circumstances. Moreover, disputes among the old elite and leftist socialist/indigenous movements over the redistribution of rents drove renationalization in Latin America (Venezuela, Bolivia, Ecuador) as did diverging interests over depletion and exploitation strategies as in the case of Argentina.

For the OECD and its multinational oil (and gas) companies the time of cheap and easily accessible conventional oil is over. Because of restrictive policies in many energy-abundant states, but also because of depletion paths in the OECD world, the IOCs have to go to geologically and geographically ever more challenging areas. All this together, is driving oil (and energy) prices. The costs escalation is remarkable: in the 1990s, the costs for an oil field development from exploration until the start of production lay between US$500 million and US$1 billion; nowadays this amounts to US$5–10 billion (Oldag 2011).

As a result of renationalization and state intervention we have seen fragmented markets developing. Whereas the EU has followed a path of markets, competition and integration based on clear rules and regulation to provide for a level playing field, other countries have gone in the opposite direction following a "regions and empires approach" (Clingendael International Energy Programme 2004).[10]

Last but not least, it has to be emphasized that oil price developments have not only been a source of conflict but have also driven remarkable achievements in international governance. When it comes to dealing with the well-known pig-cycles in raw materials and the shift in the supply and demand balances, 2008 has been a watershed that has brought energy governance further but not far enough: the peak in prices of about US$147 per barrel (Brent) and the price decrease by almost US$100 per barrel in less than six months raised the consciousness for mutual vulnerability. Both consumers and producers are exposed to the damage volatile prices can cause. Oil price levels and oil price volatility have been a constant source of concern for all relevant actors and have resulted in new or reinforced international governance initiatives such as the Joint Organisation Data Initiative (JODI) of the International Energy Forum. JODI has been built up as a major initiative to step up dialogue and cooperation between consuming and producing countries. The year 2008 has been a watershed for cooperation in the IEF because the decrease of oil prices by almost US$100 per barrel proved to be equally damaging to producing and consuming countries. Since then, the world has witnessed intents to improve transparency on the oil markets as well as in the financial markets, both under the umbrella of the IEF as well as under the roof of the G20.

## 5  CONTROVERSIAL VIEWS ABOUT THE CAUSES OF AN UNSTABLE "ENERGY SECURITY"

Thus, as has been shown above, the more recent conversation about energy security has evinced consensus among researchers and analysts that supply is no longer being guaranteed for the unforeseeable future. Import dependency of major consumer countries will grow – though for the mid-term future, the United States will be exempted from this competition. As early as 2004 the Clingendael "Study on Energy Supply Security and Geopolitics" concluded that energy will become a determining factor for international relations in the future and that "energy relations will become increasingly politicized" (Clingendael International Energy Programme 2004, 16). While gas and also renewable energy have some implications for energy security, there is also consensus that oil dominates the concern about energy security, that "(o)il plays a strategic role in both economic and social development and is one the greatest sources of world power" (Palazuelos 2012, 301). And, moreover, the question of oil will also determine whether we will see "cooperation (or conflict) in international relations" (Palazuelos 2012, 301). But that is where the consensus stops. As already evident from the discussion above, there are controversial views of the causes of this unstable situation, but they have different explanatory analytical power if applied to particular regions and case studies.

Michael T. Klare's analysis of "Great Power Games" makes sense for two conflict areas, one of which threatens to result in open military conflict in the near future: the conflict over the East and South China Seas; the other is over the oil in the Arctic.[11] It is these areas where exploration and production open new frontiers, and where borders are unclear or disputed.

## 5.1 Great Power Game in the East and South China Seas

Three conflict lines are identifiable here: between China and the US, China and Japan and China and its neighbors to the South and East. All this reflects the territorially disputed areas of the East and South China Seas, the Gulf of Thailand and the waters surrounding Indonesia. The Obama Administration's "pivot" from the Middle East to the Asian-Pacific region has to be seen in this context, manifesting itself by sending US Marines to Australia, combat ships to Singapore, the opening of a naval base in the Philippines and the redeployment of an aircraft carrier into the South China Sea[12] – leading Klare to suggest that "after a decade-long-hiatus-cum-debacle on the Eurasian continent, the Great Game v. China is back on".[13]

It is obvious that several Southeast Asian countries have deepened their military ties with the US and Japan in response to the Chinese build up of its naval capabilities over the last 15 years and China's public announcement in 2010 that its South China Sea claims were "among its 'core interests'" (Chang 2012, 21). These military deployments to this region can certainly not directly be linked only and exclusively to energy, since the issue is also about securing the Straits of Malacca against piracy and the terror of Islamic groups (Favennec 2011, 129).

But, in general, they can be understood as a commitment by the US to support its allies in the region. According to Klare "(b)ecause so many countries have advanced overlapping claims to all or part of these maritime regions – China, Japan, and Taiwan in the East China Sea; Brunei, China, Malaysia, the Philippines, Taiwan, and Vietnam in the South China Sea – it has proven nearly impossible to establish definitive offshore boundaries . . ." (Klare 2012, 67). The summer and fall of 2012 saw a conflict heating up between Japan and China over the Senkaku islands.[14] As Japan's ally, the US would be expected to come to her help, but so far the US has remained silent. Nor are there any regional organizations in place to mitigate any conflicts, hence Klare's analysis seems legitimate.

### 5.2 Russia's Hegemonic Policy in the Caspian Sea Basin and Central Asia

Even more than the East and South China Seas, it is Central Asia and the Caspian region which are generally regarded as the quintessential locus of inter-imperial rivalry[15] and for Klare they are even the "cockpit for a twenty-first-century energy version of the imperial 'Great Game' of the nineteenth century" (Klare 2008,115). But according to our analysis the picture is more of a mixed bag. A militarized "geo-strategization" is not identifiable in the case of Russia's policy in the Caspian, rather a policy of a former hegemon that has tried to preserve its geopolitical influence in the "near abroad" and has aimed to monopolize the gas flows to Europe, thereby exerting influence on the Caspian Basin energy sector. Yet, this strategy has faced limitations by the United States and China.

The region is landlocked, which forms a decisive barrier to the development of the vast resources, notably because of the complexities of financing and constructing pipelines across several countries. Alternative pipelines that bypass Russian territory had to be built. This constitutes long-term relations. In the Caspian Sea Basin and in Central Asia geopolitics and geo-economics are at play amongst those territorial disputes, such as the Russia-Georgia conflict in August 2008, the unclear legal status of the Caspian Sea, policy reversals as regards depletion strategies, and upstream access, export routes and domestic energy use. Political risk in the region is considered high and these perceptions have preempted larger export projects. With the dissolution of the Soviet Union, and most visibly illustrated in Ukrainian-Russian gas disputes of 2006 and 2009, transit became one, if not *the* security of supply issue for Europe.

Russia's priority from the very beginning has been to maintain the monopoly for oil and gas flows from the region, and to trade and re-sell these hydrocarbons with high windfall profits. Transit issues have been prominently addressed in the Energy Charter Treaty (ECT) and its related Transit Protocol (Westphal 2011a). For that reason, Russia had never ratified the ECT as it stepped in as a trader of Central Asian gas and was not willing to offer free transit. The Russian gas company Gazprom feared the loss of its strategically important position as the narrow gateway to Central Asian gas. The decisive point here is that Gazprom has bought and resold Central Asian gas instead of simply providing transit services. Russia has put Turkmenistan in a position of a swing gas supplier, exposing the country twice to two sharp collapses in deliveries to Russia because of disputes over prices and volumes (1997–1998 and 2009). It has also instrumentalized the uncertain legal status of the Caspian Sea as well as the other littoral states for their own interest. This unclear legal status has so far prevented a Trans-Caspian pipeline and large scale off-shore projects of Turkmenistan. Russia has made significant windfall profits over the past 20 years in re-selling Central Asian gas. And more than that, it has pressured Kazakhstan and Turkmenistan into the situation of swing suppliers: with the falling gas demand in Europe, Russia passed the loss in volumes down to the Central Asian producers.

The US strategy of getting a hold in the Caspian energy sector was the set up of a pipeline system that allows for a substitute export route for West Caspian oil that does not pass through Russian territory, the Baku-Tbilisi-Ceyhan pipeline. Operational since 2006 this pipeline transports the oil directly from Baku in Azerbaijan to Ceyhan in Turkey.[16] In addition, China started to tap into the Caspian resources since Central Asia's oil can be directly delivered to China by land and domestic pipelines, thus the vulnerability of the oil and gas's transportation can be curbed (Klare 2008, 133–135).

The export option to China is strategically of the utmost importance for the real independence of Kazakhstan and Turkmenistan from Moscow, and the Chinese presence has almost put an end to Russian dominance in price negotiations and play of volumes China is investing predominantly in Kazakhstan and Turkmenistan in oil and gas pipeline systems and upstream projects. With the commissioning of the Turkmenistan – China pipeline gas export has started to diversify.

For the EU, that has repeatedly claimed to diversify its imports, it seems very likely that gas exports from Shah Deniz Phase II will find its way through a Southern Corridor into Europe. This might not necessarily be realized through the Nabucco Pipeline but rather with a smaller project that is being developed first, and with the possibility of being upgraded later. In any case, however, these long, overland and multi-country projects make the reliability of exports contingent on a long chain of political arrangements, frameworks and circumstances.

### 5.3 US Hegemony by a Militarized Globalization Strategy

In contrast to Klare's analysis of the "great power game", Doug Stokes and Sam Raphael argue – on the basis of neo-Gramscian theories – that the United States strives for global power and hegemony over the world's oil rich regions by coercing them to open up their oil wealth to a "free marketization" (Stokes and Raphael, 2010, 2).

But unlike Klare they interpret the hegemonic policy of the US government in the international oil system not as a result of US "resource nationalism", but rather as motivated by the intent to maintain a functioning global economic system based on free trade of oil from which all states, including non-Western ones, should profit. This capability on part of the US to control the conditions under which all core powers receive their oil from the South again contributes to the hegemonic power of the US. Throughout the world, the US secures control of the global oil supplies primarily through "transnationalizing" the elite of these oil rich countries of the South by buttressing their power through a strategy of "military coercion". Thus these coercive and military deployments in the "Global South" are not intended to threaten the big rivals, China and Russia, in these regions, but rather against the impoverished population in the oil rich areas who might demand a fair share of the oil wealth produced by their authoritarian governments. Since Stokes and Raphael believe that "rival centers of power opt primarily to work under the American strategic umbrella" their conclusion is distinct from that of Klare in that "overt competition for the world's oil stocks will continue to be overwhelmingly pacified" (Stokes and Raphael 2010, 51).

The Iraq war is also interpreted by Stokes and Raphael as the beginning of neo-liberalism in the Gulf region with IOCs starting to have a greater role to play. Thus by taking recourse to Robinson they argue that the US invasion cannot be interpreted as a "'US imperialist plan to gain the upper hand over French, German, and Russian competition' by monopolizing Iraq's crucial oil reserves",[17] but rather as an opening up to the "investment by global capital" (Stokes and Raphel 2010, 96). If we look at the result of the Iraq War, Stokes and Raphael seem to be right. Only two US companies gained a big share of the bid (ExxonMobil and California's Occidental Petroleum). Also five OECD, but non-US, companies (British BP, Dutch/British Shell, Italian ENI, Norwegian Statoil and the Korea Gas Corporation) became involved in Iraqi oil, but large shares of the attractive Iraqi energy sector also went to Chinese, Russian,

Malaysian and Angolan companies.[18] The United States' tacit approval of Chinese and Russian companies' involvement supports strongly the thesis of the neo-Gramscian camp that the US interest was to bring the oil to the international market, and in the long run will profit from having re-strengthened its control of the international energy market.

## 6 CONCLUSION: REVISITING THE GEOPOLITICAL PARADIGM

It is impossible to deny that oil, given newly emerging energy insecurities, can (no longer) be purely seen as a "commodity". Too many voices have warned that oil is getting scarce, from CEOs of big oil companies to the IEA, though with different levels of urgency. How you frame it depends on the lens you are looking through. But with the return of energy security, traditional geopolitical interpretations are very much in vogue and "sexy" again. Among all players analyzed here only China with its authoritarian governmental system might oblige the US to come to the support of its allies in the instable region of the East and South China seas.

But, while "China's resource undertaking is global and among the most aggressive in history" (Moyo 2012, 3), it can still be interpreted as using "soft power"[19] in its global search for control of energy – as long as the search is not in its own neighborhood.[20] As for Russia it "only" strives to maintain hegemony over the former Soviet republics and Eastern Europe. More importantly in energy terms it is a fact that this strategy has fallen together with the economically and commercially driven pursuit to preserve, retain or expand its market shares in the respective region. There are also no signs of a "geo-strategization" to achieve its goal.

In general, these geopolitical narratives tend to ignore developments in exploration and production of unconventional resources. Moreover, they perceive the state as the main device and blind out the economic and corporate interests of private companies. Most importantly, the most probable possibility that the resource game is fought via pricing of fossil fuels is more or less ignored. The geopolitical storyline is about access, and tends to ignore the "rest" of the supply chain. Theoretical concepts inspired by institutional economics (North 1990) focus on other levels e.g. substate, transnational etc. and on other actors, such as companies. The analysis directs toward corporate strategies, price mechanisms and interdependencies. These theories help to understand that the energy supply chain in most cases stretches across borders: from the production (upstream), to transport and trade (midstream) to downstream (processing, retailing and marketing), sometimes this production chain is vertically-integrated, in most cases it is not. Consequently, the analytical focus was not only on the supply side, but also on the demand side. Energy security is not only defined as supply security, but further takes into account interdependencies, vulnerability and sensitivity (Keohane and Nye 2001). Thus, resilience of a system plays a major role.

Apart from some select deviating views[21] there still seems to be consensus among analysts that it is the European Union which is epitomizing the "markets and institutions"[22] approach since it is very much embedded in the spirit of globalization and neo-liberalism, trusting exclusively the forces of the market and its institutions to provide energy security for its region. But with the US shale gas and oil revolution in full swing, it remains to be seen to what extent there will be an awakening among EU politicians that from now on they might themselves have to take care of the region's "energy security". As for the US we learnt that its "intellectuals of statecraft" pursue a three-pillar strategy: while pursuing autarky based on with their domestic shale gas and oil revolution, they are at the same time preserving its strategy of globalizing the international oil market without renouncing on a geo-strategization which takes recourse to military means if necessary. Since even with a scenario of autarky the US needs to watch global oil prices, the Europeans might be lucky and might not experience a situation in which they find themselves bewildered in front of a folded US "strategic umbrella". Future research will have to analyze the implications and consequences of this new configuration in the new international energy system, in particular the effects on transatlantic relations.

## NOTES

1. One prominent figure of the US inner circles who confirmed this kind of interpretation was Alan Greenspan, Former Chairman of the US Federal Reserve Bank, see Peter Beaumont and Joanna Walters, "Greenspan admits Iraq was about oil, as deaths put at 1.2m", *The Observer*, 16 September 2007.
2. For the German case Klinke observes that the "debate is based on an unhelpful binary: the life or death of geopolitics", 2011, p. 709.
3. Tuathail (2004, 93), quoting from Tuathail and Agnew (1992), "Geopolitics and Discourse: Practical Geopolitical Reasoning and American Foreign Policy", *Political Geography*, 11: 190–204.
4. See also Peters (1999, 30–31).
5. See Moran and Russell (2009), who claim that "(t)he idea of peak oil is already becoming established as a subtext or unspoken assumption among strategists and policy-makers", 4. For a detailed discussion of "peak oil" see Peters (2004).
6. See in more detail Westphal (2011b).
7. Own calculations based on EIA (2011) and BP (2012, 8).
8. See for example Smith Stegen (2012).
9. For Saudi-Arabia the figures are as follows: 57 per cent of the crude oil exports go to

the Far East and 50 per cent of the refined product exports, too (Energy Information Administration 2011, 6).
10. See the 2004 Clingendael study which distinguishes between the "regions and empires" and "markets and institutions" approach (Clingendael International Energy Programme 2004, 84).
11. See the chapter on the Arctic in Klare (2012, 70–99).
12. *Spiegel* online, 17 October 2012.
13. Klare, Michael (2012), "Oil Wars on the Horizon", at *TomDispatch.com*, accessed at 19 September 2012.
14. *Economist*, 22 September 2012.
15. See also Kleveman (2003).
16. See Klare (2008, 125) and Favennec (2011, 160).
17. Robinson, William (2004), *A Theory of Global Capitalism: Production, Class and State in a Transnational World*, Baltimore, MD: John Hopkins University Press, p. 140, quoted in Stokes and Raphael (2010, 96–97).
18. *Aljazeera*, "Western Oil Firms remain as US exits Iraq", 7 January 2012. For a more detailed analysis of the Iraq oil sector see Kadhim (2012).
19. Kinchen, David (2012), "Book Review: 'Winner Take All': China uses 'Soft Power' to Gain Control of Mineral Resources, Other Valuable Commodities", *HuffingtonNews.Net*, 26 June.
20. For a balanced analysis see also Downs (2010).
21. For a contrary view, see Bosse (2011).
22. The 2004 Clingendael study sees the EU firmly embedded in the "markets and institutions" approach. See also McGowan (2011) who argues that the "debates and actions of European policy makers are better understood as being played out in a framework of politicization rather than securitization", p. 488.

# BIBLIOGRAPHY

Agnew, John and Stuart Corbridge (1995), *Mastering Space. Hegemony, Territory and International Political Economy*, London and New York: Routledge.
Benes, Jaromir, Marcelle Chauvet, Ondra Kamenik, Michael Kumhof, Douglas Laxton, Susanna Mursula and Jack Selody (2012), "The Future of Oil: Geology versus Technology", IMF Working Paper 12/109.
Bosse, Giselle (2011), "The EU's Geopolitical Vision of a European Energy Space: When 'Gulliver' meets 'White Elephants' and Verdi's Babylonian King", *Geopolitics*, 16: 512–535.
BGR (2009), *Energy Resources 2009. Reserves, Resources, Availability*, Hannover: BGR.
BP (2011), *BP Energy Outlook 2030*, London: BP.
BP (2012), *Statistical Review of World Energy 2012*, London: BP.
Bradshaw, Michael (2010), "Global Energy Dilemmas: A Geographical Perspective", *The Geographical Journal* 176(4): 275–290.
Chang, Felix K. (2012), "China's Naval Rise and the South China Sea: An Operational Assessment", *Orbis*, Winter: 19–38.
Clingendael International Energy Programme (CIEP) (2004), "Study on Energy Supply Security and Geopolitics", Final Report, *Institute for International Relations* "Clingendael", The Hague, the Netherlands.
Darbouche, Hakim and Bassam Fattouh (2011), *The Implications of the Arab Uprising for Oil and Gas Markets*, MEP 2, The Oxford Institute for Energy Studies.
Downs, Erica (2010), "Who's afraid of China's Oil Companies?", in Carlos Pascual and Jonathan Elking (eds), *Energy Security. Economics, Politics, Strategies, and Implications*, Washington D.C.: Brookings Institution Press, 73–102.
Energy Information Administration (2011), Saudi Arabia, Country Analysis Briefs, January 2011.

Favennec, Jean-Pierre (2011), *The Geopolitics of Energy*, Paris: Editions Technip.
Friedman, Thomas (2006), "The First Law on Petropolitics", *Foreign Policy*, May/ June: 28–36.
Götz, Roland (2012), "Mythen und Fakten, Europas Gasabhängigkeit von Russland", *Osteuropa*, 62(6–8/2012): 435–458.
IEA (2010), *World Energy Outlook 2010*, Paris: OECD/IEA.
Kadhim, Abbas (2012), "Beyond the Oil Curse. Iraq's Wealthy State and Poor Society", in Robert E. Looney, *Handbook of Oil Politics*, London and New York: Routledge, 249–261.
Keohane, Robert O. and Joseph S. Nye (2001), *Power and Interdependence*, New York: Longman.
Klare, Michael (2008), *Rising Powers, Shrinking Planet. How Scarce Energy is Creating a New World Order*, Oxford: Oneworld Books.
Klare, Michael T. (2012), *The Race for What's Left. The Global Scramble for the World's Last Resources*, New York: Metropolitan Books.
Kleveman, Lutz (2003), *The New Great Game. Blood and Oil in Central Asia*, London: Atlantic Books.
Klinke, Ian (2011), "Geopolitics in Germany – the Return of the Living Dead?", *Geopolitics*, 16(3): 707–726.
Lesage, Dries, Thijs Van de Graaf and Kisrten Westphal (2010), *Global Energy Governance in a Multipolar World*, Farnham: Ashgate.
Maugeri, Leonardo (2012), "Oil: The Next Revolution", Discussion Paper 2012-10, Belfer Center for Science and International Affairs, Harvard Kennedy School, June 2012.
McGowan, Francis (2011), "Putting Energy Insecurity into Historical Context: European Responses to the Energy Crisis of the 1970s and 2000s", *Geopolitics*, 16: 486–511.
Mommer, Bernard (2002), *Global Oil and the Nation State*, Oxford: Oxford University Press.
Moran, Daniel and James A. Russell (2009), *Energy Security and Global Politics. The Militarization of Resource Management*, London and New York: Routledge.
Moyo, Dambisa (2012), *Winner Take All. China's Race for Resources and What it Means for the World*, New York: Basic Books.
North, Douglass (1990), *Institutions, Institutional Change and Economic Performance*, Cambridge University Bridge.
Oldag, Andreas (2011), "Eine unbequeme Wahrheit", *Süddeutsche Zeitung*, September 12, p. 20.
Palazuelos, Enrique (2012), "Current Oil (dis)order: Players, Scenarios, and Mechanisms", *Review of International Studies*, 38(2): 301–319.
Peters, Susanne (1999), "The West against the Rest: Geopolitics after the end of the Cold War", *Geopolitics*, 4(3): 29–46.
Peters, Susanne (2004), "Coercive Western Energy Security Strategies: 'Resource Wars' as a New Threat to Global Security", *Geopolitics*, 9(1): 187–212.
Profant, Tomáš (2010), "French Geopolitics in Africa: From Neocolonialism to Identity", *Perspectives. Review of International Affairs* 18(1): 41–61.
Popvici, Vlad (2011), "Europe's new energy frontier", *European Energy Review*, October 27.
Roberts, John (2011), "The Arab Revolution of 2011", *Energy Economist*, Issue 353, March 3. Pp. 3.
Simmons, Matthew (2005), *Twilight in the Desert. The Coming Saudi Oil Shock and the World Economy*, New Jersey, Canada: John Wiley and Sons, Inc.
Smith Stegen, Karen, Patrick Gilmartin and Janetta Carlucci (2012), "Terrorists versus the Sun: Desertec in North Africa as a Case Study for Assessing Risks to Energy Infrastructure", *Risk Management*, 14(1): 3–26.
Stokes, Doug and Sam Raphael (2010), *Global Energy Security and American Hegemony*, Baltimore: The John Hopkins Press.
Tuathail, Gearóid Ó. (2004), "Geopolitical Structures and Cultures: Towards Conceptual Clarity in the Critical Study of Geopolitics", in Lasha Tchantouridze (ed.), *Geopolitics. Global Problems and Regional Concerns*, Centre for Defense and Security Studies, Winnipeg, Canada, 75–102.

Yergin, Daniel (2006), "Ensuring Energy Security", *Foreign Affairs*, March/April: 69–82.

Westphal, Kirsten (2011a), *The Energy Charter Treaty Revisited. The Russian Proposal for an International Energy Convention and the Energy Charter Treaty*, SWP Comments C08, March.

Westphal, Kirsten (2011b), "Energy in an Era of Unprecedented Uncertainty: International Energy Governance in the Face of Macroeconomic, Geopolitical, and Systemic Challenges", in David Koranyi (ed.) (2011), *Transatlantic Energy Futures: Strategic Perspectives on Energy Security, Climate Change and New Technologies in Europe and the United States*, Washington: Center for Transatlantic Relations, 1–26.

Westphal, Kirsten (2012), "Be Prepared. The Four Great Challenges for the European Gas Market", *European Energy Review*, 2 July.

Westphal, Kirsten (2013), *Unconventional Oil and Gas – Global Consequences*, SWP Comments C12, March.

# PART III

# SECURITY OF ENERGY SUPPLY

# 6. Securing energy supply: strategic reserves
*Elspeth Thomson and Augustin Boey*

## INTRODUCTION

Strategic Energy Reserves (SERs) are meant to help safeguard a country's economic growth and provide a buffer to external price/supply turbulence. They are intended to dampen domestic price volatility following an energy supply/price shock and concomitant financial losses incurred as a result of production stoppages and/or the necessity of making short-term purchases of energy commodities (typically oil, oil products, natural gas or coal) from the spot market.

Energy supply/price shocks can be extremely harmful to developed and developing economies alike, potentially disrupting all sectors and exacerbating existing weaknesses in economic structures. They can cause GDP to plummet and also trigger inflation and unemployment.

However, not all people believe that creating and maintaining strategic energy reserves (SERs) is useful. Some believe there is little point in having them because they are costly to build and maintain, and if a supply/price shock is not resolved within a short period of time, the stockpiling is for naught. Moreover, many regard energy supplies as fungible commodities.

While the advantages and disadvantages of strategic petroleum reserves (SPRs) have been discussed by governments around the world for some 40 years, recently, governments have also been considering the pros and cons of maintaining reserves of other forms of energy, hence, SERs, namely gas and coal. Perhaps in future, governments may also examine the possibility of maintaining reserves of biomass used in both electricity generation and transport fuel production.

## RATIONALE FOR AND OPERATION OF SERs

Besides discouraging the use of energy supplies as a political weapon, SERs also provide a number of other benefits for stockpiling states. Firstly, private sector energy corporations typically do not have sufficient economic incentive to maintain energy stocks beyond levels that are operationally optimal. The operational stocks cannot provide a sufficient supply buffer in the event of a supply disruption (Strait, 2009). Thus, in this respect, SERs can be viewed as public goods as they enhance the energy security and economic resilience of a state. SERs also give stockpiling countries the ability to buy time and to hold off more drastic measures whilst pursuing diplomatic solutions and other policy measures in the event of an energy supply shock (Yergin, 2006).

Secondly, though SERs cannot completely mitigate the price spikes and other economic impacts associated with an energy supply or price shock, releasing stocks can soften potential macroeconomic impacts, calm energy markets and prevent panic buying. The reserves are especially crucial for countries that have little in the way of domestic energy resources, i.e., they are forced to import most of their energy requirements and are unavoidably extremely vulnerable to any change in international energy markets. SERs help ensure that there are adequate supplies in the short term, and that the market price of these energy reserves remains the same or stabilises during price increases.

## THE HISTORY AND EVOLUTION OF STRATEGIC ENERGY RESERVES

The events surrounding the Yom Kippur War in 1973 were the catalyst for the establishment of the International Energy Agency (IEA) in 1974.[1] In response to the United States' support of the Israeli military, exports of oil to the United States were embargoed by the Organization of Arab Exporting Countries (OAPEC) in October 1973. Oil prices to America's West European allies were raised by 70 per cent. This led to severe supply shortages and caused the price of crude oil to soar. The key specific goal behind the IEA's establishing the Strategic Petroleum Reserves (SPRs) was 'to help countries co-ordinate a collective response to major disruptions in oil supply through the release of emergency oil stocks to the markets.'[2]

The use of petroleum as a political 'weapon' during what came to be known as the 'first oil crisis' was a turning point in governments' energy security planning and the objectives of energy security came to be commonly defined as 'assuring adequate, reliable supplies of energy at reasonable prices and in ways that do not jeopardize major national values and objectives' (Yergin, 1988). Following the first oil crisis and a second in 1979,[3] a number of energy security-related initiatives were set in motion to boost the resilience of national energy systems and, by the same token, to prevent or mitigate the impacts of energy supply and price shocks. Energy supplies and prices at any given time are a function of a combination of political, economic, social, military, transport and geological factors. A supply/price 'shock' can occur as a result of loss in equilibrium by any one or combination of these factors.

The cost of building and maintaining these reserves is seen as being justified against the far greater economic and political costs of an oil crisis (Paik et al., 1999). SPRs are believed to serve an important function in the strategic capacity and short-term energy security and economic health of stockpiling nations. It must be emphasised, however, that they constitute but one part of a comprehensive suite of short-term and longer-term energy security measures. The requirements for ensuring a country's energy security are context-dependent and the various component policies and measures, including SPRs, vary in importance across countries.

SPRs are typically in the form of crude oil due to the lower overall costs of acquisition, storage and transportation (East-West Center, 2005). However, some are in the form of refined products.[4] They can be held either by the government, private companies, stockholding agencies or a mix of these. The calculation of a country's total SPR stocks can include oil or refined products held in operation, distribution or transport, namely, in tankers or pipelines en route to refineries or consumers. When and how a country's SPRs are released is typically debated and enacted by the highest level of government.

## OECD (IEA) COUNTRY SPRs

Net oil importing countries within the IEA are legally obliged under the International Energy Programme (IEP) to hold emergency oil reserves equivalent to 90 days of the prior year's net imports. This obligation was enacted in an attempt to balance the OECD grouping against the oil-producing OPEC countries' power to cause serious economic harm. Canada, Denmark, Norway and the United Kingdom are exempt from this requirement as they are net oil exporting members. However, Denmark and the United Kingdom are required to hold stocks under consumption-based European Union regulations. About two-thirds of the IEA's strategic oil stocks are held by the oil industry. The IEP also provides a framework for rapid response and decision-making in the event of oil supply disruptions. It is supplemented by the Co-ordinated Emergency Response Measures (CERM), which are an additional set of measures established in 1984 that may come into force in the event of an oil supply disruption that is too small to trigger the IEP emergency measures.

The IEA member states currently together hold around 1.5 billion barrels of oil in their strategic reserves and around 4.1 billion barrels of oil stocks in total.

The United States has the largest SPR, which in March 2012 held about 695 million barrels. This represented around 45 per cent of the total volume of the IEA's member states' SPRs. The United States' SPR will be expanded to 1 billion barrels of crude oil under the requirements of the Energy Policy Act of 2005. The second largest SPR belongs to Japan, which holds around 320 million barrels. Germany has the third largest SPR with approximately 180 million barrels. Japan and Germany together hold about 40 per cent of the IEA member states' SPRs. The USA's SPRs are almost entirely in the form of crude oil, with the exception of a much smaller heating oil reserve. The formal agreement to build them took place in December 1975, but filling them did not begin until 1977. They were controversial right from the start. Some observers feared soil contamination and other environmental implications in the immediate area, as well as high construction and maintenance costs.

The US's reserves are contained in subterranean salt domes, thousands of feet underground, at four locations along the Gulf of Mexico. These natural caverns are by far the least costly and most secure types of storage.[5] Within the various countries, SPRs are held in different types of storage sites. Other stockpiling methods include aboveground storage tanks, partially buried storage and offshore floating platforms. These are, of course, much more vulnerable to terrorist attack.

Most IEA member governments offer various forms of financial support to oil companies holding stocks which could be instantly released when given the decree. Some impose a penalty on oil companies that default on their stockholding requirements.

**The Release of OECD Reserves**

As noted, the SPRs in the United States have been controversial right from the start. Whenever there is some event, either a natural disaster or a large-scale war effort, the pundits always debate vigorously about whether or not to release some of the reserves. Some argue that the SPRs are meant strictly for US war efforts (either on US soil or abroad), while others contend they should be used exclusively to buffer the economic impacts of high prices. The latter group also justify this stance by saying that a recession in the US could easily cause other countries, indeed the world economy, to slide into recession.

To date, the IEA has released oil stocks from its member states' SPRs three times: during the 1991 Gulf War,[6] in the aftermath of the 2005 Hurricanes Katrina and Rita (Yergin, 2006) and in 2011 during the Libyan revolution.[7]

## SPRs Held by Non-OECD Countries

Though the IEA/OECD member countries have the most developed SER programme, several other countries have demonstrated interest in developing their own strategic energy reserves. Most of these are still in the planning and capacity-building stage. Of note are the measures taken by various rapidly developing Asian economies which must import rapidly increasing quantities of oil in tandem with their economic growth (Paik et al., 1999; Downs, 2004). These governments are offering similar economic incentives as the IEA member countries to maintain SPRs designed to avoid the economic impacts of higher expenditures on oil imports and the loss of GDP during oil shocks. The need for SPRs as an emergency measure is likely to increase as Asian economies become more vulnerable to oil shocks due to their increasing reliance on the politically unstable Middle East for oil, the reduction in spare stocks and spare production capacity in the private sector, increasing oil flows through potential oil transit chokepoints and the short-run price inelasticity of oil (East-West Center, 2005; Paik et al., 1999).

The IEA posts data on its website indicating its member states' SPR stocks. However, most other countries do not publish this information. They are not required to report their SPR stockpiles and are far less transparent with respect to their SPR programmes. The specific types, volumes and locations of strategic energy reserves are considered a matter of national security and are typically not fully disclosed. When non-IEA countries make announcements about current or planned additions to their SPR capacities, the actual scale of the reserves is open to inference as SPR capacity does not usually indicate base stock levels.

### China

China is currently building the largest SPR outside of the OECD. It has had three phases of oil reserve construction since the launching of its strategic oil reserve plan in 2003, with the bases in the first phase holding 30 days' worth of China's oil imports (Yan, 2010). Construction of the storage tanks began in 2004, and filling began in 2005. The decision to start building the reserves was not unanimously endorsed by the leadership. There was disagreement, dating back to at least 1992, over what volume ought to be stockpiled and when and how to finance it. The IEA was behind China's development of SPRs. When the US and UK were preparing to invade Iraq, the IEA feared that the war could reduce or halt the export of oil from the Middle East region and that China, with its very rapidly growing demand for oil, could worsen the already fragile world oil markets.

The four completed SPRs on the eastern coast have a capacity of over 100 million barrels of storage, and there are plans to increase SPR capacity to 500 million barrels by 2020 (IEA, 2012: 9). In addition to the current strategic oil reserves that are stored in the form of crude oil, there is discussion about expanding the reserves to encompass refined oil supplies as well (IEA, 2012). In addition to the SPRs, China requires state-owned and medium-sized oil companies to maintain their own oil reserves (Chen and Lim, 2008). This became mandatory when the energy law was passed by the State Council in 2007.

China is potentially very vulnerable to supply and price shocks. At the same time, as the second largest oil consumer and importer in the world, how it behaves in the international oil market is watched very closely around the world. In the 1980s, China was a major exporter of crude oil. However, by 1993 it had become a net importer of oil. Today, the country must import over 50 per cent of its oil requirements. Other factors of key significance are the fact that most of China's oil imports come from regions suffering from political instability (Africa and the Middle East) some 60 per cent of the imports must pass through the Malacca and Singapore Straits, a critical geographic chokepoint which some analysts believe could be blockaded, albeit temporarily. Since the 1970s, sudden price hikes in transport fuel have caused political unrest in several Asian countries. Thus another reason for China's establishing SPRs is to prevent similar riots from occurring in China.

**Other non-OECD countries**

The second largest non-IEA SPR will be in Russia, where up to 15 million tons of SPR stocks are being amassed. Russia initiated the creation of an SPR in 2011 mainly for short-term economic reasons. The government's stated goals for its SPR were to 'influence oil exports and gain profits from price fluctuations' (Izvestia, 2012). Taiwan, India, Chile, Thailand and Vietnam have SPRs that have been confirmed by official announcements and/or government legislation. Rwanda, Uganda, Yemen, Kenya and Zambia have been cited as possessing SPRs. The Philippines has announced intentions to build an SPR. The reasons given by these governments for building SPRs are the same as those given by IEA member states: namely, short-term energy security via emergency preparedness against oil supply disruptions.

Thailand's government announced in June 2012 that it intends to increase the amount of SPRs held by private firms from 44 days worth of imports to 90 days, citing the fact that Thailand imports around 80 per cent of its oil from the politically volatile Middle East (Praiwan, 2012). These reasons also provide justification for the SPRs in Vietnam and Taiwan

(Energy Information Administration, 2008; REEEP, 2009). The Indian government announced plans to establish SPRS in 2004. Comprised of both crude oil and oil products, the goal was to amass the equivalent of 15 days of consumption for contingency plans and also to lower market prices of oil in the event of price fluctuations (Ma and Sharma, 2011).

The Philippines also announced its intention to create an SPR to provide oil security in the event of an oil crisis and also to mitigate oil import price volatility (PhilNews.com, 2011). Regarding energy security-related stockpiling as a non-priority in such countries provides a possible explanation for the relatively small size of their SPRs, as they are not meant to substitute for consumption on a scale similar to the IEA members' SPRs.

The justification and operation of SPRs in these countries is thus based on a wider definition of 'strategic' compared to countries with SPRs managed in accordance with the IEA SPR policy. While the IEA countries use their SPRs to enhance energy security and economic resilience, the strategic importance in these particular non-IEA countries centres around their SPRs' potential to function as a hedge against oil price fluctuations. Drawdowns from the SPRs of such countries will likely be far more frequent to fulfil their objectives of maximizing short-term economic gains, whereas the IEA's SPRs may be drawn upon only in the event of an actual physical shortage of oil.[8]

## PROBLEMS WITH SPRs

Maintaining SPRs is costly and also requires the good will of the private sector in terms of keeping precise records and transparency (Bohi and Toman, 1996: 126). At any given time, the government must know exactly how much oil is available. There must be clear guidelines within the countries pertaining to the release of the reserves, especially if some are held by private companies.

SPRs may well ease a short-term crisis. However, if the political problem causing the supply/price disruption dragged on for months or longer, the SPRs would soon be exhausted.

## STRATEGIC GAS RESERVES IN OECD COUNTRIES

Compared to strategic oil reserves, global strategic gas reserves are much more modest, with both the stockpiling regulations and actual physical stocks still in nascent stages of development. Strategic gas reserves are commonly stored underground or aboveground as LNG, with the latter method being smaller in size and more costly. The methods of underground gas storage include oil field reservoirs, aquifers (EIA, 2004), as well as caverns percolated out of salt layers (GDF SUEZ, n.d.). At present, there are 12 countries that have gas storage equivalent to at least 10 per cent of their annual demand (EIA, 2004).

In the United States, the 2005 hurricane season resulted in a 10 per cent fall in natural gas production in the Gulf of Mexico. Following this, policymakers began to consider the benefits of building a natural gas strategic reserve (Sandia National Laboratories, 2006). In 2006, H.R. 5048 was introduced to direct the Secretary of Energy to undertake a study of the need for and feasibility of establishing a strategic natural gas reserve (Opencongress.org, 2006).

The Danish government has access to about 515 million cubic metres (mcm) of strategic storage capacity. This includes an amount directly reserved by Energinet.dk[9] and volumes made available from shippers' storage filling requirements. In Italy, the Ministry of Economic Development Communiqué sets strategic stock volumes each year based on assumptions of import reduction through the system's major entry points. Strategic stocks belong to storage companies. Stocks should cover for 60 days or a 50 per cent disruption of peak capacity at the main national entry point. In 2010, Italy had around 5.1 bcm of strategic stocks. During the winter of 2005–2006, Italy faced a gas shortage due to reduced supply from Russia. In response, the Italian Administration decided to release 1.5 bcm from the strategic reserves (IEA, 2010).

As of December 2011, Hungary was the only IEA country with strategic gas reserves under government control. These were created in the wake of the gas crisis of January 2006. Stocks reached the initial planned level of 1.2 bcm in early 2010, covering 40–45 days of average demand. As of July 2011, this had been reduced to 0.92 bcm, equivalent to around 30 days of demand (IEA, 2012b).

In Poland, compulsory gas stocks are held at the disposal of the Minister of Economy. These stocks may be released by operators after receiving permission from the Minister of Economy (IEA, 2011a). In Portugal, mandatory gas reserves are mixed with commercial stocks. The average stock of mandatory gas reserves was estimated to be around 234 mcm in 2008, equivalent to 20 days' imports. The volume of commercial stocks was estimated to be equivalent to 3.5 days of imports (IEA, 2011b). In Spain, according to Royal Decree 1766/2007, natural gas and LPG operators are obliged to hold minimum stocks equivalent to 20 days' consumption, consisting of 10 days of strategic reserves and 10 days of operational reserves (IEA, 2011b).

## STRATEGIC GAS RESERVES IN CHINA AND INDIA

Due to critical gas shortages in 2009, construction of three gas stock facilities in China began in 2010, with a total working capacity of 1.4 bcm. Another 10 facilities are being planned (IEA, 2012a). These reserves are regarded as crucial for China given its aim to double the consumption of gas from about 4 per cent of total energy consumption to 8 per cent by 2015. China's natural gas consumption grew 20 per cent in 2010 to reach 106 bcm, and is expected to reach 260 bcm by 2015 (Oliver, 2012). The main reason for the rapid expansion in the use of gas is the fact that the carbon emissions released in the burning of gas are far lower than those of oil or coal.

The Indian government is currently studying ways to construct 15 days' worth of strategic gas reserves to ensure that supplies will not be disrupted. This is in anticipation of increased demand for natural gas in the future. India's gas consumption is projected to increase from 65 mcm per day to 300 mcm by 2025 (Silicon India, 2012).

## PROBLEMS WITH GAS SPRs

Although strategic gas reserves can help to reduce price volatility and smooth variations in supply, the costs and limitations of storing natural gas are significantly higher than for oil storage. According to the IEA, the costs of underground gas storage are five to seven times higher than for underground oil per ton of oil equivalent stored. The variable costs for maintaining gas in storage are also significant. The variable cost of maintaining 90 days' imports in reserves across the IEA was USD 5.4 billion per year in 2010. Variable costs are determined by economic factors such as interest rates, maintenance and cost of personnel, and also gas leakage,[10] a factor specific to natural gas. The main reasons for the high costs of gas storage compared to oil are twofold: the gaseous nature of natural gas and its low energy density.[11] In order to prevent the gas from escaping, natural gas must be fully contained at all times. Its low energy density also means that its energy density must be increased so as to make storage economical. As a result, natural gas must be stored as a liquid, either via compression at high pressures or through condensation at low temperatures, both of which result in significant equipment and energy costs (IEA, 2012d).

Gas reserves are also less effective than oil reserves in mitigating supply risks. Compared to oil, gas has a less robust and interconnected infrastructure. For instance, disruptions in a gas pipeline will affect the entire downstream distribution of gas. On the other hand, repairs to an oil pipeline are not only cheaper, alternative transport options in the form of oil trucks or tankers can also be utilised. As a result, this calls for multiple gas reserves to be sited close to corresponding consumption centres throughout the country, and this further increases the cost of storage (IEA, 2012d).

Due to its high cost and lower effectiveness compared to oil stocks, few countries have strategic gas reserves. The IEA recommends using gas storage as a part of a wider suite of options such as fuel switching and interruptible contracts (IEA, 2012d).

## STRATEGIC COAL RESERVES

To date, there are no countries with established strategic coal reserves. The only country with such plans is China, which aims to create a strategic coal reserve by 2015 (China Daily, 2009; Yu, 2008). Nearly 80 per cent of the electricity generated in China is from coal. The country has a staggering appetite for coal, with over 80 per cent of the global increase in coal demand coming from China alone. China's share of global demand increased from 27 per cent in 2000 to 47 per cent by 2010, with coal use more than doubling to 2,350 million tons of oil equivalent. In 2009, China consumed more coal than the next 16 largest consumers combined (IEA, 2012c).

China's total coal reserves (proven reserves ready for extraction) in 2011 stood at 1.021 gigatons (Gt), with a reserve-to-production ratio of 70 years. Most of these are underground, with only around 100 Gt of coal reserves able to be extracted from existing mines. Domestic output continues to increase. However, as the country has not been able to meet the continuously increasing demand, the country became a net coal importer in 2009. A second reason for importing is coal quality. Under strong international pressure, the government is trying hard to minimise carbon emissions. To this end, it is importing coals which are of higher quality than the local coals. A third reason is insufficient railway capacity. There are not enough railway cars available for loading at the mine sites, and the railway lines emanating from the main coal mining areas in the interior of the country simply cannot handle more coal traffic to the power plants in the eastern region. This has been a continuous problem since the 1950s but hitherto has been resolvable (Thomson, 2003).

A strategic coal reserve will help balance market demand with supply and keep coal prices down. Coal reserve legislation was put forward in September 2008 under a scheme to establish a Chinese legal system for coal which aims to ensure sustainable development of the industry (Yu, 2008). The strategic coal reserve will include rarer types of coal, and will be a mixture of spot and resource reserves.

In 2009, the National Development and Reform Commission (NDRC) commissioned China Shenhua Energy Co, the nation's top coal producer, to build 10 storage facilities for coal throughout the nation. These reserve sites will hold a total of between 100 to 200 million tons of coal. Zhejiang and Shandong, two major coal producing provinces, have already starting building pilot coal reserve bases. China aims to keep the equivalent of 10 per cent of its domestic consumption in reserve (Letzing, 2009).

It was reported in 2011 that the State Council had approved a programme for emergency state coal reserves, which would begin storing 5 million tons of coal in the same year. At the provincial level, Shandong announced plans to build up to eight coal reserve plants with total reserves of 6 millions tons by 2015 (Chinamining.org, 2011a). Dafeng Port in East China's Jiangsu province has also been selected as one of China's coal handling bases for strategic reserves. The port has a throughput of 3.6 million tons of coal per annum and its warehousing and logistics businesses will be further developed to serve coal mining enterprises in neighbouring areas such as Qinhuangdao (Chinamining.org, 2011b).

## PROBLEMS WITH COAL STORAGE

Similar to natural gas, coal has a lower energy density compared to oil. This therefore leads to high transport costs and high storage requirements, which translates into high maintenance costs. In addition, there are also safety issues to consider as steam coal is prone to spontaneous combustion and efflorescence (China Daily, 2009).

## CONTROVERSIES SURROUNDING THE RELEASE OF SERs

SERs are created with the intention of addressing supply disruptions. Strategic stocks, by definition, are different from commercial stocks. The latter are held by private companies or private consumers to guarantee the smooth functioning of their equipment, or in anticipation of financial gain when prices could rise in the future. Commercial stocks are therefore determined by the requirements of the stockholder, the size of available storage facilities and expectations about future prices. On the other hand, strategic stocks are meant to deal with extraordinary situations, when a security threat is perceived. At the extreme, this could refer to open warfare, but as one moves away from the extreme towards more nuanced situations, it becomes increasingly difficult to differentiate commercial and strategic threats, and hence when it is appropriate to draw down strategic reserves (Luciani and Henry, 2011).

A physical shortfall and major change in price are usually related, since a drop in supply will raise prices. However, it is difficult to define a historical point in time when a physical shortfall has resulted in a threat to energy security. Although political unrest and turmoil can disrupt a country's output, global supply consists of the sum of declines in some fields and increases in others. Therefore, it is often the case that any shortfalls are made up by increased production from other countries.

According to the IEA, the most important oil supply disruption occurred during the Iranian Revolution, when 5.6 million barrels per day were lost for six months. However, this loss was compensated for by increased production in other countries, even leading to an increase in global oil production between 1978 and 1979. As such, situations in which there is a clear disruption in physical supply are extremely rare. A definition of a threat based on high oil prices would thus be much clearer than a definition based on physical supply. However, doing so would mean that there is little difference between strategic and commercial stocks (Luciani and Henry, 2011).

This conundrum is visible in the history of the United States' SPR, where there have been differences in opinions as to what constitutes a "severe energy supply interruption" (Energy Security Analysis, Inc, 2003; Styles, 2012). This includes increases in prices of oil products independent of crude prices and also increases in crude prices without any measurable shortages in crude supply. A glaring example of the thin line between strategic and commercial stocks can also be seen in the example of the Clinton Administration selling 7 million barrels of oil to help finance the SPR programme in FY 1996. This set a precedent for three further sales in FY1996 for budgetary reasons (Bamberger, 2009).

The proper use of SPRs has thus been a perennial subject of contention. There has been increasing flexibility in the latest coordinated drawdowns from the IEA's SPRs. During the Gulf War, for instance, the IEA coordinated a release of SPR stocks to mitigate an anticipated price increase associated with fears of an impending crude oil shortfall. The use of the SPR to mitigate an anticipated price increase, instead of a "severe energy supply disruption", was controversial and was the subject of debate. This event revealed the complexities inherent in governments' decision-making processes in controlling their oil stockpiles. This debate was revisited during the 2011 SPR drawdown, triggered by the Libyan revolution, where the release of SPR oil stocks was justified as a pre-emptive measure to prevent price spikes and the consequent economic impacts. According to the IEA, the Libyan drawdown was justified as a means to ensure an adequate supply of oil in the market in order to protect the global economy from 'unnecessary damage when it is in a fragile state'.[12]

## WHY SOME GOVERNMENTS HAVE NO PLANS TO BUILD SERs

Several non-OECD countries choose not to maintain strategic energy reserves (Taylor and Van Doren, 2005). Just as some IEA member countries are not required to maintain SPRs because they are net exporters of oil, net exporters of oil outside the IEA, such as Papua New Guinea, Brunei and Mexico, do not to maintain SPRs. As noted above, Russia has recently chosen to create an SPR, but for commercial rather than energy security-related reasons.

The reasons why net oil-importing countries choose to not maintain SPRs are less clear as information from official sources is sparse. In countries such as Zambia, Tanzania and Uganda, where plans for SPRs have been either aborted or left incomplete, the poor progress of their SPR projects can be reasonably, or at least partially, attributed to the inability of their economies to provide adequate funding and also the instability of their political regimes.

Some inferences can be drawn from China's SPR experience. China is in the process of constructing the largest SPR outside the OECD but progress on it has not been without opposition. Even after it was officially endorsed, there has still been a strong anti-SPR faction in China that has contributed to delays in implementing the project (Downs, 2004). The economic rationale cited by opponents of China's SPR include the contention that China cannot afford to build an SPR that is large enough to be effective for energy security and its scarce resources would be better allocated to more important and pressing priorities (Downs, 2004). Another reason cited is the fact that China's electricity is generated mainly by coal and thus creating an SPR is unwarranted, though this overlooks the rapidly growing demand for oil in the transport sector (Downs, 2004). The opponents of China's SPR also contend that creating an SPR is unnecessary as China's energy security would be better enhanced through other means, such as energy diversification and acquiring stakes in overseas oil fields (Downs, 2004).

Other countries may have other reasons not to maintain an SER. Malaysia's lack of an SPR and decision to enhance energy security by reducing oil consumption through a fuel diversification policy can be seen to reduce the need for an SPR by focusing upon long-term energy security measures and reducing oil import dependency. Singapore, which

130  *International handbook of energy security*

maintains substantial commercial oil reserves but lacks an SPR, likely chooses not to maintain an SPR due to the lack of land and also because the government has, in the event of an oil shock, guaranteed access rights at market prices to commercial oil stocks stored by the country's large petrochemical industry.

## CONCLUSION

Most governments build strategic energy reserves to help keep their economies on an even keel in the event of an external price/supply shock. The aim is to reduce domestic price volatility and financial losses resulting from production stoppages and/or the imperative to make short-term purchases of energy commodities from the spot market. They are also often seen as contributing to *international* price stability in the long run. Some governments also build reserve capacities of oil, gas or coal for commercial reasons, i.e., to take advantage of price variations.

The 28 OECD countries, through the International Energy Agency, have the most sophisticated SER programmes. All must have oil reserves equivalent to 90 days of net oil imports, and many are also establishing, or planning to establish, strategic natural gas reserves.

Several non-OECD/IEA countries have begun, or are considering, building SERs, mainly for genuine energy security-related reasons. However, the perceived utility of SERs is not universal. Many governments do not believe the potential gains are greater than the considerable costs of building and maintaining them. No doubt, many governments have keenly watched the confused political wrangling in the United States, where the precise purpose of the SPRs has been frequently debated, most notably just before elections.

## NOTES

1. Currently, the IEA's 28 members are: Australia, Austria, Belgium, Canada, Czech Republic, Denmark, Finland, France, Germany, Greece, Hungary, Ireland, Italy, Japan, Luxembourg, Netherlands, New Zealand, Norway, Poland, Portugal, Republic of Korea, Slovak Republic, Spain, Sweden, Switzerland, Turkey, United Kingdom and United States. Full details of the IEA's oil stockpiling guidelines are available at: http://www.iea.org/topics/energysecurity/ (accessed August 2012).
2. Full details of the IEA's oil stockpiling guidelines are available at: http://www.iea.org/topics/energysecurity/ (accessed February 2013).
3. The 1979 oil crisis was caused by the fall of the Shah in Iran.
4. Much of this information is from Elspeth Thomson (2005), "China's Construction of

Strategic Petroleum Reserves: How Urgent?", *East Asian Institute Background Brief*, no. 243, 11 May.
5. Salt domes are also used in Alsace, France and Lower Saxony, Germany.
6. Thirty-four million barrels were released at the start of the war, though only half were actually delivered.
7. Thirty million barrels were released to offset the lower quantities of oil exported from Libya due to the revolution that ended Colonel Gaddafi's regime.
8. The USA and the UK recently discussed releasing SPR stocks in order to mitigate high gasoline prices, allegedly to boost President Obama's electoral support in the 2012 presidential elections. This planned drawdown did not have the support of the IEA. See (Falloon and Mason, 2012).
9. Energinet.dk is the Danish national transmission system operator for electricity and natural gas.
10. In underground storage, gas leaks from high pressure into the atmosphere which is of lower pressure.
11. Energy density refers to the energy content per unit mass or volume of a fuel.
12. See IEA Collective Action: Frequently Asked Questions: http://www.iea.org/topics/oil/oilstocks/ieacollectiveaction/ (accessed 1 March 2013).

# REFERENCES

Bamberger, R. (2009). 'The Strategic Petroleum Reserve: History, Perspectives and Issues', Congressional Research Service.
Bohi, D.R. and M.A. Toman (1996). *The Economics of Energy Security*, 1st ed., Kluwer Academic Publishers, Norwell, MA.
Chen, S. and T.S. Lim (2008).' China's Strategic Petroleum Reserves: An Update (No. 371)', East Asian Institute Background Brief.
China Daily (2009). 'Shenhua to set up coal reserves', *People's Daily Online*, 11 August.
Chinamining.org (2011a). 'China's Cabinet Approves Emergency State Coal Reserves Program-China Mining', at http://www.chinamining.org/News/2011-03-24/1300947 250d44098.html (accessed 18 February 2013).
Chinamining.org (2011b). 'Jiangsu Dafeng Port Approved as China's Strategic Reserve Coal-handling Base-China Mining', at http://www.chinamining.org/News/2011-12-09/1323398996d52244.html (accessed 18 February 2013).
Downs, E.S. (2004). 'The Chinese Energy Security Debate', *The China Quarterly*, 177, 21–41.
EIA (2004). 'The Basics of Underground Natural Gas Storage', at http://www.eia.gov/pub/oil_gas/natural_gas/analysis_publications/storagebasics/storagebasics.html (accessed 18 February 2013).
Ellison, J., A. Kelic and T. Corbet (2006). 'Is A Natural Gas Strategic Reserve for the US Neccessary? A System Dynamics Approach', Paper presented at the 25th International Conference of the System Dynamics Society, Boston, USA, at www.systemdynamics.org/conferences/2007/index.htm (accessed 1 March 2013).
Energy Information Administration (2008). 'Country Analysis Briefs: Taiwan', at www.eia.doe.gov (accessed 18 February 2013).
Falloon, M. and J. Mason (2012). 'Exclusive: Obama, UK's Cameron discussed tapping oil reserves: sources', *Reuters*, 15 March.
GDF SUEZ (n.d.). 'Storage of natural gas', athttp://www.gdfsuez.com/en/businesses/gas/infrastructure-management/storage-natural-gas/ (accessed 18 February 2013).
IEA (2010). 'Oil and Gas Security: Emergency Response of IEA Countries: Italy', at www.eia.doe.gov (accessed 18 February 2013).
IEA (2011a). 'Oil and Gas Security: Emergency Response of IEA Countries: Poland', at www.eia.doe.gov (accessed 18 February 2013).

IEA (2011b). 'Oil and Gas Security: Emergency Response of IEA Countries: Portugal', at www.eia.doe.gov (accessed 18 February 2013).
IEA (2012a). 'Oil and Gas Security: Emergency Response of IEA Countries: People's Republic of China', at www.eia.doe.gov (accessed 18 February 2013).
IEA (2012b). 'Oil and Gas Security: Emergency Response of IEA Countries: Hungary', at www.eia.doe.gov (accessed 18 February 2013).
IEA (2012c). 'Facing China's Coal Future- Prospects and Challenges for Carbon Capture and Storage', at http://www.iea.org/publications/insights/chinas_coal_future.pdf (accessed 1 March 2013).
IEA (2012d). 'Natural Gas Market Review 2007', at http://www.iea.org/publications/freepublications/publication/Gasmarket2007.pdf (accessed 1 March 2013).
Izvestia (2012). 'Russia creates strategic petroleum reserve', 10 January. at http://en.rian.ru/papers/20120110/170694865.html (accessed 1 March 2013).
Letzing, J. (2009). 'China reportedly to create strategic coal reserve', *Market Watch*, 12 November.
Luciani, G. and F.-L. Henry (2011). 'Strategic Oil Stocks and Security of Supply (No. 353)', CEPS Working Document.
Ma, W. and R. Sharma (2011). 'China and India Stock Up on Oil', *Wall Street Journal*, 21 December.
Oliver, C. (2012). 'China needs strategic natural-gas reserves: report', at http://articles.marketwatch.com/2011-04-12/news/30895624_1_natural-gas-natural-gas-reserves (accessed 18 February 2013).
Opencongress.org (2006). 'H.R.5048: To direct the Secretary of Energy to undertake a study of the need for and feasibility of establishing a strategic natural gas reserve', at http://www.opencongress.org/bill/109-h5048/show (accessed 18 February 2013).
Paik, I., Paul Leiby, Donald Jones, Keiichi Yokobori and David Bowman (1999). 'Strategic oil stocks in the APEC region', in Proceedings of the 22nd IAEE Annual International Conference, International Association for Energy Economists, Presented at the 22nd IAEE Annual International Conference, Rome.
PhilNews.com (2011). 'Philippines to create a strategic petroleum reserve'. 16 April.
Praiwan, Y. (2012). 'Strategic oil plan to take reserves to 90-day supply', *Bangkok Post*, 28 June.
REEEP (2009). 'Policy DB Details: Vietnam', at http://www.reeep.org/index.php?id=9353&text=policy-database&special=viewitem&cid=39 (accessed 18 February 2013).
Sandia National Laboratories (2006). 'Is A Natural Gas Strategic Reserve for the US Necessary? A Systems Dynamics Approach', at http://www.systemdynamics.org/conferences/2007/proceed/papers/ELLIS216.pdf (accessed 1 March 2013).
Silicon India (2012). 'India plans to have gas strategic reserves', at http://www.siliconindia.com/shownews/India_plans_to_have_gas_strategic_reserves___-nid-23423-cid-3.html (accessed 18 February 2013).
Strait, A.L. (2009). 'Strategic Petroleum Reserve', Hauppauge, New York: Nova Science Pub Incorporated.
Styles, G. (2012). 'Exports Raise the Bar for US Strategic Petroleum Releases / The Energy Collective', at http://theenergycollective.com/geoffrey-styles/107046/exports-raise-bar-us-strategic-petroleum-releases (accessed 18 February 2013).
Taylor, J. and P. Van Doren (2005). 'The case against the strategic petroleum reserve', Cato Institute Policy Analysis no. 555, at http://www.cato.org/publications/archives/studies/policy-analysis (accessed 1 March 2013).
Thomson, E. (2003). 'Chinese Coal Industry – An Economic History', RoutledgeCurzon, London.
Yan, P. (2010). 'China accelerates filling strategic oil reserves', China.org.cn, 21 July.
Yergin, D. (1988). 'Energy Security in the 1990s', *Foreign Affairs*, 67, 110–132.
Yergin, D. (2006). 'Ensuring Energy Security', *Foreign Affairs*, 85, 69–82.
Yu, T. (2008). 'Coal reserve plan a bid to tame prices', *China Daily*, 2 September.

# 7. Securing energy supply II: diversification of energy sources and carriers
*Kas Hemmes**

## INTRODUCTION

The world energy supply is largely based on fossil fuels: oil, gas and coal. Several scenarios for the future energy supply distribution have been made and reported by official institutions, such as the International Energy Agency. Depending on assumptions, a faster or slower growth of renewable energies is depicted. These pictures are overall scenarios for the world energy supply. However distinctions have to be made between global and local, centralized and decentralized systems. Moreover a distinction also has to be made between electricity supply and other forms of energy needed such as heat and transport fuel. It should also be kept in mind that there is a competition on fossil fuels for use as an energy source and for use as a bulk chemical in the chemical process industry. All these issues are relevant in the transition to a more sustainable energy system. However, in the process of restructuring the energy system the security of supply has top priority. It is rather like when renovating a shop the owner puts up a sign which says: "Business as usual during renovation". In this chapter we will explore the renovation of the energy sector, what kind of innovations are taking place and can be expected, and how that could improve the security of supply, while taking into account issues of scale and different time horizons.

## ENERGY SOURCES AND ENERGY CARRIERS

### Energy Carriers

The present energy system is characterized by several energy carriers, partly the fossil fuels themselves – coal, natural gas, oil, gasoline – but also electricity. Reducing the number of energy carriers can increase security of supply if it is accompanied by more interconnections in the energy system and if we are able to convert one form of energy into the other.

Because we are so used to fossil fuels the distinction between energy

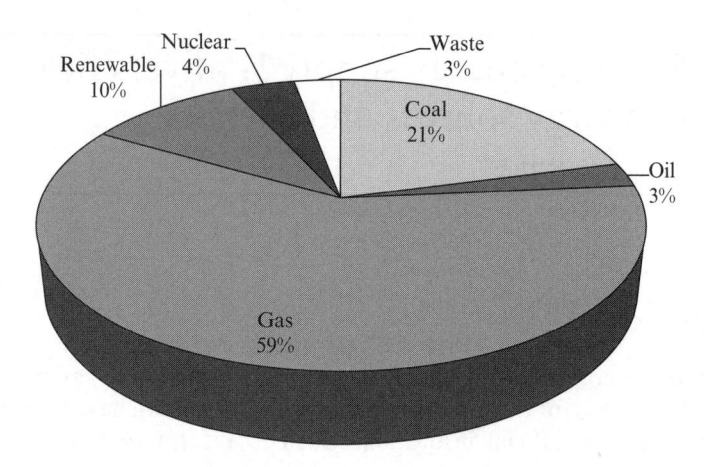

[Pie chart: Renewable 10%, Nuclear 4%, Waste 3%, Coal 21%, Oil 3%, Gas 59%]

*Source:* http://www.cbs.nl/en-GB/menu/home/default.htm?Languageswitch=on (accessed March 2013).

*Figure 7.1 Distribution of energy sources for the production of electricity in 2011 for the Netherlands*

source and energy carrier has often not been very strict. For example natural gas is an energy source as well as an energy carrier. The same holds for coal. With oil it is a bit more difficult. Although oil tankers still transport the bulk of the energy contained in oil across the oceans, in the refineries oil is converted into several other bulk chemicals and fuels such as Nafta and gasoline. We can consider oil the source but locally gasoline is the energy carrier. For electricity the distinction is quite clear. Electricity is not produced in nature (except perhaps in lightning, which we do not and cannot harvest on a significant scale). So electricity only is an energy carrier and the energy source can be quite diverse, as it is currently. In Figure 7.1 we see the distribution of energy sources for the production of electricity in the Netherlands.

**Analysis of Energy Systems**

In the next figures we see a number of possible energy systems in which we can supply energy to the consumer; indicated by D (demand) and we supply it from a source Y (Yield of energy). In general we have to convert the energy form of the yields into a form that we can transport and after transportation sometimes we have to convert it into another form as demanded by the consumer. This is called the energy chain.

However, we must be aware that there are several interconnected chains

*Diversification of energy sources and carriers* 135

Y = Yield, supply
T = Transport
C = Conversion
S = Storage
D = Demand

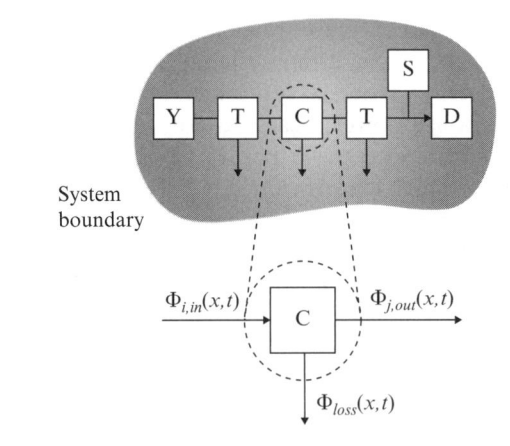

$\Phi_{i,in}(x,t)$ $\Phi_{j,out}(x,t)$ $\Phi_{loss}(x,t)$

*Figure 7.2  Supply chain in an energy system*

in real life energy systems – certainly on longer scales. They can be interconnected in different ways at different points depending on the use of converters and what type of energy carrier is dominant.

Multisource operation on a national scale increases energy security, but if the supply systems (the energy chains) are separate, failure in one chain will disrupt the whole supply chain. As a hypothesis we can claim that the security of supply increases if more interconnections exist in the energy system between the different supply chains. Interconnections can exist on different scales and in different forms. For instance we can have fuel blending in which we mix coal with biomass and use that mix in a coal-fired power plant. Other examples of fuel blending are the blending of bio-diesel into fossil fuel diesel or ethanol mixed with gasoline into E-10 and other fuel blends. There have also been studies on the blending of hydrogen into the natural gas grid as a way of decarbonizing the natural gas fuel. This is also a form of blending fuel. When we blend different fuels of course the conversion devices should be suitable for – or adapted to – the new blends of fuels. They are preferably continuously adoptable to a wide range of mixing ratios.

In general energy conversion leads to losses, but these can be minimized by technical improvements of the efficiency, but also by using a waste stream as in combined heat and power. Combined heat and power is a form of making interconnections in the energy system since the waste heat is coupled into the energy system as an energy carrier and used downstream. If combined heat and power is not applied the supply chains are not interconnected and heat is wasted into the environment (air or cooling water of the power plants). So by interconnections total conversion efficiency can increase. Some systems can produce three products and are called tri-generation systems. Although the name can refer to any three energy carriers as a product, often the name is reserved for systems that produce power, heat and cold. Below as an example the so-called Superwind concept will be described. In short Superwind is a concept based on a high temperature fuel cell capable of coproduction of hydrogen and power with some waste heat with natural gas or biogas or a mixture of both as the energy source. Here efficiencies even further increase even without counting heat as a useful product in the efficiency definition.

**Issues of Scale**

As discussed elsewhere, the issue of scale in energy supply is important. It refers to time (short-term versus long-term) as well as place (local versus global). Security supply can be analyzed on different timescales and length scales. On the timescale we can distinguish between short-term, medium-term and long-term issues. Considering the short-term security of supply normally we call it differently namely: 'reliability'. We can link security of supply on the medium term to the flexibility of our system to cope with temporary disruptions in one form of energy and therefore switching to another form of energy as our main source. Security of supply on the long term refers to our sources of energy in the long run when fossil fuel reserves will be diminished.

When we consider the scale length or the size of the system, security of supply on smaller systems such as a household or a neighborhood or an industrial installation refers to the security of supply in the sense of reliability of our system to perform what it is supposed to perform. Security of supply on median-length scales could refer to the energy supply for a region or country whereas security of supply on a long-length scale would refer to security of supply for the whole earth or security of supply globally. If you put these in a matrix, length scales versus timescales, we can distinguish nine areas with different issues and sometimes also different names for what we in general might call security of supply (see Table 7.1).

**Flexibility**

Obviously when the supply of one sort of energy is hampered either by the fact that it is running out (long-term issue) or that the transport is blocked by technical (gas pipeline rupture) or geopolitical issues, the security of energy supply is increased if one can replace one source of energy easily by another source of energy that still is available. The energy system thus becomes more robust if it is flexible in its sources of energy. This holds for

*Diversification of energy sources and carriers* 137

Table 7.1 Main security of supply issues on different length scales and time scales

| SoS main issues | Short term | Medium term | Long term |
| --- | --- | --- | --- |
| Single Unit – eco park | Reliability | Economic feasibility of particular energy source | Availability of particular energy form needed |
| Region – country | Geopolitics and technical disruptions in the supply chain | Transition management | Backcasting |
| Global | Market price fluctuations | Choice of energy carriers and its transport | Energy scenarios |

example for a small energy system like a car (single unit) that can run on gasoline as well as on LPG and can easily switch between them, as well as for a country that can switch between different fuels on a national scale. The latter can be achieved with either multi-fuel appliances like the car in the example above, or by multiple appliances that can be chosen to operate depending on the type of fuel available. Another approach is the choice of energy carrier and associated infrastructure.

The dual fuel or even multi-fuel appliances are to be preferred in general because less capital is needed for the infrastructure and the same appliance can be kept running on one fuel or the other depending on its price and availability. In the case of separate appliances they will stand idle if the particular fuel is not available or is too expensive.

**Fuel Blending**

Another form of flexibility in energy sources is mixing fuels. This is done on a large scale in the transport sector with the introduction and mixing of bio-diesel in ordinary diesel and for example bio-ethanol with gasoline. The European Renewable Energy Directive introduces a binding target of 10 per cent renewable energy in transport by 2020.

On a much smaller scale synthetic natural gas (SNG) is produced from biomass and introduced into the natural gas grid. With this European objective of blending biofuel with fossil fuel for the transport sector huge amounts of biomass will be reserved for this application and probably little will be left for synthetic natural gas production. But the principle is blending of fuels and thereby increasing security of supply in two ways: firstly decreasing the use of fossil fuels maintaining their reserves and secondly broadening the supply base with introducing a second supply source, SNG in this case.

A simple form of integration is the co-firing of biomass in a coal-fired power plant. Other simple forms of such integration are the preparation of blended fuels obtained by mixing bio-ethanol or bio-diesel into the respective fossil fuels. However, the co-firing of biogas from a biomass gasifier in a natural gas fired power plant would be a form of integration of sources going one step deeper, because it is not just a mix of fuels entering the same equipment that would need only small adaptations to the new fuel mix. Sometimes a biomass gasifier is needed to first covert one of the input streams before it is mixed with the other conventional main stream.

**Multisource Multiproduct Energy System**

There is a new development called multisource multiproduct (MSMP) energy systems. These systems differ from dual fuel appliances in the sense that in dual fuel appliances either one or the other fuel is used like in the example of the car running on gasoline or LPG, while in multisource multiproduct energy systems the system in principle can run on two or more energy sources simultaneously in a flexible way and produce more than one form of energy output. Multisource multiproduct systems are defined on the level of the conversion devices on the small length scale. Of course we can see the whole energy system for a country as a huge multisource multiproduct system but we want to exclude that from the definition, because it has no added meaning.

MSMP systems introduce flexibility needed in energy systems that have to cope with fluctuating demand and with the introduction of renewable energy sources like solar and wind also have to deal with fluctuating and not controllable supply. In general we will illustrate a topology of integration levels in multisource multiproduct energy systems. There are different forms of integration from loosely coupled through producing electricity for a common grid, to more intimately coupling in coal and biomass combustion for example or blending of biofuel into gas, gasoline and diesel:

- Single input single output conversion systems
- Combined heat and power
- Tri-generation.

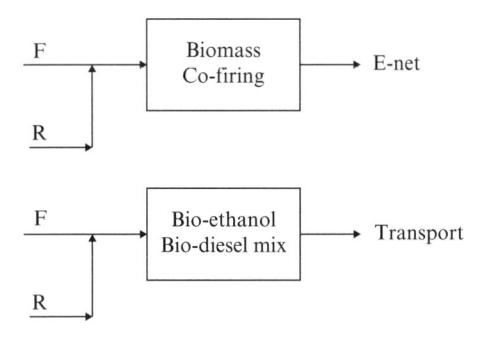

*Figure 7.3  Simple form of multi-input systems combining fossil energy with renewable energy by blending*

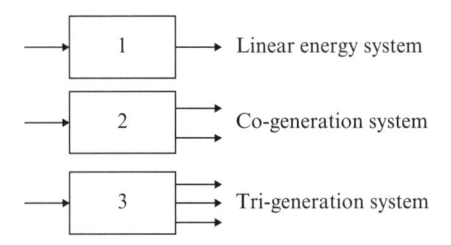

*Figure 7.4  Classification of energy systems*

The principle of cogeneration is well-known and widely applied for example in combined heat and power applications. The term tri-generation has been proposed for systems capable of the co-production of electricity, heat and cold, but also for the co-production of electricity and heat in combination with a chemical product such as hydrogen. This illustrates that a linear system as in Figure 7.2 is not generic, but only a simple representation of an example system. In general, often more than one product is produced. The other products are seen as a byproduct or even just as waste.

However, more than one source can also be applied on the input side, while the degree of integration may differ strongly. This can be illustrated by the familiar example system, called 'House'. Often the house is connected to three sources of energy: natural gas, electricity and passive sunlight for space heating, but those sources are not controlled and integrated but largely function independently.

Hence we must consider integration of sources into multisource multiproduct energy systems.

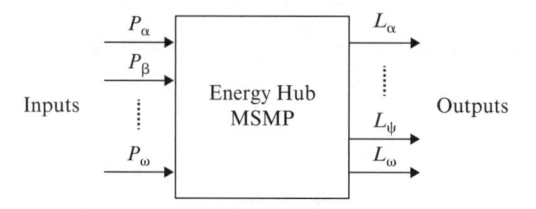

*Figure 7.5  Schematic illustration of a multisource multiproduct energy system (or energy hub)*

We will give several examples of such multisource multiproduct energy systems:

- Superwind
- Direct carbon fuel cell producing CO
- Hybrid systems of CSP and gas.

**Superwind**
Problems in practice are often different from those first anticipated. In the case of wind energy the problem is often phrased as the unpredictable and fluctuating character of the produced electricity. The solution generally proposed is to store the energy in batteries or to convert the wind electricity into hydrogen by electrolysis and subsequently store the hydrogen for later use in a fuel cell to produce electricity when there is little wind. This conventional 'hydrogen' storage solution suffers from a number of problems. Because a series of conversion steps are needed, overall efficiency becomes very low. For a rough order of magnitude calculation we may assume an efficiency for electrolysis of 80 per cent, 20 per cent losses in hydrogen storage and 40 per cent efficiency for the fuel cell. These indicative figures lead to an overall efficiency of only about 25 per cent! (0.8 * 0.8 * 0.4 = 0.256). Moreover, this conventional solution requires three types of equipment: electrolysis, hydrogen storage and fuel cells each with associated high capital costs. In addition, this expensive equipment is not used all of the time. On the contrary, it is intended to be used only when there is a surplus of electricity that absolutely cannot be used directly or fed into the grid. In an economic evaluation this will evidently lead to a poor return on investment.

However from the perspective of an individual wind turbine owner or wind corporation, the problem is to deliver electricity as promised on the "one day ahead" market (for example the APX, Amsterdam Power eXchange). When the wind fails to deliver what is predicted by weather models and subsequently offered to the market, the owner faces financial penalties from the electricity traders. So wind turbine owners are looking for a technology that "fills in the gaps" rather than for a technology that "shaves the peaks". The conventional solution is a typical exponent of our way of thinking about energy systems, being linear and either fully renewable or completely based on fossil energy. The idea of producing pure renewable ("green") electricity or hydrogen is very appealing to policymakers and the public. Large projects are carried out or planned to demonstrate this option that lacks rationality when seen from a system perspective as argued above. As long as wind energy is not the main source of electricity, storage is often not absolutely necessary as was shown in a recent study for the Dutch government. Instead, flexible production by complementary systems can be considered and of course demand side management, whenever that is possible. Yet part load operation of conventional technology will lead to decreased efficiencies, and a temporary stop of production will lead to a decreased economic efficiency. For fuel cell technology however, the situation is quite different. In principle the efficiency of fuel cells increase when operated in part load. But even more interesting the flexible co-production of hydrogen and electric power by a high temperature internal reforming fuel cell fueled by natural gas, offers the possibility to continuously produce valuable economic products all the time (Hemmes et al., 2008). If less electricity is needed the operation can be changed to produce more hydrogen. This concept that we called Superwind is a promising alternative if a hydrogen market is developing in the future, for example for the automotive sector (Hemmes et al., 2007). Flexibility of this Superwind concept can also be used to adapt gradually to a growing demand for hydrogen when more and more hydrogen fuel cell vehicles appear on the roads and the hydrogen demand increases. Applying the Superwind concepts allows the operator to produce more hydrogen when needed, without additional investments in the production capacity while in the beginning at low hydrogen demand this installation is still economically producing valuable goods namely electricity and heat.

Also, large-scale industrial energy conversion systems such as coal gasification units combined with integrated gasification combined cycle technology should be studied for their possible suitability for the flexible coproduction of hydrogen and electric power.

In many places natural gas and sometimes biogas or even hydrogen is available and these can be applied in combination with wind and solar in order to compensate for, or better, to be complementary to these fluctuating renewable energy sources. So, more research should be focused on the development of flexible and dynamic electricity production methods. The research on large-scale production and storage of hydrogen produced by electrolysis from wind and solar energy does not seem to be necessary at

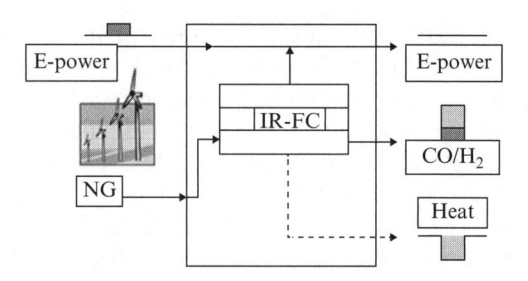

*Figure 7.6   Schematic representation of the Superwind concept*

the moment. The situation might change if wind and solar are applied at such a large scale that significant overproduction may exist at a significant fraction of the time.

**Direct carbon fuel cell producing CO**
In earlier work we have proposed an exciting, but still theoretical, possibility of the electrochemical gasification of (pure) carbon into carbon monoxide with the simultaneous conversion of (solar) heat and reaction enthalpy into electric power (Peelen et al., 1998). Because carbon monoxide can be converted with steam into hydrogen and carbon dioxide in the so-called shift reaction, we have in fact obtained a fuel cell that produces hydrogen instead of consuming it, and converts heat into power instead of dissipating heat. This would be a true MSMP system in which carbon and solar energy are converted into hydrogen, electricity and heat (Hemmes, 2004). Additionally solar energy can first be used to convert natural gas into carbon and hydrogen. And subsequently the carbon can be electrochemically converted into carbon monoxide with the help of solar energy input, while this carbon monoxide can be a second source for hydrogen through the shift reaction with steam. Needless to say that this still requires further simulation and feasibility studies and if these are positive a huge technological development is needed. A most suitable location for the implementation of the concept would be North Africa (Algeria, Egypt) which has a large potential for solar energy and resources of natural gas. The hydrogen produced or hydrogen methane mixtures as proposed in the European research project NATURALHY, can be transported to Italy by pipeline to supply Europe with an increasing amount of renewable energy.

**Hybrid systems of CSP and gas**
A further integration of energy sources is for example achieved by the performance of endothermic chemical reactions like steam reforming of

natural gas with the aid of the heat from concentrated solar power or nuclear energy. This example also illustrates the possibility of the integration of a fossil energy sources with a renewable energy source or even nuclear power. An alternative would be the thermal decomposition of natural gas (methane) with the aid of the heat from concentrated solar power. The clean hydrogen produced is easily separated from the solid carbon contrary to steam reforming where it has to be separated from a gas mixture containing carbon monoxide, carbon dioxide and steam using PSA (Pressure Swing Absorption) or other costly separation technologies. Obviously one has to find useful applications for the carbon, or carbon black as it is often called. There are applications for carbon in printers and in car tires, but not for the quantities in which it will be produced in the future if these technologies are applied on a large scale. This would require a demand for much larger quantities. A number of them are proposed in the literature by Muradov et al., one of them being Direct Carbon Fuel Cells which are presently under development (Muradov and Veziroglu, 2005).

In cooperation with the University of Perugia we have calculated that in a solar natural gas decomposition setup a ratio of 80/20 is typical for the relative contribution of fossil energy versus solar energy (Cinti 2011).

**Hydrogen as a Universal Energy Carrier**

Since the 1970s hydrogen has also been proposed as an energy carrier, with the idea that fossil fuels will eventually run out and have to be replaced by renewable energy sources such as wind energy and solar energy. These renewable energy sources mainly produce electricity but contrary to fossil energy these are fluctuating and do not include a storage function. Another well-known drawback of solar and wind related to this is the fluctuating nature of these sources on different time scales (seconds to days, weeks and even seasonal fluctuations). Some are very predictable such as day/night patterns for solar, but some are less predictable such as clouds for solar and wind speed forecasts for wind energy. As a solution, storage in the form of hydrogen has been proposed. But hydrogen can also be used in the transport sector and does not need to be converted back into electricity. With the development of a hydrogen economy as it was called other sources for the production of hydrogen can also be developed. In fact one of the main advantages proposed by the promoters of the hydrogen economy is that hydrogen can be made in so many ways, from a variety of sources, thereby providing or at least increasing energy security.

In the early days of the ideas on the hydrogen economy and sometimes even now, one can observe a rather naïve attitude towards real-life energy systems, leading to severe criticism on the hydrogen economy (e.g. by Bossel 2006). Since hydrogen is not an energy source but only an energy carrier, it must always be made by conversion from another form of energy and – although often ignored – a source of hydrogen atoms, for example water. Conversions are accompanied by losses and conversion units cost money, so hydrogen in general will be more expensive than the original source.

Secondly hydrogen can be stored, as is often claimed, but not so easily, nor on a large scale. Also the required purity of hydrogen for different applications is an issue that is not often considered in general concepts of the hydrogen economy. Thirdly a quantitative argument holds that as long as electricity production from renewable energy is not abundant, it makes much more sense not to convert renewable energy but to keep it in the form of electricity as the energy carrier. Also converting it on a large scale would require a huge capacity in the form of electrolyzers. Electrolyzers are devices similar to fuel cells and similarly priced.

The conversion efficiency argument can be addressed by MSMP systems with inherent higher efficiency. This can be illustrated with the very ambitious plans of the Dutch government. The 6 GW offshore wind energy parks would only provide about 2 GW continuously, providing 1.6 GW of hydrogen when assuming 80 per cent efficiency of the electrolyzers. If we compare this with the 40 billion normal cubic meters of natural gas consumed in the Netherlands on a yearly basis, the hydrogen thus produced from offshore wind energy would only suffice to provide 3.5% of total Dutch natural gas consumption. (1.6 GW/(40.10$^9$ Nm$^3$/year 10 kWh/Nm$^3$). (1/(8760 h/year)) x 100%)

## CONCLUSIONS

An attempt is made to systemize energy systems in terms of building blocks, supply chains and their interconnectivity. Building blocks are yield (the source), transport, conversion, storage and end use (demand). Interconnections are seen in fuel blending, combined heat and power, tri-generation systems and in MSMP energy systems. Interconnections also follow from reducing the number of energy carriers of which the hydrogen economy is an extreme case, in which hydrogen and probably also electricity are the only energy carriers.

Disruption in one energy source (Yield) does not lead to the disruption of the whole supply chain, because other sources from a redundant chain can take over via a common hydrogen energy carrier. This disruption need not only be on a large scale over a longer period of time, but can also be fluctuations in, for example, renewable energy sources like wind energy and solar energy. MSMP energy conversion systems especially are shown to be highly efficient and flexible thereby improving the system's security of supply and overall efficiency at the same time. Moreover, interconnections between different supply chains often make the use of storage unnecessary.

## NOTE

\* This research has been partly financed by a grant of the Energy Delta Gas Research (EDGaR) program. EDGaR is co-financed by the Northern Netherlands Provinces, the European Fund for Regional Development, the Ministry of Economic Affairs and the Province of Groningen. The author would like to thank the section "Technology Dynamics and Sustainability" of the Faculty of Technology Policy and Management, Delft University of Technology, for constant support.

## REFERENCES

Bossel, Ulf (2006), "Does a Hydrogen Economy Make Sense?", *Proceedings of the IEEE*, 94(10), October.

Cinti, G. and K. Hemmes (2011), "Integration of direct carbon fuel cells with concentrated solar power", *International Journal of Hydrogen Energy*, 36(16): 10198–10208.

Hemmes, K. (2004), 'Fuel Cells', in R.E. White, B.E. Conway and CG. Vayenas (eds), *Fuel Cells; in Modern Aspects of Electrochemistry*, Modern Aspects of Electrochemistry, Vol 37, New York, Kluwer.Academic/Plenum Publishers, pp. 131–251.

Hemmes, K., L.M. Kamp, A.B.H. Vernay et al. (2007), "A multi-source multi-product internal reforming fuel cell energy system as a stepping stone in the transition towards a more sustainable energy and transport sector", *International Journal of Hydrogen Energy*, 36(16): 10221–10227, DOI: 10.1016/j.ijhydene.2010.11.017.

Hemmes, K., J.L. Zachariah-Wolff, M. Geidl et al. (2008), "Towards multi-source multi-product energy systems", 2nd European Hydrogen Energy Conference, Zaragoza, Spain, 22–25 November 2005, *International Journal of Hydrogen Energy*, 32(10–11): 1332–1338, DOI: 10.1016/j.ijhydene.2006.10.013.

Muradov, N. Z. and T.N. Veziroglu (2005), "From hydrocarbon to hydrogen-carbon to hydrogen economy", *International Journal of Hydrogen Energy*, 30(3): 225–237.

Peelen, W.H.A., K. Hemmes and J.H.W. De Wit (1998), "Carbon a major energy carrier for the future? Direct carbon fuel cells and molten salt coal/biomass gasification", *High Temperature Material Processes*, 2(4): 471–482.

Vernay, A.L. G. Steenvoorden and K. Hemmes (2008), "Superwind: A feasibility study. Integrating wind energy with internal reforming fuel cells for flexible coproduction of electricity and hydrogen", final report for Senternovem, November 2008, Project number: NEOH 02010 Senternovem. 2008, Delft, TU Delft.

# 8. Energy security assessment framework and three case studies
## Aleh Cherp and Jessica Jewell

## INTRODUCTION

The interest in measuring energy security results not only from its rising prominence but also from its increasing complexity. In the past, energy security concerns were no less acute than they are today. Consider, for example, the importance of access to oil for nations engaged in major wars of the 20th century. Yet, to strive for energy security in such cases did not require complex measures because policy makers were directly engaged and closely familiar with these immediate and pressing issues. In contrast, today's energy security problems often overlap national, institutional and sectoral boundaries stretching the cognitive abilities of experts and policy makers to deal with diverse situations and challenges which may not be directly familiar or predictable. One approach to cutting through this complexity is relating energy security to a common yardstick that would allow comparing it across different countries, at different points in time or to other policy priorities, in other words quantitatively *measuring* energy security.

The challenge of measuring energy security is not only to see through natural, technological, and economic complexities and uncertainties, but also to address the fact that it has different meanings for different groups (Chester, 2009). No single set of metrics is suitable for assessing energy security for all purposes in all situations. Instead energy security should be measured through application of an assessment framework sufficiently systematic to ensure scientific rigor and sufficiently flexible to account for specific circumstances and perspectives (Cherp and Jewell, 2011a). This chapter outlines such a framework and illustrates its application in the following three cases:

- The **International Energy Agency's Model of Short-term Energy Security (MOSES)** (Jewell, 2011). The purpose of MOSES was to depict the energy security landscape of the 28 IEA member countries by characterizing their energy security profiles and grouping together countries with similar energy profiles.

- The **Global Energy Assessment** (GEA) (GEA, 2012), a major international effort to evaluate energy challenges and construct long-term scenarios for meeting these challenges. The purpose of GEA's energy security assessment (GEA Chapter 5, Cherp et al., 2012) was to "identify common energy security concerns (in over 130 countries) affecting significant parts of the world's population".
- A set of recent studies of **energy security in future scenarios** based on the methodology originally proposed by (Jewell, 2010) and subsequently used in Chapter 17 of GEA (Riahi et al., 2012) as well as in Jewell et al. (2012) and Cherp et al. (2013). The purpose of these studies has been to analyze energy security in long-term (up to the year 2010) scenarios of transformation of global energy systems.

## METHODOLOGICAL CHOICES

Any quantification of energy security requires certain methodological choices. Making such choices is a difficult task because of the multitude of interpretations of energy security (see overviews in Cherp and Jewell, 2011b; Sovacool, 2011; Chester, 2009; Winzer, 2012). The two most fundamental methodological choices in energy security assessments are (1) the choice between perceptions and facts in deciding what constitutes a significant energy security concern and (2) the choice between the specific and generic in deciding on what is the appropriate level of detail of the assessment. As we explain in the next section these choices need to be made with respect to vital energy systems, their vulnerabilities, and selection and interpretation of indicators.

The first choice in deciding what constitutes an energy security concern and whether such a concern is significant is between facts and perceptions. Focusing on facts means conceptualizing energy security as an objective property of energy systems which makes it easier to quantify and compare (e.g. by Le Coq and Paltseva, 2009 and Gupta, 2008). This approach, however, sometimes fails to explain the actual energy security policy priorities influenced by such hard-to-quantify factors as history, culture, politics and psychology.

On the other end of the epistemological spectrum are perceptions. For example, Sovacool and Mukherjee (2011) solicit views of various stakeholders to arrive at a set of "dimensions" and indicators of energy security. However, stakeholders can be biased, manipulative or poorly informed. They may either use security rhetoric or ignore obvious concerns to advance their own interests. As a result, an assessment guided

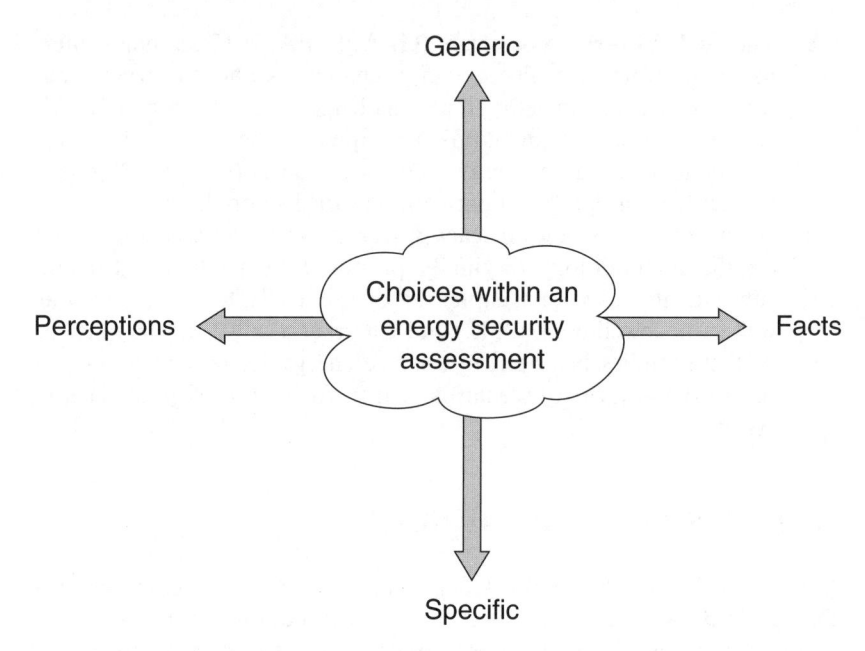

*Figure 8.1  Major methodological choices in measuring energy security and the energy systems approach*

by such a survey would risk not being policy relevant. Thus, an energy security analyst should be aware of biases and try to reduce them while still remaining policy relevant. Cherp (2012) argues that perceptions can be useful in framing energy security assessments only if they are solicited from a relevant group of stakeholders and in such a way that forces prioritization of various concerns.

The three case studies discussed in this chapter strive to combine analyses of energy systems and insights from energy security policies to arrive at findings that are both scientifically rigorous and reflective of policy concerns. MOSES was conducted under the oversight of IEA member countries and in direct and continuous dialogue with policy-makers. The GEA analysis frames its quantitative findings with an analysis of energy security policies. Finally, the analysis of future energy scenarios derives its approach from the careful study of the evolution of energy security policy paradigms over last century to distill generic concerns which can be plausibly valid for the next 100 years.

The second choice is between the generic and the specific in choosing the scope, focus and tools for an assessment. The three case-studies discussed in this chapter feature various degree of specificity. MOSES uses approaches specific to energy supply of developed market economies. GEA's approach is more generic as it needs to be applicable to over 130 countries. Finally, the analysis of energy security of future scenarios needs to deal with energy systems which are widely different from those of today and thus uses the most generic approach of the three cases.

In summary, an effective energy security assessment is specific enough to reflect context-specific issues and yet generic enough to enable sufficiently wide comparison. Likewise, it is based on hard facts, not opinions while still responding to perceptions and policy priorities. Finding such trade-offs is the science and art of energy security assessments. There is no blueprint for achieving this balance, but in all the three cases it has been guided by an energy systems approach. This approach proceeds from the premise that the term "energy" in "energy security" designates not a black box with amorphous content, but rather a set of interlinked systems each consisting of elements connected to each other and to the outside world and each with their own sets of vulnerabilities.

Thinking in such systems terms can support methodological choices within energy security assessments. For example, perceptions of energy stakeholders, especially policy makers, can be structured in accordance with three fundamental security questions: What to protect? From which risks? And by which means? Answers to these questions reflect the way policy makers perceive energy systems which can be related to objective facts about them.

The energy systems approach can also support the choice between the specific and the generic in energy security assessments. Specific approaches developed for particular situations work better when the assessment compares similar energy systems (e.g. the change in energy security of a particular country from one year to another). However, the wider the difference in energy systems that require comparison is (for example involving many diverse countries or addressing the situation in a distant future), the more generic the energy security assessment methodology should be. The range of addressed concerns may need to be wider, the indicators more universal and their interpretation involve stronger qualitative elements.

The methodological choices in an energy security assessment should be systematic rational and transparent. They should reflect the configuration of energy systems (real and perceived), justified based on the purpose on the assessment and clearly explained for the intended audience. The proposed energy security assessment framework presents an approach for guiding such choices through several stages as explained in the following section.

## ENERGY SECURITY ASSESSMENT FRAMEWORK

The energy security assessment framework includes five stages:

1. defining energy security for the purpose of the assessment;
2. delineating vital energy systems;
3. identifying vulnerabilities of vital energy systems;
4. selecting and calculating indicators for these vulnerabilities;
5. interpreting the indicators to answer the questions posed by the assessment.

### Defining Energy Security

Because there is no universal definition of energy security (Chester, 2009; Winzer, 2012), any energy security assessment should start with choosing or operationalizing an appropriate definition. For example, the analysis of energy security in future energy scenarios uses the most generic definition of energy security as *low vulnerability of vital energy systems*. It covers a wide variety of situations and at the same time provides a clear direction of operationalizing it for a specific context by narrowing down the concepts of "vulnerability" and "vital energy systems". The GEA defines energy security as *uninterrupted provision of vital energy services*. The focus on energy services reflects GEA's emphasis on energy's role in human welfare and sustainable development. As we shall see later, the GEA's actual approach to measuring energy security covers not only energy end-uses but also sources and carriers linked to those services. MOSES proceeds from the IEA definition of energy security as *the uninterrupted physical availability at a price which is affordable, while respecting environmental concerns*. MOSES focuses only on the short-term physical availability of energy referred to in the first part of the definition.

### Vital Energy Systems

As already mentioned, energy security is fundamentally a systemic notion. What is secure for a particular system may not be secure for its sub-system(s) and vice-versa. Thus, evaluating energy security entails clearly and explicitly defining the boundaries of the energy systems, which are being evaluated. The choice of these systems is not arbitrary. In addressing the *What to protect?* question, energy security policies are focused not on some abstract "energy" but rather on protecting energy systems which are critical for societies, in other words, vital energy systems.

An early example of a vital energy system can be traced to the time when the British Navy switched from coal to oil on the eve of the First World War (Yergin, 1991). The first vital energy system critical for the survival of the British Empire consisted of a fleet of navy ships and oil wells connected by transportation lines. It formed a true system: shortfalls of oil supplies could be replaced by oil from another source but not, for example, by coal or wood. This explains why even though oil was a tiny proportion of the overall energy consumed at that time it was still at the center of its energy security concerns.

Thus, the notion of a vital energy system combines two aspects. The term "vital" means that it is critical for the functioning and stability of a society.[1] The term "system" means that it consists of resources, materials, infrastructure, technologies, markets and other elements connected to each other stronger than they are connected to the outside world. From the energy security angle, the meaning of such connections is that in the case of a disruption the elements within a system can replace each other, but the elements from outside the system – can't.

Energy systems can be delineated along geographic or sectoral boundaries. Various combinations of geographic and sectoral choices define a potentially large number of energy systems (see Figure 8.2 for an illustration). Only some of these combinations making up vital energy systems will be relevant for a particular energy security assessment.

With respect to geographic boundaries, energy security concerns are primarily articulated at the national level.[2] This is because historically it has always been the responsibility of the nation state to protect security. Even such supra-national entities as the IEA and the European Union respect national boundaries by focusing on energy security of their individual member states. Thus, MOSES focuses on national energy systems of the IEA member countries and GEA focuses on energy security of over 130 countries.

Regional and global energy systems can also be viewed as vital by energy security policies. An historic example is the US Carter doctrine which called for the protection of global oil-producing regions and transportation routes because they are linked to US "vital interests" (Carter, 1980). More recent UK and EU energy security strategies address Eurasian and global gas markets. The Australian National Energy Security Assessment (NESA) analyzes the global markets in liquid fuels and natural gas (Department of Resources, 2011). State-supported Chinese investments in overseas oil assets have been driven by concerns over the security of the global oil market (Zhang, 2012). Bridge et al. (2012) develop an elegant notion of the "global production networks" for the energy system encompassing natural gas production and trade. The GEA discusses energy security in individual regions and analyzes the global market for international traded fuels and the global nuclear fuel cycle.

The analysis of energy security in future scenarios (as well as earlier studies of future energy security such as Turton and Barreto (2006) and Costantini et al. (2007)) faces the limitation that global long-term energy models do not have national-level resolution, instead they generate scenarios for a dozen or so "global regions" (for example the Middle East and North Africa). Based on the assumption that intra-regional energy integration and trade will likely be stronger than at present, these assessments analyze energy security at the regional (as well as the global) level.[3]

With respect to sectoral boundaries of vital energy systems, some academic literature refers to "security of supply" drawing the systems boundaries around all primary energy sources. The supply-focused approach is implicit in such generic concepts of energy security as the "4 As" (availability, accessibility, affordability and acceptability) (proposed in APERC (2007)). Such an approach is based on the assumption that various primary energy sources can substitute one another, which is often not the case. In reality, different primary energy sources often have distinct vulnerabilities which need to be analyzed separately. That is why for example Le Coq and Paltseva (2009) analyze vulnerabilities of oil, gas and coal separately.

We already mentioned the historic focus of energy security analysis on oil. This focus has persisted starting from early 20th century and been fueled by such events as the two world wars[4] and the oil embargoes of the 1970s. Security of oil supply clearly remains on the global energy security agenda, however, other sources have entered the picture as well. The IEA's "comprehensive view of energy security" is reflected in MOSES' analysis of seven primary energy sources (oil, natural gas, coal, biomass and waste, nuclear energy, hydropower and geothermal energy). The vital energy systems addressed in the GEA are shown in Figure 8.2. They include biomass particularly important for developing nations.

An example of a vital global energy system examined in the GEA is the nuclear fuel cycle. The GEA analysis shows that while nuclear power plants are constructed and maintained nationally they depend upon supply of nuclear fuel, parts of nuclear reactors and nuclear fuel reprocessing organized globally. Such global systems are another example of "global production networks" (Bridge et al., 2012).

Vital energy systems may also be structured around energy carriers such as electricity analyzed in the GEA (Figure 8.4) and the future energy studies. National electricity grids and power plants represent a truly unified energy system (often backed up by international interconnections). Electricity generation usually relies on a mix of sources so that disruptions in one fuel can be compensated by increased input from another fuel.

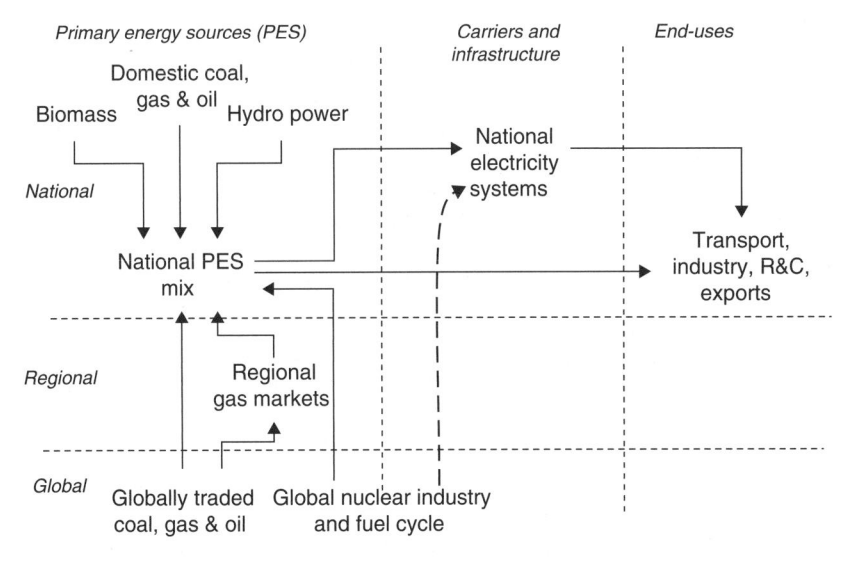

*Energy security assessment framework* 153

```
Primary energy sources (PES)        Carriers and              End-uses
                                    infrastructure
        Domestic coal,
        gas & oil
Biomass            Hydro power
                                    National
                                    electricity
National                            systems

        National PES                                      Transport,
        mix                                               industry, R&C,
                                                          exports
- - - - - - - - - - - - - - - - - - - - - - - - - - - - - - - - - - - -
Regional    Regional
            gas markets

- - - - - - - - - - - - - - - - - - - - - - - - - - - - - - - - - - - -
Global  Globally traded  Global nuclear industry
        coal, gas & oil  and fuel cycle
```

*Note:* The dotted arrow represents vulnerabilities associated with the concentration of nuclear reactor parts, nuclear fuel reprocessing and long-term nuclear waste storage.

*Source:* Adopted from Cherp et al. (2012).

*Figure 8.2   Vital energy systems covered in the GEA*

That is why many energy security policies and studies (e.g. Stirling, 1994; Awerbuch, 2006 and Grubb et al., 2006) address security of electricity. Other energy carriers include oil products (diesel, gasoline and others) and biofuels (both categories are analyzed in MOSES) or liquids fuels in general (Department of Resources, 2011; Cherp et al., 2013).

The assessment of energy security in future scenarios faces a major challenge to delineate vital energy systems of the future which might be significantly different from those of today. Thus, this energy security assessment looked into primary energy sources and energy carriers which will play a significant role in future energy systems. With respect to energy sources, it considers tradable fuels: oil, gas, coal and biofuels. With respect to energy carriers, it included synthetic fuels and hydrogen in addition to electricity and liquid fuels.

Finally, end-use sectors (sometimes called "energy services") can also be considered as vital energy systems (an example of an analysis focused on end-use services is Jansen and Seebregts, 2009). For example, one energy end-use vital for all countries is transportation. In the same way as the British Empire could not defend itself without a fleet of navy ships, a modern society cannot function without a fleet of motor vehicles. Other end-uses analyzed in both the GEA and the assessment of future scenarios include the residential and commercial sector and the industrial sector. In addition, the GEA assessment also addresses energy exports as a vital energy system for energy exporting nations, sometimes referred to as "demand security".

In summary, delineation of vital energy systems for an energy security assessment can be supported by the following checklist:

- ☑ Is this a true system? Are the elements within this system mutually substitutable? Can it be divided into sub-systems or merged with a larger system without making the assessment less meaningful?
- ☑ Is this a sufficiently significant system in terms of its size or the population using it or the economy it supports? Does this system support truly vital functions of a society?
- ☑ Is there a history or plausible scenario of disruption of this system or similar systems?
- ☑ Is this system consistently delineated and meaningful for all situations covered by the assessment?
- ☑ Are there energy security policies or discourses that address this system?

**Vulnerabilities**

Vulnerabilities of an energy system are a combination of its exposure to risks and resilience, i.e. its capacity to respond to disruptions. Some authors only look at risks (e.g. APERC, 2007; Winzer, 2012) others focus primarily on resilience (Stirling, 1994; 1998) whereas others (e.g. Kendell, 1998; Gupta, 2008) look at both risks and resilience. Energy security risks differ with respect to their time-profile (shocks or stresses) and the nature of disruptions (physical or economic). Resilience can relate to specific risks (e.g. the presence of alternative pipelines may help to reroute gas imports in case of problems in transit countries) or to more general risks categories (e.g. strategic storage can protect from shocks of supply caused by political, economic or technical factors). The distinction between risk and resilience capacities is not always observed in reality: sometimes these two can only be analyzed in combination.

Disruptions of vital energy systems come in the form of shocks (rapidly unfolding short-term disruptions) and stresses (slowly approaching and longer-lasting phenomena) (Stirling, 2010). Historically the energy security agenda was primarily shaped by shocks such as the oil crises of the 1970s, the coal miners' strikes of the 1980s, and the disruptions of natural gas supply and electricity blackouts of the 2000s. Stresses include unrelenting demand growth, resource depletion and aging of infrastructure.

The second distinction between physical and economic risks is drawn in a classic definition of energy security "sufficient supplies at affordable prices" (Yergin, 2006). Whereas "sufficient supplies" is an intuitively clear concept referring to physical risks, "affordable prices" is more of a widely debated political construct. Policy rhetoric on this issue uses such colorful but unhelpful terms as "reasonable", "true", "fair", "affordable", "cost-effective" and "competitive". A wide body of literature (among which Keppler (2007), Greene (2010) and Helm (2002) can be especially recommended) explores the economic aspects of energy security. An analysis of the policy measures from the UK, Sweden and the EU shows that despite the rhetoric their real focus (compatible with the overall idea of energy security) is on *stable* and *competitive* prices that do not threaten the operation of vitally important industries. Whatever the case, the exact meaning of economic risks to energy systems should be clarified at this stage of the assessment.

The literature proposes multiple ways to classify vulnerabilities often dividing them into economic, political, natural, technical, military, etc. (see e.g. Alhajji, 2008). However, in order to be useful for energy security assessments such classifications need to be more fundamental. Indeed, gaining insight into the causes of potential vulnerabilities of vital energy systems requires detailed understanding of how these systems function. This understanding needs to go beyond common sense and be rooted in a disciplined epistemological community and armed with an effective tool kit. Cherp and Jewell (2011b) identify three such perspectives on energy security rooted in their own historic experience and different disciplines as summarized in Table 8.1.

Historically, the energy security discourse emerged in the context of military hostilities and therefore focused on risks associated with hostile actions such as attacks on supply lines or oil fields. The risk that an adversary would attack or otherwise disrupt vital energy systems has remained high on the political agenda for the last 100 years (be it in the discourse of the "Arab oil weapon" (Paust and Blaustein, 2008), "the Russian gas weapon" (Baran, 2007), or in discussing possible "resource wars" between the US and China (Klare, 2008). The notion of targeted and intentional embargoes is now broadened and more nuanced: it includes concerns over political extortion, political stability of suppliers or collateral damage due to unrelated energy disputes. Nevertheless, all these concerns focus on risks arising from foreign control over vital energy systems. As an influential UK energy security policy document puts it: "[energy security would allow the UK to] retain independence in its foreign policy through avoiding dependence on particular nations" (Wicks, 2009:8). This *sovereignty perspective* on energy security analyzes risks in terms of interests, alliances, power balances and space for maneuver as the sovereignty perspective.

The second perspective on energy security sees the origin of risks in

Table 8.1 Three perspectives on energy security

| Perspective | Sovereignty | Robustness | Resilience |
| --- | --- | --- | --- |
| Historic roots | War-time oil supplies and the 1970s oil crises | Large accidents, electricity blackouts, resource scarcity | Liberalization of energy systems |
| Parent discipline | Security studies, international relations, political science | Engineering, natural science | Economics, complex system analysis |
| Key risks | Intentional actions by malevolent agents including politically motivated disruptions, political extortion and price manipulations | Probabilistically predictable natural, technical and economic factors. Infrastructure failures and aging, extreme natural events, depletion of resources, demand growth | Diverse and partially unpredictable factors: political instability, labor actions, terrorism, climate, economic volatility etc. |
| Resilience capacities | Competitive market arrangements, diversity of actors, trusted suppliers and reliable regimes | Emergency stocks and redundancies, spare capacities, infrastructure diversity | Diversity of energy technologies, low energy intensity, emergency preparedness, investent in research and development, etc. |
| Primary protection mechanisms | Control over energy systems and institutional arrangements preventing disruptive actions | Upgrading infrastructure, constraining demand, switching to more abundant resources | Increasing the ability to withstand and recover from various disruptions |

Source: Modified from Cherp and Jewell (2011b).

natural and technical factors rather than in hostile or intentional human actions. It puts at the center concerns such as aging of infrastructure, depletion of resources, and vulnerability of energy systems to extreme natural events. This *robustness perspective* has its roots in natural science and engineering and relies on forecasts and estimation of probabilities for risk evaluation.

The third, *resilience perspective* sees the origin of risks in increasing complexity and uncertainty of technological, social and economic factors affecting energy systems. It recognizes that many disruptions and risks cannot be accurately predicted. It shifts attention from identifying and managing risks to building resilient energy systems that are able to respond to diverse disruptions.

Among the three assessments, MOSES has the narrowest focus on short-term physical disruptions whereas the GEA and the analysis of future energy scenarios cover both shocks and stresses with both a physical and economic nature. All of the assessments seek to integrate the three perspectives on energy security though MOSES predominantly focuses on sovereignty and robustness concerns (which it classifies in external and domestic risk and resilience factors, see Table 8.2) whereas the future analysis in its current form only covers sovereignty and resilience concerns. The list of potential future vulnerabilities (see Table 8.4) is largely derived from the prioritization of the *current* vulnerabilities as identified in the GEA (Table 8.3 and Figure 8.4) and interpreted in more generic terms to be applicable to future vital energy systems.

As with vital energy systems, it is important to make systematic and transparent choices of which vulnerabilities (either listed in Table 8.1 or additional ones) to include in (and which to exclude from) the energy security assessments. The following checklist may aid identification of vulnerabilities of vital energy systems in an energy security assessment:

- ☑ Does a particular vulnerability characterize one of the vital energy systems identified at the previous stage of the assessment?
- ☑ Is the vulnerability likely to cause a significant disruption to one of the vital energy systems?
- ☑ Is the vulnerability addressed in energy security policies or rhetoric?

**Selecting Indicators**

Energy security indicators should reflect the vulnerabilities of vital energy systems identified at the earlier stages of the assessment. They can be selected from those suggested in the abundant literature or designed specifically for the purpose of a particular assessment. Selection of indicators should be guided by how well they represent a particular risk or vulnerability of a vital energy system. However, an indicator is rarely a direct measure of a risk or a resilience capacity. Rather it is a quantitative proxy, a signal of a state of a complex and dynamic energy system. A good analogy here is body temperature as an indicator of human health. As a proxy it does not exactly point to the causes, nature or extent of illness but it is still widely used and relatively reliable, especially when used in conjunction with other observations. So are energy security indicators. One indicator may signal the presence of several risks. For example, import dependency may reflect the exposure to deliberate supply cuts, disputes with transit countries, failures or sabotage of transportation lines, or price volatility. Similarly, one vulnerability can be reflected in several indicators. For example, the risk of blackouts may be reflected by their historic frequency, the age of the power plants, the spare capacity, and the diversity of electricity generation.

Some indicators can be directly found in existing statistical information and other data sources. In most cases, however, the indicators will need to be calculated based on available data. For example, MOSES used data from the IEA, the World Bank and the IAEA; the GEA used publicly available IEA and BP energy statistics as well as Platts energy database, the World Bank, and the IAEA. The analysis of energy security in future energy scenarios derived its data from the variables calculated from Integrated Assessment Models such as MESSAGE and REMIND. Calculation of indicators may use relatively simple formulas such as the reserves-to-production (R/P) ratios or a diversity index such as the Shannon-Weiner Diversity Index or the Herfindahl Hirschmann index. An example of a more complex formula is the calculation of the diversity of energy sources used in transport in future energy scenarios which reflects dozens of links inside the energy system (Jewell et al., 2012).

MOSES uses 35 indicators (see Table 8.2 for a sample, the full list is available in Jewell (2011:11) grouped into four dimensions of vulnerability for each of the primary sources and secondary fuels.

GEA uses some 30 indicators most of which are listed in Table 8.3. In contrast to MOSES, GEA addresses a wider range of energy systems and vulnerabilities and thus uses less detailed but more diverse indicators. The more general nature of the GEA indicators is also explained by the fact that the GEA analysts did not have access to as detailed information for all 134 countries as MOSES had for the IEA members.

The analysis of energy security in future energy scenarios used 20 global and five regional indicators summarized in Table 8.4 (some of these are proposed for future studies).

*Table 8.2  Indicators of crude oil supply security used in MOSES*

|  | Risks | Resilience |
|---|---|---|
| External | External risks:<br>• Import dependence<br>• Political stability of suppliers | External resilience:<br>• Number of ports<br>• Number of pipelines<br>• Diversity of suppliers |
| Domestic | Domestic risks:<br>• Share of offshore production<br>• Volatility of domestic production | Domestic resilience:<br>• Domestic storage level |

*Source:* Jewell (2011).

Selection of energy security indicators may be guided by the following checklist of questions:

- ☑ Is the indicator a characteristic of one of the vital energy systems?
- ☑ Does the indicator reflect one or more significant vulnerabilities (risks and/or resilience capacities) identified earlier?
- ☑ Does the indicator provide useful information about this risk or vulnerability in addition to that provided by other indicators?
- ☑ Are there reliable data and tools (models, etc.) available for calculating the indicator at all time points or for all situations covered by the assessment purposes?

**Making Sense of Indicators**

After the indicators have been calculated, the complex journey from the initial assessment questions to a set of numbers needs to be traced backwards: from those numbers to meaningful answers. The final task is to process, interpret and communicate the indicators in such a way that they convey accurate and relevant information cognitively accessible to the intended audiences of the assessment. There are three interrelated strategies for achieving this objective:

- interpreting individual indicators;
- reducing the number of indicators by combining them into aggregated metrics;
- presenting the indicators (individually or jointly) in a format that facilitates the assessment.

First, well-selected indicators can sometimes directly provide the answers. For example, policy makers often use indicators such as import dependency, R/P ratios, demand growth rates, blackout frequencies and

Table 8.3  Indicators used in the GEA energy security assessment

| Energy system | Energy security indicators ||
|---|---|---|
| | Shocks | Stresses |
| Globally traded fuels: oil, gas and coal | Import dependency, cost of imports | Global R/P<br>Domestic R/C<br>Growth in oil consumption |
| Nuclear | Fuel intensity of GDP | Average age of nuclear power plants<br>Start of last plant construction (reflecting the capacity to replace existing fleet) |
| Hydro | Diversity of hydro power dams ||
| Electricity | Dependency on imported fuels | Electricity demand growth rate<br>Rate of access to electricity |
| | Diversity of energy sources used in production of electricity ||
| End-use sectors: transport, industry, residential and commercial | Dependence on imported fuels | Demand growth rate in the sector |
| | Diversity of sources and carriers used in the sector ||
| Energy exports | Revenue from energy exports as share of GDP (reflecting exposure to price fluctuations) | R/P ratios of exported fuel |
| National Energy Systems | Overall energy import dependency<br>Cost of energy imports compared to GDP<br>Cost of energy imports compared to export earnings | Energy demand growth |
| | Diversity of PES; Energy intensity ||

*Source:* Adopted from Cherp et al. (2012).

*Energy security assessment framework* 161

Table 8.4 *Vulnerabilities of future energy systems and related indicators*

| Energy systems | | Perspectives: | |
| --- | --- | --- | --- |
| | | Sovereignty | Resilience |
| Primary energy sources | Total Primary Energy Supply (TPES) | Global energy trade (absolute and relative to the total TPES) Net import dependency* | Diversity of TPES Energy intensity |
| | Oil, gas, coal, biofuels | Global fuel trade Fuel import dependency* | |
| | | Regional diversity of fuel production | |
| Carriers | Hydrogen, electricity | Global trade in carrier Regional diversity of carrier production Reliance on imported fuels in carrier production* | Diversity of PES used in carrier production |
| End-use sectors | Transport, industry, residential and commercial | Reliance on imported fuels in end-use sector* | Diversity of PES used in the end-use sector Energy intensity of end-use sector |

*Notes:* *Regional level indicator.

*Source:* Jewell et al. (2012).

the age of power plants. Interpretation of individual indicators may involve comparison between countries or different points in time or relating them to some reference values such as the baseline. For example, the ranking of indicators for crude oil supply used in MOSES is shown in Table 8.5. Each indicator is assigned to a band of low, medium or high vulnerability on the basis of the indicator's values for IEA countries.

The GEA uses simple indicators to demonstrate that oil is the most vulnerable among the globally traded fuels because it has the lowest global R/P ratio, the highest proportion of international trade in global production, the largest number of people living in countries with major oil import dependency and the highest concentration of global production. The assessment reaches these conclusions by comparing indicators for global and national oil vulnerability with those for coal and natural gas.

Interpretation of individual indicators of future energy security is based on their comparison to the present situation and other scenarios (including business as usual development). For example, in most low-carbon scenar-

162  *International handbook of energy security*

*Table 8.5  Ranges of indicators for crude oil supply in MOSES*

| Dimension | Indicator | | Low | Medium | High |
|---|---|---|---|---|---|
| External risk | Import dependency | | ≤5% | 40–65% | ≥80% |
| | Political stability of suppliers | | <2.5 | ≥2.9 | |
| Internal risk | Volatility of production | | <20% | >20% | |
| | Share of offshore production | | <5% | >90% | |
| External resilience | Diversity of suppliers | | >0.8 | 0.30–0.8 | <0.30 |
| | Import infrastructure (entry points) | Ports | 0–1 | 2 | 3–4 | ≥5 |
| | | Pipelines | 1–2 | 3–4 | 5–8 | ≥9 |
| Internal resilience | Storage levels | | ≤15 | 20–50 | ≥55 |

*Source:* Simplified from Jewell (2011:16).

ios the global energy trade decreases in comparison to the present situation and the diversity of fuels used in the most vulnerable transport sector increases. At the same time in the business-as-usual scenarios the levels of global energy trade significantly rise and the diversity of transport fuels rises much slower. This leads to a conclusion that most low-carbon energy transition scenarios are beneficial to energy security at the global level.

In many cases, however, direct interpretation of individual indicators is not sufficient. Policy makers often need to see an integrated picture of energy security as reflected in several indicators. However, the more indicators that come into the picture the more difficult it is to make sense of them, especially if each tells a different story. Thus, the second strategy is aggregating indicators into energy security "indices" using one of the many methods proposed in the academic literature (Gupta, 2008; Scheepers et al., 2007). The rationale for such indices is that they can reduce the amount of information and thus make the results of an assessment more understandable.

However, policy-maker's enthusiasm for compound indices has been varied. The problem is not that they have an aversion to aggregation as such: in fact even the most simple, straightforward and much used energy security indicators are already to some extent aggregated. For example, the most widely used indicator of import dependence aggregates imports at different periods of time (usually across a year) from different suppliers, at different prices, by different routes and for different purposes. There is even more aggregation involved when import dependence is calculated not for an individual fuel or a carrier (such as LNG or gasoline) but for "oil products", "fossil fuels" or total "energy".

In systematic energy security assessments energy security indicators should be aggregated at the level of vital energy systems and their vulnerabilities (e.g. Le Coq and Paltseva (2009) aggregate vulnerabilities of individual fuels). If the initial identification of systems and vulnerabilities correctly accounts for policy perspectives, policy makers are comfortable with such aggregation, because it corresponds to their familiar boundaries of energy systems and their ideas of vulnerabilities. If, on the other hand, the methods of aggregation (or calculation of complex indicators in the first place) produce a disconnect between the intuitively familiar systems and vulnerabilities and the numbers resulting from the assessment, policy-makers are likely to feel much less comfortable. In this latter case, the aggregated metrics designed to make the results more understandable achieve exactly the opposite: they complicate and obscure the message of the assessment.

Thus, any aggregation must strike a very delicate balance between on the one hand reducing the amount of data and on the other hand staying true to the systems and vulnerabilities which were identified as important at earlier stages. In line with the energy systems approach, the aggregation of indicators should to the extent possible correspond to how energy systems function. Aggregation makes more sense when the indicators relate to the same vital energy systems and/or to vulnerabilities which can potentially interact. For example, it may take into account how particular risks may exacerbate one another and how particular resilience capacities may mitigate specific risks. Such aggregation preserves the focus of the assessment on key energy systems and their vulnerabilities and thus facilitates achieving the purpose of the assessment. In contrast, aggregating indicators which relate to different and disconnected energy systems or to vulnerabilities which reflect different perspectives on energy security or different types of risks and resilience capacities is usually counterproductive.

The first step of aggregation is closely connected to interpretation of individual indicators that we discussed above. As a result of such interpretation, indicators may be normalized or related to a non-dimensional scale (e.g. ranking) making them comparable. Once indicators are normalized, the methods of aggregation can be based on simple semi-quantitative matrices as shown in Table 8.6 illustrating semi-quantitative aggregation of two external resilience indicators for crude oil in MOSES. The aggregation in MOSES proceeds through several similar stages until arriving at the final results (illustrated in Table 8.7). MOSES does not aggregate results across fuels and carriers because energy officials guiding this process perceived that important information might be lost as a result of such aggregation.

The energy systems approach used in MOSES and GEA allows aggregating not only vulnerabilities related to one and the same energy system, but also indicates the proliferation of vulnerabilities from one energy

Table 8.6  *Aggregating indicators for external resilience of crude oil supply in MOSES*

|  |  | Import infrastructure |  |  |
|---|---|---|---|---|
|  |  | Low | Medium | High |
| Diversity of suppliers | Low | Slovakia | Finland |  |
|  | Medium | Ireland | Sweden |  |
|  | High | Austria | Turkey | Japan |

*Note:* As a result of combining these two indicators, the countries are divided into four groups indicated by different shades, the lighter shades indicating more resilience. One country is listed for every group as an example.

*Source:* simplified from Jewell (2011:17).

system to another. For example, MOSES accounts for the aggregate security of crude oil supply in calculating the vulnerability of oil products. The GEA takes into account for concerns associated with individual primary energy sources in calculating vulnerability of electricity systems and end-uses that rely on those sources.

In the quest for an "objective" evaluation of energy security, many studies use mathematical operations to aggregate indicators into a combined index. Scheepers et al. (2007) use relatively arbitrary (but transparently defined and explained) weights to aggregate indicators throughout the energy system into the "S/D index" for EU countries (Scheepers et al. 2007, 31). Gupta (2008) analyzes oil security by using principal component analysis to remove correlation between indicators to avoid double-counting vulnerabilities.

Aiming for a strictly objective evaluation of energy security is futile. All methods for interpreting and aggregating indicators require some form of human judgment, implicit or explicit, on the relative importance of energy systems or their vulnerabilities. For example, in MOSES expert judgments are used to determine the "safe" levels of risks or "adequate" resilience capacities (see Table 8.5). Some of the aggregation methods solicit such judgment in a more formal and sophisticated way. Badea et al. (2011) use the idea of risk aversion to prioritize energy security concerns in cases a country ends up at the bottom of the list with respect to a particular indicator.

Though complex manipulations of indicators can be very thoughtful and elegant they always involve a lot of assumptions and a risk that they might conceal rather than highlight truly important information. Therefore if the main reason for aggregating indicators is to reduce their number, two alternative approaches may be tried. Firstly, it is important

*Energy security assessment framework* 165

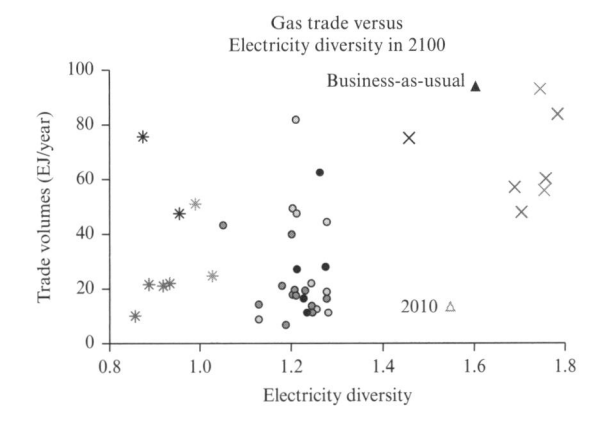

Gas trade versus
Electricity diversity in 2100

*Note:* Different shapes and shades of the data points represent different scenarios. Scenarios represented by lightly shaded crosses imply high energy efficiency and constraints on renewable energy penetration.

*Source:* Adopted from Jewell et al. (2012).

*Figure 8.3  Aggregate analysis of energy security in future energy scenarios using a two-dimensional plotting*

to ask whether all of the indicators are necessary in the first place. Do they all tell meaningful stories? Perhaps some of them looked promising at the stage of selecting indicators but turned out to not be sufficiently reliable or differentiating. Perhaps the focus of the assessment was initially defined too widely and it is necessary to exclude some systems or vulnerabilities for the purposes of communication.

Secondly, it may be possible to present disaggregated indicators in such a way that they are more understandable without aggregating them. For example, instead of combining two independent indicators they can be presented on a two-dimensional scatterplot as shown in Figure 8.3, giving an example of analysis of future energy scenarios. The analysis does not combine two unrelated indicators of electricity diversity and the gas trade into a single index but instead presents the two most prominent vulnerabilities identified in the assessment in a two-dimensional plot. It clearly shows that low trade and high diversity (the optimal conditions for energy security) are only possible in certain scenarios.

There are other techniques for visualizing multiple numerical data which can be successfully used in communicating assessment results. Since energy security is very much about context and perceptions it is useful to consider methods of communication which have a clear qualitative aspect such as narratives or visuals. For example MOSES summarizes

*Table 8.7    Results of the crude oil analysis for MOSES*

| Group | Countries that: | No. of countries |
|---|---|---|
| A | Export crude oil or import ≤15% of their crude oil consumption. | 5 |
| B | Import 40–65% of their crude oil consumption or Import ≥80% of their crude oil consumption and have<br>• ≥5 crude oil ports, high supplier diversity and ≥55 days of crude oil storage. | 4 |
| C | Import ≥80% of their crude oil consumption and have:<br>• ≥5 crude oil ports, high supplier diversity, and <50 days of crude oil storage or<br>• 2–4 crude oil ports, high supplier diversity and >20 days of crude oil storage. | 9 |
| D | Import ≥80% of their crude oil consumption and have:<br>• 2–4 crude oil ports, high supplier diversity, and ≤15 days of crude oil storage or<br>• 2 crude oil ports or 3 crude oil pipelines, low supplier diversity, and ≥15 days crude oil storage or<br>• 1–2 crude oil pipelines or 1 crude oil port and have either:<br>  ○ medium to high supplier diversity and ≥15 days of crude oil storage or<br>  ○ low supplier diversity and ≥55 days of crude oil storage. | 6 |
| E | Import ≥80% of their crude oil consumption and have:<br>• 1–3 crude oil pipelines or 1 crude oil port and ≤15 days of crude oil storage or<br>• 1–2 crude oil pipelines, low supplier diversity and <50 days of crude oil storage. | 3 |

*Source:*   Jewell (2011:18).

its results in terms of "profiles" of energy security of individual countries which together form a "landscape" of energy security in the IEA Member Countries. The terms "profile" and "landscape" convey clear qualitative images. The results of MOSES convey holistic stories about countries (divided into groups according to their vulnerability profiles) as shown in Table 7 for the case of crude oil.

The GEA messages are also expressed in a narrative and qualitative form. Thus GEA summarizes one of its main messages as follows (note how quantitative indicators and depiction of energy systems which span end-uses and primary energy sources are woven into the narratives):

Oil is at the center of contemporary energy-security concerns for most nations, regions, and communities. Oil products provide over 90% of transport energy in almost all countries. Thus, disruptions of oil supplies may have catastrophic effects, not only on personal mobility, but also on food production and distribution, medical care, national security, manufacturing, and other vital functions of modern societies. At the same time, conventional oil resources are increasingly concentrated in just a few regions. The concerns over political stability affecting resource extraction and transport add to uncertainty. Moreover, the global production capacity of conventional oil is widely perceived as limited. Furthermore, the demand for transport fuels is steadily rising, especially rapidly in emerging Asian economies. Thus, for most countries, an ever higher share of their oil, or even all of it, must be imported. More than three billion people live in countries that import more than 75% of the oil and petroleum products they use. An additional 1.7 billion people live in countries with limited domestic oil resources (including China) which are likely to experience similarly high levels of import dependence in the coming decades.

In summary, interpretation of indicators can use the following approaches:

☑ Individual indicators may be interpreted by comparing them across the systems (or points in time) covered by the assessment or with meaningful reference values;
☑ Several indicators may be aggregated into a compound index. Such aggregation makes sense if it:
- Combines indicators related to systems or vulnerabilities that potentially interact with or affect each other;
- Uses techniques which reflect such interaction;
- Does not obscure or conceal important choices and trade-offs that are meant to be highlighted by the assessment.

☑ Other methods for making sense of a large number of indicators and data points include various visual techniques and qualitative narratives;
☑ Subjective judgments are an inevitable part of interpreting indicators and should be made in a transparent way consistent with the overall purpose of the assessment.

## CONCLUSIONS

This final section recaps the main messages of the chapter and outlines the agenda for further development and application of the energy security assessment framework. In contrast to the mainstream tradition the framework does not place indicators at the center of measuring energy security. Instead it focuses on how to make transparent and informed choices at five distinct stages of an energy security assessment as schematically shown at Figure 8.4.

# Generic vs. Specific

## Energy systems approach

Operational definition → Delineating vital energy systems → Identifying vulnerabilities → Selecting indicators → Interpretation and communication

## Energy systems approach

# Facts vs. Perceptions

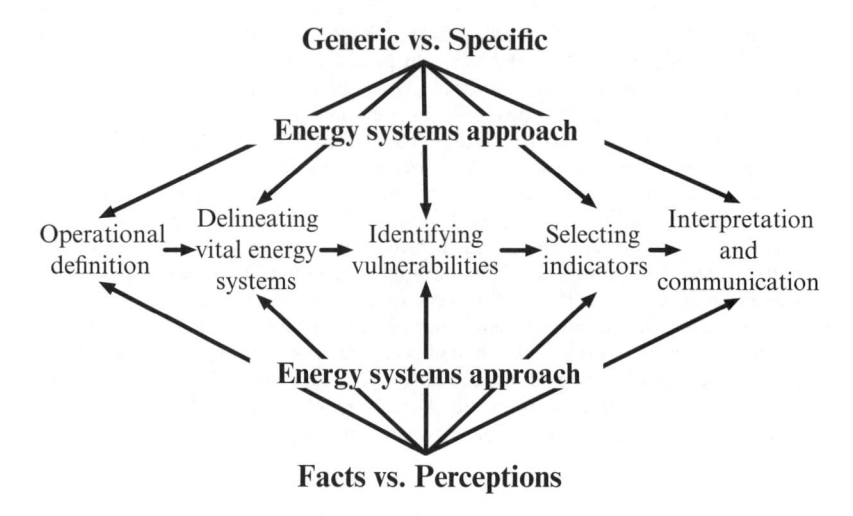

*Figure 8.4  Energy security assessment framework*

The first set of choices reflects the idea that energy security is as much about perceptions as it is about the hard realities of energy systems. The second set of choices reflects the fact that energy security is a highly contextualized characteristic of energy systems which nevertheless should be rendered generic for the purpose of comparison. A good assessment strikes the right balance between these major choices at each of its five stages:

- At the first stage, it selects a definition of energy security acceptable to the audience of the assessment and sufficiently operational with respect to all energy systems analyzed.
- At the second stage it delineates vital energy systems, in a manner that is meaningful and consistent for all points of comparison, with reference to both policy concerns and the realities of energy flows.
- At the third stage, it identifies the vulnerabilities of these vital energy systems. Existing policy concerns are a good starting point, however, human perception of risks can be severely biased towards higher-profile, particularly dreaded events, especially resulting from actions of hated adversaries rather than "Acts of God". This bias may need to be adjusted by an objective analysis.
- At the fourth stage, it selects energy security indicators that reflect (but not necessarily measure!) the identified vulnerabilities. It is usually easier to start with the metrics already used in policy-making because (a) they will be more familiar and easier to interpret

and communicate; and (b) there will usually be data available for such metrics. More complicated calculation and data mining may be required to obtain indicators for vulnerabilities which for one reason or another are not on the political agenda. Such complex indicators will also require more efforts to interpret and communicate. While indicators should be relevant for a particular situation, they should also be comparable across all situations covered by the assessment.
- At the fifth and final stage, the indicators are interpreted and presented in a form that facilitates answering the original questions posed by the assessment. This may require aggregating indicators quantitatively into compound indices or qualitatively into narratives. Perspectives of the audiences of the assessment need to be taken into account in this process. They should, however, not distort the rigor of the assessment or obscure its main messages.

As indicated in Figure 8.4, the energy systems approach helps making informed choices at each stage of the assessment. It means that at every stage analysts should work not with a black box of amorphous "energy" but with actual energy systems. Vital energy systems should be delineated based on an understanding of energy flows and their significance for societies. Vulnerabilities should be identified based on how energy systems might respond to disruptions. Indicators are selected based on their abilities to serve as proxies for such complex system behavior. Finally, indicators should be interpreted, processed and presented to reflect the way actual energy systems function.

This chapter illustrates the application of the proposed framework in three case studies summarized in Table 8.8. Despite the fact that all the studies make different choices about the definition of energy security, vital energy systems, key vulnerabilities, indicators and approach to their interpretation they all systematically move through the five stages and apply the principles of the energy systems approach.

There are several ways in which the proposed framework can be further developed and used. This research agenda can also be structured in line with the key stages of the assessment, as follows. There should be better understanding of different types of vital energy systems; for example, more research is needed to understand the vulnerabilities of nuclear energy, renewable energy sources and traditional biomass. There should be better methods to explore vulnerabilities of vital energy systems, for example through researching their reaction to possible disturbances in dedicated modelling exercises. Based on this understanding new indicators of energy security may be developed, e.g. based on indicators used to characterize

Table 8.8 Cases of energy security assessments presented in this chapter

| Energy security assessment | IEA model of short-term energy security (MOSES) | Global Energy Assessment (*Energy and security chapter*) | Future energy security |
|---|---|---|---|
| *Scope and purpose* | Develop energy security profiles of 28 IEA countries | Identify the most prominent energy security challenges at present and in the near future affecting the world as a whole (134 countries) | Examine energy security implications of long-term energy transition scenarios |
| *Vital energy systems* | | | |
| Geographic and political boundaries | National | Global and national, qualitative regional discussions | Global and regional |
| Sectoral boundaries | Seven primary energy sources and secondary fuels | Primary energy sources, electricity, key end-uses | Globally traded fuels and carriers, electricity, selected end-uses |
| *Vulnerabilities* | Short-term physical disruptions | Stresses and shocks both physical and economic | Primarily physical shocks and stresses, sovereignty risks and resilience factors |
| *Indicators* | 35 indicators | 34 indicators | 20 global and five regional indicators |
| *Interpretation and aggregation method* | Comparative ranking between IEA countries, then semi-quantitative aggregation to characterize their "energy security profiles" and the IEA "energy security landscape" | Narratives to describe the most significant vulnerabilities | Narratives to describe the dynamics of future energy security |

resilience of ecological systems and social networks. In order to make sense of the new and existing indicators, large and consistent data sets will need to be created spanning a range of energy systems and time points for monitoring and comparison. Energy security assessments should go hand in hand with developing a toolkit for energy security policy analysis, in which policies and vulnerabilities of energy systems are understood as interacting and co-evolving.

## NOTES

1. This is in line with the classic definition of the objective of energy security by Daniel Yergin (1988:112): 'The objective of energy security is to assure adequate, reliable supplies of energy at reasonable prices and in ways that *do not jeopardize major national values and objectives.*' (emphasis added)
2. It is also not uncommon, especially for larger countries, to address energy security of sub-national regions (e.g. regional electricity grids in Sweden or the US or regional gas markets in Australia).
3. In the case of some regions this approach is a good proxy of assessing national energy security. This concerns highly integrated and homogenous regions (e.g. the European Union) and those that are dominated by a single major country (e.g. North America by the US, South Asia by India and Centrally Planned Asia by China). In other cases such as Africa, Latin America and the former Soviet Union the results of the assessment using this method are likely to be very different from an assessment from national perspectives which unfortunately cannot be conducted when dealing with long-term radical energy transformation scenarios.
4. We have already mentioned the importance of crude oil for the British Navy in World War I. The importance of oil products for the USSR during the World War II is vividly described by (Matvejchuk, 2012).

## REFERENCES

Alhajji, A. F. (2008). "What Is Energy Security? Economic, Environmental, Social, Foreign Policy, Technical and Security Dimensions", *Oil, Gas & Energy Law Intelligence*, 6(3).

APERC (2007). *A quest for energy security in the 21st century: Resources and Constraints*, Japan: Institute of Energy Economics, pp. 1–113.

Awerbuch, S. (2006). "Portfolio-Based Electricity Generation Planning: Policy Implications For Renewables And Energy Security", *Mitigation and Adaptation Strategies for Global Change*, 11(3), 693–710, doi:10.1007/s11027-006-4754-4.

Badea, A., C. M. Rocco, S. Tarantola and R. Bolado (2011). "Composite indicators for security of energy supply using ordered weighted averaging", *Reliability Engineering and System Safety*, 96(6), 651–662, doi:10.1016/j.ress.2010.12.025.

Baran, Z. (2007). "EU Energy Security: Time to End Russian Leverage", *The Washington Quarterly*, 30(4), 131–144.

Bridge, G., M. Bradshaw, S. Bouzarovski, J. Dutton and G. C. K. Leung (2012). "Globalising Gas", Working Paper No. 1, The Geopolitical Economy of Global Gas Security and Governance: Implications for the UK, Manchester, UK.

Carter, J. (1980). The State of the Union Address, *The American Presidency Project*, Delivered Before a Joint Session of the Congress.

Cherp, A. (2012). "Defining energy security takes more than asking around", *Energy Policy*, 48, 841–842, doi:10.1016/j.enpol.2012.02.016.

Cherp, A. and J. Jewell (2011a). "Measuring energy security: from universal indicators to contextualized frameworks", in B. K. Sovacool (ed.), *The Routledge Handbook of Energy Security*, Oxon, UK and New York: Routledge, pp. 330–355.

Cherp, A. and J. Jewell (2011b). "The three perspectives on energy security: intellectual history, disciplinary roots and the potential for integration", *Current Opinion in Environmental Sustainability*, 3(4), 202–212, doi:10.1016/j.cosust.2011.07.001.

Cherp, A., A. Adenikinju, A. Goldthau, L. Hughes, J. C. Jansen, J. Jewell, M. Olshanskaya et al. (2012). "Energy and Security", *Global Energy Assessment: Toward a Sustainable Future, Global Energy Assessment*, Cambridge, UK; and New York, USA: Cambridge University Press, and the International Institute for Applied Systems Analysis, Laxenburg, Austria, pp. 325–384.

Cherp, A., J. Jewell, V. Vinichenko, N. Bauer and E. DeCian (2013). "Energy security in scenarios modelled by REMIND and WITCH", *Climatic Change*.

Chester, L. (2009). "Conceptualising energy security and making explicit its polysemic nature", *Energy Policy*, 38(2), 887–895, doi:10.1016/j.enpol.2009.10.039.

Costantini, V., F. Gracceva, A. Markandya and G. Vicini (2007). "Security of energy supply: Comparing scenarios from a European perspective", *Energy Policy*, 35(1), 210–226, doi:10.1016/j.enpol.2005.11.002.

Department of Resources, Energy and Tourism of Australia (2011). "National Energy Security Assessment 2011", Government of Australia. Canberra.

Greene, D. L. (2010). "Measuring Energy Security: Can the United States Achieve Oil Independence?" Energy Policy 38 (January 27): 1614–1621. doi: 10.1016/j.enpol.2009.01.041.

Grubb, M., L. Butler and P. Twomey, P. (2006). "Diversity and security in UK electricity generation: The influence of low-carbon objectives", *Energy Policy*, 34(18), 4050–4062, doi:10.1016/j.enpol.2005.09.004.

Gupta, E. (2008). "Oil vulnerability index of oil-importing countries", *Energy Policy*, 36(3), 1195–1211, doi:10.1016/j.enpol.2007.11.011.

Helm, D. (2002). "Energy policy: security of supply, sustainability and competition", *Energy Policy*, 30, 173–184.

Jansen, J. C. and A. J. Seebregts (2009). "Long-term energy services security: What is it and how can it be measured and valued?", *Energy Policy*, 38(4), 1654–1664, doi:10.1016/j.enpol.2009.02.047.

Jewell, J. (2010). "Sustainable Energy Pathways: Are they secure?" Young Scientists Summer Program Final Report. Laxenburg, Austria: International Institute for Applied Systems Analysis (IIASA).

Jewell, J. (2011). *The IEA Model of Short-term Energy Security (MOSES)*, Paris: OECD/IEA.

Jewell, J., A. Cherp and K. Riahi (2012). "Energy security indicators for use in Integrated Assessment Models. LIMITS deliverable 4.1. Accessed at: http://www.feem-project.net/limits/docs/limits_d4-1_iiasa.pdf.

GEA (2012). Global Energy Assessment – Toward a Sustainable Future, Cambridge University Press, Cambridge, UK and New York, NY, USA and the International Institute for Applied Systems Analysis, Laxenburg, Austria.

Kendell, J. M. (1998). *Measures of oil import dependence*, US, EIA, Washington, D.C. http://www.eia.doe.gov/oiaf/archive/issues98/oimport.html.

Keppler, J. H. (2007). "Energy supply security and nuclear energy: concepts, indicators, policies", Background Study in the Context of the Nuclear Development Committee's Security of Supply Project. OECD, Paris.

Klare, M. T. (2008). *Rising powers, shrinking planet: the new geopolitics of energy* (illustrated), New York: Metropolitan Books.

Le Coq, C. and E. Paltseva (2009). "Measuring the security of external energy supply in the European Union", *Energy Policy*, 37(11), 4474–4481, doi:10.1016/j.enpol.2009.05.069.

Matvejchuk, A. (2012). "The 'burning' problem of Soviet Aviation", *Oil in Russia*, 5, 120–125.
Paust, J. J. and A. P. Blaustein (2008). "Arab Oil Weapon – A Threat to International Peace", *American Journal of International Law*, 68, 1–31.
Riahi, K., F. Dentener, D. Gielen, A. Grubler, J. Jewell, Z. Klimont, V. Krey et al. (2012). "Energy pathways for sustainable development", in *Global Energy Assessment: Toward a Sustainable Future, Global Energy Assessment: Towards a More Sustainable Future*, Cambridge, UK; and New York, USA: Cambridge University Press, and the International Institute for Applied Systems Analysis, Laxenburg, Austria, pp. 1203–1306.
Scheepers, M., A. J. Seebregts, J. de Jong and H. Maters (2007). *EU Standards for Energy Security of Supply*, Petten, Netherlands: ECN, Energy Research Center of the Netherlands.
Sovacool, B. K. (2011). "Introduction: Defining, measuring, and exploring energy security", in B. K. Sovacool (ed.), *The Routledge Handbook of Energy Security*, Oxon, UK and New York: Routledge, pp. 1–42.
Sovacool, B. K. and I. Mukherjee (2011). "Conceptualizing and measuring energy security: A synthesized approach", *Energy*, 36(8), 5343–5355, doi:10.1016/j.energy.2011.06.043.
Stirling, A. (1994). "Diversity and ignorance in electricity supply investment: Addressing the solution rather than the problem", *Energy Policy*, 22(3), 195–216, doi:10.1016/0301-4215(94)90159-7.
Stirling, A. (1998). "On the economics and analysis of diversity", Science Policy Research Unit (SPRU), Electronic Working Papers Series, Paper, 28, 1–156.
Stirling, A. (2013). "From Sustainability, through Diversity to Transformation: towards more reflexive governance of technological vulnerability", in A. Hommels, J. Mesman and W. Bijker (eds), *Vulnerability in Technological Cultures: New Directions in Research and Governance*, Cambridge MA: MIT Press, in press.
Turton, H. and L. Barreto (2006). "Long-term security of energy supply and climate change", *Energy Policy*, 34(15), 2232–2250.
Wicks, M. (2009). *Energy Security: A national challenge in a changing world*. London: Malcolm Wicks MP
Winzer, C. (2012). "Conceptualizing energy security", *Energy Policy*, 46(C), 36–48, doi:10.1016/j.enpol.2012.02.067.
Yergin, D. (1988). "Energy Security in the 1990s", *Foreign Affairs*, 67(1), 110–132.
Yergin, D. (1991). *The Prize: The Epic Quest for Oil, Money, and Power*, New York: Simon & Schuster.
Yergin, D. (2006). "Ensuring energy security", *Foreign Affairs*, 85(2), 69–82.
Zhang, Z. (2012). "The overseas acquisitions and equity oil shares of Chinese national oil companies A threat to the West but a boost to China's energy security?", *Energy Policy*, 48(C), 1–4, doi:10.1016/j.enpol.2012.05.077.

# 9. National energy strategies of major industrialized countries
*Stephan Schott and Graham Campbell*

## 1 INTRODUCTION

Energy security in major industrialized countries has traditionally been concerned with securing the supply of particular energy commodities at reasonable cost. For many countries, this has meant a secure supply of fossil fuels in order to support a manufacturing base, to enable increasing living standards based on large centralized electricity grids, and to support fossil fuel based transportation systems. Energy strategies, therefore, typically consisted of ensuring availability of energy commodities and avoiding disruption to energy supplies. This situation has drastically changed in the last two decades as a consequence of the integration of the former Eastern bloc countries into Europe, the ever more visible threat of climate change, new emerging global energy suppliers like Norway, Australia, Canada, China and Russia, and new rapidly growing large scale energy consumers like China and India, which greatly expand opportunities for sale of primary energy commodities to meet their rapidly growing demand.

More recent energy strategies are more concerned about securing the supply of energy services, rather than energy commodities per se, moving towards a more diverse suite of energy generation sources, such as nuclear energy and renewable energy technologies. There is a growing trend in weaving energy strategies with either economic development strategies (as in the case of Russia and Norway), economic, regional development and industrial strategies (as in the case of Canada, Australia and Norway), or climate change policies (as in the case of the European Union). Certain industrialized countries illustrate this broadened interest in energy security. For example, the United States is trying to rely more on domestic or secure energy supplies and sees energy security as an employment, innovation and economic rejuvenation strategy. Exporting countries, such as Norway, Russia, Canada and Australia, recognize the importance of energy exports to their long-term economic development prospects.

Our approach to examining developments in energy strategies is to select a suite of industrialized countries, which are characterized on the basis of their import/export dependence in primary energy commodities.

Countries can be characterized under this approach as being: net energy exporters, concerned primarily about the security of demand for their commodities (such as Russia, Norway, Australia and Canada); countries that are both importers, say of crude oil, but are also emerging as major exporters, in part due to recent technological advancements (such as Canada as a result of advancements in oil sands technologies, potentially the EU with a more integrated electricity grid and the United States due to the recent progress in shale gas extraction); and countries which are almost totally reliant on limited domestic production and a major share from imports for their energy supplies (as is the case in Japan). Countries can change from one category to another suddenly. For example, a natural gas revolution (Voser 2012) arising from the exploitation of gas resources in shale formations, has changed the global supply-side picture dramatically, facilitating the potential of establishing a global natural gas market featuring new players.

Many international institutions such as the IEA, the G8, the G20 and many others are addressing energy security issues. But recent failures in the world financial systems and the Euro crisis question the ability of world institutions to coordinate actions and to successfully implement common objectives. Furthermore existing institutions need to be overhauled and strengthened to secure energy flows between net exporters and importers (Goldthau et al. 2010). As a result there have been calls for a new institution, an Energy Stability Board similar to the Financial Stability Board (Victor and Yueh 2010).

On a more limited scale, the G8 countries declared seven principles of energy security in St. Petersburg in 2006, and annually measure progress for each G8 country. The principles are aimed at increasing transparency, predictability and stability of global energy markets, improving the investment climate in the energy sector, enhancing energy efficiency and energy saving, diversifying energy mix, ensuring physical security of critical energy infrastructure, reducing energy poverty and at addressing climate change and sustainable development. In the latest G8 Summit at Camp David (The White House 2012a) the countries agreed to pursue a comprehensive energy strategy that focuses on safe, efficient and environmentally sustainable development of a mix of energy sources. The action plan also encourages the facilitation of access to energy resources in order to enhance energy security, decrease price volatility and to provide for a favorable investment climate in the energy sector. The G8 also promote the sustainable deployment of renewables and addressing climate change by reducing short-term pollutants.

Each of the selected countries, with the exception of Canada, has developed a long-term national energy strategy with their own particular focus and characterization of energy security. In this chapter, we examine the structure of national energy strategies in the G8 countries and Norway, and evaluate and contrast different components of individual strategies. We chose Norway in addition to the G8 countries because of its unique strategic position in Europe and the Arctic and its progressive views on many environmental and energy related polices. We hope to contribute to the better understanding of the importance and meaning of national energy strategies, differences in the definition of and approach to energy security, and the compatibility of strategies with the emerging world energy order. We start by assessing current energy mixes and flows (imports and exports) of these major industrialized countries. We then discuss stated energy strategies in all these countries and regions and contrast them with respect to a number of identified criteria. We examine future desired energy profiles and the role of trade links and markets, and the role and definition of energy security in each case. Finally we analyze the compatibility of energy strategies and the potential for national, regional and global energy security.

## 2 CURRENT ENERGY MIX AND IMPORT DEPENDENCE

The countries referenced in this chapter are among the leading global industrialized nations. Each country's perspective on the issue of energy security is influenced by the national situation for supply and demand for the principal energy commodities. We examine the picture for energy security from the standpoint of countries characterized by security concerns dominated by supply of energy commodities, and those concerned about the security of demand for their energy commodities from foreign markets (see Table 9.1).

Countries with a rich resource base are more likely to be economically dependent on the sale of their energy commodities into foreign markets; exceptions are countries which have large unsatisfied domestic energy demand arising from a large population with increasing expectations for energy services, such as China and India for example. The export-dependent countries typically orient their energy security strategies to developing indigenous resources to maintain export revenues. For example, Norway exports over 89 per cent of its oil and gas production to markets in Europe. Other countries have a more balanced picture between exports and imports, and their energy security strategies will be concerned with both domestic supply issues and foreign demand issues. While the economic development and growth of these countries is dependent on

*Table 9.1  Export-import dependence of selected countries*

| Country | Coal<br>Exp/ (prod + Imps) | Crude Oil<br>Exp/ (prod + Imps) | Natural gas<br>Exp/ (prod + Imps) |
|---|---|---|---|
| **High Export Dependence** | | | |
| Norway | 71% | 80% | 94% |
| **Balanced Export-Import Dependence** | | | |
| RU | 41% | 51% | 28% |
| CA | 42% | 51% | 52% |
| AU | 76% | 34% | 39% |
| **High Import Dependence** | | | |
| EU | 7% | 8% | 14% |
| US | 6% | 1% | 4% |
| JA | 1% | 0% | 0% |

*Source:* International Energy Agency (2009).

sales of energy commodities into foreign markets, these countries also face issues around domestic security of supply. Russia, Canada and Australia are in this situation.

Countries, which rely on domestic production and imports to support their economic development and growth, are primarily concerned with energy supply issues, lacking a significant export trade in energy commodities. The high level of dependence on imports is expected to result in a dominant focus on security of supply issues in the formulation of their energy strategies. We will evaluate national energy strategies in more detail next and will assess the driving factors of energy security issues.

## 3  FEATURES OF NATIONAL ENERGY STRATEGIES: A COMPARATIVE APPROACH

### 3.1  Australia

A balanced, commodity-dependent import/export picture characterizes Australia (Table 9.1). Australia enjoys a significant resource endowment in important energy commodities: oil, gas, and coal. Coal resources, located primarily in the southeast quadrant, are a major export commodity, that contribute a significant share of Australia's export revenue (31 per cent of total national export revenue). Natural gas markets are well established in the southeast, southwest, and northwest quadrants; LNG exports

178  *International handbook of energy security*

*Table 9.2  Characteristics of national energy strategies in countries with high export and balanced export-import dependence*

| Country/Region | Energy Strategy (Name and Date) | Focus of Energy Strategy | Components of Energy Strategy | Definition of Energy Security |
|---|---|---|---|---|
| Australia | *Draft Energy White Paper: Strengthening the Foundations of Australia's Energy Future National Energy Security Assessment 2011* | To build a secure, resilient and efficient energy system that: (1) provides accessible, reliable and competitively priced energy for all Australians (2) enhances Australia's domestic and export growth potential (3) Delivers clean and sustainable energy | Priority areas: (1) strengthening the resilience of Australia's energy policy framework (2) reinvigorating the energy market reform agenda (markets and energy productivity) (3) developing Australia's critical energy resources – particularly Australia's gas resources (4) accelerating clean energy outcomes. | Adequate, reliable and competitive supply of energy to support the nation's economic and social development |
| Norway | For petroleum, an industry for the future – Norway's petroleum activities (Report: white paper for the Norwegian Parliament) | For petroleum (includes oil and gas), to ensure the profitable production of oil and gas in a long-term perspective, to improve the quality of life in Norway for years to come. For electricity, increase ability to manage | For petroleum, (1) improved recovery from existing fields, (2) development of discoveries not yet on production, (3) proving up undiscovered resources, in areas open for exploration now, and in areas to be opened in the near future | Open and transparent world markets with clear price signals (Norwegian Minister of Petroleum and Energy, Ola Borten Moe in a speech in the U.S. in March of 2012) |

National energy strategies of major industrialized countries 179

| Projected Energy consumption mix | Institutions Governing Energy Coordination/ Strategy | Flexibility/ Adaptability | Reliance on unconventional energy sources | Reliance on int. cooperation/ markets |
|---|---|---|---|---|
| Total energy produced: 18,000 PJs in 2011 to 43,000 PJs in 2035 Coal: 10,500 PJs in 2011 to 18,700 PJs in 2035 (93% exported) Natural Gas: 2400 PJs in 2011 to 8,300 PJs in 2035 (67% exported) Uranium: 3,200 PJs in 2011 to 14,500 PJs in 2035 (100% exported) (Department of Resources, Industry and Tourism (2011)) | For Commonwealth government: Department of Resources, Energy and Tourism, Energy Security Council (to be created) – For States: Territories: their ministries, own resources, responsible for production, transport, land-use, mineral rights, and environ-mental assessments. | Periodic situation and policy reviews, and assessments. Sound awareness of global and domestic developments and trends. Largely self-sufficient in energy resources, other than crude oil and products. Expectations for large growth in AU's exports of coal, uranium and natural gas. | Recent development of coal seam gas in Victoria, offshore gas production well-established (since 1989) on NW shelf. | For security of supply, relies entirely on international oil market, with concern re implications of future political instability and future price increases. For security of foreign markets, relies on Asia for purchases of coal, Uranium and LNG, expecting very large growth in sales between now and 2034–2035 (coal to double, natural gas by five times) |
| Oil: 2.1 mbopd in 2009 to 1.2 mpopd in 2030 (90% exported) Gas: 130 Bcmpd in 2009 to 130 bcmpd in 2030 (95% exported) Ministry of Petroleum and Energy (2011) | Ministry of Petroleum and Energy | abundant potential oil and gas resources underlying the northern shelf and in the far Northern offshore. Also, revenues from development of offshore natural gas resources will offset | Unconventional resources are not being pursued in Norway. The focus is on exploration and develop-ment of conventional offshore | Norway derives the bulk of its national revenue from sale of oil and gas production to western Europe and the UK. This source of revenue is seen as the key economic factor in Norway's future. In this sense, "energy security" in |

*Table 9.2* (continued)

| Country/Region | Energy Strategy (Name and Date) | Focus of Energy Strategy | Components of Energy Strategy | Definition of Energy Security |
|---|---|---|---|---|
| Norway | | shortfalls in supply from hydropower in times of low reservoir capacity. Strong commitment to energy efficiency. | For electricity, (1) increase share from other renewables (primarily wind) (2) improve reliability and resilience of electricity system | |
| Russia | Energy Strategy for the Period up to 2030 (Ministry of Energy 2010) | Creation of innovative and efficient energy sector, meeting foreign economic interests, socio-economic development objectives | (1) Energy security (2) Energy efficiency (3) Budget efficiency of the energy sector (4) Environmental safety of the energy sector | National security, ability to withstand threats to reliable supply of fuel and energy and threats caused by external factors (geopolitical, macro-economic and market) |
| Canada | n/a | n/a | n/a | Global energy security with a free choice of energy mix (Foreign Affairs and International Trade, Canada) |

| Projected Energy consumption mix | Institutions Governing Energy Coordination/ Strategy | Flexibility/ Adaptability | Reliance on unconventional energy sources | Reliance on int. cooperation/ markets |
|---|---|---|---|---|
| | | the ongoing decline in revenues from oil production. | oil and gas resources. | Norway means secure access to European and UK markets, supported by pipeline infrastructure. For electricity supply, Norway relies on imports from the neighbouring countries via the NordPool electricity market. |
| 2030: 46% Nat. Gas 22% Liquid fuels 18.5% Solid fuels 13% Non-fuels | Ministry of Energy, Institute of Energy Strategy | Built-in adjustment of development paths | Not a strong focus | Stability of foreign demand, security of transit |
| | Federal Minister of Natural Resources, Council of the Federation, National Energy Board, Canadian Nuclear Safety Commission | Limited flexibility due to provinces' authority over energy policy | Oil sands and shale gas exploration in some provinces (moratorium on shale gas in Québec) | Heavy reliance on international markets particularly the U.S. market, strong urge to diversify market access |

make up a growing share of national export revenue, originating from the Northwest shelf in 1989, and more recently from the development of coal seam resources in the southeast quadrant. As domestic oil production dwindles, Australia is becoming progressively more reliant on imports, largely through the Singaporean transportation corridors.

Given the importance of energy commodities to the national economy, Australia has periodically developed an energy strategy, supported by a national assessment of energy security risks associated with each primary energy commodity. Most recently, the draft Energy White Paper has been issued and is entitled "Strengthening the Foundations for Australia's Energy Future" (Department of Resources, Energy and Tourism, Government of Australia 2011), complemented by the most recent National Energy Security Assessment (December 2011). Taken together, these documents provide an objective, comprehensive picture of how Australia is coming to grips with energy security issues. Four priority areas are explored in depth in the draft Energy White Paper: resource development to meet both domestic needs and to benefit from providing primary energy commodities to the export market; the safe and sustainable development of energy resources through community engagement and proper environmental regulation of resource development activities and management of competing pressures associated with diverse, multiple interests; investment across the energy supply chain in new and aging infrastructure; and prudently addressing the increasing energy costs which arise from investments in new energy technologies and infrastructure and which directly affect consumers and industry alike.

The objectives of the energy security framework are to ensure an adequate, reliable and competitively priced supply of energy to support Australia's ongoing economic and social needs. The White Paper lists several factors which have a direct bearing on Australia's energy security: the ability to efficiently and sustainably develop energy resources; domestic and global political conditions which influence the operation of international energy markets and financial markets; the efficiency and robustness of domestic and international energy markets; the degree of integration with international energy markets and supply chains; and the impact of domestic and global energy prices.[1] The policy framework has seven key elements:

1. promoting timely and efficient development of Australia's energy resources and developing robust energy markets that provide reliable and efficient supply-side and demand-side outcomes;
2. ensuring effective and resilient governance and regulatory institutions with effective and transparent market monitoring arrangements;

3. ensuring energy and related resource policies are appropriately and efficiently integrated with climate change and other environmental policy frameworks;
4. developing new generation and end-use technologies to improve performance (including environmental outcomes), diversify the energy system and reduce critical dependencies;
5. providing a stable, attractive and open economy that facilitates international trade and investment;
6. undertaking effective international engagement and collaboration with key trading partners; and
7. international energy organizations.

### 3.1.1 National Energy Security Assessment

The Energy White Paper is supported with the outcomes from the periodic National Energy Security Assessment (NESA), most recently issued in December 2011 (Department of Resources, Energy and Tourism, Government of Australia 2011). With respect to particular energy commodities, Australia has analyzed the overall risks and opportunities associated with crude oil, natural gas, and electricity. Each commodity is given a categorization of the security risks (high security, moderate security, or low security[2]) over three time periods (2012, to 2016, and to 2035). The assessment of Australia's security in liquid fuels indicates *high* security with respect to adequacy and reliability through the medium term of the analysis, and *moderate* security throughout the analysis period due to uncertainties in global oil prices, in part due to pressures from growing international demand for oil and higher production costs associated with new sources of supply. Natural gas security is rated as *moderate* for adequacy and reliability[3] through 2018. Affordability ranges from *moderate* in the short term, due to the current market situation where supply is adequate at moderate market prices, to *low* in 2018 due to anticipated tighter domestic supply conditions and completion between domestic and international market prices as the global market for LNG expands in the near future.

An interesting component of the security assessment is consideration of the role of critical infrastructure and cyber security. As noted in the NESA 2011, "... the rise of more interactive and technology-connected energy systems creates an emerging area of vulnerability through cybersecurity threats."[4] Cyber security threats are rated as *moderate* in keeping with the experience gained through known cyber security incidents and the fact that there is a growing concern for further vulnerabilities to attacks from remote sites since proximity is no longer a requirement to cause interruptions to energy supply or damage to infrastructure.

184　*International handbook of energy security*

The global energy scene is characterized by ever more rapid changes, which could arise as a result of political upheavals, the deployment of new energy technologies which permit access to previously-untapped components of the resource base, or new policies which are developed to address environmental needs or market failures. As a result, energy security policies should be kept under constant review by governments so that updated assessments can be issued frequently. The Government of Australia has committed to undertaking four-year strategic reviews of national energy policy, starting in 2016, and to a biennial national energy security assessment starting in 2014. In addition, and in clear recognition of the divided responsibilities for energy matters in Australia, the Commonwealth government of Australia will undertake a risk preparedness audit of the country's energy sector, in collaboration with the six states and territories, with industry, and with energy market bodies. With respect to the outlook for domestic consumption of energy commodities, Australia foresees steady growth in domestic use of natural gas, matched by gradual decline in use of coal, and significant decline in domestic oil production.

### 3.2　Norway

The energy security situation in Norway differs significantly from the other countries selected for study in this chapter. Norway is one of the world's leading energy suppliers, based on its rich endowment of oil and gas resources, which are the source of current and future petroleum production. Due to its secure and reliable source of petroleum, Norway provides a significant contribution to the security of supply for countries in Western Europe and to the UK.

Fields developed to date are located in basins beneath its quadrant of the North Sea and in the southern offshore areas. Norway's overall energy production is dominated by oil (52 per cent) and gas (41 per cent)[5] (International Energy Agency 2011). Geological assessments have indicated significant undiscovered oil and gas resources beneath the central and northern offshore areas, and farther north adjacent to the islands of Svalbard. These resources position it to be a major oil and gas supplier for decades to come. The bulk of Norway's petroleum production is exported to Western Europe and the UK; 80 per cent of oil production and 94 per cent of natural gas production (see Table 9.1). Globally, Norway is the third largest oil exporter, after Saudi Arabia and Russia.

The revenues from petroleum exports contribute a correspondingly large share to Norway's total revenues: 22 per cent of national value creation[6] in 2009 (Norwegian Petroleum Directorate 2010). The reliance on petroleum export revenues as the basis for Norway's economic future means that "energy security" translates into continuation of long-term oil and gas production, the availability of secure and reliable infrastructure for transportation of production to foreign markets, and open and secure access to foreign markets without import barriers and at international prices.

### 3.2.1 Norway's strategies for maintaining and growing petroleum production

Norway recognizes that production from oil fields is currently declining, and that there is a significant opportunity to increase natural gas production in the future. According to data from the Ministry of Petroleum and Energy, oil production peaked in 2001 at 3.3 million barrels of oil per day (mbopd) and had declined to approximately 2.1 mbopd by 2011; projected production by 2030 is 1.2 mbopd (International Energy Agency 2011). The situation for natural gas is an increasing trend, from approximately 26 billion cubic metres (bcm) in 1985, to 105.9 bcm in 2009; with projections to up to 130 bcm by 2020 depending on the timing of new production projects (International Energy Agency 2011). Norway's strategy to ensure continuation of this favorable, export-oriented situation for petroleum has key elements aimed at sustaining oil production and growing natural gas production over the long term. In brief, the elements are:

1. improved recovery from existing fields based on tapping undeveloped reserves;
2. development of discoveries not yet on production; and,
3. proving up undiscovered resources, in areas open for exploration now, and in areas to be opened in the near future.

### 3.2.2 Electricity: security of supply

In contrast to Norway's strong situation as an energy supplier of oil and gas, the situation is dramatically different for electricity generation. Norway obtains over 96 per cent of its electricity from its hydropower facilities (International Energy Agency 2011), a very favorable situation with respect to clean energy generation. However, the amount of generation is directly dependent on reservoir capacity; fluctuations in rainfall and reservoir inflow result in significant year-to-year variations in annual generation. "Since 2000, hydropower generation has ranged from a low of 106 Twh in 2005 to an all-time high of 140 Twh in 2008." (International Energy Agency 2011) Norway is addressing this electricity security issue through actions by the country's transmission system operator, Statnett, which uses market-based practices and back-up generation to ensure sufficient supply and power quality. The first recourse is to draw on the Nord

Pool spot market, based on the quantities and prices available at that time. Statnett attempts to have access to at least 2000 MW of capacity in the balancing market every hour. If the bids from producers and consumers in the balancing market are insufficient to meet peak power requirements, Statnett can conclude contracts for reserve capacity with producers and power consumers so that a prudent balance can be maintained. Lastly, the TSO also has two mobile gas turbine plants (150 MW capacity) to provide reserve generation capacity (International Energy Agency 2011).

Looking ahead over the long-term, Norway's security strategy with respect to electricity is intended to address the country's vulnerability to fluctuations in reservoir levels, and stresses:

1. an increased share from other renewables (primarily wind);
2. improving reliability and resilience of the electricity system.

### 3.3 Russia

Russia's latest energy strategy is documented in great detail in "Energy Strategy for the Period up to 2030" (Ministry of Energy of the Russian Federation 2010). Its main objectives are the effective use of natural resources, the potential of the energy sector to sustain economic growth, and to strengthen the foreign economic position of the country. Russia aims to create an innovative and efficient energy sector that meets foreign economic interests and socio-economic development objectives. The national energy strategy identifies three implementation phases. The first phase (that is supposed to end between 2013 and 2015) aims to overcome the economic and energy crisis in the country. It is also recognized as a renewal phase and an opportunity to modernize the Russian fuel and energy complex. In the second phase (2016–2020/2022), an array of cutting-edge, highly efficient innovations and technologies are to be introduced: greenfields are to become operational and new capital-intensive energy projects in Eastern Siberia and Far East, on the continental shelf of the Arctic seas and the Yamal peninsula are to be fully operational. The latter projects are supposed to be implemented in the first period. In the final period of 2021/2023–2030, considerably improved energy efficiency coupled with enhanced use of non-fuel energy sources (nuclear, solar, wind, etc.) are expected to gradually transition Russia's energy sector and economy. The state is supposed to have an important role in guiding investments and steering the economy in the right direction (Ministry of Energy of the Russian Federation 2010).

It seems somewhat surprising that energy security is the first of the four main strategic components of the energy strategy. Energy security is defined as being provided and determined by resource sufficiency, as well as economic availability and ecological and technological acceptability of energy sources. Resource sufficiency refers to the deficit-free supply of energy to the Russian population. Economic availability relates to the profitability of energy supply and the determination of appropriate market prices. The Russian energy strategy is concerned about improving energy efficiency and energy conservation and aims to utilize market signals to a larger extent. Russia also realizes, however, that there are ecological and technological limits that need to be respected to guarantee operational safety of energy facilities and to meet environmental standards.

Russia also has experienced regional disparity in energy supplies, a high degree of fixed asset depreciation and low levels of investments in the fuel and energy complex. The national energy strategy, however, realizes Russia's heavy dependence on natural gas and natural gas markets in Europe and Asia. Energy security is, therefore, also expressed as the ability to withstand threats to reliable supply of fuel and energy and threats caused by external factors (geopolitical, macroeconomic and market). Russia aims at strengthening partnership with the leading energy producers and developing active dialogue with the consumers and transit countries in accordance with the principles of energy security adopted by the G8 Summit in Saint Petersburg in 2006 (Shadrina 2010). The other three major components of the energy strategy are the improvement of energy efficiency in all sectors, budget efficiency of the energy sector to become more competitive with world markets and environmental safety of the energy sector. Budget efficiency refers to the predictability of public revenues from business entities of the energy sector and necessary state investments to further develop the energy sector.

Russia predicts a reduction in the share of natural gas in the primary energy consumption from 53 per cent in 2008 to 46 per cent in 2030. This will be offset with an increase in the proportion of liquid fuels and non-fuels. Energy exports are expected to rise by around 13 per cent over 2008 figures and natural gas exports to far-abroad countries are estimated to rise by 65 per cent (Ministry of Energy of the Russian Federation 2010).

Based on these figures Russia is still planning to be a fuel-based energy exporter in 2030. It is, of course, very uncertain what the political and economic situation in 2030 will be. Since the energy strategy was crafted we have already seen drastic changes in the world energy situation. The natural gas revolution might result in world markets for natural gas similar to the trade in oil, particularly if countries invest into liquefied natural gas terminals and other continents develop production from shale gas reservoirs on a larger scale. The Russian economy needs to be flexible and cannot just count on the current trade in natural gas based on a

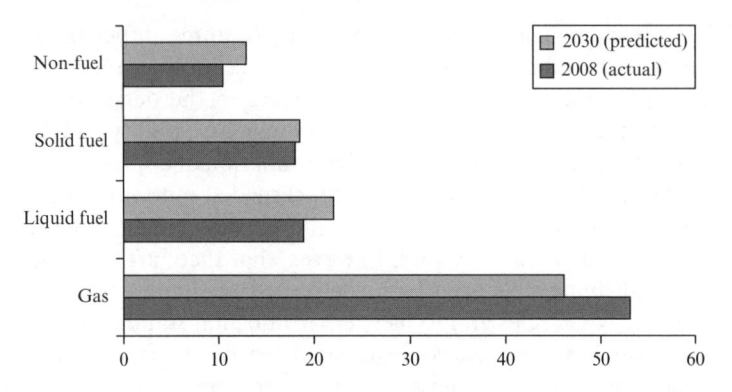

188   *International handbook of energy security*

[Bar chart showing Russia's primary energy consumption by fuel, comparing 2030 (predicted) and 2008 (actual) values:
- Non-fuel: ~13 (2030), ~11 (2008)
- Solid fuel: ~18 (2030), ~18 (2008)
- Liquid fuel: ~22 (2030), ~19 (2008)
- Gas: ~46 (2030), ~53 (2008)]

*Source:*   Ministry of Energy of the Russian Federation (2010).

*Figure 9.1   Russia's primary energy consumption by fuel as a percentage of total consumption*

monopoly position and long-term contracts rather than spot market and future markets. The Russian energy strategy recognizes this to some extent and, therefore, has different development adjustment paths. It is more difficult for a country like Russia that is still in a transition to pinpoint in what direction its economy will be heading. The abundance of natural resources could very well fuel a more value-added manufacturing sector. The Russian Ministry of Energy and the central government is very much in charge of the direction Russia is taking. Coordination between energy sectors will not be an issue in Russia. The downside is that state energy policies determine to a large extent what investments will be made and what investments and technological innovations will occur. This could stifle innovation and the entry of new energy players.

### 3.4   Canada

Canada is the only country that does not currently have a formal energy strategy even though the current Prime Minister, Stephen Harper, has referred to Canada as an emerging "energy superpower". Canada has vast resources of bitumen in the oil sands, shale gas, uranium, hydropower, coal, oil and natural gas, and a vast potential of alternative and renewable resources in the form of wind, solar, biomass and tidal power. The country is a net exporter of energy with 99 per cent of all energy exports going to the United States (Foreign Affairs and International Trade Canada. While Europe is dependent on a single supplier of natural gas for much of its energy demands, Canada relies on a single buyer of its resources. Due to the geographic separation of Canada's oil supply basins and the major areas of demand, about 50 per cent of the crude oil used by Canadian refiners to meet consumers' oil product needs comes from imports. About 44 per cent of those imports come from OPEC countries and 37 per cent come from the North Sea. Because of this import and export dependency and tremendous potential benefits from further development of the vast renewable resource potential, it seems astounding that Canada has no official energy strategy and vision. There are a number of reasons that have contributed to this. The Dominion of Canada is a federal state, but one of the most decentralized ones in the world. Provinces have authority over crucial sectors such as energy, education, health and natural resource management. This complicates coordinated decision-making at the federal level or between the provinces. Although the federal government has a proscribed role to play, it has the sole responsibility for development of national strategies (see also the latest report, Standing Senate Committee on Energy 2012), depending to a large degree on political will and leadership of the federal government in power. Under Pierre Elliot Trudeau as Prime Minister in the 1980s, the federal government implemented the National Energy Programme (NEP) that divided the country. Some of the main objectives of the NEP were to promote oil self-sufficiency for Canada, maintain the oil supply for the Eastern Canadian manufacturing base, promote oil exploration in Canada, and to increase government revenues from oil sales through taxes, other instruments and agreements between the federal government and the provinces. The NEP also capped the price of oil for Canadian producers. The province of Alberta and Western Canadians were, therefore, deprived of cashing in on escalating prices arising from the international oil crisis. Instead they had to subsidize the manufacturing sector of the Eastern part of the country. Canada's western producing provinces objected strongly to the intrusion of the NEP into their jurisdiction over resources. For these reasons, no federal government after Trudeau's has attempted to devise a national energy strategy.

Subsequently, Western Canadians have seized the opportunity to accelerate development of the provinces rich resource base of bitumen in oil sands and shale gas resources for domestic and export markets. In the last 10 years Alberta (and soon also Saskatchewan) has experienced an ongoing oil boom (with small interruptions through the world economic crisis). The problem they face today is the lack of pipeline capacity to deliver bitumen to existing and new export markets. Due to monopsonistic power by the United States and a glut of supply in the Western U.S., Alberta oil is receiving heavy discounts. The Natural Resources Minister is estimating that the price differential to global markets costs Canadian producers over

$ 20 million a day (Plecash (2012)). Alberta is, therefore, very keen on diversifying its export markets. In addition, under the free trade agreement with the U.S. and Mexico (NAFTA), article 605 requires that Canada continue the amount exported to the U.S. and Mexico as a ratio of Canadian production and not to disrupt natural supply channels. The commitment to new pipelines to the United States could, therefore, seriously impede Canadian energy security and could tie them to lower prices in the U.S. market. Pipeline companies such as Enbridge and TransCanada have recently applied to the Canadian National Energy Board to have an oil pipeline between Sarnia and Montreal reversed in order to provide a means to flow western Canadian bitumen to Canadian customers, refineries, and potentially export markets. This will open the door for access to world markets outside of North America. In order to expand and diversify Canadian energy markets, Canadian provinces need to coordinate their support for pipelines, which pass through different provinces in order to cross the Canada-US boundary or to reach export terminals.

In addition, Canada must properly account for environmental damages, particularly carbon emissions from bitumen extraction, transportation and the electricity sector in order to be seen as environmentally responsible and to win support for exporting energy commodities and products to European markets. Canada and the EU are in the process of signing a free trade deal, which would open up new markets for Alberta oil. Trans Canada Pipeline is considering conversion of the existing natural gas pipeline to enable transportation of bitumen from Alberta through Québec and New Brunswick to a future east coast export terminal.

Because all these energy developments require substantial coordination and policies at the national level, it is no surprise that the current Alberta premier Alison Redford is pushing for a national energy strategy. The current federal government has, however, chosen to not get directly involved and instead defers to the Annual Meeting of Federal, Provincial and Territorial Mines and Energy Ministers to make progress on this matter. The 2011 Meeting of Provincial and Territorial Premiers meeting resulted in a declaration of commitment to a "Pan-Canadian" approach to developing the country's natural resources based on economic development, energy efficiency and innovation (Plecash 2012). Alberta Premier Redford refers to an action plan from 2007 that included improvements in energy efficiency and conservation, the acceleration of clean-technology deployment, clean energy for the future, a modernization of the energy transmission infrastructure (including an East-West grid), streamlining the regulatory process, developing human resources in the energy sector, and permitting the provinces and territories to participate in international energy negotiations.

The making of energy policy in Canada is a divided responsibility between the federal and provincial governments. Provincial governments own the energy resources within their boundaries and develop independent energy policies for the development, management, and taxation of their resources. At present, the federal government has not taken an interest in developing a national energy strategy. Any strategy and any future changes would require the approval of 13 provincial and territorial premiers. At the most recent meeting of provincial premiers, all provinces except British Columbia agreed to proceed to develop a Canadian energy strategy. The province of British Columbia (B.C.) is currently boycotting the development of a national energy strategy because of the risk associated with the controversial Northern Gateway pipeline proposal that is planned through pristine areas of B.C. and many aboriginal lands in order to transport Alberta bitumen to Asian markets through B.C. ports. The B.C. premier is not willing to take the risks of pipeline leaks and shipping accidents in the pristine marine environment without being properly compensated from oil revenues.

### 3.5 The European Union

The European Union (EU) has a long-term (European Commission 2011) and a short-term (European Commission 2010) energy strategy. The EU is committed to reducing greenhouse gas emissions to 80–95% below 1990 levels as stated in its "Roadmap for moving to a competitive low-carbon economy in 2050". In its long-term energy strategy the EU is exploring ways to achieve its decarbonization objective while at the same time ensuring security of energy supply and competitiveness. The short-term and long-term energy strategies of the EU are, therefore, guided by the decarbonization policies. Consequently the EU strategy heavily relies on the transition to renewables and significant improvements in energy efficiency. In addition it relies on natural gas imports, and in contrast to the United States, is cautious on the development of internal market shale gas deposits. Instead it counts on a diversification of energy fuels and a doubling of electricity in final energy demand. Electricity will be generated with a mix of nuclear, renewables, coal and natural gas. The latter two are planned to be effectively connected to CCS technologies by 2030 in order to control carbon emissions from the energy sector.

The EU framework is technology neutral in the sense that it does not dictate specific technologies on individual member countries but instead attempts to provide incentives through carbon pricing, as for example through the EU Emission Trading System (ETS). Since member states have very different views on certain technologies such as nuclear, renewables or unconventional natural gas, there is large uncertainty about the role of various energy sources in 2050. One EU objective does stand out

Table 9.3  Characteristics of national energy strategies in countries with high import dependence

| Country/Region | Energy Strategy (Name and Date) | Focus of Energy Strategy | Components of Energy Strategy | Definition of Energy Security |
|---|---|---|---|---|
| United States | Blueprint for a Secure Energy Future (The White House (2011)) | Increasing domestic energy supply, reducing energy imports, reduction of energy costs, energy efficiency and R&D for clean energy technologies | (1) Expansion of Domestic Oil and Natural Gas Development and Production (2) Increasing energy supplies worldwide and promotion of switch to cleaner technologies (3) Fuel efficiency standards for vehicles (4) Energy efficiency in buildings (5) Domestic clean energy promotion and support of R&D | Affordable, safe and reliable supply of energy |
| EU | Energy Roadmap 2050 (15/12/11) Energy 2020 (10/11/10) | Decarbonization, security of energy supply and competitiveness, develop a long-term European technology-neutral framework | (1) Energy efficiency measures (2) Drastic increase in the share of renewables (3) CCS utilization as of 2030 (4) Nuclear part of the mix but uncertain what exact role it will take (5) Doubling share of electricity | Diversification of fuels, sources of supply and transit routes, protection of EU and foreign investments in energy producing countries, safety and security of oil, natural gas pipelines |

*National energy strategies of major industrialized countries* 193

| Projected Energy consumption mix | Institutions Governing Energy Coordination/ Strategy | Flexibility/ Adaptability | Reliance on unconventional energy sources | Reliance on int. cooperation/ markets |
|---|---|---|---|---|
| 2035:<br>15% RES and Liquid Biofuels<br>25% Nat. Gas<br>32% Oil<br>20% Coal<br>9% Nuclear | Federal Energy Regulator's Commission (FERC) | Limited flexibility, no drastic changes in primary energy mix, continued reliance on coal and oil, shale gas might add flexibility, reliance on cost improvements for solar and advancements in CCS for coal | Large reliance on shale gas and oil reserves | Reducing oil demand while increasing the supply of oil worldwide and the trade in natural gas, new international framework for nuclear energy |
| 2050:<br>50% RES, 25% Natural Gas, 25% Oil, Nuclear and solid fuels | ACER, ENTSO-G, ENTSO-E | Flexible resources in the power system (e.g. flexible generation, storage, demand management), further electricity and gas market integration, more | Cautious consideration of shale gas | Large reliance, especially interconnectedness in the EU internal market, increased reliance on electricity grids, LNG and natural gas pipeline projects |

194   *International handbook of energy security*

*Table 9.3*   (continued)

| Country/ Region | Energy Strategy (Name and Date) | Focus of Energy Strategy | Components of Energy Strategy | Definition of Energy Security |
|---|---|---|---|---|
| EU | | | in final energy demand (6) Smart technology, storage and alternative fuels | and related production and transport infrastructure |
| Japan | Basic Energy Plan (2010) | Three prime areas of focus: – energy security – environmental protection – efficient energy supply. Start with significant decrease in energy demand, by 13% by 2030 | (1) Securing resources and enhancing supply stability (2) Independent and environment-friendly energy supply structure (3) Realizing a low carbon energy demand structure (4) Building next-generation energy and social systems (5) Developing and diffusing innovative energy technologies | To secure sufficient energy supply at reasonable prices for the achievement, pursuit and maintenance of maximizing economic/ social welfare and sustainable development of national economy and citizens (2006 BES presentation) |

clearly, which is to drastically reduce the dependence on oil by half to 20 per cent by 2050 (see Figure 9.2).

Because increasing electrification of the energy system plays such a strong role in the EU, the EU energy strategy pushes further electricity

### National energy strategies of major industrialized countries 195

| Projected Energy consumption mix | Institutions Governing Energy Coordination/ Strategy | Flexibility/ Adaptability | Reliance on unconventional energy sources | Reliance on int. cooperation/ markets |
|---|---|---|---|---|
| | | integrated view on transmission, distribution and storage | | |
| Energy consumption (no exports) (million kilolitres) Energy: 589 in 2007 to 514 in 2030 Petroleum: 240 in 2007 to 142 in 2030 Coal: 130 in 2007 to 85 in 2030 Gas: 106 in 2007 to 80 in 2030 Nuclear: 60 in 2007 to 122 in 2030 Renewables: 35 in 2007 to 67 in 2030 (Ministry of Economy, Trade and Industry (2010)) | Agency for Natural Resources and Energy (ANRE), a part of Ministry of Economy, Trade and Industry (METI) Nuclear and Industrial Safety Agency (NISA), a part of ANRE | Entirely import dependent, no flexibility re development of domestic resources. Major effort to decrease demand, and, thereby, to decrease export dependency. Major effort on energy efficiency and conservation, to reduce demand. Focus on increasing share of nuclear and renewables | Nil, negligible domestic resource endowment | Reliant on imports for over 80% of primary energy. Forging closer ties with energy commodity suppliers |

market integration and more integrated views on transmission, distribution and storage. The EU already has institutions that coordinate the electricity market such as the Agency for the Cooperation of Energy Regulators (ACER) and the European Network of Transmission Grid Operators for

Graph 1: EU Decarbonisation scenarios – 2030 and 2050 range of fuel shares in primary energy consumption compared with 2005 (in %)

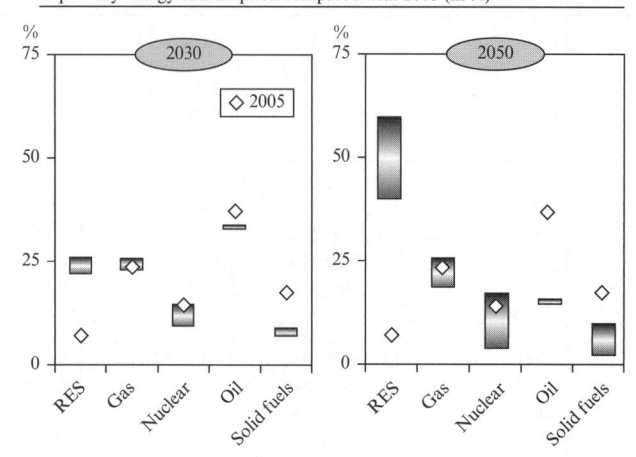

*Source:* European Commission (2011).

*Figure 9.2   Energy mix outlook for the EU 2030 and 2050*

Electricity (ENTSO-E). These agencies already have experience in the coordination between countries and will make it easier to find additional improvements in the efficiency of electricity use and the development of smarter grids. The increased reliance on electricity for energy supply in the internal EU market is one aspect of energy security for the EU. In addition, the EU relies on further diversification of fuels, sources of supply and transit routes, and the protection of EU and foreign investments in energy producing countries. Securing additional natural gas pipelines and related production and transport infrastructure will be a key part of the energy strategy particularly as conventional natural gas plays a crucial role in the transition to a low carbon economy, and as a backup for the increasing addition of intermittent renewable energy sources. Increased trade in liquefied natural gas could be an important additional source of natural gas for the EU, which could have the effect of reducing the use of coal for electricity generation and of significantly dropping energy costs in the EU.

### 3.6   The United States

The Unites States latest energy strategy entitled "Blueprint for a Secure Energy Future" (The White House 2011) is very different from the EU strategy. The focus of the strategy is not on decarbonizing the economy and transitioning to a low carbon economy. Instead the U.S. Blueprint

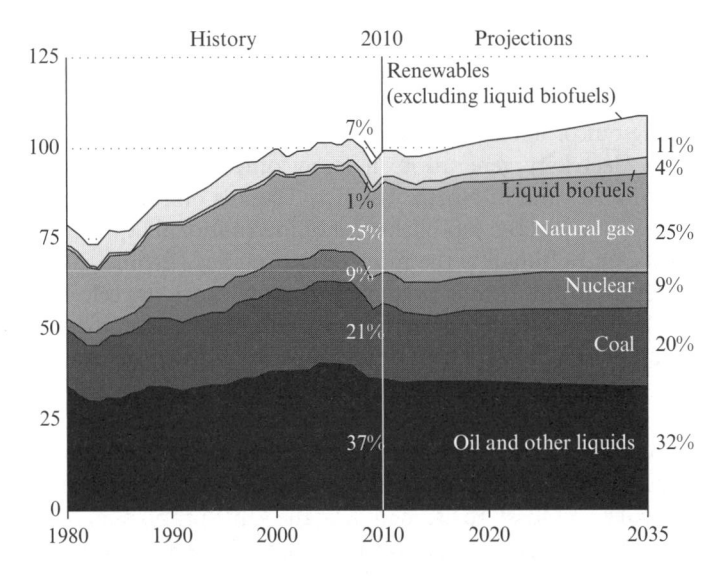

*Figure 9.3* U.S. primary energy consumption by fuel (quadrillion btu per year)

emphasizes the reduction of energy costs and the reduction of energy imports. The U.S. tries to achieve this by significantly increasing domestic energy supplies, by increasing energy efficiency, and by investing in R&D for cleaner technologies and biofuels. In contrast to the EU's approach that stresses carbon prices and incentives to reduce consumption and switch to cleaner technologies and modes of transportation, the U.S. targets the production or supply side. On the energy demand side of the equation, the United States has already implemented ambitious fuel efficiency standards for vehicles including heavy-duty vehicles (White House 2012b). The U.S. energy strategy also targets energy efficiency of buildings, the expansion of domestic unconventional oil and natural gas development and production. The U.S. also is actively involved in trying to increase energy supplies worldwide and promote switch to cleaner technologies, which would reduce the demand on fossil fuels. Energy security is defined as affordable, safe and reliable energy supply for Americans.

The projected energy mix for the U.S. (EIA 2012) is not very different from the energy mix in primary demand in 2010 (see Figure 9.3). A 5 per cent reduction in oil is compensated by a slight increase in renewables and a 3 per cent increase in liquid biofuels.

Nuclear energy and natural gas remain constant and coal retains 20 per cent of the mix, about twice as much as in the EU (in 2030). The development of nuclear and coal is, however, uncertain due to concerns arising from the Fukushima incident and the recent nuclear "Ausstieg" (the phasing out of nuclear) in Germany, resulting in an uncertain future for nuclear electricity generation in many other countries. The reliance on coal in the EU and in North America depends on the development of shale gas reserves and the establishment of international liquefied national gas markets, and ultimately the price of natural gas (Brooks 2012). Shale gas production in the U.S. is predicted to grow nearly threefold between now and 2035, and would represent nearly half of all gas production in the United States (EIA 2012). These predictions will only come true if natural gas prices allow shale gas producers to break even. It is, however, not until 2015 that gas futures prices are consistently above break even prices for only two out of the five major shale basins (Brooks 2012).

The U.S. energy strategy has limited adaptability. It continues to rely to a large extent on fossil fuels, including the same proportion of coal and a slightly reduced proportion of oil in the primary energy mix. Shale gas may add some flexibility but it is quite uncertain to what extent it can be relied on for energy security and affordable consumer and industry prices since it only becomes lucrative at significantly higher natural gas prices. The Energy Blueprint also assumes substantial reductions in the cost of solar power (one quarter of 2009 costs by 2030) and in the cost of electric vehicle batteries. It is also quite optimistic on the deployment of carbon storage and capture (CCS) technologies within 10 years of the blueprint (by 2021). On the institutional side, similar to the EU, the U.S. has the federal energy regulators commission (FERC) that coordinates between states. For example, it regulates the transmission and wholesale sales of electricity in interstate commerce, it regulates the transmission and sale of natural gas for resale in interstate commerce, regulates the transportation of oil by pipeline in interstate commerce, approves the siting and abandonment of interstate natural gas pipelines and storage facilities, ensures the safe operation and reliability of proposed and operating LNG terminals, and licenses and inspects private, municipal, and state hydroelectric projects. FERC has its own strategic plan with the two primary goals to ensure that rates, terms and conditions are just, reasonable and not unduly discriminatory or preferential, and to promote the development of safe, reliable and efficient energy infrastructure that serves the public interest.

### 3.7 Japan

Among the countries covered in this study, Japan is unique due to its almost complete reliance on energy commodities from foreign sources.

Japanese governments have squarely addressed this vulnerable situation through legislation established in 2002. The Basic Act on Energy Policy (BEAP) focused on three central energy issues of importance to Japan: securing stable supply of energy; ensuring environmental sustainability; and, utilizing market mechanisms to control energy pricing and flows.

Security of supply has been the central focus of a succession of three energy plans, all based on the BEAP in 2002, in 2003, in the New National Energy Strategy in 2006, and most recently in 2010, in the Strategic Energy Plan of Japan: Meeting global challenges and securing energy futures (SEPJ) (Ministry of Economy, Trade and Industry, Government of Japan 2010).[7] The SEPJ was issued nine months prior to the nuclear incident at the Fukushima Daiichi nuclear power plant in March 2011. The significant reliance on an increasing share of electricity from nuclear power plants as a means to achieve a significant share of the SEPJ's objectives is now a matter of public debate and opposition. As described below, while the general directions and principles remain sound and commendable, the means to achieve Japan's laudable goals is in serious question.

### 3.7.1 The strategic energy plan of Japan

Three "basis points of view" set the context for the 2010 SEPJ. In keeping with previous Japanese energy plans, the three primary thrusts of energy policy are energy security, environmental protection, and efficient energy supply. Two additional policy directions were introduced in 2010: energy-based economic growth and reform of the energy industrial structure; and, a fundamental overhaul of the energy supply and demand system by 2030. The plan sets five ambitious targets between 2010 and 2030. Three of the targets are directly related to energy security: doubling the energy self-sufficiency ratio (renewables and nuclear), from the present 18 per cent and doubling the self-developed fossil fuel supply ratio (coal, natural gas, oil and LPG produced by Japanese companies); raising the share of zero-emission power share from 34 per cent to 70 per cent; and maintaining and enhancing energy efficiency in the industrial sector at the highest level in the world.[8] Highlights of each of the measures, which are expected to achieve these targets, illustrate the significant level of effort, which will be required. With respect to energy supply, the BEPJ calls for assisting the expansion of renewable energy sources through expanded feed-in tariffs, support for R&D, and power grid stabilization in order to effectively handle an increase in variable generation. Nine new nuclear plants are to be built by 2020 with overall utilization rates of 85 per cent, and 14 more plants by 2030 with utilization rates of 90 per cent by 2030. It is also planned to reduce $CO_2$ emissions from coal-fired plants to levels comparable with IGCC plants, and to accelerate the implementation of carbon capture and storage technology by 2030. Reducing energy demand is an effective pathway towards improved energy security and reduction of GHG emissions. The BEPJ calls for a suite of energy efficiency initiatives in each of the end-use sectors of the Japanese economy. Incorporating improved grid-based technologies, such as smart grid, smart meters, and equipment to control electricity demand in real-time are identified as a means to improve the reliability, efficiency, and stability of the electricity system (Ministry of Economy, Trade and Industry, Government of Japan 2010). The impacts of the suite of measures on total primary energy supply and on the shares of each energy commodity are significant. First, primary energy supply is projected to decrease dramatically by 13 per cent by 2030. Japan is unique in setting this aggressive target to reduce overall energy demand. Second, the shares estimated to come from nuclear power and renewable energy technologies are expected to more than double over the 20-year projection period.

One significant factor in Japan's long-term energy future is the role that nuclear power will play, post-Fukushima. The BEPJ indicates that if these estimates are realized, that the share of Japan's electricity from nuclear power will be approximately 50 per cent of generated electricity in 2030. LNG is projected to contribute approximately 10 per cent. In light of the impact of the Fukushima incident, the subsequent shutdown of the bulk of nuclear generation, and the expressions of public anxiety around nuclear energy in general, the energy security picture for Japan is highly uncertain. One option available to Japan is to increase the generation capacity from plants using natural gas turbines, from IGCC plants, and from cogeneration facilities. Rather than a projected reduction of the contribution from natural gas (see Table 9.4), the role of natural gas supplied via international LNG markets creates one viable, and prompt, pathway for maintaining electricity generation. The negative impact on the hoped-for

*Table 9.4  Dramatic supply shifts in Japan by 2030*

| Energy Source | Actual Fiscal 2007 | Estimated 2030 |
| --- | --- | --- |
| Petroleum | 41% | 28% |
| LPG | 3% | 3% |
| Coal | 22% | 17% |
| Natural Gas | 18% | 16% |
| Nuclear Power | 10% | 24% |
| Renewable Energy | 6% | 13% |
| Total Primary Energy | 589 (m kilolitres) | 514 (in kilolitres) |

*Source:* (Ministry of Economy, Trade and Industry, Government of Japan 2010).

reduction in GHG emissions will be one unfortunate consequence from the expansion in natural gas generation capacity.

## 4 WHAT EXPLAINS DIFFERENCES IN NATIONAL ENERGY STRATEGIES? FEATURES, RECOMMENDATIONS AND "BEST PRACTICES"

The observations from our selection of countries, ranging from those largely dependent on imported energy commodities to those that rely on sales of their resources to export markets, demonstrate the multi-faceted and complex nature of energy strategies. Each of the countries studied, with the exception of Canada, has an official national, or – in the case of the EU – economy-wide, energy strategy that provides energy directions at least until 2030. The focus of most strategies is on the availability and affordability of domestic energy supplies, clean energy, technological innovation, conservation and sustainability. Only Japan and the EU explicitly mention energy security as one of the main focuses, while the U.S. indirectly refers to energy security as a main focus of their energy strategy. Energy security is not a component of the energy strategies of Australia and Norway. Canada also has no explicit energy strategy and hence no energy security policy. Although Norway, Australia and Canada are rich in energy resources, it is surprising that they do not address export market or demand security as a threat to their overall economic future. Only in a recent speech by Norway's Energy Minister to industry delegates in the United States was energy security defined for Norway. Norway is concerned (as Canada, Australia and Russia should be) about markets that are not open, transparent and that do not have clear, competitive price signals. Without open market operation, exporters of energy markets will have difficulty accessing their existing and new export markets. Energy resources will continue to be dictated by geopolitical processes.

The definitions of energy security of all of the G8 countries have many similarities. All refer to securing reliable, affordable and sufficient energy resources. Only Canada and Norway stress global energy security since both countries depend on maintaining access to stable world energy markets. Canada's approach is to reduce its dependence on one client (the U.S.) and Norway wants to maintain and expand its export markets in Europe. It supplies Europe with petroleum and natural gas and has bidirectional electricity trade with neighboring countries via the NordPool electricity market. Russia, on the other hand, has access to European natural gas markets through pipeline connections, which are, however, being challenged by alternative pipeline projects and liquefied natural gas terminals in the near future. This means Russia will need to rely more on access to open, transparent and competitive markets.

As stated by Chester (2010): "The meaning of energy security differs over the short, medium and long term because the probability, likelihood and consequences of different risks or threats to supply will vary over time. Thus we will never reach an end-state of energy security as such." Changes can occur quickly in energy markets, due to geopolitical events, the advent of new more efficient energy technologies, or the development of new ways to access previously untapped components of a country's national resource base. Such developments will typically have a direct influence on energy security considerations. Examples are the potentially negative impact of the Fukushima incident on Japan's expectations for an increasing share of nuclear power and the United States' recent opportunity to possibly become a LNG exporter as a consequence of growth in gas supply from shale gas reservoirs. This calls for regular reviews of energy security strategies, and also a readiness to revisit strategies in situations in which fundamental supply-demand factors have changed. Japan's practice of issuing updated comprehensive basic energy plans every four years is a good "best practice", in fulfillment of legislative provisions. The Fukushima incident will precipitate a thorough review of Japan's goals and strategies. Japan needs to review regularly because it does not have a lot of built-in flexibility in its energy system. The EU intends to have more flexible resources in the power system that could advance electricity integration and the electrification of European energy and transportation systems. The United States is relying heavily on unconventional and some alternative resources. It is not drastically changing its primary energy consumption mix and increasingly relies on domestic sources. This limits flexibility and adaptability of its energy system and makes the U.S. vulnerable. The U.S. energy strategy illustrates, however, how technology advancements directly interplay with energy strategies through potentially facilitating access to previously-untapped components of the resource base, bringing down the cost of new energy technologies, and opening up completely novel ways of providing energy services. Examples are the use of new production techniques to access the natural gas in shale reservoirs in the United States and Canada, the dramatic fall in the cost of solar cells, and emerging technologies, which will facilitate electrification of transportation. In fact, technological advances can have a profound effect on national energy security, illustrated by the emerging possibility of natural gas LNG exports from the United States to European markets and the development of highly-efficient natural gas turbines for electricity generation, as well as potential solutions for cleaner coal production through carbon capture and storage technologies. Nuclear is the elephant in the room and creates large uncertainty as countries are either undecided about its future (e.g. in Japan or in Canada) or cannot agree among themselves (e.g. in the EU).

In addition, policy developments in related areas have played a crucial role in the energy path that nations choose, the institutions that develop to govern energy developments, and the strategies that are chosen. The EU for example had agreed on a climate change strategy that had a direct influence on the characteristics of its energy strategy. In fact, one could argue the EU devised an energy strategy in order to fulfill its climate change aspirations. The opposite holds for the United States where energy strategy and energy security always trumps climate change policy. The U.S. always had a very explicit and clear energy strategy but to date has not been able to pass a climate change bill through Senate. In Canada experience with the NEP has stalled national discussions about a national energy strategy and has strengthened the authority of provinces over energy and environmental policy, and now that provinces are ready for an energy strategy, there is a lack of federal authority to pull it through. Historic events such as Harrisburg/Three Mile Island (1979), Chernobyl (1986) or most recently Fukushima (2011) critically influence public opinion and energy strategies. Due to these events Germany took a quick stand on a long debated question to phase out nuclear, and Japan is struggling to develop a revised energy plan. Even the U.S. and understandably Russia have become more cautious with nuclear technologies. None of the G8 countries (with the exception of Germany) makes a clear or firm commitment to nuclear energy, but also do not step away from it, and leave at least a small proportion in their energy mix.

## 5 CONCLUSION

Energy security issues for major industrialized nations are changing. It is no longer merely a question of securing adequate supplies of energy to fuel an ever-growing manufacturing base. The major industrialized nations are focusing more on becoming major energy exporters of conventional oil (Norway and Russia), coal (Australia), conventional natural gas (Norway and Russia), unconventional oil (Canada) and unconventional natural gas (potentially the United States). In the meantime the EU intends to set up a more integrated and sophisticated electricity system with a large proportion of renewables and might act as a major importer and exporter of electricity. Only Japan is left on its own, being dependent on imports for securing energy for its manufacturing and consumption base. These new developments are demanding new energy strategies that create a vision about supply and demand security, integration of markets and electricity systems and open access to global energy markets. In addition, decisions in these G8 countries and Norway on climate change commitments and nuclear technology investments will determine to a large extent the new world energy order. The potential for a nuclear renaissance will be determined by some of the G8 countries that then will export their technologies to the major emerging energy markets in China, India, Brazil and developing nations. Climate change negotiations have to advance sooner rather than later and will critically influence the path of energy developments. All this will require more regular adjustments to national energy strategies.

## NOTES

1. Draft Energy White Paper, p. 65
2. 'High energy security' is defined as meeting Australia's economic and social needs. 'Moderate energy security' means that needs are being met but with emerging issues that will need to be addressed to maintain this level of security. 'Low energy security' means that needs are not being, or might not be, met.
3. In the Draft White Paper, the term "adequacy" is defined as the provision of sufficient energy to support economic and social activity, and "reliability" is the provision of energy with minimum disruptions to supply.
4. NESA, p. 78
5. Electricity production from hydro facilities makes up the remaining 7 per cent.
6. The term "national value creation" is synonymous with gross domestic product.
7. A thorough review of Japan's new basic energy plan has been provided by Duffield and Woodall (2011).
8. The two remaining targets are to reduce the emissions from the residential sector to one half of current levels, and to maintain or obtain top-class shares of global markets for energy-related products and systems.

## REFERENCES

Brooks, Allen (2012), "Optimistic NPC report could point US energy strategy in wrong direction", *Energy Strategy Reviews*, **1**: 57–61.
Chester, Lynne (2010), "Conceptualizing energy security and making explicit its polysemic nature", *Energy Policy*, **38**: 887–895.
Department of Resources, Energy and Tourism, Government of Australia (2011), "Draft Energy White Paper: Strengthening the Foundations for Australia's Energy Future". www.energywhitepaper.ret.gov.au
Department of Resources, Energy and Tourism, Government of Australia, "National Energy Security Assessment" December 2011, www.ret.gov.au (accessed February 2013).
Duffield, John S. and Brian Woodall (2011), "Japan's new basic energy plan", *Energy Policy*, **39**: 3741–3749.
EIA, U.S. Energy Information Administration (2012), "Annual Energy Outlook 2012 – Early Release Overview", at http://www.eia.gov/forecasts/aeo/er/pdf/0383er%282012%29.pdf (accessed February 2013).
European Commission (2010), "Communication from the commission to the European

Parliament, the Council, the European Economic and Social Committee and the Committee of the Regions – Energy 2020: A strategy for competitive, sustainable and secure energy", at http://eurlex.europa.eu/LexUriServ/LexUriServ.do?uri=CELEX:5201 0DC0639:EN:HTML:NOT (accessed February 2013).

European Commission (2011), "Communication from the commission to the European Parliament, the Council, the European Economic and Social Committee and the Committee of the Regions – Energy Roadmap 2050", at http://ec.europa.eu/energy/energy2020/roadmap/doc/com_2011_8852_en.pdf (accessed February 2013).

Eurostat Statistical Books (2012), "Edition-Energy balance sheets 2009–2010", at http://epp.eurostat.ec.europa.eu/portal/page/portal/release_calendars/publications (accessed February 2013).

Foreign Affairs and International Trade Canada (2012), at www.international.gc.ca, (accessed July 19, 2012).

Goldthau, A., W. Hoxtell and J.M. Witte (2010). "Global Energy Governance: The Way Forward", in A. Goldthau and J.M. Witte (eds) *Global Energy Governance: The New Rules of the Game*, Global Public Policy Institute: Berlin, Chapter 16.

International Energy Agency (2009), Export-Import Dependence of Selected Countries-Energy Balances, at http://www.iea.org/stats/balancetable.asp?COUNTRY_CODE=NO (accessed February 2013).

International Energy Agency (2011), "Norway 2011 Review", Energy Policies of IEA Countries, at http://www.iea.org/publications/freepublications/publication/Norway2011_web.pdf (accessed February 2013).

Ministry of Economy, Trade and Industry, Government of Japan (2010), "The Strategic Energy Plan of Japan – Meeting global challenges and securing energy futures [Summary]", June 2010, at http://www.meti.go.jp/english/press/data/pdf/20100618_08a.pdf (accessed February 2013).

Ministry of Energy of the Russian Federation (2010), "Energy Strategy of Russia for the Period up to 2030", at http://www.energystrategy.ru/projects/docs/ES-2030_%28Eng%29.pdf (accessed February 2013).

Ministry of Petroleum and Energy (2010) "Norway's oil history in 5 minutes", (updated November 2010).

Norwegian Petroleum Directorate (2010). "The petroleum sector – Norway's largest industry", at http://www.npd.no/Templates/OD/Article.aspx?id=2944&epslanguage=en (accessed February 2013).

Plecash, Chris (2012). "Ottawa and provinces must collaborate on resource development: Energy Minister Oliver", *The Hill Times*, August 6, 2012.

Shadrina, Elena (2010), "Russia's foreign energy policy: norms, ideas and driving dynamics", Electronic Publications of Pan-European Institute 18/2010, at www.tse.fi/pei (accessed February 2013).

Standing Senate Committee on Energy, the Environment and Natural Resources (2012), "Now or Never-Canda Must Act Urgently to Seize its Place in the New Energy World Order", July.

Victor, D.G. and L. Yueh (2010). "The New Energy Order-Managing Insecurities in the Twenty-first Century", *Foreign Affairs*, Jan./Feb: 61–73.

Voser, P. (2012). "The Natural Gas Revolution", *Energy Strategy Reviews*, **1**: 3–4.

The White House (2011). "Blueprint for a Secure Energy Future", Washington, March 30, at http://www.whitehouse.gov/sites/default/files/blueprint_secure_energy_future.pdf (accessed February 2013).

The White House (2012a). "Fact Sheet: G-8 Action on Energy and Climate Change", May 19, at http://www.whitehouse.gov/the-press-office/2012/05/19/fact-sheet-g-8-action-energy-and-climate-change (accessed February 2013).

The White House (2012b). "The Blueprint for a Secure Energy Future: Progress Report", Washington, March 30, at http://www.whitehouse.gov/sites/default/files/email-files/the_blueprint_for_a_secure_energy_future_oneyear_progress_report.pdf (accessed February 2013).

# 10. Developing world: national energy strategies
*Sylvia Gaylord and Kathleen J. Hancock**

## 1 INTRODUCTION

In July 2012, India experienced the worst blackout in its history, plunging 670 million people – about 10 percent of the world's population – into darkness and bringing energy security into headlines around the world. India's inability to guarantee energy supply is seen as a major factor hindering its development. Companies depend on reliable energy and are thus hesitant to invest in countries where security of supply is questionable (Sharma et al. 2012; Harris and Bajaj 2012). News coverage of the blackout linked energy access to long-term development. If India's system is vulnerable to such a massive shut-down, how can domestic and international companies be expected to invest and thrive in India, and without this investment, how will India raise the standard of living for its citizens?

While developed states also have electricity blackouts, India's energy disaster highlighted the severity of energy security problems in the developing world and invited debate on the best strategies for enhancing energy access. In this chapter, we focus on energy issues and related strategies of particular relevance to the developing world. These tend to emphasize increasing electricity access for the larger population and long-term economic development for the state. These primary concerns, along with limited financial and military resources for energy security, shape the strategies developing countries can and do employ.

The term "developing countries" encompasses most of the world, including a broad range of income levels, from the least-developed states to large emerging economies, and population sizes, all the way from tiny island states to the two most populated states in the world. Since examining strategies in all of the states covered by the term is not possible, we look at general trends in the three main world regions where lower per-capita-income states are the majority: Southeast and East Asia (including China and India), sub-Saharan Africa, and Latin America.

We begin by briefly discussing the concept of energy security (see earlier chapters for more detail) and then elaborate on energy security goals for developing states. We then discuss in detail six energy strategies:

(1) nationalize energy companies, (2) privatize energy companies to attract foreign direct investment (FDI), (3) increase and diversify domestic supplies through nuclear or renewable energy sources, (4) diversify imports, (5) build and maintain strategic petroleum reserves, and (6) increase efficiency. Under each strategy, we discuss the issues involved and summarize some associated activities in developing states. We conclude the chapter with a number of observations about issues surrounding these strategies in the developing world and suggestions for future research.

## 2   ENERGY SECURITY ISSUES

While there is significant debate about how to define energy security, most current discussions include four key concepts: availability, reliability, affordability, and sustainability (APEC 2007; UNDP 2004; Xu 2012; Chester 2010). For example, the United Nations Development Program defines energy security as "the availability of energy at all times in various forms, in sufficient quantities and at affordable prices without unacceptable or irreversible impact on the environment" (UNDP 2004, 42). The two primary energy goals for the developing world – electricity for the general population, and long-term economic development for the state – relate to all four elements of this definition.[1]

One in five people in the world (1.3 billion) lacks access to electricity in their homes and businesses (IEA 2011b, 1) and lives in conditions broadly labeled as "energy poverty" (IEA 2012). Nearly 40 percent rely on wood, coal, charcoal, or animal waste to cook their food – all sources of toxic smoke – causing lung disease that kills nearly 2 million people a year, most of them women and children. The Advisory Group on Energy and Climate Change (AGECC) states that "current energy systems are inadequate to meet the needs of the world's poor and are jeopardizing achievement" of the United Nation's Millennium Development Goals (AGECC 2010). Modern energy infrastructure allows people in remote areas to participate in economic activity, facilitates the delivery of health, education and other services, and the provision of clean water and safer food.

The AGECC breaks energy needs into three levels: (1) basic human needs (lighting, health, education, communication, community services and modern fuels for cooking and heating), (2) productive uses (electricity and other energy services to improve productivity–for example, in agriculture, water pumping for irrigation, fertilizer, and mechanized tilling), and (3) modern society needs (domestic appliances, private transportation, etc.) (AGECC 2010). The IEA estimates that meeting basic human needs (level 1) alone will require $35 billion/year for electricity access plus $2–3 billion/year for modern fuels access (AGECC 2010). Developing states are primarily focused on the first two levels.

While developing countries are broadly labeled as "energy poor", access varies significantly between and within regions and between rural and urban areas within states. In 2009, 91 percent of the population in East Asia and the Pacific region had access to electricity (World Bank 2012). Despite this high regional rate, access in some states, such as Myanmar, is as low as 13 percent. In the region, annual commercial energy production grew over 6 percent in the decades from 1980–2002, well above the global average (ESCAP 2006, 78). Average electrification for Africa as a whole is 67 percent in urban areas and 23 percent in rural areas (Yépez-García et al. 2010, 32; Niez 2010; World Bank 2008). Sub-Saharan Africa has the lowest electricity access rate, estimated between 24 and 32 percent; rural electricity access is only 8 percent. Despite growth of 70 percent in electricity generation in the period 1998–2008, 85 percent of the population relies on traditional biomass (mostly wood) (UNEP 2012, 9; World Bank 2012; Eberhard et al. 2011). In Latin America, 93 percent of the population had access to electricity in 2009 (World Bank 2012). Average rural electrification stood at 70 percent, with the lowest being Haiti at 12 percent (Yépez-García et al. 2010, 32). While coverage is poorer in rural areas, rapid growth in urban centers has put considerable pressure on infrastructure in urban areas as well (Fay and Morrison 2005, 2; Calderón and Servén 2010).

The second primary goal for developing countries is long-term sustainable development, similar to the AGEEC's level 2 focus on energy for productive purposes. The effect of India's inadequate electricity supply on long-term growth, illustrated by the 2012 blackout, is substantial, equating to an estimated loss of two percentage points in GDP growth (India Reels From Second Power Grid Collapse 2012). As with India, most developing countries need to overhaul energy and transportation infrastructure to provide immediate basic needs but also to attract the investment necessary to meet long-term development goals.

The U.S. Energy Information Agency (EIA) estimates energy demand in developing countries will drive future world energy consumption. In 2035, developing country demand is expected to be 85 percent higher than in 2008, with consumption projected to rise the most in Asia (117 percent) and the least in Africa (67 percent). The industrial sector is expected to lead the growth in demand, at an annual rate of 2 percent, followed by the transportation, services, and residential sectors. This projection assumes annual average growth in world GDP of 3.4 percent, with developing economies averaging 4.6 percent, and China and India averaging 5.7 percent and 5.5 percent, respectively (EIA 2011, 10–11). The projections also show the gap in energy consumption between developed and developing countries increasing dramatically, with developing countries collectively consuming 67 percent more than developed ones in 2035, up from a gap of 7 percent in 2008. The growth in the energy demand of the industrial sector will also be higher in developing countries because they tend to focus on less-efficient and more energy-intensive products (EIA 2011, 15). Exxon Mobil estimates that developing countries' combined energy demand will increase by 60 percent between 2010 and 2040, keeping pace with population growth (ExxonMobil 2012; BP-British Petroleum 2011). These projections do not take into account potential improvements in technology, improved energy efficiency or climate control policies.

To achieve energy security, both in terms of providing for people's basic needs and creating the conditions for long-term sustainable economic growth, developing countries need energy resources and infrastructure. The following section provides an overview of energy strategies for developing countries to attain these goals.

## 3 ENERGY SECURITY STRATEGIES

Developing countries employ a variety of energy security strategies. For each strategy, we include definitions of concepts, an evaluation of the debates surrounding the strategy, and examples of projects related to these strategies in the three regions on which we focus: East Asia and the Pacific, sub-Saharan Africa, and Latin America. There are numerous initiatives, not all of which can be covered here. As such, the summaries of projects are illustrative only. Many articles that define energy security and elaborate on strategies do not distinguish between levels of development (Winzer 2012; Sovacool and Mukherjee 2011). Of the few that do, most focus on the industrialized states (Löschel et al. 2010) while others evaluate strategies for specific countries, many of which we cite in the following sections. We focus here on strategies that a number of states are pursuing and seem most likely to address energy poverty and longer-term development concerns.

States concerned with the security of energy supply generally choose one or more of the following strategies: nationalize energy companies, privatize energy companies to attract foreign direct investment (FDI), increase and diversify domestic supply by developing nuclear and renewable energy sources, build and maintain strategic petroleum reserves, protect energy infrastructure, and increase energy efficiency. Some of these strategies are available to all developing states, such as increasing efficiency, while others demand significant financial resources, technological development (for nuclear energy, for example) and assets (such as military equipment for protecting sea lanes) that are out of reach for all but the largest emerging states of China, India, and Brazil.

### 3.1 Nationalize Energy Companies

Energy nationalism has recently re-emerged as a strategy for Latin American developing countries seeking to protect domestic energy resources and secure the supply of energy from other countries (Vivoda 2009). In addition, Asian countries have been using national oil companies (NOCs) since the late 1970s to guide their economic development and secure greater control over energy supply, pricing, and distribution. While energy nationalism is not new – Mexico nationalized its oil industry in 1938 (Macalister 2007) and Thailand's NOC was formed in 1978 – it has recently become a major source of debate. The present wave of energy nationalism is driven by a confluence of factors. High oil prices, the emergence of new large consumers, such as India and China – which is giving energy producing nations increased bargaining power and fomenting fear of future resource scarcity (Hughes 2011; Vivoda 2009) – disenchantment with privatization (Steinberg 2012), and the perception that international firms are taking resources that may one day be needed for domestic use without paying adequate compensation (Stevens 2008).

The term energy nationalism encompasses a variety of actions, carried out via presidential or legislative mandates. States can take ownership by purchasing assets from private owners or by seizing assets without compensation (Monagas 2012). States can also renegotiate contractual terms, imposing tougher conditions on private companies (Rigobón 2008; Hoyos 2006). For many developing countries, NOCs play an important role in enhancing energy security by acquiring overseas investments. These aggressive investment strategies from ascending countries, such as China, India and Brazil, are seen as evidence of growing energy nationalism (Herberg 2010).

National oil companies are receiving considerable attention on two counts: their growing presence in world markets in terms of control over reserves and production, and their active participation in international markets in direct competition with the major private oil firms (Chen and Jaffe 2007). NOCs control an estimated 90 percent of the world's oil reserves and 75 percent of production of oil and gas (Tordo et al. 2011a, ix; Economist 2006). In 2010, NOCs were responsible for 19 percent of global investments in mergers and acquisitions in oil and gas and for the first time outspent the major private oil firms, including investments in the U.S. and Canada (P&GJ 2011). While the growing presence of NOCs in the energy sector is often interpreted as evidence of growing nationalism (Dirmoser 2007), some point out that in developing countries states tend to intervene across broad segments of the economy and energy strategies anchored in NOCs should be understood in the context of broader development programs (Hughes 2011). Others note that not all NOCS are the same in terms of management and investment capacity (Marcel 2009), or in terms of the degree of government influence over investment decisions, and that despite the depiction of NOCs in the press as a threat to world energy supplies, they are, on average, at a disadvantage relative to private firms in terms of technological and managerial expertise (Herberg 2007; Agashe 2010).

Asia and Latin America have been at the center of the surge in energy nationalism, each region exhibiting its own brand of nationalism. In Asia, energy nationalism is considered a dominant regional trait (Vivoda 2010). According to the National Bureau of Asian Research, Asia's energy nationalism is driven by the region's limited energy resources and dependence on imports to sustain trends in economic growth and safeguard the region's political stability (Herberg 2011). A number of Asian states, both net importers and exporters of oil and gas, have formed NOCs over the last few decades. China's two most significant NOCs – China National Petroleum Corporation (CNPC) and China Petroleum and Chemical Corporation (Sinopec) – are among the world's largest oil companies. In 2009, China began dramatically increasing its overseas acquisitions through its third most important NOC, the National Offshore Oil Corporation (CNOOC), spending $47.59 billion in 2009 and 2010. The seven countries where China is most invested are Kazakhstan (23 percent of China's overseas equity), Sudan (15 percent), Venezuela (15 percent), Angola (14 percent), Syria (6 percent), Russia (4 percent), and Tunisia (3 percent). While China's NOCs are government-owned, they operate somewhat independently (Jiang and Sinton 2011).

India's Oil and Natural Gas Corporation Ltd. is a dominant player in the domestic market (though in recent years it has lost ground to private firms) and participates in overseas investments in 19 countries (Booz Allen Hamilton Inc. 2007; Herberg 2007; Madan 2007). Petronas, the national oil company of Malaysia, is the second largest multinational company based in a developing country. Petronas has exclusive rights to the development of the country's oil reserves, as well as considerable interests in South Africa's refining and retail fuel markets, among other countries (Lewis 2010). The Petroleum Authority of Thailand (PTT), established in 1978, integrates the various segments of the energy value chain, playing a central role in Thailand's national energy strategy (Kennedy 2010; Tordo et al. 2011b).

While the NOC strategies used in Asia are similar to those pursued by

Brazil's Petrobrás (Wertheim 2012; Sennes and Narciso; Fletcher 2009), recent energy nationalism in Latin America has mostly involved the nationalization of foreign owned assets and operations, and renegotiation of contracts with private firms (Rigobón 2008). Bolivia, Argentina, and Venezuela have recently attracted international attention with energy nationalizations. In 2006, the Bolivian government passed a decree increasing taxes, mandating a renegotiation of contracts, and returning formal ownership of underground resources to the national oil company YPFB, after having privatized them in the 1990s. The nationalization did not involve seizing assets and private firms maintained control of their operations (EIA 2012). The decision to nationalize was driven by the view that foreign firms (including Petrobrás of Brazil) were reaping excessive profits (Gjelten 2012) and effectively defining policy for the sector (Bolivia Asserts 2007; Loftis et al. 2007; Cayoja 2012; Zissis 2006). In 2012, Bolivia also nationalized the main hydroelectric plant and the power grid operator, which controls 74 percent of the electricity lines in the country (Azcui 2012). In 2012, Argentina renationalized a controlling share of its oil company, YPF, after having privatized it in the 1990s. Venezuela, since nationalizing its oil company PDVSA in 1976, has moved between periods of intense state control and greater reliance on private operators. In the last decade, the government has increased taxes and royalties, and under threat of nationalization, forced private firms to renegotiate their contracts with PDVSA (García 2011; Fletcher 2009).

Energy nationalism is among the strategies pursued by developing countries to meet growing demand and respond to developments in international markets. In Latin America, energy nationalism has emerged as a response to an earlier wave of privatizations and discontent with the performance of private energy firms. In Asia, energy nationalism is more outward-oriented and more overtly competitive. In both regions, the renewed interest in state-driven development has encouraged energy nationalism. The perception that energy nationalism is spreading to vast segments of the developing world and is a strategy with the potential to upset the world's energy scenarios stems from the changing balance of power in energy markets, as growing energy demand from the Asia-Pacific region has decreased the leverage of historically large consumers, such as the U.S. and Japan (Stanley Foundation 2006; Hughes 2011), and new competition is emerging in the form of NOCs (Agashe 2010; Brune 2010).

### 3.2 Privatize to Attract FDI

Attracting private investment is a key strategy of several developing countries facing rapidly growing energy needs. The IEA calculates that

$9 billion of the capital invested in energy in 2009 provided access to "modern energy" to previously unserved populations, and that $48 billion in investment per year would be necessary to provide universal access by 2030 (IEA 2011a, 7). In the last two decades (1990–2011), investment in energy accounted for 30 percent of the private participation in infrastructure (which also includes telecom, transport, and water and sewage) in low- and middle-income countries. South Asia led with 44 percent of infrastructure investment going to energy projects, followed by East Asia and the Pacific (39 percent), Latin America (35 percent), MENA (22 percent) and Sub-Saharan Africa (10 percent) (authors' calculations from the Private Participation in Infrastructure Database (World Bank Group (n.d.)).

Private investment in developing countries began to grow in the 1990s, peaking in 1997 and again in 2007 (Farquharson et al. 2011, 1; PWC 2012a). The global financial crisis has brought a decline in private source of financing, though not all regions have been affected equally; Africa is expected to bear the greatest impact of the crisis (PWC 2012b). Predictions of future energy investment flows recognize that while the majority of cross-border energy investment still take place within developed countries (North America and Europe received 85 percent of all cross-border energy investments in 2011 (PWC 2012a)), the focus for the future is on expanding further into China and other developing countries (McKinsey & Co 2012; PWC 2012b).

Private investment in energy infrastructure takes two forms: privatization and greenfield investments. Privatization entails transfer of total or partial ownership and control of state assets and firms to the private sector (OECD 2002). Privatization can also entail a concession to operate a public asset, like a transmission line or a port. Between 1988 and 1999, there were 221 energy privatizations (including gas, petroleum, and electricity) in 39 developing and middle-income countries for a total of $47 billion. From 2000 to 2008, there were 84 energy privatizations in 26 developing and middle-income countries for a total of $69 billion in investments (World Bank n.d.).

In the 1980s, Latin American countries, starting with Chile and then Argentina and Brazil, began to privatize electricity, a process that entailed restructuring the sector into separate generation, transmission, and distribution companies. This model was later adopted by Peru, Bolivia, Colombia, Guatemala, and others (Hall 2005). Ten countries in the region privatized a total of 116 electric distribution utilities, now serving over 60 percent of electricity subscribers in Latin America, up from only 3 percent in the early 1990s (Foster et al. 2006, 2).

Throughout the developing world, states are continuing to privatize electricity, though at a slower pace, particularly following the 2008 financial crisis. The Philippines, Jordan, Russia, Poland, Turkey and Romania recently sold state-owned electricity companies to private firms. In 2009, India launched the sale of the National Hydroelectric Power Corporation (Kikeri and Perault 2010). Turkey privatized 53 state-owned enterprises in the utilities and oil and gas sectors in 2010, and 29 in 2011 (PWC 2011; Shehadi 2002).

Petroleum privatization went the furthest in Argentina and Bolivia (though it has been partly reversed in both countries; see the section on Nationalism). Brazil opted for a state-controlled deregulation scheme, with Petrobrás maintaining a dominant presence in the sector but under the regulation of an independent agency. Mexico and Venezuela remained in complete state ownership (Palacios 2002). Colombia ended its state monopoly over petroleum in the 2000s and became one of the countries most open to foreign investment in oil (Economist 2012), and along with Brazil, is one of the largest recipients of investment in the hydrocarbons in the region (ECLA 2011).

Greenfield investment entails constructing new infrastructure where none previously existed. In the period 1990–2011, 64 percent of private infrastructure investment in developing countries was in greenfield projects, with the rest going to divestitures, concessions, and management and lease contracts. In this period, there were over 1,200 greenfield energy projects in low- and middle-income countries for a total value of $173 billion, with the East Asia and Pacific region accounting for 45 percent of the projects and 59 percent of the investment (authors' calculations from the Private Participation in Infrastructure Database (World Bank Group (n.d.)).

Private investment in energy reversed many decades of state-sponsored infrastructure development and required legal, and in some cases, constitutional reforms that ended state monopolies, permitted divestiture and unbundling of assets, and removed restrictions on private and foreign participation. Privatization was part of a broader agenda of market-oriented reforms strongly motivated by developing country debt rescheduling and conditionality schemes in the 1980s and 1990s (Vazquez 1996; Sachs 1989; Cramer 1999; Edwards 1995). During the 1980s, public investment in infrastructure lagged as countries struggled to cut spending to service their debts, which also encouraged developing country governments to surrender ownership and control of large segments of their energy sectors to private and foreign investors (Davis et al. 2000; Edwards 1995). As of 2004, approximately 51 percent of developing countries had an independent regulatory agency for electricity and 47 percent had private participation in electricity generation (Estache and Goicoechea 2005, 4).

Despite successes in improving investment, profitability, and performance of formerly state-owned and operated companies (Megginson and Netter 2001; La Porta and Lopez-De-Silanes 1999; Frydman et al. 1999; Ehrlich et al. 1994), privatization remains controversial and increasingly unpopular (Nellis 2006). Developing countries that carried out energy privatizations have had difficulty in reconciling investor's profit expectations and public expectations of better, more accessible service at low cost (Trebing and Voll 2006). Many scholars identify the inadequate sequencing of reforms, which weakens competition under private ownership (Zhang et al. 2005), and the lack of adequate regulatory capacity as critical factors undermining the success of privatization (Boubakri et al. 2008; Durakoglu 2011; Parker and Kirkpatrick 2005). The weakness of local capital markets and legal environments are also important in explaining the mixed record in energy privatization (World Bank 2004; Bayliss and McKinley 2007).

Privatization is unpopular in many developing countries because of the perception that it is unfair, having not met the expectations of the public while providing opportunities for local elites and foreigners to capture public wealth (Birdsall and Nellis 2002; Chong and López-de-Silanes 2003; Ripley 2010; Roxas and Santiago 2010; Silvestre et al. 2010). In some cases, privatization worsened the distribution of income, though this pattern is less clear in the areas of electricity where access for the poor has increased (Kundu and Mishra 2011; Birdsall and Nellis 2002). These problems have been most noticeable in Africa where host countries have had difficulty attracting foreign investors and the focus on cost recovery has not contributed to the final goals of reducing poverty and encouraging development. As a result, privatization in Africa was significant in only a few countries, with sub-Saharan Africa receiving less than 4 percent of world private investment in infrastructure between 1990 and 2003 (Hall 2007; Bayliss and McKinley 2007).

A large number of developing countries sought private investment in energy infrastructure as a response to a sharp decline in public investment. This strategy was strongly encouraged by development agencies as part of a broader agenda of economic reform, and was expected to inject new capital, better technology and more efficient management into the energy sector and have a positive spillover effect on development (Hall et al. 2005). Developing countries continue to rely significantly on private investment as a strategy to meet future energy demand, though development agencies are advising a return to building up public investment in infrastructure as private flows have proven insufficient and conditions in many countries are not favorable to this type of investment (Fay and Morrison 2005; Bayliss and McKinley 2007).

### 3.3 Increase Domestic Supply

An obvious choice for any state to enhance its energy security is to increase the amount of fuel and infrastructure it produces domestically. Nearly all developing states are doing this to some extent, with the exact strategy depending on their natural resource endowments and the state's total GDP and level of technological development. Nationalization and privatization strategies are often designed to accomplish this goal, as discussed in earlier sections. However, for many developing states, increasing the production of fossil fuels is off the table since they do not have significant known reserves. Developing countries are also under pressure from developed countries and international organizations to limit fossil fuels which contribute to climate change. The two major options discussed in this section are nuclear energy for electricity and renewable energy for all three end uses: heating, fuel, and electricity.

#### 3.3.1 Nuclear energy

Nuclear-generated energy, experiencing a revival after decades of stagnation, is available to a few developing states (Ferguson 2011). In the past, developing countries used nuclear power to gain a seat at the table of industrialized countries. Currently, nuclear energy plans are viewed as part of a strategy of energy security (Goldemberg 2009) that can provide energy independence and contribute to the economic stability of developing countries. International institutions in particular consider nuclear energy as a valid option for developing countries to address the rapid rise in demand for electricity and the need to diversify away from fossil fuels (Embrace nuclear 2006; Bratt 2010). Despite these incentives, many developing countries balk at the investment requirements and are often less pressured to adopt nuclear technology to avoid the polluting effects of fossil fuels (Goldemberg 2007).

As of 2011, 28 percent of the world's 439 nuclear reactors, and 24 percent of the installed capacity (including those in Japan) is outside of North America and Europe (Stanculescu 2011). Excluding Japan, the rest of the world has 12 percent of the installed nuclear power capacity (Nuclear Power in Japan 2012). Among the developing states, India has the most significant commitment to nuclear energy, with 20 operating reactors and six under construction (World Nuclear Association 2012b; NEI 2012). Despite this investment, nuclear power accounts for only 2 percent of India's electricity generation (IAEA 2010). India plans an eight-fold increase in nuclear-generated power to 10 percent of the electricity supply by 2022, and to 26 percent by 2052. This works out to be an annual average growth rate of 9.5 percent, slightly less than the average global growth in nuclear generation from 1970 through 2002 (McDonald 2012). Like India, China is meeting its steeply rising demand for energy by expanding generating capacity using all possible sources, including nuclear power. China has six reactors under construction and plans nearly a fivefold increase by 2020. Because it has such low generation now, even this significant increase would result in only 4 percent of expected electricity generation. China may eventually be a significant supplier of nuclear technology and services, especially in Asia (McDonald 2012). Despite frequent discussion of nuclear power potential for Africa, South Africa has the only two operating power reactors on the continent (Sokolov and McDonald 2005). In Latin America, Argentina, Brazil and Mexico each have two reactors; Argentina has a second one under construction. No other Latin American states have plans to build nuclear reactors (McDonald 2012).

According to the International Atomic Energy Agency (IAEA), some 65 countries without nuclear power have expressed interest in developing it; however, few of these countries are actively planning to build reactors. Furthermore, over half of these states have grids of less than 5 GW which are too small to accommodate nuclear power (IAEA 2010). Most of the growth in installed nuclear energy capacity is expected to take place in countries with well-established programs (World Nuclear Association 2012a). The rate at which countries have adopted their first nuclear technology in the past decades has been fairly slow, with only three countries (China, Mexico and Romania) connecting their first nuclear power plants to the grid in the post-Chernobyl era (IAEA 2010).

Several factors discourage the adoption of nuclear energy in developing countries. Nuclear power requires strong regulations governing nuclear waste and decommissioning, working with international non-proliferation agreements and organizations, and insurance for third party damage, much of which is beyond the capabilities of many developing countries (Emerging Nuclear 2012; World Bank 1992). Other obstacles faced by developing countries include the need for governments to facilitate the financing of nuclear power in a context of reduced public investment in infrastructure (Goldemberg 2007; IAEA 1993) and the protectionism that has typically surrounded nuclear reactor technology (Bratt 2010). Thailand is likely one of many states whose interest was further diminished after the 2010 Fukushima (Japan) accident (Chotichanathawewong and Thongplew 2012; IAEA 2012).

### 3.3.2 Renewable energy

Renewable energy sources provide an alternative to traditional fossil fuel and nuclear energy, and can significantly contribute to developing country energy needs. Renewable energy (RE) includes wind power, solar power, biomass, geothermal, hydropower, and ocean energy (REN21 2012). RE can entail large, centralized electricity systems as well as mini-grids and off-grid options. Globally, renewables provide 19 percent of power generation, with hydro being the dominant form, accounting for 84 percent of renewables (IEA 2011b).

High income countries and multilateral organizations, such as the United Nations Environment Program and the World Bank, have been encouraging low-income countries to invest more heavily in RE. For example, one of the six major U.S. policy goals for sub-Saharan Africa is to "promote low-emissions growth and sustainable development, and build resilience to climate change," including "supporting the adoption of low-emissions development strategies, and mobilizing financing to support the development and deployment of clean energy" (The White House 2012).

Developing countries are more likely than high-income states to use bio-energy and to need decentralized electricity systems. Instead of waiting for an expensive grid to come to a rural area, renewable energies can be quickly deployed in even the most remote areas. Decentralized generation can also help lessen the risk of massive power outages and the resulting dependence on expensive and unhealthy diesel power, which can cost up to 5 percent of a country's annual GDP – a problem that affects 60 percent of the 48 countries in sub-Saharan Africa (Amin et al. 2012).

A plethora of RE projects is planned or underway throughout the developing world. As in so many areas of energy security in the developing world, China is taking the lead on RE, now possessing the world's largest RE capacity. Renewable sources accounted for more than 30 percent of the 90 GW that China added to its electric capacity in 2011; of that amount, non-hydro sources, mostly wind and solar, accounted for about 20 percent (REN21 2012). In Latin America, Brazil leads the way in RE, having started using sugarcane to produce ethanol in the 1970s, though it is used mainly for fueling vehicles (Valdes 2011). Argentina and Colombia also have established biofuels markets, while a number of other Latin American countries have significant potential for producing biofuels for transportation but have only recently started exploring this option (Janssen and Rutz 2011). While African states already have a very high percentage of renewables, these are mostly from traditional sources (wood and animal dung) that have high negative externalities. The push is to move African states to modern RE. Kenya is one state that has made significant commitments in this direction: by 2031 it hopes to generate about 27 percent of its electricity needs through geothermal energy. The African Development Bank recently committed $145 million to one of these projects, and the World Bank along with French and

Japanese development agencies have given $500 million in loans and grants (Manson 2012). South Africa's Department of Energy recently concluded the second round of bids for 19 RE independent power projects for 1,044 MW of power using a variety of technologies, including wind, small hydro, solar photovoltaic, and concentrated solar power (Nedbank Capital 2012). Sub-Saharan Africa, where wood and dung are the most common biofuels, has the greatest bio-energy potential due to its large areas of cropland, unused pasture, and low agricultural productivity (Watson 2008). In Bangladesh – which has installed 1.2 million rural solar home systems – solar is the "one energy-related area in which the country is making progress" (Ebinger 2011, 100). A micro-hydro plant initiative in Nepal has led to 3,850 permanent jobs since 1998 (REN21 2012, 26). Thailand has a 15-year RE development plan that includes more energy with wind, hydropower, biogas, and municipal waste as the source (Chotichanathawewong and Thongplew 2012). In 2006, India created the Ministry of New and Renewable Energy, which includes four technical institutions focused on solar, wind, biofuels and development issues. India is now the fifth-largest wind power generator (Ebinger 2011, 33, 48; Pachauri 2011, 100).

Renewable energy projects present a number of political and economic challenges and tradeoffs. Government-driven initiatives require significant research and policy making abilities, which many of the poorer states may not have. The World Bank found that even China, with its comparatively strong central government, has suffered a number of challenges with its RE projects (Xie et al. 2009; Xu 2012). In addition, RE is often expensive to develop and requires government incentives, which states may have difficulty forming and enforcing (Chotichanathawewong and Thongplew 2012). Policies to force producers to invest in renewables and to get consumers to use the RE sources – such as feed-in tariffs, quota-based incentives, and financial/tax-based incentives – can be effective, but applied improperly can be costly and inefficient. Decentralized systems (off-grid or mini-grid) combined with renewable sources can provide a lower cost alternative for bringing energy to rural and remote settlements but generally do not make economic sense for denser areas (Deichmann et al. 2011). Wind power requires suitable land that may be scarce in smaller states (Chotichanathawewong and Thongplew 2012). Bio-energy presents its own problems. Using land to grow crops for fuel rather than food can alter domestic food prices and may end up displacing small-shareholders (Deepchand 2002). In addition, land that could be used for biofuels may contain significant biodiversity that would be lost (Diaz-Chavez et al. 2010). As shown from Brazil's ethanol program, modern bio-energy can also create significant waste products (World Energy Council 2011).

Municipal waste projects must also earn acceptance from local users (Chotichanathawewong and Thongplew 2012).

Thus, while the strategy of increasing energy supply by investing in RE holds long-term promise, there are a number of financial and technical challenges that must be met, as well as tradeoffs to consider.

### 3.4 Diversify Import Options

While diversifying energy sources can include a broad range of issues (Stirling 2011), in this section we focus exclusively on diversifying the number of states from which a developing state imports fossil fuels. Many developing states do not have significant imports, in part because they lack sufficient energy infrastructure and have relatively few private vehicles per capita compared to developed states. Others who have substantial energy demands, notably China and India, are diversifying their energy supplies to enhance security. One of the most important investments is in financing oil and natural gas pipelines to states that export fossil fuels.

China's energy security issues are significant and growing and have had a global impact. In oil trade, China accounted for about two-thirds of growth from 2011 to 2012, with its net imports rising by 13 percent (BP-British Petroleum 2012). Until 1993, China was self-sufficient in energy and therefore faced few risks (Bambawale and Sovacool 2011). China now imports about half its oil, mostly from the Middle East and Africa (Zhou et al. 2010). To increase its options and lower the risks that arise from shipping through the various Straits (see "Protect Energy Infrastructure" below), China has invested in, or is planning to invest in, oil pipelines linking it directly to Kazakhstan, Russia, Myanmar and Pakistan (Leung 2011). In 2006, despite wide-spread skepticism from petroleum analysts, China and Kazakhstan announced the opening of the first pipeline to deliver oil directly to China (Hancock 2009). Other investments include projects in Angola, Ecuador, Myanmar, Nigeria, Peru, Sudan, and Venezuela (Janardhanan 2010; Jaffe and Medlock 2005). As discussed under the "Nationalize Energy" section, China has significantly invested in foreign oil and gas projects. China has also been developing technology and infrastructure to create liquid natural gas which can then be transported on ships, opening up new avenues for energy imports.

While not as invested in this strategy as China, India's increasing energy demands – which some expect will account for 12 percent of global energy demand by 2030 (compared to 27 percent for China) (Thavasi and Ramakrishna 2009) – may push India to further diversity its options. Dependent on Gulf oil and the fluctuations that political issues in the Gulf sometimes bring, India has reportedly been "energy hunting" in West

Asia, the Persian Gulf, Central Asia, South Asia and Asia-Pacific (Misra 2007). Brazil has also been actively pursuing energy sources in neighboring countries. It built a pipeline to transport natural gas from Bolivia to the southern part of Brazil (World Bank 2003), and in 2010 concluded an agreement with Peru to build multiple hydroelectric plants mainly to serve demand in the south of Brazil, though actual construction has been halted by protests (Brazil and Peru 2010; Outrage 2012). While this strategy has been highly effective for the emerging economies, it is out of reach for most of the developing world for two reasons: most states do not use enough fossil fuels to warrant the significant investment required to build pipelines and lack the domestic financing and international investment interest to pay for them.

### 3.5 Build and Maintain Strategic Petroleum Reserves

Stockpiling petroleum is increasingly being considered by developing countries as a viable energy security strategy. Strategic Petroleum Reserves (SPRs) were first established in response to the 1973–1974 Arab oil embargo as a buffer against oil supply shortages that may threaten economic stability (Andrews and Pirog 2012; DOE 2012). While the largest SPRs are in Western states, a number of developing countries have established, or are planning to establish, reserves of their own. China has the largest reserve with total capacity of nearly 500 million barrels (Bahgat 2011, 71). India plans to accumulate about 40 million barrels (14 days' worth of imports) (Hook 2011). Albeit on a much smaller scale, the Philippines, Vietnam, Rwanda, Uganda, Kenya and Zambia also have plans for some reserves (Castillo 2011; Kojima 2009; Do and Sharma 2011, 5774).

For many developing countries, cost and management challenges are serious obstacles to realizing their plans for SPRs. Building fuel stockpiles is expensive – it is estimated that a 40-day reserve for India would cost $6.3 billion (Dietl 2004) – and implies an opportunity cost that can negatively affect GDP growth in the short term and outweigh the long-term benefit of using SPRs to smooth oil price fluctuations (Wei et al. 2008). Building SPRs makes more sense for large countries, like China and India, while smaller developing states may be better off implementing energy strategies based on diversifying resources and suppliers, or limiting consumption by reducing speed limits, banning weekend driving, reducing public transportation fares, and increasing gas taxes to reduce the demand for oil (IEA 2005).

SPRs can help developing countries maintain stable access to energy resources in periods of short-term supply disruption. However, the costs of creating and maintaining SPRs may not serve the long-term interests of developing states (Yergin 2006, 1990; Goolsbee 2012).

### 3.6 Protect Energy Infrastructure

Another aspect of energy security is protecting one's energy infrastructure. This issue has recently gained enough attention to warrant a database on such attacks (Giroux and Burgherr 2012). Critical energy infrastructure includes electricity grids, hydroelectric dams, nuclear plants, petroleum production facilities (oil and gas fields, wells, platforms, and rigs), refineries, transportation facilities (such as pipelines, terminals, and tankers), distribution sites, and even corporate offices (Koknar 2009). Threats to infrastructure may originate from domestic or international terrorists, pirates, local populations seeking direct access to energy resources for consumption or trade with separatist or other anti-state political activists (Chaturvedi and Samdarshi 2011; Jaffe and Lewis 2002; Giroux and Burgherr 2012).

Energy infrastructure is also vulnerable to technical problems unrelated to political issues (Yu and Pollitt 2009; Hines et al. 2008; Lukszo et al. 2010). Technical strategies, such as increased redundancy, backup generators and early warning and load distribution systems address these threats (Cherp and Jewell 2011). Since this is an extensive literature which moves away from the political and economic focus of this chapter, we do not further address technical vulnerabilities.

Petroleum products are the main energy sources transported via massive physical infrastructure (Bateman 2003). In 2011, total world oil production was about 88 million barrels per day, nearly half of which was transported via tankers and one-third via land-based pipelines (EIA 2012). Maritime security remains a considerable threat to transportation and economic stability (Komiss and Huntzinger 2011; Dannreuther 2011). Two of the most important chokepoints for energy transportation are the Strait of Hormuz and the Strait of Malacca. About 17 million barrels of crude move through the Strait of Hormuz every day, while Malacca moves 15.2 million barrels a day (EIA 2012).

Energy infrastructure attracts three forms of non-government organized attacks: piracy, smuggling, and terrorism. The International Maritime Organization defines piracy as "any illegal act of violence or detention, or any act of depredation committed for private ends by the crew or passengers of a private ship or a private aircraft," and directed at another ship or aircraft at sea (IMO 2012). For a number of economic and political reasons, pirate attacks are on the rise (Chalk 2008; Sinai 2004). Pipelines have long been targets of smuggling operations. Crude smuggling in Nigeria and Iraq offer examples of heightened sophistication in both method and organization, and the burdens it places on host governments and corporations (Ikelegbe 2005, 221; Onuoha 2008). In Nigeria, Shell reported losing 100,000 barrels a day in 2003, translating to a nearly 10 percent reduction of Nigeria's total production and costing companies and the host country millions of dollars (Oduniyi 2003). In Iraq, some reports suggest that nearly 40–50 percent of the revenue from smuggled oil in 2006 was pocketed by insurgents (Wahab 2006). Since 1986, persistent attacks on Colombia's Caño Limón pipeline – 950 attacks as of 2004 – have cost the government $2.5 billion in revenues and have cut supplies for months (Parfomak 2004, 4). Terrorism magnifies threats to energy infrastructure (Chalk 2008, 24–25; Luft and Korin 2004). Cyberterrorism presents an additional threat, particularly against critical outdated infrastructure, which is more common in the developing world (Sheldon et al. 2004; Oliveira 2010). Environmental disasters caused by infrastructure accidents can also damage long-term economic growth, with potentially lasting effects on ecological systems (O'Rourke and Connolly 2003; Amin and Gellings 2006). Finally, massive political instability and even coups can threaten infrastructure, such as in Libya, Syria, Egypt and Yemen (Giroux 2012).

Strategies to counter energy infrastructure threats can be as complex as regional cooperation or diversifying maritime transportation through development of land-based transportation (Blank 2009), and as simple as providing locks and antipiracy training (Rosenberg 2009, 47–48). To avoid piracy and international conflict disruptions, China is transiting from maritime imports to land-based imports from Central Asia (Sheives 2006; Chung 2004). Military build-up is also an option, but this strategy can threaten the tenuous balance of maritime sovereignty (Cole 2008) and power stability (Kaplan 2009; Ball 1993–1994). Indonesia, Singapore and Malaysia have attempted to increase cooperation in combating piracy, and Japan and India have proposed larger-scale naval patrols of the Strait of Malacca. Naval intervention, however, breaches the sensitive issues of (imperial) militarism in the region, as well as unresolved territorial issues (Mo 2002). Africa faces similar complications in collective attempts to address piracy and smuggling, especially with regards to commitments to international law (Gibson 2009; Baker 2011). Some of these threats are addressed by strategies discussed elsewhere, such as using RE sources to avoid relying on maritime and pipeline deliveries.

### 3.7 Increasing Efficiency

Many developing countries include energy efficiency as a critical strategy for reducing demand, or keeping it from growing as fast as it might otherwise, and thus enhancing security (Chotichanathawewong and Thongplew 2012). In China, where energy efficiency has "been on the agenda" for decades, improved efficiency enabled the state to only double energy demand while GDP quadrupled since 1990 (Bambawale and Sovacool 2011). Brazil has an extensive energy efficiency labeling program to inform consumers about long-term electricity costs for various products. A review of 119 projects covering nine manufacturing sub-sectors in developing states found substantial efficiency gains (Alcorta et al. 2012). However, while about two-thirds of the available opportunities for energy efficiency investments are in the developing world (Farrell and Remes 2009), many companies likely remain uninformed about the opportunities (Alcorta et al. 2012). In addition, two factors that are a significant challenge in the developing world are seen as critical to the success of energy efficiency strategies: generating consumer demand for efficient products, and building consumer confidence and trust in labeling schemes (Ellis et al. 2009). Access to local financing is considered another major obstacle to greater efficiency in developing states (ESMAP 2006). For many states, such as India, price subsidies discourage investments in fuel-efficient technologies (Ebinger 2011, 51). Finally, states and international organizations need to improve data collection to better understand consumption patterns and to develop state-of-the-art indicators to better inform policy making (Trudeau and Taylor 2011).

## 4 CONCLUSIONS

Developing states – a very broad category in terms of population, territorial size, income levels, and status in the international system – pursue a variety of strategies meant to enhance their energy security in terms of access to supply. In this chapter, we have summarized a variety of strategies that appear most frequently in the literature on the developing states and energy issues. To illustrate these strategies, we mention cases throughout the developing world providing some historical background and highlighting some of the issues states face as they pursue these strategies. We draw a number of conclusions from our research.

First, nearly all of the strategies we discuss are actively employed by developed states as well as developing states. This does not mean that all developing states are equally equipped–in terms of financing means and state capacity–or even committed to implementing the individual strategies. In addition, developing states often have a different focus than the highly industrialized, wealthy states due to the high percentages of people who lack access to electricity and the states' long-term focus on industrial development. For example, scholars caution that energy efficiency may be low on the agenda for many developing states, whereas it has become more important in developed regions.

Second, the energy mix in the developing states looks very different from that in developed states. For example, in much of Africa and parts of Asia, RE comprises a majority of the states' energy resources. However, much of the RE is from traditional sources – wood and sometimes animal dung. The challenge is thus to move to modern sources, such as solar and wind, and even more advanced and/or sustainable forms of biomass, such as industrial waste or plentiful cassava.

Third, the energy needs of developing countries will require massive injections of capital that cannot be obtained without private sector participation. However, the experience of the last three decades with FDI and privatization has led developing country governments and international development agencies to reconsider the role of the state in steering private investment towards fulfilling long-term development goals. In many countries, private investment flows have proven insufficient or inadequate to achieving these goals. As a result, some countries have partially returned to state controlled energy strategies. Given conditions of high energy prices and private financing constraints we should expect energy nationalism to remain an attractive strategy for larger countries with active NOCs as well as smaller countries with resources to protect.

Fourth, some of the strategies appear to be endorsed and advocated far more by international organizations (such as the World Bank and the United Nations Development Program), Western states (particularly the US and the EU), and industry groups (such as the World Alliance for De-centralized Energy) than by the developing states themselves.

Fifth, while it is often not discussed in the literature, particularly by those advocating for the policies, the lack of state capacity remains a significant issue for many developing states. Adequate regulation, designing long-term energy policies, and orchestrating new investments while keeping energy at affordable prices require significant government capacity, which not all developing states can supply.

Sixth, as the only major rising power with global reach and ambitions, China's strategies often have more in common with the developed than the developing states. India and Brazil also have options open to them that are out of reach for others. A repeated theme in the strategies section is the cost of these strategies, many of which are beyond the means of most developing states and raise issues of opportunity costs. If states invest in nuclear energy, what other basic human needs might be sacrificed? Furthermore, are there more cost effective measures? For example, when one compares the costs of building and maintaining a strategic petroleum reserve to reducing fuel demand through taxes and lowering the speed limit, the former strategy seems a poor use of funds.

Finally, we identify areas for future research. In the existing literature, few authors discuss developing state strategies as distinct from developed state strategies and few compare cases within the developing world. Research that focuses on groups of countries according to their size (both territorial and economic), energy resources, and level of development would provide a better mapping of energy issues and strategies. Research that includes local print sources and interviews of government officials and individuals in the private energy sector of developing countries is also necessary to distinguish the agenda of developing states from that of external actors, in particular with respect to the pursuit of alternative sources of energy, such as renewables and nuclear. Another promising avenue of research is on the role of sub-national governments and non-state actors in promoting the development of alternative sources of energy. State strategies at the national level tend to focus on large infrastructure projects while most of the off-grid and alternative energy projects are sponsored by international agencies, NGOs, and community-level organizations with goals and agendas that coexist but can also be at odds with those of the national government. This situation is specific to developing countries, where infrastructure is not fully developed and integrated and the presence of the state throughout a country's territory can be uneven. Research in this area would help us better understand the conditions under which particular energy strategies are formulated and deployed.

## NOTES

* The authors thank Margaret Albert for her excellent research support for this chapter.
1. For a detailed discussion on defining energy security, including an analysis of 45 definitions, see Sovacool (2011).

## REFERENCES

Agashe, Geeta (2010). "Big oil faces global competition from national oil companies". *Pipeline & Gas Journal* 237 (6): 42.

AGECC (2010). *Energy for a Sustainable Future*. New York: United Nations Advisory Group on Energy and Climate Change.

Alcorta, Ludovico, Morgan Bazilian, Giuseppe De Simone and Ascha Pedersen (2012). *Return on Investment from Industrial Energy Efficiency: Evidence from Developing Countries*. Vienna, Austria: United Nations Industrial Development Organization.

Amin, Adnan, Achim Steiner and Kandeh K. Yumkella (2012). "Lighting the Dark

Continent." Project Syndicate, http://www.project-syndicate.org/commentary/lighting-the-dark-continent (accessed February 2013).
Amin, S. Massoud, and Clark Gellings (2006). "The North American power delivery system:Balancing market restricting and environmental economics with infrastructure security." *Energy* 31 (6–7): 976–999.
Andrews, Anthony and Robert Pirog (2012). *The Strategic Petroleum Reserve: Authorization, Operation, and Drawdown Policy*. Washington DC: Congressional Research Service.
APEC (2007). *A Quest for Energy Security in the 21st Century: Resources and Constraints*. Tokyo: Asia Pacific Energy Research Centre.
Azcui, Mabel (2012). "Evo Morales nacionaliza la filial de Red Eléctrica en Bolivia." *El País online*, http://economia.elpais.com/economia/2012/05/01/actualidad/1335887717_799794.html/ (accessed February 2013).
Bahgat, Gawdat (2011). *Energy Security: An Interdisciplinary Approach*. West Sussex, UK: John Wiley & Sons, Ltd.
Baker, Michael (2011). "Toward an African Maritime Economy: Empowering the African Union to Revolutionize the African Maritime Sector." *War College Review* 64 (2): 39–62.
Ball, Desmond (1993–1994). "Arms and Affluence: Military Acquisitions in the Asia-Pacific Region." *International Security* 18 (3): 78–112.
Bambawale, Malavika Jain and Benjamin K. Sovacool (2011). "China's energy security: The perspective of energy users." *Applied Energy* 88 (5): 1949–1956.
Bateman, Sam (2003). "Sea Lane Security." *Maritime Studies* 128: 17–27.
Bayliss, Kate and Terry McKinley (2007). "Providing basic utilities in sub-Saharan Africa: why has privatization failed?" *Environment* 49 (April).
Birdsall, Nancy and John Nellis (2002). "Winners and Losers: Assessing the Distributional Impact of Privatisation." Center for Global Development Working Paper. Washington, DC: Center for Global Development.
Blank, Stephen (2009). "Chinese Energy Policy in Central and South Asia." *Korean Journal of Defense Analysis* 21 (4): 435–453.
Bolivia Asserts (2007). "Bolivia asserts oil sovereignty: an interview with Carlos Villegas." *Multinational Monitor* 29 (4).
Booz Allen Hamilton Inc. (2007). India's National Oil Companies Emerging as Commercial Actors in the International Energy Market, http://www.nbr.org/Downloads/pdfs/ETA/ES_Conf07_MacDonald.pdf (accessed February 2013).
Boubakri, Narjess, Jean-Claude Cosset and Omrane Guedhami (2008). "Privatisation in Developing Countries: Performance and Ownership Effects." *Development Policy Review* 26 (3): 275–308.
BP-British Petroleum (2011). "BP Energy Outlook 2030." London: British Petroleum.
BP-British Petroleum (2012). "Statistical Review of World Energy 2010." London.
Bratt, Duane (2010). "Re-igniting the Atom: The Political Consequences of the Global Nuclear Revival." *The Whitehead Journal of Diplomacy and International Relations* Summer/Fall: 59–74.
Brazil and Peru (2010). "Brazil and Peru sign agreement for energy integration." *Bank Information Center*, http://www.bicusa.org (accessed March 2013).
Brune, Nancy (2010). "Latin America: A Blind Spot in US Energy Security Policy." *Journal of Energy Security* (July), http://www.ensec.org (accessed February 2013).
Calderón, César and Luis Servén (2010). "Infrastructure in Latin America." Policy Research Working Paper no. 5317. Washington, DC: The World Bank.
Castillo, Lorelei (2011). "Establishment of National Strategic Petroleum Reserve Sought." Congress of the Philippines, Public Relations and Information Bureau.
Cayoja, Mario Roque (2012). Cómo funcionan las nacionalizaciones en Bolivia," infobae.com, http://america.infobae.com/notas/49523-Cmo-funcionan-las-nacionalizaciones-en-Bolivia (accessed February 2013).
Chalk, Peter (2008). "The Maritime Dimensions of International Security: Terrorism, Piracy, and Challenges for the United States." Santa Monica, CA; Arlington, VA; Pittsburgh, PA: RAND Corporation.

Chaturvedi, A. and S.K. Samdarshi (2011). "Energy, economy and development (EED) triangle: Concerns for India." *Energy Policy*, 39 (8): 4651–4655.

Chen, Matthew E. and Amy M. Jaffe (2007). "Energy Security: Meeting the Growing Challenge of National Oil Companies." *The Whitehead Journal of Diplomacy and International Relations*, Summer/Fall: 9–21.

Cherp, Aleh and Jessica Jewell (2011). "The three perspectives on energy security: intellectual history, disciplinary roots and the potential for integration." *Current Opinion in Environmental Sustainability*, 3: 202–212, doi: 10.1016/j.cosust.2011.07.001.

Chester, Lynne (2010). "Conceptualising energy security and making explicit its polysemic nature." *Energy Policy*, 38: 887–895.

Chong, Alberto and Florencio López-de-Silanes (2003). "The Truth about Privatization in Latin America", Research Network Working Paper, Washington, DC: Inter-American Development Bank.

Chotichanathawewong, Qwanruedee, and Natapol Thongplew. 2012. Development Trajectory, Emission Profile, and Policy Actions: Thailand. Tokyo: Asian Development Bank Institute.

Chung, Chien-peng (2004). "The Shanghai Co-operation Organization: China's Changing Influence in Central Asia." *China Quarterly*, 180: 989–1009.

Cole, Bernard (2008). *Sea Lanes and Pipelines*, Westport, CT: Praeger Security International.

Cramer, Chris (1999). "Privatisation and the Post-Washington Consensus: Between The Lab And The Real World?", Discussion Paper. Centre for Development Policy & Research (CDPR), http://eprints.soas.ac.uk/7376/1/DiscussionPaper0799.pdf (accessed February 2013).

Dannreuther, Roland (2011). "China and global oil: vulnerability and opportunity." *International Affairs*, 87 (6): 1354–1364.

Davis, Jeffrey, Rolando Ossowski, Thomas Richardson and Steven Barnett (2000). "Fiscal and Macroeconomic Impact of Privatization," Occasional Paper No.194, Washington, DC: International Monetary Fund.

Deepchand, K. (2002). "Promoting equity in large-scale renewable energy development: the case of Mauritius." *Energy Policy*, 30 (11–12): 1129–1142.

Deichmann, Uwe, Craig Meisner, Siobhan Murray and David Wheeler (2011). "The Economics of Renewable Energy Expansion in Rural Sub-Saharan Africa." *Energy Policy*, 39 (1): 215–227.

Diaz-Chavez, Rocio, Steven Mutimba, Helen Watson, Sebastian Rodriguez-Sanchez, and Massaër Nguer. 2010. "Mapping Food and Bioenergy in Africa," in *Ghana: Forum for Agricultural Research in Africa*. Accra, Ghana: Forum for Agricultural Research in Africa.

Dietl, Gulshan (2004). "New threats to oil and gas in West Asia: Issues in India's energy security." *Strategic Analysis*, 28 (3): 373–389.

Dirmoser, Dietmar (2007). *Energy Security: New Shortages, The Revival of Resource Nationalism, and the Outlook for Multilateral Approaches.* Berlin: Compass 2020.

Do, Tien Minh and Deepak Sharma (2011). "Vietnam's energy sector: A review of current energy policies and strategies." *Energy Policy*, 39 (10): 5770–5777.

DOE (2012). "Strategic Petroleum Reserve – Profile," U.S. Department of Energy (2012), http://www.fossil.energy.gov/programs/reserves/spr/index.html (accessed February 2013).

Durakoglu, S. Mustafa (2011). "Political institutions of electricity regulation: The case of Turkey." *Energy Policy*, 39 (9): 5578–5587.

Eberhard, Anton, Orvika Rosnes, Maria Shkaratan and Haakon Vennemo (2011). *Africa's Power Infrastructure*, Washington, DC: The World Bank.

Ebinger, Charles K. (2011). *Energy and Security in South Asia: Cooperation or Conflict?* Washington, D.C.: Brookings Institution Press.

ECLA (2011). "Foreign direct investment between the European Union and Latin America and the Caribbean," in The Economic Commission for Latin America (ed.) *Foreign Direct Investment in Latin America and the Caribbean 2011*, Santiago, Chile.

Economist, The (2006). "Really Big Oil; National oil companies," *The Economist (US)*, 380.

Economist, The (2012). "Colombia's oil industry: Gushers and guns," *The Economist (UK)*, http://www.economist.com/node/21550304 (accessed February 2013).
Edwards, Sebastian (1995). *Crisis and Reform in Latin America: From Despair to Hope*, World Bank and Oxford University Press.
Ehrlich, Isaac, Georges Gallais-Hamonno, Zhiqiang Liu and Randall Lutter (1994). "Productivity Growth and Firm Ownership: An Empirical Investigation." *Journal of Political Economy*, 102: 1006–1038.
EIA (2011). "International Energy Outlook 2011," U.S. Energy Information Administration, http://www.eia.gov/forecasts/ieo/pdf/0484%282011%29.pdf (accessed February 2013).
EIA (2012). "World Oil Transit Chokepoints," U.S. Energy Information Administration [cited July 25, 2012], http://www.eia.gov/countries/regions-topics.cfm?fips=WOTC (accessed February 2013).
Ellis, M., I. Barnsley and S. Holt (2009). "Barriers to maximizing compliance with energy efficiency policy," Paper read at ECEEE 2009 Summer Study proceedings.
Embrace nuclear (2006). "Should developing nations embrace nuclear energy?," SciDevNet, http://www.scidev.net/en/editorials/should-developing-nations-embrace-nuclear-energy.html (accessed February 2013).
Emerging Nuclear (2012). "Emerging Nuclear Energy Countries," World Nuclear Association, http://www.world-nuclear.org/info/inf102.html (accessed February 2013).
ESCAP (2006). "Enhancing Regional Cooperation in Infrastructure Development Including that Related to Disaster Management," in *62nd Commission Theme Study*, New York, NY: United Nations Economic and Social Commission for Asia and the Pacific.
ESMAP (2006). "The Energy Efficiency Investment Forum: Scaling Up Financing in the Developing World," in *Workshop Proceedings*, Washington, DC: Energy Sector Management Assistance Program.
Estache, Antonio and Ana Goicoechea (2005). "How widespread were private investment and regulatory reform in infrastructure utilities during the 1990s?" World Bank Policy Research Working Paper 3595, Washington, DC: The World Bank.
ExxonMobil (2012). "2012 The Outlook for Energy: A View to 2040," http://www.exxonmobil.com/Corporate/files/news_pub_eo.pdf (accessed February 2013).
Farquharson, Edward, Clemencia Torres de Mastle, E.R. Yescombe and Javier Encinas (2011). *How to Engage with the Private Sector in Public-Private Partnerships in Emerging Markets*, Washington, DC: The World Bank.
Farrell, Diana and Jaana Remes (2009). "Promoting Energy Efficiency in the Developing World," *McKinsey Quarterly Economic Studies*.
Fay, Marianne and Mary Morrison (2005). *Infrastructure in Latin America & the Caribbean*, Washington, DC: The World Bank.
Ferguson, Charles D. (2011) Think Again: Nuclear Power. *Foreign Policy*, 189 (Nov): 49–53.
Fletcher, Sam (2009). "Special Report: Pemex, PDVSA, Petrobras: how strategies, results differ," *Oil&Gas Journal*, http://www.ogj.com (accessed February 2013).
Foster, Vivien, Luis Andrés, and José Luis Guasch (2006) "The Impact of Privatization On The Performance Of The Infrastructure Sector: The Case Of Electricity Distribution In Latin America Countries," World Bank Policy Research Working Paper 3936. Washington, DC: The World Bank.
Frydman, Roman, Cheryl Gray, Marek Hassel and Andrzej Rapaczynski (1999). "When Does Privatization Work? The Impact of Private Ownership on Corporate Performance in Transition Economies." *Quarterly Journal of Economics*, 114: 1153–1191.
Gambill, Gary (2001). "Syria's Foreign Relations: Iraq." *Middle East Intelligence Bulletin*, 3 (3).
García, Julián Cárdenas (2011). "Rebalancing oil contracts in Venezuela." *Houston Journal of International Law*, 33 (2): 235–301.
Gibson, John (2009). "Maritime security and international law in Africa." *African Security Review*, 18 (3): 61–70.
Giroux, Jennifer (2012). "Energy Infrastructure Attacks Examined: An Emerging Research

Area," in United States Institute of Peace, *International Network for Economics and Conflict*, http://inec.usip.org/blog (accessed February 2013).

Giroux, Jennifer and Peter Burgherr (2012). "Canvassing the Targeting of Energy Infrastructure: The Energy Infrastructure Attack Database." *Journal of Energy Security*, July.

Gjelten, Tom (2012). "The dash for gas: the golden age of an energy game-changer." *World Affairs*, 174, January/February.

Goldemberg, José (2007). "The Limited Appeal of Nuclear Energy." *Scientific American*, 297 (1).

Goldemberg, José (2009). "Nuclear energy in developing countries." *Daedalus*, 138 (4): 71–80.

Goolsbee, Austan (2012). "There's Too Much Crude in the Strategic Petroleum Reserve," http://online.wsj.com/article/SB10001424052702303772904577335372708364592.html (accessed July 2012).

Hall, David (2005). *Electricity privatisation and restructuring in Latin America and the impact on workers*, London, UK: Business School University of Greenwich.

Hall, David (2007). "Energy privatisation and reform in East Africa," Public Services International Research Unit, University of Greenwich, http://www.psiru.org (accessed February 2013).

Hall, David, Emanuele Lobina and Robin de la Motte (2005). "Public resistance to privatisation in water and energy." *Development in Practice*, 15 (3&4): 286–301.

Hancock, Kathleen J. (2009). *Regional Integration: Choosing Plutocracy*. New York: Palgrave.

Harris, Gardiner and Vikas Bajaj (2012). "As Power is Restored in India, the 'Blame Game' over Blackouts Heats Up," *New York Times*, August 1.

Herberg, Mikkal (2007). "The Rise of Asia's National Oil Companies." *NBR Special Report*, 14.

Herberg, Mikkal (2010). "The Rise of Energy and Resource Nationalism in Asia," in Ashley Tellis, Andrew Marble and Travis Tanner (eds), *Strategic Asia 2010–11: Asia's Rising Power and America's Continued Purpose*, Washington, DC: The National Bureau of Asian Research.

Herberg, Mikkal (2011). "Introduction," in Gabe Collins, Andrew Erickson, Yufan Hao, Mikkal Hergerg, Llewelyn Hughes, Weihua Liu and Jane Nakano (eds), *Asia's Rising Energy and Resource Nationalism: Implications for the United States, China, and the Asia-Pacific Region*, Washington, DC: National Bureau of Asian Research.

Hines, P., J. Apt and S. Talukdar (2008). "Trend in the History of Large Blackouts in the United States," IEEE Power and Energy Society General Meeting-Conversion and Delivery of Electrical Energy in the 21st Century. Pittsburgh, PA.

Hook, Leslie (2011). "Asia moves to shore up strategic oil reserves," *Financial Times*, March 2.

Hoyos, Carola (2006). "Nationalist politics muscle back into world energy," *ft.com*, May 4.

Hughes, Llewelyn (2011). "Resource Nationalism in the Asia-Pacifi Region: Why does it Matter?" in Gabe Collins, Andrew Erickson, Yufan Hao, Mikkal Herberg, Llewelyn Hughes, Weihua Liu and Jane Nakano (eds), *Asia's Rising Energy and Resource Nationalism: Implications for the United States, China, and the Asia-Pacific Region*, Seattle, WA: The National Bureau of Asian Research.

IAEA (1993). *Financing Arrangements for Nuclear Power Projects in Developing Countries – A Reference Book*, Vienna, Austria: International Atomic Energy Agency.

IAEA (2010). "International Status and Prospects of Nuclear Power," International Atomic Energy Agency.

IAEA (2012). "Starting Right: Developing Countries Make Progress Toward Nuclear Power," http://www.iaea.org/newscenter/news/2012/startingright.html (accessed February 2013).

IEA (2005). "Saving Oil in a Hurry," International Energy Agency, http://www.iea.org/publications/freepublications/publication/savingoil.pdf (accessed February 2013).

IEA (2011a). "World Energy Outlook, Executive Summary," International Energy Agency, http://www.iea.org (accessed February 2013).
IEA (2011b). "World Energy Outlook, Executive Summary," http://www.iea.org (accessed February 2013).
IEA (2012.) "Energy poverty," http://www.iea.org/topics/energypoverty/ (accessed February 2013).
Ikelegbe, Augustine (2005). "The Economy of Conflict in the Oil Rich Niger Delta Region of Nigeria." *Nordic Journal of African Studies*, 14 (2): 208–234.
IMO (2012). "Piracy and armed robbery against ships," International Maritime Organization, http://www.imo.org/ourwork/security/piracyarmedrobbery/ (accessed February 2013).
India Reels From Second Power Grid Collapse (2012). *Dow Jones News*, http://www.4-traders.com/POWER-GRID-CORPORATION-OF-9059859/news/India-Reels-From-Second-Power-Grid-Collapse-14439883/ (accessed March 2013).
Jaffe, A.M. and K.B. Medlock (2005). "China and Northeast Asia," in J.H. Kalicki and D.L. Goldwyn (eds), *Energy and security, toward a new foreign policy strategy*, Washington DC: Woodrow Wilson Center Press, pp. 267–289.
Jaffe, Amy and Steven Lewis (2002). "Beijing's Oil Diplomacy," *Survival*, 44 (1): 115–134.
Janardhanan, N. (2010). "Global energy geopolitics and the strategy of Chinese oil corporates," in V.R. Raghavan (ed.), *Emerging challenges to energy security in the Asia Pacific*, Chennai, India: Centre for Security Analysis, pp. 70–86.
Janssen, Rainer and Dominik Damian Rutz (2011). "Sustainability of biofuels in Latin America: Risks and opportunities." *Energy Policy*, 39 (10): 5717–5725.
Jiang, Julie and Jonathan Sinton (2011). "Overseas Investments by Chinese National Oil Companies. Assessing the drivers and impacts," International Energy Agency. Information Paper (February). Paris, France: OECD/IEA.
Kaplan, Robert (2009). "Center Stage for the Twenty-first Century: Power Plays in the Indian Ocean." *Foreign Affairs*, 88 (2): 16–32.
Kennedy, Andrew (2010). "Rethinking energy security in China." East Asia Forum 2012, http://www.eastasiaforum.org/2010/06/06/rethinking-energy-security-in-china/ (accessed March 2013).
Kikeri, Sunita and Matthew Perault (2010). "Privatization Trends," Note Number 322, Washington, DC: The World Bank.
Kojima, Masami (2009). "Government Response to Oil Price Volatility: Experience of 49 Developing Countries," Extractive Industries for Development Series #10, World Bank, http://siteresources.worldbank.org/INTOGMC/Resources/10-govt_response-hyperlinked.pdf (accessed July 2012).
Koknar, Ali M. (2009). "The Epidemic of Energy Terrorism," in Gal Luft and Anne Korin (eds), *Energy Security Challenges for the 21st Century: A Reference Handbook*, Santa Barbara: ABC-CLIO, LLC.
Komiss, William and LaVar Huntzinger (2011). "The Economic Implications of Disruptions to Maritime Oil Chokepoints," CNA Analysis and Solutions, http://www.cna.org/sites/default/files/research/The Economic Implications of Disruptions to Maritime Oil Chokepoints D0024669 A1.pdf, mar-2011 (accessed July 2012).
Kundu, Goutam K. and Bidhu B. Mishra (2011). "Impact of reform and privatization on consumers: A case study of power sector reform in Orissa, India." *Energy Policy*, 39 (6): 3537–3549.
La Porta, Rafael and Florencio Lopez-De-Silanes (1999). "The Benefits of Privatization: Evidence from Mexico." *Quarterly Journal of Economics*, 114: 1193–1242.
Leung, Guy C.K. (2011). "China's energy security: Perception and reality." *Energy Policy*, 39: 1330–1337.
Lewis, Ian (2010). "Petronas goes back to NOC basics," Petroleum Economist, http://www.petroleum-economist.com/Article/2731082/Petronas-goes-back-to-NOC-basics.html (accessed July 2010).
Loftis, James, Adrianne Goins and Miranda-Lin Gong (2007). "Latin America: arbitration overview," http://www.GlobalArbitrationReview.com (accessed February 2013).

Löschel, Andreas, Ulf Moslener and Dirk T.G. Rübbelke (2010). "Indicators of energy security in industrialized countries." *Energy Policy*, 38: 1665–1671.
Luft, Gal and Anne Korin (2004). "Terrorism Goes to Sea." *Foreign Affairs*, 83 (6): 61–71.
Lukszo, Zofia, Geert Deconinck and Margot P.C. Weijnen (2010). *Exploring the Risks of Information and Communication Technology in Tomorrow's Electricity Infrastructure*, Dordrecht, Heidelberg, London, New York: Springer.
Macalister, Terry (2007). "IEA warns against 'resource nationalism'." *The Guardian*, http://www.guardian.co.uk/business/2007/jul/09/2 (accessed February 2013).
Madan, Tanvi (2007). "India's ONGC: Balanciing Different Roles, Different Goals," Rice University: James A. Baker III Institute for Public Policy Rice University.
Manson, Katrina (2012). "Geothermal: Kenya pins its hopes on steam power," *Financial Times*, http://www.ft.com/cms/s/0/f08f0064-b55a-11e1-ad93-00144feabdc0.html – ixzz24n PBpsaK (accessed February 2013).
Marcel, Valerie (2009). "States of play: national oil companies control ever more of the world's oil. But they're not all alike." *Foreign Policy*, Sept/Oct.
McDonald, Alan (2012). "Nuclear Power Global Status, IAEA Bulletin," International Atomic Energy Agency, http://www.iaea.org/Publications/Magazines/Bulletin/Bull492/49204734548.html (accessed February 2013).
McKinsey & Co. (2012). "Winning the $30 trillion Decathlon: going for gold in emerging markets," http://www.mckinsey.com (accessed February 2013).
Megginson, William and Jeffrey Netter (2001). "From State to Market: A Survey of Empirical Studies on Privatization." *Journal of Economic Literature*, 39: 321–389.
Misra, Ashutosh (2007). "Contours of India's energy security: Harmonizing domestic and external options," in Michael Wesley (ed.), *Energy security in Asia*, London and New York: Routledge, pp. 68–87.
Mo, John (2002). "Options to Combat Maritime Piracy in Southeast Asia." *Ocean Development & International Law*, 33 (3–4): 343–358.
Monagas, Yessika (2012). "U.S. property in jeopardy: Latin American expropriations of U.S. corporations' property abroad." *Houston Journal of International Law*, 34 (2): 455–498.
Nedbank Capital (2012). "African Renewable Energy Review," www.nedbank.co.za (accessed March 2013).
NEI (2012). *Global Nuclear Power Development: Major Expansion Continues*, Washington, DC: Nuclear Energy Institute.
Nellis, John (2006). "Privatization – A Summary Assessment," Working Paper Number 87, Center for Global Development, http://www.cgdev.org (accessed February 2013).
Niez, Alexandra (2010). *Comparative Study on Rural Electrification Policies in Emerging Economies*, Paris, France: OECD/IEA.
Nuclear Power in Japan (2012). "Nuclear Power in Japan," Nuclear World Association, http://www.world-nuclear.org/info/inf79.html (accessed February 2013).
O'Rourke, Dara and Sarah Connolly (2003). "Oil Production and Consumption." *Annual Review of Environment and Resources*, 28: 587–617.
Oduniyi, Mike (2003). "Crude Oil Theft: Bunkerers Get More Daring," Legal Oil, http://www.legaloil.com/NewsItem.asp?DocumentIDX=1056729417&Category=news;27-ma7-2003 (accessed July 2012).
OECD (2002). "Glossary of Statistical Terms," Organisation for Economic Co-operation and Development, http://www.Oliveira, Daniela (2010). "Cyber-terrorism and critical energy infrastructure vulnerability to cyber-attacks." *Environmental and Energy Law and Policy Journal*, 5 (2): 519–526.
Onuoha, Freedom (2008). "Oil pipeline sabotage in Nigeria: Dimension, actors and implications for national security." *African Security Review*, 17 (3): 99–115.
Outrage (2012). "Outrage over Peru-Brazil Energy Agreement," International Rivers, http://www.internationalrivers.org/resources/outrage-over-peru-brazil-energy-agreement-3756 (accessed February 2013).
P&GJ (2011). "Keep Eye on Unconventional Plays, Aggressive NOC Spending, IOC Restructuring." *Pipeline & Gas Journal*, 238 (4).

Pachauri, Shonali (2011). "The Energy Poverty Dimension of Energy Security," in Benjamin K. Sovacool (ed.), *The Routledge Handbook of Energy Security*, New York and Oxon (United Kingdom): Routledge.

Palacios, Luisa (2002). "The Petroleum Sector in Latin America: Reforming the Crown Jewels." *Les Etudes du CERI*, 88.

Parfomak, Paul (2004). "Pipeline Security: An Overview of Federal Activities and Current Policy Issues," CRS Report for Congress. Congressional Research Service.

Parker, David and Colin Kirkpatrick (2005). "Privatisation in developing countries: a review of the evidence and the policy lessons." *Journal of Development Studies*, 41 (4): 513–542.

PWC (2011). "Energy Deals: Merger and acquisition activity in Turkey's energy market, 2011 Annual Review," Price Waterhouse Coopers, http://www.pwc.com/tr/energy (accessed February 2013).

PWC (2012a). "Power Deals: 2012 outlook and 2011 review," Price Waterhouse Coopers, http://www.pwc.com (accessed February 2013).

PWC (2012b). "The shape of power to come: Investment, affordability and security in an energy-hungry world," in 12th PwC Annual Global Power & Utilities Survey, http://www.pwc.com/utilities (accessed February 2013).

REN21 (2012). "Renewables 2012 Global Status Report," Paris: REN21 Secretariat.

Rigobón, Roberto (2008). *Dealing with Expropriations: General Guidelines for Oil Production Contracts*, Sloan School of Management, MIT and NBER.

Ripley, Charles (2010). "The Privatization of Nicaragua's Energy Sector: Market Imperfections and Popular Discontent." *Latin American Policy*, 1 (1): 114–132.

Rosenberg, David (2009). "The Political Economy of Piracy in the South China Sea." *Naval War College Review*, 62 (3): 43–58.

Roxas, Fernando and Andrea Santiago (2010). "Broken dreams: Unmet expectations of investors in the Philippine electricity restructuring and privatization." *Energy Policy*, 38 (11): 7269–7277.

Sachs, Jeffrey D. (1989). "Conditionality, Debt Relief, and the Developing Country Debt Crisis," in Jeffrey D. Sachs (ed.), *Developing Country Debt and Economic Performance, VI*, Chicago: University of Chicago Press.

Sennes, Ricardo and Thais Narciso (forthcoming). "Brazil as an International Energy Player," in Lael Brainard and Leonardo Martinez-Diaz (eds), *Brazil as an Economic Superpower? Understanding Brazil's Changing Role in the Global Economy*, Washington, DC: Brookings Institution Press.

Sharma, Amol, Saurabh Chaturvedi and Santanu Choudhury (2012). "India's Power Grid Collapses Again," July 31, http://online.wsj.com/article/SB10000872396390444405804577560413178678898.html (accessed February 2013).

Shehadi, Kamal S. (2002). "Privatization: Considerations for Arab States," Sub-regional Resource Facility, http://arabstates.undp.org/contents/file/LessonsinPrivatization.pdf www.nedbank.co.za (accessed March 2013).

Sheives, Kevin (2006). "China Turns West: Beijing's Contemporary Strategy towards Central Asia." *Pacific Affairs*, 79 (2): 205–224.

Sheldon, F, T. Potok, A. Krings and P. Oman (2004). "Critical Energy Infrastructure Survivability, Inherent Limitations, Obstacles, and Mitigation Strategies." *International Journal of Power and Energy Systems*, Special Issue on Blackouts: 86–92.

Silvestre, Bruno, Jeremy Hall, Stelvia Matos and Luiz A. Figueira (2010). "Privatization of electricity distribution in the Northeast of Brazil: the good, the bad, the ugly or the naive?" *Energy Policy*, 38 (11): 7001–7013.

Sinai, Joshua (2004). "Future Trends in Worldwide Maritime Terrorism." *The Quarterly Journal*, 31 (1): 49–66.

Sokolov, Y.A. and A. McDonald (2005). "The Nuclear Power Options for Africa." *ATDF Journal*, 2 (2): 12–18.

Sovacool, Benjamin K. (2011). "Introduction: Defining, measuring, and exploring energy security," in Benjamin K. Sovacool (ed.), *The Routledge Handbook of Energy Security*, New York and Oxon (United Kingdom): Routledge.

Sovacool, Benjamin K. and Ishani Mukherjee (2011). "Conceptualizing and measuring energy security: A synthesized approach." *Energy*, 36: 5343–5355.

Stanculescu, Alexander (2011). "Overview of Nuclear Energy: Present and Projected Use," Idaho National Laboratory: U.S. Department of Energy.

Stanley Foundation (2006). "China's Energy Security and Its Grand Strategy," http://www.stanleyfoundation.org/publications/pab/pab06chinasenergy.pdf (accessed February 2013).

Steinberg, Federico (2012). "Nacionalismo energético argentino," Infolatam, http://www.infolatam.com/2012/04/17/nacionalismo-energetico-argentino/ (accessed February 2013).

Stevens, Paul (2008). "National oil companies and international oil companies in the Middle East: Under the shadow of government and the resource nationalism cycle." *Journal of World Energy Law & Business*, 1 (1): 5–30.

Stirling, Andy (2011). "The Diversification Dimension of Energy Security," in Benjamin K. Sovacool (ed.), *The Routledge Handbook of Energy Security*, New York and Oxon (United Kingdom): Routledge.

Thavasi, V. and S. Ramakrishna (2009). "Asia energy mixes from socio-economic and environmental perspectives." *Energy Policy*, 37 (11): 4240–4250.

The White House (2012). "U.S. Strategy Toward Sub-Saharan Africa," http://www.whitehouse.gov/sites/default/files/docs/africa_strategy_2.pdf (accessed March 2013).

Tordo, Silvana, Brandon Tracey and Noora Arfaa (2011a.) *National Oil Companies and Value Creation*, Vol. I, Washington, DC: The World Bank.

Tordo, Silvana, Brandon Tracey and Noora Arfaa (2011b.) *National Oil Companies and Value Creation*, Vol. II, Washington, DC: The World Bank.

Trebing, Harry M., and Sarah P. Voll (2006). "Infrastructure deregulation and privatization in industrialized and emerging economies." *Journal of Economic Issues*, 40 (2): 307–315.

Trudeau, Nathalie and Peter G. Taylor (2011). "The Energy Efficiency Dimension of Energy Security," in Benjamin K. Sovacool (ed.), *The Routledge Handbook of Energy Security*, New York and Oxon (United Kingdom): Routledge.

UNDP (2004). *World Energy Assessment: Overview 2004 Update*, New York: United Nations Development Program.

UNEP (2012). "Financing renewable energy in developing countries. Geneva, Switzerland: United National Environment Program," http://www.unepfi.org/fileadmin/documents/Financing_Renewable_Energy_in_subSaharan_Africa.pdf (accessed March 2013).

Valdes, Constanza (2011). "Brazil's Ethanol Industry: Looking Forward," U.S. Department of Agriculture, http://www.ers.usda.gov (accessed February 2013).

Vazquez, Ian (1996). "The Brady Plan and Market-Based Solutions to Debt Crises." *Cato Journal*, 16 (2): 233–243.

Vivoda, Vlado (2009). "Resource nationalism, bargaining and international oil companies: challenges and change in the new millennium." *New Political Economy*, 14 (4): 517–534.

Vivoda, Vlado (2010). "Evaluating energy security in the Asia-Pacific region: A novel methodological approach." *Energy Policy*, 38 (9): 5258–5263.

Wahab, Bilal (2006). "How Iraqi Oil Smuggling Greases Violence." *Middle East Quarterly*, XII (4): 53–59.

Wei, Y.M., Go Wu, Y. Fan and L.C. Liu (2008). "Empirical analysis of optimal strategic petroleum reserve in China." *Energy Economics*, 30 (2): 290–302.

Wertheim, Peter (2012). "Oil Giant Petrobras Strategies Abroad," Latin Trade, http://latintrade.com/2012/07/oil-giant-petrobras-strategies-abroad (accessed August 2012).

Winzer, Christian (2012). "Conceptualizing energy security," *Energy Policy*, 46: 36–48, doi:10.1016/j.enpol.2012.02.067.

World Bank (n.d.). "Privatization Data," http://data.worldbank.org/data-catalog/privatization-database (accessed March 2013).

World Bank (1992). Guidelines for Environmental Assessment of Energy and Industry Projects. World Bank Technical Paper No. 154. Environmental Assessment Sourcebook, Vol III.

World Bank (2003). "Brazil – Gas Sector Development Project, Sao Paulo Natural Gas

Distribution Project, and Hydrocarbon Transport and Processing Project," Washington, DC: The World Bank.
World Bank (2004). "The Challenge of Financing Infrastructure in Developing Countries," in *Global Development Finance 2004: Harnessing Cyclical Gains for Development*, Washington, DC: The World Bank.
World Bank (2008). "The Welfare Impact of Rural Electrification: A Reassessment of the Costs and Benefits," Washington, DC: IBRD/The World Bank.
World Bank (2012). "WDI Online: World Development Indicators 2012," http://data.worldbank.org/data-catalog/world-development-indicators (accessed March 2013).
World Bank Group (n.d.). "Private Participation in Infrastructure Database," at http://ppi.worldbank.org/ (accessed August 2012).
World Energy Council (2011). "Assessment of Country Energy and Climate Policies," London.
World Nuclear Association (2012a). "Emerging Nuclear Energy Countries," http://www.world-nuclear.org/info/inf102.html; 25-jun-2012 (accessed August 2012).
World Nuclear Association (2012b). "Nuclear Power in India," http://www.world-nuclear.org/info/inf53.html (accessed August 2012).
Xie, Jian, László Pintér and Xuejun Wang (2009). *China: Promoting a Circular Economy*, Washington, DC: World Bank.
Xu, Yi-chong (2012). "Energy and Environmental Challenges in China," in Luca Anceschi and Jonathan Symons (eds), *Energy Security in the Era of Climate Change: The Asia-Pacific Experience*, United Kingdom: Palgrave Macmillan, pp. 91–110.
Yépez-García, Rigoberto, Todd Johnson and Luis Andrés (2010). *Meeting the Electricity Supply/Demand Balance in Latin America & the Caribbean*, Washington, DC: The World Bank.
Yergin, Daniel (1990). *The Prize: The Epic Quest for Oil, Money, and Power*, New York, NY: Free Press.
Yergin, Daniel (2006). "Ensuring Energy Security." *Foreign Affairs*, 85 (2): 69–82.
Yu, W. and M. Pollitt (2009). "Does liberalization cause more electricity blackouts? Evidence from a global study of newspaper reports," EPRG Working Paper, Electricity Policy Research Group.
Zhang, Yinfang, David Parker and Colin Kirkpatrick (2005). "Competition, regulation and privatisation of electricity generation in developing countries: does the sequencing of the reforms matter?" *The Quarterly Review of Economics and Finance*, 45 (2–3): 358–379.
Zhou, W., B. Zhu, S. Fuss, J. Szolgayova, M. Obersteiner and W. Fei (2010). "Uncertainty modeling of CCS investment strategy in China's power sector." *Applied Energy*, 87 (7): 2392–2400.
Zissis, Carin (2006). "Bolivia's Nationalization of Oil and Gas," Council on Foreign Relations, http://www.cfr.org/economics/bolivias-nationalization-oil-gas/p10682 (accessed February 2013).

# PART IV

# SECURITY OF ENERGY DEMAND

# 11. Energy demand: security for suppliers?
*Tatiana Romanova*

## 1 INTRODUCTION

The vast majority of papers on energy security look at it from the consumer's point of view, making security of supply the key preoccupation. However, this volume is an excellent illustration that energy security is an 'umbrella term', covering 'many concerns linking energy, economic growth and political power' (Westminster Energy Forum 2006, p.9). Preoccupations of suppliers are not new; they emerged in the 1980s and have grown in scope ever since. These interests and concerns are expressed by both individual producing countries and by organizations that bring them together.

Demand security is frequently portrayed as the other side of the energy security medal, the side which was overlooked for a long time. Most decision-makers and analysts currently agree about the interconnectedness of supply security and demand security. A vicious circle emerges, for example, when high prices lead to the decline in demand and, therefore, limit the will of producing countries to invest in new production facilities and in infrastructure bottlenecks. 'The disincentive to invest then creates the roots of the next oil price shock once oil demand recovers' but the capacities to feed the demand remain limited (Fattouh and van der Linde 2011, p.12). Another vicious circle emerges when prices are too low, they constrain opportunities for bringing into operation new oil and gas fields or ways of their transportation. As a result, the increase in the demand for oil and gas cannot be met and this leads to a price hike.

Cooperation of producing and consuming countries is needed to convert these vicious circles into virtuous ones. In other words, energy security is about cooperation and interdependence rather than about confrontation and zero-sum games. Liberalism and institutionalism are more appropriate perspectives for energy demand security than international relations realism. However, understanding producing countries' objectives and their feasibility is a pre-requisite for the application of liberal and institutional perspectives. This chapter seeks to achieve that understanding through analysis of key documents (energy strategies, doctrines, concepts) and speeches and declarations of representatives of key oil and gas exporting countries (see Table 11.1).

Table 11.1  Key oil and gas exporting countries[1]

|  | Net oil export, billion t | | | Net natural gas export, billion m$^3$ | |
| --- | --- | --- | --- | --- | --- |
|  | 2010 | **2011** |  | 2010 | **2011** |
| Saudi Arabia | 343.4 | **398** | Russia | 174.8 | **182.4** |
| Russia | 376.2 | **375.4** | Qatar | 96.3 | **130.6** |
| Kuwait | 103.7 | **121** | Norway | 102.3 | **97.4** |
| UAE | 102.5 | **119.6** | Canada | 64.9 | **55.7** |
| Iran | 117.3 | **118.8** | Algeria | 57.3 | **50** |
| Venezuela | 105.6 | **101.3** |  |  |  |

This chapter will first look at how the concept of demand security came about and how it evolved. It will then examine in more details requests of consuming countries. Finally, the chapter will analyze the means that producing countries use to ensure demand security. Both the requests and the means to meet them are classified in two groups (economic and political). The term 'political' is preferred to 'geopolitical', which is frequently used in this context, because it embraces not only geopolitical ambitions of producing countries but also such instruments of traditional international relations like dialogues among producers or between producers and consumers, as well as efforts to induce new international legislation. In sum, 'political' for us includes both realist and liberal institutionalist approaches.

## 2 HOW DID THE CONCEPT OF DEMAND SECURITY DEVELOP?

The peculiar needs of producing countries had already been recognized by the experts' community in the 1970s, at the start of the current studies of energy security, although they did not immediately become the focus of political interaction. The needs of exporting countries were defined then as sovereignty over resources and granted access to consumers abroad (see, for example, Willrich 1975).

OPEC members first made requests for demand security in the 1980s. That was the result of the two oil crises of the 1970s, which encouraged European consumers to drastically reduce their energy consumption. Demand security in the 1980s boiled down to price stability and amounts to be exported. It was also limited to oil, a commodity which at that time was globally traded. Demand security has been present in both political agendas and experts' deliberations from that time on.

The 1990s made the arguments and concerns of oil producing countries more sophisticated. Firstly, instead of arguing only about a fair price, they made a link between the cash flow and their propensity to invest in upstream business as well as in transportation and oil processing. In other words, they connected concerns of exporting countries and the long-term supply stability of consumers. Moreover, rather than arguing about a fair price, consumers and producers introduced at the Third International Energy Conference in 1994 the notion of price stability, which is mutually beneficial for both sides. They underlined that:

> price stability is a key concern for the energy security from the point view of both consumers and producer countries. It is therefore necessary to enhance the study of the limits of that reasonable price level in order to identify the range that would provide for common benefits and at the same time, avoid the risks of price volatility for both consumer and producer countries. (Fattouh and van der Linde 2011, p. 65)

In other words, an effort was made to make energy security a cooperative game as opposed to relying on a zero-sum approach.

Secondly, environmental arguments moved in the forefront of the discussion in the 1990s. On the one hand, traditional sources were undermined due to their environmental impact; in particular, producing countries objected to energy taxation, which discriminated oil as a polluting source. On the other hand, European countries progressively encouraged development of renewable sources of energy, deemed more climate-friendly and at the same time decreasing their external dependence. It is, therefore, noteworthy that the Saudi Minister for Petroleum and Mineral Resources defined producers' security then as 'continued access into the markets of oil importing countries, the steady share of oil in total energy consumption over the long term, and fair and stable prices that allow for their sustainable development over the lifetime of the resource' (Fattouh and van der Linde 2011, p. 61).

Besides, the 1990s also saw the emergence and beefing-up of the International Energy Forum (IEF). It brought together key oil exporting and importing countries and enabled their dialogue on oil trade. Hence, the first institutional structure to promote permanent policy interaction between consumers and producers emerged; it also became the venue, which strived to find the balance between the two sides of the energy security medal (between supply and demand concerns). Ultimately it facilitated an approach to energy security as a cooperative game, which takes into consideration the demands of suppliers and consumers.

Finally, towards the end of the 1990s the discussions on demand security broadened to gas due to its liberalization, which started in the US and UK and then made its way to Europe. Liberalization in consuming countries raised producers' concern due to unbundling, which undermined their ownership of some pipelines beyond their territory, as well as due to the changes in long-term contracts, which consuming countries requested. Gas demand security debates, however, remained fairly tacit until the turn of the millennium because initial liberalization meant only information transparency whereas unbundling (division in production, transportation and distribution) was implemented through the division of management and through clear costs' allocation. In 2002 gas discussions were incorporated into the agenda of the IEF.

The turn of the new millennium ushered further developments of the demand security debates. The IEF expanded the linkage between prices and their volatility, investments and a fair share of hydrocarbons in the market. These debates 'were given more content' because they included 'issues of policy uncertainty, data transparency, human capital shortages, the IOC-NOC relationship and the role of technology' (Fattouh and van der Linde 2011, p. 116). However, environmental issues remained relatively low-profile, presumably because 'parties wanted to avoid confrontational topics and focus more on themes that can bring them closer together' (Ibid).

At the same time, a new line of debates emerged outside of the IEF. Oil producing countries increasingly applied development arguments to justify high prices and guaranteed volumes of export. On top of economic pragmatism (mutually beneficial for exporters and importers price stability and guaranteed investments), they applied a normative justification. They argued that revenues from the sale of hydrocarbons are crucial for their development, for social inclusion and stable employment. It is telling, for example, that the head of the Energy Studies Department of OPEC Secretariat, Mohamed Hamel, interpreted that energy is a two-way street in the following way:

> On the one hand, oil is important to the economic growth and prosperity of consuming/importing countries, but on the other hand it is also crucial to the development and social progress of producing/exporting countries. For example, while net oil imports in OECD countries account for around 60 per cent of their total demand, oil exports from OPEC's Member Countries account for no less than 77 per cent of their total exports, and for some of them, more than 90%. (Hamel 2007)

This line of arguments can be seen as offsetting the normative environmental logics of developed countries, which mostly happen to be consumers (Norway provides a noticeable exception here).

However, the most tangible development in the concept of demand security that came in the new millennium is not from the oil sector but from that of natural gas. Two developments facilitated it. Deepening of liberalization was the first one: as the time passed unbundling in key consuming markets was increasingly about organizational independence of upstream, midstream and downstream businesses, and eventually about legal and ownership separation. That development meant that gas suppliers progressively were stripped of the guaranteed consumption, which underlined the very development of the industry. Secondly, the first decade of this century also saw a drastic decrease in the costs of liquefied natural gas (LNG). Its significance lies in the fact that it makes redundant costly pipeline infrastructure and converts previously closed and separated regional gas markets into one global one.

Given these developments, gas moved into the forefront of the demand security debates. It is noteworthy that the IEF held a specialized ministerial forum on gas in November 2008. But even before, at the start of the century, key gas producing countries beefed up their efforts to promote demand security, and Russia has so far been the most active player in this game. This later development is not surprising given the fact that Moscow is by far the largest producer and exporter of natural gas in the world. The country also relies on the revenues from the sale of oil and gas to carry out domestic reforms and fulfill its numerous other social obligations.

Russia's urge to assume the leading role among gas producers was also fed by its ambitions to stay among key world players at the time when some prerequisites for it (in particular, economic strength and ideological, normative appeal) were missing. It is worth mentioning, for example, that President Vladimir Putin has repeatedly stated the goal to achieve Russian leadership in the context of global energy security (see Putin 2005). Current Russian energy strategy notoriously stresses the need 'to ensure the contribution of the energy sector into improvement of foreign economic activities and to reinforcement of Russia's positions in the world economic system' (Russian Federation 2010, p. 14–15).[2]

Russian activity served to increase the perception that energy security is an issue of mutual dependence because it has emphasized cooperation and shared responsibility.[3] Firstly, it involved a larger array of international, non-energy specialized organizations (G-8, G-20, OSCE, UN, APEC). It meant that the number of arenas for the discussions on demand security was increased and awareness on this issue grew. Moreover, it was introduced into the agenda of developed countries, the majority of which were net consumers. Gradually most fora recognized the 'mismatch between security of supply and security of demand' (OSCE 2010).

Secondly, Russia contributed to the change of the definition by putting in the agenda the term 'global energy security' during its G-8 presidency in

2006. The new discussion stressed the 'interdependence between producing, consuming and transiting countries', the indivisibility of energy security into that of exporters and that of importers (G-8 2006). It also made central the 'development of transparent, efficient and competitive global energy markets' as well as 'enhanced dialogue on relevant stakeholders' perspectives on growing interdependence, security of supply and demand issues' to achieve energy security (Ibid).

The G-8, of course, lacks legitimacy as a forum for the full-scale dialogue between consumers and producers (primarily because many suppliers are absent but also because it is not an international organization but rather a discussion club). Therefore, it made several references to the IEF as a venue for further elaboration.

The reason for Russia introducing the global energy security as a point of G-8 discussions was not tactical, 'to recover its tattered reputation' following the 2005–2006 transit conflict with Ukraine and interruption of supply (Van de Graaf and Westphal 2011, p.22) but rather strategic. Russia used the new term (global energy security) to challenge existing legal obligations (i.e. Energy Charter Treaty), which it believed to favour consumers over suppliers. It shifted the discussion from the contrasting of two terms (supply security and demand security) to the holistic idea of global energy security, to a more cooperative game. This shift enabled Russia to introduce a new conceptual approach in 2009 and a draft convention on energy security in 2010.

Thirdly, Russia has sought to modify some international legal instruments to take into account the needs and priorities of producers. It famously refused to ratify the Energy Charter Treaty, which it perceived as the document privileging consumers over producers. In 2009 President Dmitry Medvedev suggested an alternative legal regime (Medvedev 2009), which became the basis for the draft Convention on global energy security. This later document defined global energy security as:

> the state of world energy system, which guarantees steady and uninterrupted supply of consumers with energy materials and products on the conditions, which satisfy all market participants with minimal environmental impact and pursuing the goal of sustainable socio-economic development of the global community. (Shtilkind 2010).

The draft also contained such principles as interdependence, equal responsibility of producers, consumers and transit countries, fair distribution of risks and balance of all interests (Ibid). It, therefore, incorporates the interests of producing and consuming countries. Russian activity also prompted modernization of the Energy Charter process (the relevant process was launched in 2010 and reflected the ideas, promoted by

Moscow, including the concept of global energy security (Energy Charter Secretariat 2010)).

In sum, Russia contributed greatly to raising awareness about demand security. It also diversified arenas of this discussion as well as the ways of promoting this concept. Interestingly, however, Russia has never attempted either to expand the normative definition of demand security, which OPEC countries worked out (contribution to social and development goals) or to change it. Rather, in its internal documents (Russian Federation 2003, 2010) Moscow adhered to western normativity (meaning, environmental protection and energy efficiency) and argued that although energy sector is important in terms of revenues it brings to the budget, the ultimate goal is modernization and a drastic reduction of the share of oil and gas in the GDP. This attitude, in fact, reflects the ambiguous position of Russia, which strives to straddle the two worlds (that of oil and gas producers and another one, composed of developed states). It is also an excellent illustration of the fact that Russian normativity is embodied in an immediate and pragmatic economic action rather than in abstract issues.

The increasing concern about demand security for natural gas also encouraged key producing countries to coordinate their activities. In 2001 they set up an informal Forum of Gas Exporting Countries (FGEC), which was transformed into a fully-fledged international organization in 2008.

Thus at the time of writing, in 2012, the concept of demand security has been quite developed and was integrated into the concept of global energy security. It included both oil and natural gas, both economic and political aspects. These later elements are analyzed in more details in the next section.

## 3 WHAT DO SUPPLIERS WANT WHEN THEY TALK ABOUT DEMAND SECURITY?

Two observations are to be made before we discuss bits and pieces of demand security. Firstly, producing countries are different in the extent they promote various parts of demand security. At least three factors influence their propensity to defend demand security. These are the degree of their integration in the western world (or the wish to be there); the dependence on oil and gas revenues; and the ability to supply resources to global market. For example, Norway promotes environmentally-sound energy sectors and stresses that stable supply is its international responsibility rather than defends demand security. Russia has always restated western arguments on energy efficiency and environmental safety. At the same time the western concern about the supply stability and credibility of Moscow stimulated the latter to underline the importance of export markets stability. OPEC countries, Canada and Algeria have also been extremely vocal in defending their right to access consuming markets. On the other hand, Iran has been practically silent due to its increasing domestic demand, which challenges its ability to export as well as the increasing number of international sanctions against it.

Secondly, some experts contrast oil and gas producers' reasons to seek demand security. It is true that gas is more dependent on fixed infrastructure than oil (the amount of LNG is still limited and it presents a commercially viable option only on long distances), and hence redirection of gas is more complex. Furthermore, oil 'producers can add value by refining oil but gas-value is location specific' (Ghilès 2009). Finally, while oil producers use demand security to justify low investments in production capacities (see Jesse and Van der Linde 2008), gas producers are mainly interested in maintaining their share in the market and fear the competition from other fuels (El-Katiri 2010). These variations are worth pointing out as they shape the specificity of demand security concerns in oil and gas sectors. However, the elements of demand security are the same for oil and gas sectors. They are structured below in economic and political groups.

### 3.1 Economic Concerns

Five economic aspects stand out in the demand security, requested by producing countries. *Price stability* is the first, and, possibly, the best known. Current discussion is not so much about a fair price but rather about contained price volatility. This latter is the effect of both physical problems with supply and demand (due to economic recession, availability, infrastructure disruptions, etc.) and the intricate work of financial markets, which are frequently 'detached from supply and demand fundamentals' (Salem El-Badri 2011, p. 2).

Both excessively high and excessively low prices are harmful, as the introduction to this chapter pointed out. The first affect the demand and encourage consumers to look for alternative sources. Moreover, the higher the prices for oil and gas, the more attractive are renewable sources and energy efficiency technologies. Low oil prices, on the other hand, question the security of supply because they discourage producers from investing in new fields and transit capacities. They also raise the concern of whether idle capacities are needed and who should pay for them. On the consumers' side they also undermine 'conservation and climate change agenda' (Fattouh 2012).

A particular concern of gas exporting countries is the price-setting mechanism. For a long time gas prices were linked to oil prices; the historical reason is that gas was not traded freely and globally but was viewed as a substitute to oil. However, the development of spot markets for gas as well as the use of LNG for price arbitrage between various regions of gas consumption have created preconditions for independent gas pricing. Currently most long-term gas-supply contracts presuppose an oil-related price formula whereas short-term trade is based on spot prices, the latter most of the time being lower compared to the long-term prices. Hence, consumers would like to change the price mechanism whereas producers (especially, Russia and Algeria) would prefer to maintain the existing practice.

*Stability of oil and gas consumption* is another key challenge for today's producers. It is affected by a number of factors. One is temporary and relatively short-term: the results of 2008 financial crises are still felt; the lack of growth means, in turn, a depressed energy demand. Another trend, which affects the stability of oil and gas consumption, is more long-term. It stems from the efforts to instill the so-called green economy, which consists of increases in energy efficiency and the development of renewable sources of energy.

The benefits of green economy are two-fold. They reduce the dependence of importers on foreign supply and hence, from the classical realist point of view, increase their national security. On the other hand, they also mean new local, knowledge-intensive production and, by consequence, additional employment possibilities. Moreover, 'the renewable energy market is flexible, with greater opportunities for smaller, independent producers to phase into the energy markets using smaller scale investments' (Boëthius 2011). In sum, the advantages are diverse and numerous, but not for producers of traditional sources of energy.

There is also a technical concern here. It is particularly acute for the natural gas sector: once a gas field is put into operation it is impossible to stop the flow. Hence, producers can interrupt the supply for as long as they have storage capacities. Then, if consumers are still not interested in their gas, exporters can either divert the flow or start burning it. As a result, demand insecurity makes the supplier a hostage of the consumer.

The good news for oil and gas exporters is that renewable sources of energy cannot fully substitute oil and gas (with the possible exception of small Scandinavian economies). More good news lies in the low acceptability of nuclear energy today, which is the result of both the Chernobyl and the Fukusima disasters. Nor can nuclear power stations be fully substituted with alternative sources of energy.

However, the stability of consumption remains uncertain. It is for this reason that they request that future policies 'ensure that all sources of energy, including oil and gas, are part of a balanced future energy portfolio' (Hussain 2012).

The third concern, which is linked to the discussion on the stability of demand is that of *(un)just taxes* in importing countries. In part, they are used to level price fluctuations for final consumers. However, many producing countries also believe that as a result exporters reap a part of their rent, depriving them of the resources, which could be used for the development and social cohesion. Russia has also argued against import taxation because it results in higher prices for final consumers, which producers are ultimately blamed for. It is for this reason that Russian draft convention on global energy security suggested inter alia to ban all import taxes and duties (but also to preserve all export ones).

Furthermore, some taxes are also introduced to encourage consumers to shift from one energy source to another. This strategy is frequently used to subsidise renewable sources of energy with the revenues, collected from dirtier (in terms of green-house gases and other environmental impacts) traditional sources of energy. Producers frequently argue, however, that 'increased use of fossil fuels is consistent with the protection of the environment, through the development and dissemination of advanced cleaner fossil fuel technologies, and in particular the promising technology of carbon capture and storage' (Hamel 2007). Furthermore, Moscow has also suggested – to no avail – that natural gas environmental soundness is to be recognized and upheld in the EU's tax regime.

*Stability of regulation* on consumers markets is another source of concern for producing countries. It is especially pronounced for gas exporters, who increasingly face the results of gas liberalization, especially in the European markets. Russia inter alia voiced its unease about the ownership rights on the infrastructure, which Russian (or Soviet) companies have constructed or still plan to build. Suppliers have consistently tried to prove that as a result the EU will soon face the deficit of infrastructure investments (the Californian crisis is frequently cited as an example). Moscow also questioned the legitimacy of unconditional third-party access, which undermines both its ability to sell its gas and, ultimately, the security of supply of consumers in the European market.

Finally, suppliers (especially in gas) are *uncomfortable about competition* in their export markets. They would much prefer exclusive contracts and a guaranteed access to final consumers. However, what is even more annoying to them is unfair competition. This later emerged as a result of discriminative taxation or instability in the regulation in their export markets.

These five economic concerns of exporting countries frequently make them classified as supporters of the Beijing consensus, emphasizing the role of the state in economic relations as opposed to more liberal consuming countries, oriented towards Washington consensus and ultimately free markets (see, for example, van der Linde 2005). The reality is, however, more complex. Greater state interference in the economy of the majority of producing countries is the fact of life. However, what producing countries would like to see is not a similar ownership / regulation structure in consuming countries. Rather, they encourage shared responsibility and markets, which are not distorted by the normative logics of the west (i.e. by the wish to limit their supply dependence and to promote renewable sources of energy at the expense of traditional sources of energy).

Another interesting argument, which has been raised by representatives of the OSCE, Energy Charter Secretariat, is that supply security requires only technical adjustment by governments (in the form of 'binding multilateral rules on investment, transit, resources' (Dickel 2009)) whereas demand security is something that governments cannot guarantee because it is up to the markets to define the overall demand, the role of various resources in the energy mix as well as access to markets (Ibid). The reality is, however, that these market choices are distorted by the deliberate government regulation (including taxation). This is not to say that it is a wrong choice but rather to underline that both demand security and supply security are a combination of market choices and government regulation and, therefore, can be equally affected by public authorities.

### 3.2 Political Concerns

Political concerns of energy exporting countries, linked to demand security, are a logical continuation of economic concerns. They have already been mentioned before, and therefore deserve only a brief summary here:

1. Demand security guarantees *stable revenues for national budgets*: this money is mostly used to finance the development of producing countries, their improved social cohesion, education of local population, and modernization of their economies. Thus the revenues have a goal, which goes beyond the economic one. At the end of the day the very stability of exporting countries hinges on the stability of oil and gas money flows.
2. Suppliers are interested in the *development of a new legal regime* (or adjustment of the present one) in a way that takes their interests into consideration: as depicted in the first part of this chapter, Russia has been particularly vocal and active in pursuing this way.
3. Finally, supplying countries are *against any form of politicization of energy relations*: Russia has consistently argued for more business-type

relations in the field of gas and perceived the wish of the EU to reorient its contracts towards alternative (and more expensive) suppliers of natural gas as a political rather than an economic step. Alexei Miller, the CEO of Gazprom, has famously argued that his company 'is not afraid of competition' but that he feels that 'competition should be honest, rational, without any discrimination, not distorted by any political, bureaucratic and ideological factors' (Miller 2012). Other countries (like Iran) have also voiced their concern about boycotts as a political weapon to put pressure on their ruling regime.

In sum, political concerns are a continuation of the economic ones. They are not independent, rather they are meant to support the economic concerns and thus perform an auxiliary function.

## 4 WHAT INSTRUMENTS DO SUPPLIERS APPLY?

The means that producers use to further demand security and that can be applied by consuming countries can also be structured in economic and political groups. Again, the classification is more for the purpose of clarity because political tools just elaborate and enhance economic ones.

### 4.1 Economic Means

At least five economic leverages are used by suppliers to defend their interests. *Storage facilities* for both oil and natural gas and *spare capacities* to produce oil are the first, rather technical ways that producers use. These instruments are helpful to offset short-term (daily and seasonal) fluctuations in demand, sharp hikes and falls in prices and so on. Some of these facilities go beyond the needs of an individual supplier and acquire a global dimension. This is particularly the case with the Saudi spare capacities. As an unnamed industry official of this country said, 'Saudi Arabia will always play a balanced role in supplying global demand' (Reuters 2011). The regulation of volumes of export has a proved record of success, although the exact correlation between volumes and price fluctuations remains a topic of discussion.

Both storages and idle capacities represent considerable investments that do not pay back immediately and rather represent a global public good (particularly when they are used to smooth price volatility). Hence, the question has always been who is to pay for them and how to share the price between consumers and producers in a fair manner (for a recent discussion see, for example, Bodro Irawan 2012).

Another classical way to guarantee a predictable price in the field of natural gas is a *long-term contract*. It came into being in the 1970s and ever since it has served the basis for the majority of trade in gas. The attractiveness of long-term contracts lies in the fact that they guarantee certain consumption for a period of about 20 years. If an importing country does not need that amount of gas, it still has to pay for it (this is the notorious take-or-pay obligation). As a result long-term contracts guarantee a predictable cash-flow, which is essential to make costly investments in upstream business and transportation. These contracts also served as a guarantee for credits, used to develop new fields and construct pipelines and other infrastructure. Russia is a particularly vocal supporter of these contracts, which is reflected in its current energy strategy (Russian Federation 2010) as well as in its conceptual approach to global energy security and draft convention on global energy security (see Medvedev 2009, Shtilkind 2010).

Some experts suggest that individual countries (big and prospective consumers like China or India) can also use this practice of long-term contracts in oil business and 'index the price to signals generated elsewhere in the world' (Luciani 2011, p. 13).

*Control of transportation routes* is yet another way to secure the demand. This strategy mostly works for natural gas, which relies on fixed infrastructure, too costly to duplicate. For example, Russia has effectively denied third-party access to its gas pipelines (both in bilateral negotiations and in the talks on the Energy Charter Treaty and on its accession to the WTO). This allowed Moscow to regulate access of Central Asian resources to the European market. Another interesting strategy was adopted by Russia when negotiations on Nabucco (a gas pipeline that would link Central Asian gas to Europe bypassing Russia) started. Moscow did its best to contract the majority of resources in the region thus making investments in the Nabucco pipeline commercially unattractive.

Producers regularly try to offset or *correct the changes in the regulation of consuming countries*, which disadvantage them. An array of various measures is applied for that. Most recent activities have taken place in the gas domain with the view to adapt to the liberalization process. Two approaches have been applied.

The first is to defend the interests of relevant national companies in the new legal environment. Both Russia and Algeria have actively searched an access to EU's consumers. It guarantees them export markets, acts as a safeguard against prospective increase of competition and heightens their profit (the largest margin of profit is believed to be reaped in the so-called last-mile of supply). In order to get this access they bid for distribution networks but also invite energy companies of importing countries to participate in the development of new gas fields on their territory in exchange for their share in the distribution market (the strategy of assets-swap). They also frequently apply the concept of reciprocity in the access to reserves and consumers to defend their position. The purchase of distribution networks by gas-developing companies is, however, increasingly against the legislation, prescribing unbundling in the majority of consumers.

The second is to encourage importers to fine-tune their legislation in a way which is favourable to producers. International legislation is mostly used for that. Russia has also exploited Energy Charter Treaty's clause on the protection of investors to claim that consuming countries cannot force foreign companies to sell their stakes in the transit and distribution business. These clauses were also included in the Russian suggestions on how to improve global energy security (Medvedev 2009, Shtilkind 2010).

Finally, all producers actively apply *the strategy of diversification* to hedge their stakes and to balance the desire of consumers to decrease their dependence on oil and natural gas. One variant is *to diversify export markets*. For example, at the 2006 G-8 meeting in St. Petersburg Russia made known its intention 'to increase oil exports to Asian countries from 3% to 30% by 2020 and gas exports from 5% to 25%' (RIA Novosti 2006). This goal was repeated in the 2009 energy strategy of the country, although the indicators were adjusted to 22–25 per cent and 19–20 per cent for oil and gas respectively (Russian Federation 2010). Interestingly, Russia also tried to encourage Qatar to move to Asia to diversify its markets. This is, however, a strategy, which aims not so much at the increase of profit of the largest LNG producer but rather at the protection of Russian share in the European gas market (see Hulbert 2012).

Asian markets clearly present a very lucrative opportunity for both oil and gas producers due to their growing consumption, which by far outstrips that of European states. However, there is a fly in the ointment, and that is much lower prices in the most populated part of the world, compared to the revenues that can be reaped in Europe.

The aim of the geographical diversification, from the producers' point of view is to return to the situation when consumers compete for resources as opposed to the competition among exporters. In a way, this diversification race has already brought some results:

> Already, Europe has to compete for LNG flows from other gas-rich regions. Moreover, Europe is now effectively competing for new gas developments in Russia with other consuming countries because Gazprom does not have an unlimited capacity and capability to develop very many large projects all at once. (Van der Linde 2007, p. 17)

Another approach is *to diversify energy production*. A curious trend is the interest of oil and gas producers to renewable sources of energy. For example, Algerian Sonatrach 'launched a business line to harvest power from the sun' (El-Katiri 2010, p. 15); the aim is to become a world leader in the production and export of the solar energy. The decision is clearly based on a good understanding of today's energy trends. It is also noteworthy that the CEO of New Energy Algeria (partly owned by Sonatrach) underlines that his country's 'potential in thermal solar power is four times the world's energy consumption', which allows it to 'have all the ambitions' in the field (see El-Katiri 2010, p. 15). In a similar way Saudi Arabia successfully bid to become the headquarter of the new International Renewable Energy Agency (IRENA), which allows the country to closely follow all the trends in the field.

Another energy production diversification is to move away from the export of raw energy materials to oil- and gas-processing on the national territory and to the export of the production of refineries and gas factories. This is the strategy which Russia has tried to pursue, at least according to its energy strategies. The problem, however, has been in the legal uncertainty for business in Russia, in the low propensity to long-term investments and clash of interests of established energy companies, exporting raw materials and those interested in the export of processed energy goods. Similar processes can also be identified in other exporting countries.

Finally, a trend to *diversify away from energy* goods can be identified. It manifests itself at the national and companies level. For example, Russian 2009 energy strategy presupposes a drastic reduction of the energy sector in both economy and revenues of the state budget (Russian Federation 2010). Moreover, various oil and gas exporting states now set up special funds, which they fill with oil and gas revenues. This approach allows them to remove superior oil and gas revenues (and therefore fight inflation), cushion external shocks (like the 2008 financial crises) with the help of this money, set aside, and use them to finance modernization of the country (for details also see Austvik 2009).

Similarly big oil and gas companies moved to other (energy-intensive) sectors (like metal or chemical production) to secure the demand and at the same time gain profit from alternative sources.

### 4.2 Political Means

Again, as in the situation with political concerns, political means are a continuation of the economic ones. In most of the cases they perform an auxiliary function rather than attempt to use energy as a political weapon. However, we would like to differentiate five types of activities of producing countries:

1. Producers have attempted to use *international legislation* to limit the freedom of consumers to change their internal legislation. Russia has been particularly skillful in this domain. At one instance it tried to apply the Energy Charter Treaty (which Moscow did not ratify but applied provisionally) to challenge the EU's right to demand ownership unbundling of gas pipelines from production and distribution. Later Russia presented a Conceptual Approach to regulate international energy relations (Medvedev 2009) and a draft convention on the same subject (Shtilkind 2010) to amend current legal norms. It inter alia attempted to make the demand more transparent and predictable, to limit unbundling, abolish export duties and limit the clause on regional integration, which exempted the EU from the application of universal transit norms. As mentioned, the end result was the launch of the reform of the Energy Charter Treaty (Energy Charter Secretariat 2010).
2. *Producers have increasingly cooperated with each other.* OPEC became the first – and probably, still the best known – of these clubs.[4] More recently gas exporting countries took this practice on board by forming the FGEC. Unlike OPEC, FGEC has never had the goal of price control because of the still regional character of gas markets and because of broad specter of interests among its limited members (pipeline users vs producers of LNG). Rather the driving idea has been to exchange information on production and transportation, relevant technologies and practices. However, confidence-building measures between producers and consumers have moved to the front of the FGEC agenda.
3. As described in the first part of this chapter, producers have relied to a great extent on soft measures, like *dialogues between producers and consumers*. There are both global initiatives (like the IEF) and regional ones (i.e. between the EU and its suppliers like Norway, Russia and the Gulf Countries). Some, relatively modest, role was accorded to voluntary efforts to increase the transparency of the relations (notably, through the IEF JODI system). Although these instruments have only limited results, the uncertainty in the world energy market encourages their mushrooming. Russia, for example, in June 2012 announced that it was organizing a Russian International Energy Forum in 2013, which will bring together representatives of governments, business community and experts.
4. Both oil and gas producers have invested a lot in *improving their reputation of stable suppliers*. The fears that they have to deal with have varied. For the majority of OPEC countries the concern has been

about their stability and openness to cooperation. Russia has dealt with the accusation that it uses energy to pursue not so much business relations but rather its geopolitical ambitions. Producing countries have used various measures to calm their partners (among them are speeches of political leaders and business representatives, increased transparency, hiring western PR agencies to present their strategies). The best way is of course to improve their legislation and openness to the world economy but activities in this domain have been patchy and slow.
5. Russia has entertained the *ideas of integration* with the European Union where it would bring its energy complex while the EU will guarantee consumption. Vladimir Putin made the most famous presentation of this kind in 2010 in *Sueddeutsche zeitung* (Putin 2010). The crux of the idea is that integration of economies of Russia and the EU (its key consumer) will lead to the elimination of threats to demand security because the very position of Russia will change (from an external to an internal supplier). The plan also has wider political and economic implications but it has so far fallen on deaf ears in the EU.

## 5 CONCLUSION

Demand security is a more recent concept compared to supply security. However, since its inception in the 1980s it has developed considerably in scope, issues and instruments. It does, indeed, represent the other side of the coin and, together with supply security, it forms global energy security.

Despite numerous efforts to instill mutually beneficial cooperation between producers and consumers, however, energy security remains a combination of liberal efforts to cooperate and realist attempts of both consumers and producers to decrease dependence on each other. It has not become an area of constructive, liberal cooperation and it will remain in the present state for the foreseeable future. In this game producers will be increasingly more vulnerable. Their strategies to minimize their exposure will grow in sophistication.

As for many years before, a liberal agenda remains a goal on the horizon, which moves away every time one tries to approach it. However, further policy efforts and expert studies are needed to find the way for true global energy security, which takes into consideration the needs of consumers and producers and makes operational all the pieces of governance emerging in the area.

## NOTES

1. Calculated by the author on the bases of the statistics, provided by the BP (BP 2012).
2. This phrase is one of the most controversial in the document and is frequently cited to prove political rationale of Russian energy policy, which Moscow has mostly sought to deny.
3. We stop short of using the term 'liberal' here because to be recognized as liberal, Russia has to have the credibility of a liberal actor, which it currently lacks. Therefore, most Moscow initiatives, which do have a neoliberal institutionalist character, are met with such disbelief and criticism.
4. In theory the International Gas Union was the first because it was established in 1931. However, it remained rather loose, and included too many members with diverse interests. Moreover, its power until very recently was constrained by the nature of the gas market.

## REFERENCES

Austvik, Ole Gunnar (2009), 'EU Natural Gas Market Liberalization and Long-term Security-of-supply and -Demand', in Gunnar Ferman (ed.), *The Political Economy of Energy in Europe: Forces of Integration and Fragmentation*, Berlin, pp. 85–118.
Bodro Irawan, Puguh (2012), 'Understanding the global rivalry between OPEC and IEA. Oil production management vs. Strategic petroleum reserves arrangements', Vienna, 18 April, available at www.scribd.com/doc/89977373/Understanding-the-Global-Rivalry-Between-OPEC-and-IEA-Puguh-B-Irawan-18-04-2012 (accessed 20 August 2012).
Boëthius, Gustav (2011), 'Demand Security IV – Instabilities in the Global Energy Markets', Middle East Insight, No. 38, 6 September.
BP (2012), 'BP Statistical Review of World Energy', June, available at http://bp.com/statisticalreview (accessed 20 August 2012).
Dickel, Ralf (2009), 'Energy Interdependence between EU and Russia: Security of Supply/Security of Demand', Speech, Brussels: CEPS, 29 September.
El-Katiri, Mohammed (2010), 'Sonatrach: An International Giant in the Making. Defence Academy of the United Kingdom', Special Series, No 10/05.
Energy Charter Secretariat (2010), 'Decision', Brussels, 24 November.
Fattouh, Bassam and van der Linde, Coby (2011), 'The International Energy Forum. Twenty years of producer-consumer dialogue in a changing world', available at www.clingendael.nl/publications/2011/2011_IEF_History_of_IEF_Clinde_BFattouh.pdf (accessed 15 August 2012).
Fattouh, Bassam (2012), 'The Consumer-Producer Dialogue: Themes and Prospects', Speech, Kuwait City, 14 March.
G-8 (2006), 'Global Energy Security', St. Petersburg, 16 July, available at http://en.g8russia.ru/docs/11.html (accessed 15 August 2012).
Ghilès, Francis (2009), 'Algeria: A Strategic Gas Partner for Europe', 19 February, available at http://www.ensec.org/index.php?option=com_content&view=article&id=176:algeria-a-strategic-gas-partner-for-europe&catid=92:issuecontent&Itemid=341 (accessed 20 August 2012).
Hamel, Mohamed (2007), 'OPEC: Dialogue between Producers and Consumers', Speech delivered to the Side Event of Vienna-based intergovernmental organisations at the UNCSD-15, New York, 7 May, available at www.opec.org/opec_web/en/press_room/870.htm (accessed 15 August 2012).
Hulbert, Matthew (2012), 'The Vital Relationship: Why Russia needs Qatar (and Qatar could use Russia)', available at http://www.europeanenergyreview.eu/site/pagina.php?id=3607 (accessed 20 August 2012).
Hussain, Hani (2012), 'Technology, Environment &Policies Supporting Oil Industry',

Vienna: 5th OPEC International Seminar, 13 June, available at www.scribd.com/doc/89977373/Understanding-the-Global-Rivalry-Between-OPEC-and-IEA-Puguh-B-Irawan-18-04-2012 (accessed 20 August 2012).

Jesse, Jan–Hein and Van der Linde, Coby (2008), 'Oil Turbulence in the Next Decade: An Essay on High Oil Prices in a Supply-constrained World', The Hague: Cligendael.

Lenexpo (2012), 'Russian International Energy Forum was highly appreciated by experts', 18 June, available at http://lenexpo.ru/en/node/56811 (accessed 22 August 2012).

Luciani, Giacomo (2011), 'The Functioning of the International Oil Markets and Its Security Implications', CEPS Working Document, No. 351, May, available at www.princeton.edu/~gluciani/pdfs/WD 351_Secure_Luciani_on_Oil Markets.pdf (accessed 22 August 2012).

Medvedev, Dmitry (2009), 'Conceptual Approach to the New Legal Framework for Energy Cooperation (Goals and Principles)', Moscow, 21 April, available at http://archive.kremlin.ru/eng/text/docs/2009/04/215305.shtml (accessed 21 August 2012).

Miller, Alexei (2012), 'peech at the Conference 'Energeticheskaya bezopasnost i novye vozmozhnosti dlia prirodnogo gaza', 31 May, available at www.gazprom.ru/press/miller-journal/966597/ (accessed 20 August 2012).

OSCE (2010), 'OSCE Special Expert Meeting on Assessing the OSCE's Future Contribution to International Energy Cooperation', Vilnius, 14 September.

Putin, Vladimir (2005), 'Vstupitelnoe Slovo Presidenta RF V.V. Putina na Zasedanii Soveta Bezopasnosti "Rol Rossii v Obespechenii Mezhdunarodnoi Energeticheskoi Bezopasnosti"' ('Introductory speech of President Vladimir Putin at the Session of the Security Council "The Role of Russia in Guaranteeing International Energy Security"'), Moscow, 22 December, available at http://president.kremlin.ru/text/appears/2005/12/99294.shtml (accessed 10 August 2012).

Putin, Vladimir (2010), 'Plädoyer für Wirtschaftsgemeinschaft. Von Lissabon bis Wladiwostok' ('Towards a Common Economic Market from Lisbon to Vladivostok'), *Sueddeutsche zeitung*, 20 November, available at www.sueddeutsche.de/wirtschaft/putin-plaedoyer-fuer-wirtschaftsgemeinschaft-von-lissabon-bis-wladiwostok-1.1027908 (accessed 20 August 2012).

Reuters (2011), 'Factbox – Saudi Arabia's stable energy policy', 15 February, available at http://af.reuters.com/article/commoditiesNews/idAFLDE71913020110215 (accessed 22 August 2012).

RIA Novosti (2006), 'Possibilities of Energy Dialogue', Moscow, 24 April, available at http://en.rian.ru/analysis/20060425/46902706.html (accessed 20 August 2012).

Russian Federation (2003), 'Energeticheskaya strategiya Rossii do 2020 goda' ('Energy Strategy of Russia to 2020'), Decree of the Government of the Russian Federation, No 1234-r, Moscow, 28 August.

Russian Federation (2010), 'Energy Strategy of Russia for the Period up to 2030', Decree of the Government of the Russian Federation, No1715-r, Moscow, 13 November 2009.

Salem El-Badri, Abdalla (2011), 'Foreword', *OPEC World Oil Outlook*, Vienna: OPEC.

Shtilkind, Theodor (2010), 'O proekte Konventsii po obespecheniu mezhdunarodnoi bezopasnosti' ('On the Draft Convention on Ensuring International Energy Security'), Vilnius: OSCE Special Expert Meeting, 13–14 September., available at www.osce.org/ru/eea/71274 (accessed 15 August 2012).

Van de Graaf, Thijs and Westphal, Kirsten (2011), 'The G8 and G20 as Global Steering Committees for Energy: Opportunities and Constraints', *Global Policy*, vol. 2, September, pp. 19–30.

Van der Linde, Coby (2005), 'Energy in a Changing World', Inaugural Lecture as Professor of Geopolitics and Energy Management at the University of Groningen, Clingendael Energy Papers, No. 11, March.

Van der Linde, Coby (2007), 'The Geopolitics of EU Security of Gas Supply', *European Review of Energy Markets*, vol. 2, no. 2, pp. 1–24.

Westminster Energy Forum (2006), *The New Energy Security Paradigm*, Geneva: Westminster Energy Forum.

Willrich, Mason (1975), *Energy and World Politics*, New York: The Free Press.

# 12. Oil producers' perspectives on energy security
*Gawdat Bahgat*

## 1 INTRODUCTION

For centuries energy has played a major role in the evolution of human civilizations. In the last two centuries fossil fuels (coal, oil, and natural gas) were crucial for the birth and development of the Industrial Revolution and global economic prosperity. Energy products are certain to maintain their character as the "engine" for maintaining and improving our way of life.

A major characteristic of energy is the mismatch between resources and demand. Generally speaking, major consuming regions and nations (the United States, Europe, Japan, China, and India) do not hold adequate indigenous energy resources to meet their large and growing consumption. On the other hand, major producers (i.e., the Middle East, Russia, the Caspian Sea, and Africa) consume a small (albeit growing) proportion of their energy resources. This broad global mismatch between consumption and production has made energy products the world's largest traded commodities. Almost every country in the world imports or exports a significant volume of energy products. This means the wide fluctuation of energy prices plays a key role in the balance of payments almost everywhere.

The heavy reliance on energy in conjunction with the asymmetric global distribution of energy deposits have underscored the importance of energy security. This sense of vulnerability is not new. Despite the abundance of energy resources and a favorable political and economic environment, industrialized countries started expressing their concerns over energy security as early as the first part of the twentieth century. First Lord of the Admiralty Winston Churchill's decision that the Royal Navy needed to convert from coal to oil in order to retain its dominance signaled a growing intensity of global competition over energy resources (mainly oil). This rivalry between global powers was played out in World War II when the Allies enjoyed access to significant oil deposits while Germany's and Japan's strategies to gain access to oil resources failed and led, among other developments, to their eventual defeat.

The availability of cheap energy resources played a major role in the reconstruction and development of Europe and Japan in the aftermath of World War II. This prolonged era of relative confidence in the availability of abundant and secure energy resources came to an abrupt end following the outbreak of the 1973 Arab-Israeli War. Arab oil producing countries cut their production and imposed oil embargoes on the United States and a few other countries to force a change in their political support for Israel. This use of oil by major producers to gain political leverage shattered consumers' sense of energy security. Since then, the fluctuation of energy prices (partly due to geopolitical developments and partly in response to supply and demand changes) has reinforced this sense of vulnerability.

In the last few decades there has been a growing understanding of the challenges that climate changes poses to life on earth. More people have come to realize that our way of life (i.e., human activities) contributes to and accelerates global warming and that something needs to be done to restrain this human-made environmental deterioration. This slowly growing consensus has added a new dimension to energy security. The concept is no longer limited to the availability of energy resources at affordable prices. Environmental considerations restrain the exploration and development of these resources and urge consideration of less polluting alternative sources of energy. This brief overview of energy history suggests that there is a wide variety of threats to energy security.

The 1973–1974 oil embargo served as a turning point in global and domestic energy markets. The availability of energy supplies at affordable prices was no longer taken for granted. The turmoil in the world economy focused on the disruption of supplies to consuming countries. These oil consumers have implemented several measures (individually and collectively) to mitigate the impact of such disruptions and to reduce their energy vulnerability. The measures include the creation of the International Energy Agency (IEA), the storage of oil supplies in strategic petroleum reserves, and encouraging energy conservation, among others.

Not enough attention was given to the other side of the energy equation – producing nations. The concept of "energy security" is not static. Since the mid-1970s a broader definition has emerged that addresses all the energy players' concerns. In the past few decades, while the industrialized countries have successfully diversified their sources of crude oil imports and greatly reduced their relative dependence on energy (albeit at different degrees), the major oil exporters remained dependent on oil revenues. Petroleum revenues have continued to be the principal source of income for almost all major oil exporting countries. As a result, oil exporters have as many reasons to worry about the security of their markets as importers have to worry about the security of supplies. In short, the security of demand is considered as important as the security of supply (Zanoyan 2003). Abdullah Salem El-Badri, Secretary General of the Organization of Petroleum Exporting Countries (OPEC), summed up the argument: "Energy security should be reciprocal. It is a two-way street." (El-Bardi 2008)

Within this context energy analysts have provided different definitions of energy security highlighting different aspects of the concept. Barry Barton, Catherine Redgwell, Anita Ronne, and Donald Zillman define it as a condition in which "a nation and all, or most of its citizens and businesses have access to sufficient energy resources at reasonable prices for foreseeable future free from serious risk or major disruption of service" (Barton et al. 2004, 4). Daniel Yergin underscores a number of "fundamentals of energy security." The list includes diversification; high quality and timely information; collaboration among consumers and between consumers and producers; investment flows; and research and development technological advance (Yergin 2007). Yergin argues that the experience since the early 2000s has highlighted the need to expand the concept of energy security in two critical dimensions: globalization of energy markets and the need to protect the entire energy supply chain and infrastructure (Yergin 2006, 77).

Christian Egenhofer, Kyriakos Gialoglou, and Giacomo Luciani distinguish between short-term and long-term risks. The former are generally associated with supply shortages due to accidents, terrorist attacks, extreme weather conditions or technical failure of the grid. The latter are associated with the long-term adequacy of supply, the infrastructure for delivering this supply to markets, and a framework for creating strategic security against major risks (such as non-delivery for political, economic, force majeure of other reasons) (Egenhofer et al. 2004).

Finally, a report by the IEA argues that energy security stems from the welfare impact of either the physical unavailability of energy, or prices that are not competitive or overly volatile.

Analysts at the Paris-based organization add that the more a country is exposed to high-concentration markets the lower is its energy security (International Energy Agency 2007).[1]

All these definitions underscore the fact that energy security is a multidimensional concept that incorporates cooperation between producers, consumers, and national and international companies. The experience of the last few decades indicates that the availability of clean energy resources at affordable prices cannot be addressed only at a national level. Rather, international cooperation is a necessity. Thus, energy is part of broader international relations between states. A major theme of today's energy markets is interdependence between consumers and producers. Calls for self-sufficiency or energy independence are more for domestic constituencies. Indeed, energy interdependence fosters cooperation between countries in other areas such as economic development and world peace.

Another major theme of the energy security literature is the importance of diversification of energy mix and energy sources. The less dependent a country is on one form of energy (i.e., oil, natural gas, coal, nuclear power, and renewable sources), the more secure it is. Similarly, the more producing regions there are around the world, the better.

## 2 ORGANIZATION OF PETROLEUM EXPORTING COUNTRIES (OPEC)

For a long time, energy policy was perceived as a zero-sum conflict where the interests of producers and consumers were mutually exclusive. Each side pursued a strategy to maximize its interests at the expense of the other side. Stated differently, consuming countries were interested in low oil and gas prices while producing nations sought to raise prices. The two sides realized that individual states or companies would have little leverage and creating a collective entity would make it easier to reach their respective goals. Against this backdrop, the Organization of Petroleum Exporting Countries (OPEC) was established in 1960 and the International Energy Agency (IEA) was founded in 1974. The former represents the producers' interests and the latter promotes the consumers' objectives.

The two organizations were created in the midst of price crises. In the late 1950s and early 1960s, producing nations believed that International Oil Companies (IOC) were paying them very little for their precious product. Driven by this perception and the desire to receive a "fair" price for crude oil and enhance their negotiation leverage, major oil producers founded OPEC. Following the 1973 Yom Kippur War between Israel and Arab countries, most OPEC members imposed an oil embargo on the United States and some other countries to punish them for their support of Israel. More importantly, they incrementally cut off their production. These steps led to the so-called first oil price shock in 1973–1974. The soaring oil prices drove consuming nations, led by the United States to create the IEA. The goal was to articulate a strategy on how to ensure consumers' energy security.

Under these circumstances, it was almost inevitable that the two organizations adopted conflicting strategies. Their opposing policies failed to assure global energy markets and, indeed, contributed to the wide fluctuation of oil prices and the overall instability of international economy. Little wonder that a growing consensus emerged calling for cooperation between producers, consumers, and other major energy players (i.e., IOCs). Increasingly, more producers and consumers have ceased to perceive their respective interests as mutually exclusive and have identified growing areas of common interest. In order to promote and consolidate cooperation, several dialogues, partnerships, and organizations were created. The International Energy Forum (IEF) is a prominent example of these efforts.

The roots of the IEF go back to the mid-1970s when the producers and consumers held a meeting in Paris to discuss the rise in oil prices. Positions were polarized and consequently no concrete results came out of this meeting. However, in the ensuing years it became clear that sharply fluctuating oil prices were detrimental to both producers and consumers and there could be no long-term winners in a volatile environment. A growing realization by both sides was that stable prices at a reasonable level would serve their common interests. This realization of mutual interest, coupled with the geopolitical turmoil of the 1990–91 Gulf War, furnished the ground for renewed efforts to establish a producers-consumers dialogue. The Iraqi invasion and occupation of Kuwait (both are OPEC members) highlighted the threat to global oil markets and the broader world's economic prosperity. A more cooperative framework between producers and consumers was born out of this conflict.

At the initiative of Presidents Francois Mitterrand of France and Carolos Perez of Venezuela, a Ministerial Seminar of producers and consumers was held in Paris in 1991. This initiative helped to clarify the atmosphere of mistrust that characterized the relations between producers and consumers and underscored the areas of mutual interest and potential for cooperation to address common challenges. A follow-up meeting was held in Norway in 1992. Since then, the IEF has held meetings alternately in an exporting and an importing country.

The IEF is unique not only in its global perspective and scope, but also in its approach. It is not a decision-making organization or a forum for the negotiation of legally binding settlements and collective action. Nor is it a body for multilateral fixing of prices and production levels. The informality of this framework has encouraged a degree of frank exchanges, which cannot be replicated in traditional and more formal international settings.

For most of the twentieth century the global oil markets were dominated by a few major international oil companies (IOCs), the so-called Seven Sisters: Standard Oil Co. of New Jersey (later Exxon), Standard Oil Co. of New York (originally Socony, later Mobil), Standard Oil Co. of California (Socal, later Chevron), Royal Dutch Shell, Texaco, BP, and Gulf (Bahgat 2003, 178). The OPEC oil producing countries did not participate in production or pricing of crude oil, but simply received a stream of income through royalties and income taxes as part of the concession system. OPEC countries were too weak to challenge the multinational Seven Sisters' domination of the industry.

The first move towards the establishment of OPEC took place in 1949, when Venezuela approached Iran, Iraq, Kuwait, and Saudi Arabia and suggested that they exchange views and explore avenues for regular and closer communication between them. The need for closer cooperation became more apparent when, in 1959, the major international oil companies unilaterally reduced the price of oil. This prompted the convening of the First Arab Petroleum Congress, held in Cairo. The Congress adopted a resolution calling on oil companies to consult with the governments of the producing countries before unilaterally taking any decision on oil prices. It also set up a general agreement on the establishment of an oil consultation commission (OPEC 2010).

In 1960 the IOCs reconfirmed their domination by further reducing oil prices. In response, delegates from five major oil producing nations – Iran, Iraq, Kuwait, Saudi Arabia, and Venezuela – met in Baghdad and announced on September 16, 1960 the foundation of OPEC. The goal was, and still is, to protect the interests of these major producing nations. Accordingly, in their first resolution the five OPEC members emphasized that the companies should maintain price stability and that prices should not be subjected to fluctuation. They called for companies not to undertake any change in the posted price without consultation with the host country. They pledged to establish a price system that would secure stability in the market by using various means, including the regulation of production, with a view to protecting the interests of both consumers and producers.

These ambitious goals, however, proved hard to achieve given the little leverage that producing nations had and the dominant role of the IOCs. By the early 1970s, some major dynamics of the global oil industry fundamentally changed. First, some IOCs, such as ENI of Italy and Occidental of the United States, started operations in the Middle East and elsewhere and offered more attractive financial terms than those offered by the other major IOCs. This gradual process eroded the near monopoly imposed by the multinational companies. Second, economic prosperity in the United States, Western Europe, and Japan accelerated and was fueled by growing demand for oil. These major consumers, however, lacked sufficient indigenous supplies. Thus, the growing appetite for oil was met, mainly, by production from OPEC countries. Stated differently, the world grew more dependent on OPEC supplies. The combination of these two developments left OPEC producers in a stronger bargaining position than in the early 1960s.

Building on this newly acquired bargaining power, OPEC producers sought to increase oil prices. Their demands were rejected by the multinational oil companies and negotiations between the two sides collapsed. The 1973 Arab-Israeli War provided the geopolitical and geo-economic opportunity to fundamentally alter the balance of power between OPEC members and IOCs. In addition to imposing an oil embargo on the United States and a few other countries, and incrementally cutting production, some OPEC governments stopped granting new concessions and started to claim equity participation in the existing concessions, with a few of them opting for full nationalization. Asserting their power, OPEC members decided in October 1973 to unilaterally raise oil prices independently of the multinational oil companies' participation. These developments paved the way for structural changes in the world oil industry.

In the aftermath of the first oil shock, OPEC members consolidated their control over production and prices. However, their lack of the necessary technological and financial infrastructures left them dependent on multinational oil companies. Thus, rhetoric aside, OPEC members continued to sell large proportions of their production to the old concessionaires. Political developments in the Persian Gulf including the 1979 Iranian Revolution followed by the eight-year war between Iran and Iraq had a significant impact on OPEC and the broad global oil industry. The interruption of oil supplies from Iran and Iraq triggered widespread chaos leading to soaring oil prices in what became known as the second oil price shock.

The continuing push for higher prices underscored a division within OPEC between two competing strategies. The first strategy was advocated by OPEC members with considerable proven reserves, small populations, and high per capita incomes (i.e., Kuwait, Qatar, Saudi Arabia, and the UAE). These countries sought to moderate prices in order to maintain demand over the long run. The other strategy was pursued by members with larger populations, lower oil exports per capita income (i.e., Algeria, Indonesia, Iran, and Nigeria). This second group demanded restraint on OPEC production and higher prices (Blaydes 2004).

This disagreement between the so-called hawks and doves laid the ground for an awkward situation whereby a two-tiered pricing system prevailed. Saudi leaders perceived high oil prices as harming oil producers in the long run by encouraging investment in high-cost areas outside OPEC and switching to alternative fuels. Accordingly, Riyadh refused to raise prices beyond a certain level.

Against this backdrop, OPEC adopted a quota system in March 1983 which set a production ceiling. By controlling the volume of global production, OPEC sought to influence prices. Within this framework OPEC adjusted its production upward and downward based on the level of production from non-OPEC countries. Saudi Arabia, the major oil producer and exporter, played the role of "swing producer" within OPEC.

Political turmoil and the lack of consensus among OPEC members for a unified strategy prompted IOCs to increase their investments in areas outside OPEC, most notably the North Sea and the Gulf of Mexico. The increasing supplies from outside OPEC coupled with the fall in demand as a result of high prices led to a drastic fall of OPEC's share of the global oil market. This intensive rivalry between OPEC and non-OPEC producers and within OPEC proved unsustainable. In the mid-1980s, Saudi Arabia decided to drop its system of selling oil at fixed prices and instead adopted a market-oriented pricing system. Consequently, Saudi Arabia's production started to rise quickly and by 1986 global markets became saturated. This led to a severe collapse in oil prices.

The very low prices in the mid-1980s hurt the interests not only of OPEC producers, but also those of other producers such as the North Sea, the United States, and the Soviet Union as well as the overall global economy. This broad chaos and the emergence of many suppliers and many consumers led to the development of "a complex structure of oil markets which consist of spot, physical forward, futures, options, and other derivative markets." This structure is based on formula pricing where the price of a certain variety of crude oil is set as a differential to a certain benchmark or reference price. These include Brent Blend, West Texas Intermediate (WTI), Dubai, and Nigerian Forcados, among others. One of the major characteristics of the oil market in the later part of the 1980s and most of the 1990s was the stability of the long-term oil price at a relatively low level.

From 2000 to 2008, oil prices soared and, as a result, most oil exporting countries in OPEC and non-OPEC members accumulated substantial revenues. The imbalance between supply and demand was the driving force behind the soaring oil prices. Unlike the supply-interruption oil shocks of 1973–1974 and 1979–1980, the 2000s' surge was a demand-driven one, fueled by strong Asian consumption. Furthermore, the surge reflected not only increasing demand and decreasing supply, but also broader macroeconomic and geopolitical changes such as rising exploration and production costs, the falling value of the US dollar, the re-emergence of "resources nationalism," inadequate refining capacity, and an aging labor force (Bahgat 2008). The combination of these factors had shaped the global oil market as well as regional and national ones, albeit to different degrees

After reaching a peak of $147 per barrel in 2008, prices dropped significantly and then started a slow process of recovery. OPEC members and non-OPEC producers reacted in different ways to the rise in oil prices.

There was a common assumption that in the face of high and rising oil prices, OPEC would respond by increasing supply to moderate prices and stabilize the market. Such an action would help maintain healthy growth in global oil demand and limit the entry of substitutes such as tar sands and ethanol. This view was influenced by OPEC's decision to introduce a price band in 2000, which involved production adjustments when the price moved about $28 for 20 consecutive trading days or when the price moved below $22 for 10 consecutive trading days.

OPEC members failed to put a ceiling on the price and, indeed, most members took advantage of rising prices by increasing their production and exporting as much as they could to maximize their profit. This attitude suggests that OPEC's role is not to prevent oil prices from rising above certain levels or to set a price ceiling. Rather, a key objective of the organization is to avoid oil prices from falling below a level deemed unacceptable by its members.

On the other hand, non-OPEC producers' response to the 2000 price boom was weak. They were not able to raise their production to take advantage of rising prices. This suggests that non-OPEC production has peaked or is close to reaching this stage. It is increasingly becoming more costly and technologically and environmentally challenging to maintain or increase production from non-OPEC producers. As one analyst argues, there seems to be an asymmetric response to oil prices: "A sharp rise in oil price induces a modest investment response in non-OPEC countries, while a decline in the oil price generates a sharp fall in investment and a period of underinvestment in the oil sector" (Fattouh 2010, 21). This argument is in line with the IEA's projection, which predicts that as conventional oil production in countries not belonging to OPEC peaks, "most of the increase in output would need to come from OPEC countries, which hold the bulk of remaining recoverable conventional oil resources" (International Energy Agency 2009, 42).

Three conclusions can be drawn from this brief review of the fluctuation of oil prices in the last few decades and the role OPEC played in this process. First, as with any commodity, the role of oil prices is to signal relative scarcity or abundance which in turn causes all energy players (i.e., consumers, producers, national oil companies, and IOCs among others) to adjust to the allocation of resources (Mabro 2006). Second, rhetoric aside, OPEC does not fix oil prices and does not have a direct impact on their rise and fall. Rather, given the OPEC members' substantial proven reserves and large volumes of production and exports, the organization plays a significant indirect role in influencing price formation (Mabro 2005). OPEC signals its preferred price and alters its production volume up or down. These signals are perceived by other energy players and, in turn, they respond to these signals. Within this context, it is important to point out that despite impressive improvements, the availability of accurate data on production, consumption, reserves, and other vital information is not perfect. As a result, "market psychology" plays an important role in shaping the movement of oil prices. Third, historically, exporters and importers have had divergent interests, with the former favoring higher prices and the latter favoring lower ones. These perceived opposite interests have recently changed in favor of an emerging realization that too high or too low prices do not serve anyone's interests. Too high prices encourage conservation and alternative energy, and destabilize the overall global economy. Too low prices reduce investment in new exploration and the development of oil deposits and contribute to economic and political turmoil in producing countries.

## 3 OPEC's OBJECTIVES AND PERSPECTIVES ON ENERGY SECURITY

Article 2 of OPEC's statute clearly spells out the main objectives of the organization. These include coordinating the members' petroleum policies and safeguarding their interests, individually and collectively. Another related goal is to stabilize prices and eliminate fluctuations. Most importantly, OPEC seeks to secure a steady income for its members, sufficient and reliable supplies to consuming countries, and fair return on investment by either national oil companies or international ones.

The statute aside, OPEC member countries' heads of state and government occasionally meet to deliberate and articulate broad strategic guidelines. Since its foundation in September 1960, OPEC has held three summits: Algiers, 1975; Caracas, 2000, and Riyadh, 2007. For each of these summits, the main purpose was to step back from the day-to-day activities of the international oil market and examine issues at the national leadership level, pertaining to the fundamental principles, objectives, and procedures of OPEC. The summits also examined contemporary issues confronting the world, particularly the global economy and divisions between rich and poor countries and how OPEC, individually and collectively, can help bridge this gap.

The first summit was held in 1975 in the aftermath of the first oil shock and in the midst of intense confrontation between producing and consuming countries. Not surprisingly, the summit's deliberations and resolutions reflected this international confrontation. OPEC leaders rejected allegations attributing to the price of petroleum the responsibility for the instability in the world economy. They claimed that adjustment in the price of oil did not contribute to the high rates of inflation within the economies of developed countries. OPEC leaders reminded the rest of the world that they have contributed through multilateral and bilateral channels to the development efforts and balance of payments adjustments of other developing countries, as well as industrialized nations. They reaffirmed solidarity with other developing countries in their struggle to overcome underdevelopment.

Furthermore, OPEC leaders asserted that the conservation of petroleum resources is a fundamental requirement for the well-being of future generations and urged the adoption of policies aimed at optimizing the use of this essential and non-renewable resource. On the other hand, OPEC leaders reaffirmed their readiness to ensure that supplies will meet the essential requirements of the developed countries, provided that the consuming countries do not use artificial barriers to distort the normal operation of the laws of supply and demand.

By the time the second summit was held in Caracas in September 2000, the international economic system and the global energy markets had experienced some fundamental changes; in particular, environmental issues had attracted significant attention and OPEC's share in global supply was falling in favor of supplies by other producers, such as Russia and the Caspian Sea. Finally, oil prices were stable at a relatively low level for most of the 1990s. The Caracas Summit reflected OPEC leaders' concerns over these issues, among others.

OPEC heads of state and government confirmed their commitment to provide adequate, timely, and secure supplies of oil to consumers at fair and stable prices and emphasized the strong link between the security of supply and the security and transparency of demand. They called for a fair share for OPEC in the world oil supply and for growing cooperation on a regular basis between OPEC and other oil exporting countries. On the other hand, they demanded the opening of effective channels of dialogue between oil producers and consumers. OPEC leaders also asserted their association with the universal concern for the well-being of the global environment, and their readiness to continue to participate effectively in the global environmental debate and negotiations, including the UN Framework Convention on Climate Change and the Kyoto Protocol, to ensure a balanced and comprehensive outcome.

OPEC leaders called on consuming nations to adopt fair and equitable treatment of oil by ensuring that their environmental, fiscal, energy, and trade policies do not discriminate against oil. They also expressed their concern that taxation on petroleum products forms the largest component of the final price to the consumers in the major consuming countries, and called upon them to reconsider their policies with the aim of alleviating this tax burden for the benefit of the consumers, for just and equitable terms of trade between developing and developed countries, and for the sustainable growth of the world economy.

The third summit, held in Riyadh in November 2007, came amid rising global concern over climate change and deep global dependence on fossil fuels driven by soaring oil prices in the previous seven years. OPEC leaders sought to address these concerns. Thus, the summit focused on three major themes: the stability of global energy markets; energy for sustainable development; and energy and the environment.

OPEC leaders pledged to efficiently manage and prolong the exploitation of their exhaustible petroleum resources in order to promote the sustainable development and the welfare of future generations. They re-emphasized the connection between demand security and supply security and recognized that with globalization the economies of the world and energy markets are integrated and interdependent. They urged all parties to find ways and means to enhance the efficiency of financial petroleum markets with the aim of reducing short- and long-term price volatility. They reiterated the need to continue the process of coordination and consultation with other petroleum exporting countries and the necessity to strengthen and broaden the dialogue between energy producers and consumers. They repeated their call on consuming governments to adopt transparent, non-discriminatory, and predictable trade, fiscal, environmental, and energy policies and promote free access to markets and financial resources.

OPEC leaders associated their countries with all global efforts aimed at bridging the development gap and making energy accessible to the world's poor while protecting the environment. They stated that eradicating poverty should be the first and overriding global priority guiding local, regional, and international efforts.

As major oil producers and exporters are heavily dependent on oil revenues, OPEC members have adopted a cautious stance on the climate change controversy. They reiterated that the process of production and consumption of energy resources poses different local, regional, and global environmental challenges. Meanwhile, they stressed that human ingenuity and technological development have long played pivotal roles in addressing such challenges and providing the world with clean, affordable, and competitive petroleum resources for global prosperity. OPEC leaders underscored that they share the international community's concern that climate change is a long-term challenge, and recognize the interconnection between addressing such concerns on the one hand, and ensuring secure and stable petroleum supplies to support global economic growth and development on the other hand. Finally, they stressed the importance of cleaner and more efficient petroleum technologies and demanded that all policies and measures developed to address climate change concerns be both balanced and comprehensive.

## 4 OPEC: LONG-TERM STRATEGY

The main guiding texts for OPEC are the OPEC statute, approved in January 1961; the Summit Declarations of 1975, 2000, and 2007; and the Long-Term Strategy, adopted on the organization's 45th anniversary in September 2005. This strategy, prepared over a period of two and a half years, provides a broad vision and framework for the organization's future.

The strategy identifies the uncertainties surrounding future demand for OPEC oil as a key challenge. Factors such as future world economic growth, consuming countries' policies, technology development, and future non-OPEC production levels contribute to these uncertainties regarding demand for OPEC oil. The strategy explores these uncertainties in three scenarios that depict contrasting futures of the global energy scene. These scenarios are dynamics-as-usual, protracted market tightness, and a prolonged soft market.

There are substantial uncertainties over future economic growth arising from the complex interplay of domestic and global determinants of that growth, including such diverse factors as demographics, advances in technology, capital availability, and trends in commodity prices, domestic policies, and global trade developments, regimes, environmental policies, and financial regulations. Another area of uncertainty stems from consuming countries' policies. OPEC claims that taxation of energy products is often seen not only as a means of raising revenue, but also as a means of controlling demand in addressing environmental and energy security issues. The strategy alleges that consuming countries' policies demonstrate significant discrimination against oil, involving not only higher tax rates, but also subsidies for competing fuels.

A third area of uncertainty with significant impact on oil demand is technological development. For example, in the transportation sector, conventional internal combustion engines could continue to achieve significant fuel economy improvements, while hybrid vehicles may witness an important growth. The introduction of non-oil-fueled vehicles and the use of alternative fuels, such as bio-fuels, are drivers that could affect oil demand growth patterns in the transportation sector. A fourth area of concern is the development in non-OPEC supply. A number of factors, such as oil prices, upstream legal and fiscal regimes, and investment in non-OPEC countries, technological advances, and exploration successes, will shape future scenarios regarding non-OPEC supply.

A fifth area of uncertainty is related to environmental concerns. The profile of incremental global demand is overwhelmingly for light and clean products, while incremental supply comprises significant volumes of sour, medium, and heavy crude grades. The combination of this with the move to ever-stricter product quality and environmental regulations represents a challenge for the downstream industry. Future refining capability needs to be considered in terms of both the adequacy of secondary processes – for example, to upgrade heavy streams and to meet tight targets for sulfur – and crude distillation capacity.

The combination of these uncertainties signifies a heavy burden of risk in making the appropriate investment decisions. Accordingly, the strategy calls for several measures to meet these concerns. First, the promotion of the development of technologies that address climate change concerns such as carbon dioxide capture and storage technology. Second, OPEC members should continue to play an active role in the climate change negotiations. OPEC supports the principle of common, but differentiated, responsibilities. It believes that the international community should fulfill its obligations to strive to minimize the adverse effects of policies and measures on developing countries, in particular fossil-fuel exporting developing countries (i.e., OPEC members). This could involve assistance in relation to economic diversification and transfer of technology. Third, dialogue among producers and between producers and consumers should be widened and deepened to cover all issues of mutual concern. These include security of demand and supply, market stability, upstream and downstream investment, and technology. Finally, the strategy emphasizes OPEC's commitment to support oil market stability. It recognizes that extreme price levels, either too high or too low, are damaging for both producers and consumers, and points to the necessity of being proactive under all market conditions.

To conclude, when OPEC was founded in 1960 the oil industry and the global economy were very different from what they are more than five decades later. During this time the oil industry has experienced several upheavals, OPEC policy contributing to some of them. Over the years many observers have predicted OPEC's demise. However, it has survived and become an important driving force in the global energy markets. Its members' interests are not identical, though they have managed to coordinate their policies and find common ground most of the time. Any assessment of OPEC's role in managing global oil prices would be highly controversial. The push for higher prices, which characterized the first decades of OPEC's life, has waned in favor of an emerging consensus that stable prices at a "reasonable" level would serve both the producers' and consumers' interests.

## NOTE

1. The International Energy Agency, "Energy security and climate policy," p. 422, available at http://www.iea.org (accessed May 9, 2007).

## REFERENCES

Bahgat, G. (2003) *American Oil Diplomacy in the Persian Gulf and the Caspian Sea*, Gainesville, Florida: University Press of Florida.
Bahgat, G. (2008) "Supplier-user teamwork key to stable oil prices," *Oil and Gas Journal*, **106** (32), 20–24.
Barton, B., Catherine Redgwell, Anita Ronne and Donald N. Zillman (eds) (2004) *Energy Security: Managing Risk in a Dynamic Legal and Regulatory Environment*, Oxford: Oxford University Press.
Blaydes, L. (2004) "Rewarding impatience: A bargaining and enforcement model of OPEC," International Organization, **58** (2), 213–237.
Egenhofer, C., K. Gialoglou and G. Luciani (2004) "Market-based options for security of energy supply," Center for European Policy Studies, available at http://www.ceps.be (accessed March 21, 2004).
El-Bardi, A.S. (2008) "Energy security and supply," available at http://www.opec.org (accessed February 14, 2008).
Fattouh, B. (2010) *Oil Market Dynamics Through the Lens of the 2002–2009 Price Cycle* Oxford: Oxford Institute for Energy Studies.
International Energy Agency (2007) "Energy security and climate policy," available at http://www.iea.org (accessed May 9, 2007).
International Energy Agency (2009) *World Energy Outlook*, Paris: International Energy Agency.
Mabro, R. (2005) "The international oil price regime: origins, rationale and assessment," *Journal of Energy Literature*, **11** (1), 3–20.
Mabro, R. (2006) "Introduction," in R. Mabro (ed.), *Oil and the Twenty-First Century: Issues, Challenges and* Opportunities, Oxford: Oxford University Press, 1–18.
OPEC (2010) "General Information," available at http://www.opec.org (accessed 8 January 2010).
Yergin, D. (2006) "Ensuring energy security," *Foreign Affairs*, **85** (2), 69–82.
Yergin, D. (2007) "The fundamentals of energy security, testimony before the US House of Representatives Committee on Foreign Affairs," available at http://www.cera.com (accessed March 23, 2007).
Zanoyan, V. (2003) "Global energy security," *Middle East Economic Survey*, **41** (15), 20–22, available at http://www.mees.com (accessed April 14, 2003).

# 13. Energy security governance in light of the Energy Charter process
*Andrei V. Belyi*

## 1 INTRODUCTION

Policy and academic discussions about energy security open an important debate about possibilities and limits of energy security governance (Goldthau, 2012; Kuzemko et al, 2012). Therefore, this chapter aims to assess opportunities for energy security governance in the context of the Energy Charter Treaty (ECT), and subsequent and alternative documents. Since the oil shocks of 1973, energy relations between states have been primarily viewed from a security perspective, which considers states to be in a permanent struggle for access to resources and infrastructures (Waltz, 1979, 177–179). According to this state-centred view, energy security is a factor in the unpredictability of international relations, especially for countries dependant on external energy supplies. Recently, much analysis has concentrated on EU-Russia energy relations, where the security dimension has a particular flavour (Kuzemko et al, 2012; Belyi, 2012). More particularly, the issue of the ECT has been discussed in light of EU-Russia energy security relations (Yafimava, 2011). The development of an international legal framework would aim to reduce the level of geopolitical clashes between energy producing, consuming and transit countries. Nevertheless, the emergence of the Energy Charter did not greatly diminish traditional antagonistic energy security relations, although an international framework was developed.

Significantly, a geopolitical, state-centred approach to energy security cannot ignore attempts at international energy governance, even where an international legal approach does not eliminate antagonisms in interstate energy relations. This conceptual duality leads us once again to a debate between state-centred and law-centred approaches in international relations (Keohane, 2001; Waelde, 2008). Energy relations have engendered an evolution of international legal norms, including investment law, best-practice transfer in energy policies, and even a multilateral Energy Charter Treaty (ECT). Hence, an understanding of energy security based on state-centred self-sufficiency co-exists with significant attempts to create international patterns of energy governance (Waelde, 2008; Belyi,

2009). Analyzing security in light of both law-centred and state-centred conceptions reflects various analytical perspectives on international political economy (Gilpin, 1987). In both cases, multilateralism would imply a view of collective action, as being either successful or limited. The significance of the ECT has re-emerged with the European Parliament's Communication on External Energy Policy (European Parliament, 2012), in which it appeals for expansion of the Energy Charter beyond its current members. Likewise, recent Russian proposals either to revise or repeal the ECT emphasizes the need to analyse the issues in greater depth. In doing so, a number of questions emerge about the relevance of a multilateral energy security framework and about its possible limits. The view defended in this chapter points to important steps towards international energy security governance within the ECT, whereas its limits stem from discrepancies in understandings of energy security needs, for example energy security of supply and security of demand invoke different priorities and objectives (Seliverstov, 2009). The dichotomy reflects regional and country level perspectives on the wider, global challenges of energy security. New trends imply the growing significance of regional economic blocs, protectionism and resource nationalism. Thus, the dynamics of global energy governance reflect a development of regionalisms and nationalisms. More particularly, security understandings and views on energy governance differ between the European Union (EU) and Russia, and therefore these two major actors clashed over Energy Charter perspectives.[1] In order to outline the differences, this chapter will analyse the ECT governance framework in the transit and investment fields, then outline discrepancies between the EU and Russia on the multilateral commitments, and finally outline their recent positions on international energy governance.

## 2 ENERGY CHARTER TREATY AS GROUNDS FOR ENERGY GOVERNANCE

Multilateralism involves an acceptance of norms based on a belief of common action. In this respect, the Energy Charter represents a unique attempt of a regime formation specifically in energy. However, as this regime operates in a highly politicized area, a number of conflicts about an acceptance of the regime emerged. Therefore, an analysis of the Energy Charter should include both aspects: particularities of the regime formation on one hand, and the impact on energy governance on the other. Therefore, a brief historical assessment of the international energy security governance implies an understanding of the context in which the Energy

Charter emerged. Before outlining differences in energy security governance understandings between the EU and Russia, it would be useful to understand both the political context of the Treaty and its legal implications of the text adopted in 1994.

## 2.1 European Energy Charter

The Energy Charter process started with the European Energy Charter initiated by the Dutch Presidency of the European Community in 1990. In December 1991, the European Charter Declaration was signed with a view to bringing about a new East-West economic relationship. The institutional logic at that time was marked by the 1990 Charter of Paris signed by the two former ideological blocks, and which had often been seen as a starting point for the "New Europe". The Charter of Paris was then the basis for a "steadfast commitment to democracy based on human rights and fundamental freedoms; prosperity through economic liberty and social justice".[2] Europe's new image is related to the "*security community*", where multilateralism and "seminar diplomacy" aim to integrate "academic and diplomatic discourse in practice".[3]

The European Energy Charter has been part of the "seminar diplomacy" semantics, which involves a number of *foras* and declarations since the beginning of 1990s. The spirit of the European Energy Charter declaration of 1991 is clearly influenced by the Charter of Paris semantics. The Concluding Document of the European Energy Charter (December 1991) stipulates:

> The representatives of the signatories meeting in The Hague on 16 and 17 December 1991, Having regard to the Charter of Paris for a New Europe, signed in Paris on 21 November 1990 at the summit meeting of the Conference on Security and Cooperation in Europe (CSCE); Having regard to the document adopted in Bonn on 11 April 1990 by the CSCE Conference on Economic Co-operation in Europe; Having regard to the declaration of the London Economic Summit adopted on 17 July 1991; having regard to the report on the conclusions and recommendations of the CSCE meeting in Sofia on 3 November 1989, on the protection of the environment, as well as its follow-up; Having regard to the Agreement establishing the European Bank for Reconstruction and Development signed in Paris on 29 May 1990; Anxious to give formal expression to this new desire for a European-wide and global co-operation based on mutual respect and confidence.

Unlike other "seminar diplomacy" *foras*, the European Energy Charter declared an objective to create a legally-binding regime. The European Energy Charter declares: "The signatories of the European Energy Charter undertook to pursue the objectives and principles of the Charter and implement and broaden their co-operation as soon as possible by negotiating in good faith a Basic Agreement and Protocols".

Furthermore, negotiations of the "Basic Agreement" started in order to promote "East-West industrial co-operation by providing legal safeguards in areas such as investment, transit and trade". Creating energy governance is beyond a simple "seminar diplomacy" declaration. It involves a creation of an effective regime in a highly strategic area, which can not be done without controversies.

### 2.2 Negotiations of the Treaty

In July 1991, the European Union initiated negotiations over the "Basic Agreement". Other signatories of the Energy Charter joint the efforts of negotiations. At an early stage of negotiations, controversies around the "Basic Agreement" emerged between the "Western" participants of the process. Main opposition lines emerged between the US and the EU. The US considered that the ECT should be largely incorporated into the General Agreement on Tariffs and Trade regime. By contrast, the EU adopted a flexible position, which was favourable to the participation of the non-GATT members of Eastern Europe and former Soviet Union (hereinafter FSU). A problematic point between the two actors also involved the scope of national regime, because for the US resource ownership is a competence of the states rather than of a federal state. Hence, the national regime should have been, according to Washington, allocated to the federal subjects of the country.

In addition, the US brought to the table the idea of a non-discrimination of investments at the pre-investment phase, which would aim to ease access to the investments. Hence, the US insisted on enlarging the scope of investment protection to the pre-investment phase, whereas the EU insisted on the protection of the existing investments. Nevertheless, contracting parties decided to follow with a supplementary treaty on investment protection, which would also involve access to resources and markets. The US in turn withdrew from the treaty negotiations and remained an observer (Dore, 1996). Consequently, discrepancies over international energy governance are strongly reflected in transatlantic relations. In turn, absence of the US in the framework hinders the political scope of the ECT.

In the meantime, "the East", namely Russia and other FSU countries, remained passive despite their importance in oil and gas resources. Economic depression, coupled with political instability in 1992–1993, did not allow Russia to be an active player in the negotiation process. Therefore, in the later stages, Russia considered it could legitimately review several points, which played a vital role for its energy exports. However, during the first round of the negotiations, Russia allowed the

EU to lead the process. An institutional reflection of the EU's dominance is the location of the Energy Charter Secretariat in Brussels. In addition, the EU influenced the spirit of the treaty text (Blamberger and Waelde, 2007). Furthermore the early EU initiatives influenced the Russian perception of the Treaty as a pro-EU instrument in the FSU area.

In December 1994 the treaty was signed by 49 countries representing major energy producers on the Eurasian continent (Azerbaijan, Kazakhstan, Norway and Russia), the EC as well as EU Member States, Japan and Australia. During the second phase, three more countries joined the ECT. Many producing countries, including Saudi Arabia, Iran and Venezuela, obtained observer status without signing the Treaty. New energy consuming countries in Asia, mainly China and Pakistan, became observers as well. However, the first phase of the Charter process was characterized by a EU-Russia focus. Therefore there is a confusion because the EC is seen as an EU Russia relations instrument.

However, objectives of the treaty cover a broader scope. Article 2 declares "this Treaty establishes a legal framework in order to promote long-term cooperation in the energy field, based on complementarities and mutual benefits, in accordance with the objectives and principles of the Charter." The treaty enters into a logic of promoting open international energy markets, as Article 3 stipulates: "The Contracting Parties shall work to promote access to international markets on commercial terms, and generally to develop an open and competitive market, for Energy Materials and Products."

The outcome of negotiations led to eight draft parts of the text, which represent the overall complexity of the newly emerged regime. The ECT issue-specific regime can be further divided into two major components: trade (which includes transit) and investment regimes. Both components represent a particular importance in the development of international energy governance and therefore necessitate a particular attention. Particular attention should also be paid to a very embryonic environmental regime, which has been also stipulated by the ECT. Finally, a state-to-state dispute settlement mechanism represents the core element for legal effectiveness of the ECT regime.

A definition of governance is coupled with the existence of peaceful methods of conflict-resolution, including the dispute settlement mechanisms. The general principle of dispute settlement is based on the UN Charter, which in its Article 1 aims to "bring about by peaceful means, and in conformity with the principles of justice and international law, adjustment or settlement of international disputes or situations which might lead to a breach of peace". In this respect, the ECT institutionalizes existing methods of conciliation and arbitration mechanisms (Waelde, 1996).

As mentioned above, specific dispute settlement mechanisms are foreseen within each of the main components of the regime, namely Article 29 for trade and Article 7(7) for transit. Arbitration norms under the ECT are stipulated in Article 26 for investor-to-State disputes and Articles 27 and 28 for state-to-state disputes (Happ, 2002). Our particular focus of ECT governance is the trade and transit as well as investment governance. The issue of energy efficiency promotion does have a declaratory essence, albeit representing an important normative dimension.

### 2.3 Energy Charter's Trade and Transit Governance

The ECT represents the first international regime in energy trade and transit. Specificities of energy trade and transit were not considered earlier in the GATT-based global trade governance. The reason for this lies in the embryonic status of the oil and gas cross-border trade in the late 1940s. By the beginning of the 1990s, issues of access to energy markets and infrastructures emerged in world economic relations. Increasingly, complex trade and investment structures led to a need for the creation of mechanisms for the settling of international disputes. On these grounds, we observed that the ECT responded to a changing structure in international energy markets. At the same time, the Energy Charter's trade and transit provisions are largely borrowed from GATT wording (Salem Haighighi, 2006, 162).

Unlike many other products, transport and storage of gas are capital intensive and need to flow in one direction only. Therefore, an access to cross-border capacities can become a strategic issue in both political and economic terms. Our particular interest consists in analysing the path towards a transit regime, which has been particularly hindered by the high level securitization of intra-FSU gas trade relations.

In the aftermath of the break-down of the USSR, East-West oil and gas trade became a cross-border issue, where transit of energy occurs. With the disintegration of the Soviet state, the Soviet Unified Gas System was substituted by a number of companies owned by the newly emerged states. Each of them owned pipelines, underground storages and gas equipments (Mitrova, 2009). In the aftermath of the breakdown of the USSR, Russia inherited most of the pipeline network, which also connects Central Asian gas production to the extra-FSU exports. Ukraine possesses the largest transit pipeline network with Europe-widest gas storages, which is nowadays an important factor for the security of gas supply. The Energy Charter negotiators took into consideration the new economic environment and therefore the transit of energy was integrated into the treaty. A transit regime should address a new context of energy trade. That is why Article 7(1) of the ECT iterates a *facilitation* of transit as a prerequisite:

> Each Contracting Party shall take the necessary measures to facilitate the Transit of Energy Materials and Products consistent with the principle of freedom of transit and without distinction as to the origin, destination or ownership of such Energy Materials and Products or discrimination as to pricing on the basis of such distinctions, and without imposing any unreasonable delays, restrictions or charges.

Article 7 sets two aspects of facilitating transit: access to existing networks and construction of new transport capacities on the territory of a third country. Article 7 does not require third party access to the networks, unlike, for example, EU internal market legislation, which was elaborated later.

The question raised during the ECT process is whether the domestic energy transport and cross-border transit tariffs should be uniform. Moreover, domestic legislation rarely distinguishes between domestic and cross-border transport flows, which complicates the separation between the two. Ambiguous normative provisions made Russia hesitant about the charter. Indeed, Russia is an important transit state for Central Asian gas to Europe through swap agreements: Russia purchased gas from Turkmenistan to its own domestic market and exported part of its own gas to Europe on behalf of the Central Asian state. In practice, it would mean that Russia buys Turkmen gas without setting domestic rules for transit. As the conditions for export of Turkmen gas to Europe are concluded by bilateral agreements and Russian domestic gas transport by the Russian regulator, one could consider that Russia applies a differentiated tariff mechanism for Turkmen and Russian gas flowing through Russia. The issue was brought up later during the Transit Protocol negotiations, which have been analysed in detail by many other prominent experts (Konoplyanik, 2009).

An important aspect of the ECT transit regime is the existence of a mediation mechanism foreseen by Article 7 (7). Transit dispute settlement relates to non-discrimination and national treatment with regards to transit and construction of new transport capacities. The text of the article puts forward a mediation mechanism, which provides a certain flexibility for actors in dispute. The Energy Charter Secretary General designs a conciliator, who, in turn, has to find a compromise solution within 90 days. As was proved later, an inherent difficulty of both conciliation and arbitration mechanisms consists in the slow procedures, whereas the transit conflicts are usually very short and need a fast resolution. Length of the procedures demonstrated a practical ineffectiveness of those mechanisms in the transit disputes, which occurred more than a decade after the conclusion of the treaty. It is interesting to note that the conciliation mechanisms were not accepted by Russia and Ukraine whilst they were involved in the transit conflicts (Yafimava, 2011; Belyi, 2012).

## 2.4 Investment Protection within the Energy Charter

A multilateral investment protection process can represent a positive trend towards a unification of energy investment governance. In the ECT, the investment regime includes general terms of investment protection based on positive reciprocity and MFN principle, iterates a fair compensation for investors in case of expropriation and represents a balance between a principle of national sovereignty and investors' rights.

The scope of the investment provisions (Art. 10) covers fair and equitable treatment, "most constant protection and security" for investors as well as an obligation to protect any contractual arrangement. All these aspects demonstrate a requirement for general conditions for doing business in a country (Hober, 2010). Oil and gas upstream investments have often involved conflicts between host states and international investors. Indeed, investors are attracted mainly during low oil prices, whereas states attempt to reinforce control over the oil and gas production during high oil prices.

Investors attempted to use various mechanisms of protecting the investments against a sovereign state. Clear norms on investment arbitration increases the level of predictability for an investor, which is reinforced by a broad definition of investments by the ECT. For example, investments include various forms of activities in energy, such as licensing for exploration, production and development. Hence, an upstream license holder can be considered as an investor (license being a property right) and therefore can be protected accordingly.

Furthermore, the definition of investments is directly related to the issue of expropriation of foreign assets. In the ECT, expropriations are not forbidden but are put under a number of conditions related to a non-discriminatory aspect of expropriation:

(a) for a purpose which is in the public interest;
(b) not discriminatory;
(c) carried out under due process of law; and
(d) accompanied by the payment of prompt, adequate and effective compensation.

The issue of compensation for expropriation has been addressed by various scholars (Collier and Lowe, 1999; Hober, 2010) and will therefore not be examined here. Nevertheless, the very idea of compensation for investment breach creates a stable ground for establishing business by an investor in the energy sector. In addition, evaluating damages becomes a counter-stone to the high transaction costs related to the legal and political system of the host state.

The issue of national control over resources has always been an important factor in energy producing states. The ECT attempts to balance interests of sovereign states with investors' rights. In its Article 17, the ECT stipulates on non-application of the investment provisions in certain circumstances, which are related to general political relations with another state (i.e. interrupted diplomatic relations, adoption of unfriendly measures by another state, etc). Article 17 results from a necessary protection of state sovereignty in international economic relations regarding investments. Later, the ECT investment jurisprudence demonstrated that the provisions of Article 17 apply for access to new investments whereas they cannot apply to existing investments.

A similar balance between sovereignty and international obligations is translated in Article 18, which declares national sovereignty over resources. Interestingly, Article 18 links sovereign rights over resources to the respect of international norms and agreements, which are not further specified. It would then cover other norms declaring the sovereignty over resources as well as norms of investment protection. If the investment has already occurred (post-investment phase), the investors' rights should be then respected. However, access to the resource remains a discretionary power of the states: "The Contracting Parties undertake to facilitate access to energy resources, inter alia, by allocating in a non-discriminatory manner on the basis of published criteria authorizations, licences, concessions and contracts to prospect and explore for or to exploit or extract energy resources".

Relations between the sovereignty over resources and obligations towards investment protection remained one of the most complex areas between hydrocarbon-producing states and investors. As T. Waelde has pointed out, economic cycles of oil price often contributed to the balance of interests: there was more tension between states and investors during high oil prices than during low oil prices (Waelde 2008). Therefore, states must first accept the rule of law in order to make the treaty provisions effective. At the same time, a non-acceptance of investment protection rights may stem from the securitization of the access to resource issue in hydrocarbon producing states. The major concern of the Energy Charter process has indeed been related to a non-acceptance of its norms by most of the energy exporting states. None of the OPEC countries signed up to the Treaty, whereas big OECD energy exporters (Australia and Norway) refused to ratify the text and hence to integrate it into national legislation.

## 2.5 Energy Efficiency Issues

The ECT marks the first step towards integration of sustainable and efficient development into energy cooperation, although these provisions are still not binding. Energy efficiency is recognized as an additional source of energy as it represents an improved ratio between energy consumption and economic output. ECT Article 19 stipulates: "' Improving Energy Efficiency' means acting to maintain the same unit of output (of a good or service) without reducing the quality or performance of the output, while reducing the amount of energy required to produce that output."

The ECT does not establish a legally binding regime on energy efficiency, nor does it set the targets to achieve. Instead, the ECT represents an investment protection regime, which also covers environmental investments.

It should also be noted that the ECT was signed three years before the Kyoto Protocol and therefore international actors were less cautious about a global environmental regime. The only treaty document which elaborates on the topic of energy efficiency is a non-binding Protocol on Energy Efficiency and Related Environmental Aspects.

Consequently, we can observe that the energy efficiency regime became part of a more holistic system of governance, which also included an investment protection and energy trade. For instance, it could be argued that the ECT investment regime could be used to protect environmental energy efficiency investments. And in turn, the energy saved would become commercializable. Nevertheless, interest in the ECT process from the energy efficiency perspective has paled in the context of the climate change mitigation regime.

### 2.6 Institutional Setting

The ECT established two institutions which contributed further to the process: the Energy Charter Conference and the Energy Charter Secretariat. In accordance with the treaty, the Energy Charter Conference is the decision-making institution, which takes place once a year and which is composed from governmental representatives of the Contracting Parties. The Energy Charter Secretariat was established in 1996 in order to create a regular institution. For instance, arbitration cases are registered at the Secretariat. Moreover, the Secretariat holds regular thematic working groups (i.e. on transit and trade and on investments). Although the scope of activities has been wide, the Energy Charter institutions have not been successful in decreasing the level of politicization of energy transactions. In addition, Moscow has often perceived the activities of the Secretariat as being EU-sponsored.

With an intergovernmental body and a Secretariat in place, the Energy Charter took the shape of an international organization. Nevertheless, the very institution of the Charter has not aimed at creating a new "UN

Security Council" in energy. In other words, unlike the Security Council, which is created to reinforce international security governance, the ECT does not envisage being a supranational authority. Hence, the ECT institutions (including dispute settlement mechanisms) are used only if the contracting parties wish to do so (the UN Security Council may take decisions regardless of the choice of the states). In addition, the ECT foresees a quinquennial review of the process and thereafter contracting parties can either enlarge or reduce competences of the Secretariat or even dissolve it. Institutional limits can furthermore explain a relatively passive role of the Secretariat during the gas transit crises, which are discussed below.

Consequently, the Energy Charter Conference decided to establish the monitoring of entry barriers to investment rather than a legally-binding supplementary treaty. In 2004, a consultation forum was set with an Industry Advisory Panel, which was intended to present the industry view on investment access barriers. Again, the Industry Advisory Panel consultations do not create a legally binding regime. It demonstrates however that private commercial actors, in particular energy investors, are keen to see the multilateral investment regime in place.

## 3 EU-RUSSIA CONTROVERSIES ON INVESTMENT AND TRANSIT GOVERNANCE

Although the ECT represents a very deep and overarching legal framework, important controversies occurred between the European Union and Russia vis-à-vis the international energy governance from the early existence of the Energy Charter process. Russia's assertiveness in international affairs and in its model of state capitalism impacted on its vision on international energy markets. In particular, Russia defended a concept of security of demand in the G-8 Summit hosted in St. Petersburg in 2006. Meanwhile, the EU was in the process of regionalizing energy governance, which was characterized by a conclusion of the Energy Community Treaty in 2005. This, in turn, created a foundation for the EU-based governance, which is exported beyond the Union's political borders.

Contradictions with Russia mostly emerged after the signature of the ECT. Russia's "alternative proposal" constituted a culminating moment, which also led to a termination of the provisional application by Russia by the end of 2009.[4] In geopolitical terms, in the early 1990s Russia represented an alternative supply to the Middle East; two decades later industrialized nationals preferred to diversify supplies from Russia. Moreover, the Eastern enlargement of the EU further reinforced energy dependency of the newly accessed member states on Russia (Belyi, 2003).

### 3.1 Controversies Surrounding Investment Protection

Russia's post-Soviet resource access regime has been marked by post-command economy legacies, especially regarding property rights and the maintenance of a strong link between the state and extracting industries. At the same time, Russian energy policies of the 1990s would have compared badly with the Middle Eastern and Latin American oil producers. Russia favoured a privatization of the oil sector and initially adopted investor-friendly legislation. Nevertheless, the post-Soviet understanding of property rights and of the investment differ from the spirit promoted by the Energy Charter.

From the early 2000s especially, Russian policy started to prioritize resource nationalism based on non-acceptance of international investor's rights. Russia did not proceed to large-scale contract renegotiation. However, the country established an indirect form of governmental control over extractive industries. A mineral exploration license can be revoked by a ministry without a preliminary court decision. In turn, the investor may claim its rights against the ministry after a Russian court hearing. In this case, the Russian state becomes a pivotal player for any international contract negotiation for oil and gas upstream. In other words, for the Russian state, revocation of license is a political issue rather than a legal one. And therefore reciprocity may be applied for access upstream.

Russia's market remains barely accessible for international investors. In response to the concerns of national resource control, Russia adopted legislation which allows Russia to limit or to increase the presence of a foreign investor in a strategic enterprise. The recent Strategic Sector Law[5] applies to investments above certain thresholds by non-Russian Investors. Its Article 6 lists 42 activities to which the national regime is restricted at the pre-investment phase. That clearly contradicts MFN-based reciprocity. Instead, under this quota allocation mechanism, Russia allowed a number of foreign investments into its energy sector through the inter-company package agreements. Consequently, retaliatory reciprocity became a practice in energy investment at the pre-investment phase and thus the multilateral MFN principle has been hindered.

In turn, the EU moved towards a liberalization of energy markets, which has required a fragmentation of companies. The EU third gas directive (2009) requires an unbundling of both Community and non-Community undertakings. Hence, non-European vertically-integrated companies would face an obligation to sell their assets in energy transportation, which subsequently provoked legal conflicts with non-EU investors (Willems et al, 2010). In this way, the Directive also avoided a massive sell-out of strategic EU energy assets to the external monopolies with strict certification procedures. Subsequently, investors' rights would be conditioned with respect to competition norms, which is hardly in keeping with the ECT liberal approach. The EU legislation based restrictions created grounds for the specific regional conception of a market transformation.

### 3.2 Different Views on Transit Governance

ECT transit governance has been largely associated with the Transit Protocol, which aimed at clarifying the aforementioned ECT provisions on transit. The Transit Protocol represents the longest international negotiations within the ECT framework. In 1998 at the G-8 Summit, Russia proposed to reinforce the transit regime under the ECT framework. Its main concern at that time was transit theft, which occurred frequently in the energy flows to Ukraine. In December 1999 the Energy Charter Conference was mandated to negotiate the Transit Protocol, which aimed to clarify the provisions of Article 7 (1). In parallel, the Conference adopted non-binding Transit Model Agreements. Transit Protocol negotiations entered into a murky area during the repeated transit crises, which have occurred in FSU area since 2004.

In December 2002 a final draft of the Transit Protocol was proposed. The Protocol deepens the initial objective of "facilitating" transit under Article 7(1) of the ECT. The Protocol stipulates a number of obligations regarding transit:

> (a) to ensure secure, efficient, uninterrupted and unimpeded Transit for the benefit of all Contracting Parties concerned; (b) to promote transparent and non-discriminatory access to and use of Available Capacity in present and future Energy Transport Facilities used for Transit; (c) to facilitate efficient use of Energy Transport Facilities used for Transit; (d) to facilitate the construction, expansion, extension, reconstruction, and operation of Energy Transport Facilities used for Transit; (e) to minimise harmful Environmental Impacts of Transit; (f) to promote the prompt and effective settlement of disputes relating to Transit.

The Transit Protocol defines the scope of available capacity in cross-border pipelines. The Protocol stipulates a definition of available capacity, which should be uniform for any transit agreement. For hydrocarbons, definition of available capacity includes:

> forecasted requirements, for the transportation of Energy Materials and Products which are owned by the owners or operators of the Energy Transport Facilities or their Affiliates; and (d) necessary for the efficient operation of the Energy Transport Facilities, including any operating margin necessary to ensure the security and reliability of the system.

For the electricity sector, the Transit Protocol defines capacity in terms of possibility of flow and reverse flow.

Under the terms of Article 8 of the Transit Protocol, a contracting party must justify a denial of available capacity. In addition, there is an obligation to fulfil an existing long-term supply contract and hence to renew transit agreements in order to avoid a mismatch between supply and capacity agreement.

In addition to that, the Transit Protocol iterates the prohibition of unlawful taking of energy by a transit country or its network operator (Article 6) and to apply discriminatory tariffs on different capacity users (Article 10).

In December 2002, Transit Protocol negotiations were stalled mainly due to the EU-Russia controversies (Konoplyanik, 2009). Between 2003 and 2007 the EU and Russia processed bilateral consultations over the Transit Protocol. In order to avoid a mismatch between supply and capacity agreements in accordance with Article 8 (4) of the Protocol, Russia suggested a concept of "right of first refusal", which would allow a supplier of the long-term obligation to have the right to conclude a transit agreement in the first place. Initially, European countries opposed the idea of the "right of first refusal" as such in order to promote an easier transfer of capacity agreements. In 2005 a compromise was proposed. The idea was that the "right of first refusal" could be used where other suppliers had access to the available capacity (Konoplyanik, 2009; Blamberger and Waelde, 2007). Nevertheless, the Russian position has changed since 2005 and in 2006 a *de jure* gas export monopoly over Russian gas was established by Russian legislation. The export monopoly might create difficulties for Central Asian gas to get a share of transit gas through the Russian network.

After the adoption of the second Energy Directives in 2003, the European Community declared itself a Regional Economic Integration Organisation (REIO), which is an internal market *per se* and hence does not have transit within the EU. The concept of transit may then only be applicable if gas flows cross Switzerland. The REIO clause meant that the Transit Protocol is not applicable in the EU. According to Konoplyanik (one of the best known experts in the field who until 2007 was a Deputy to the Secretary General of the Energy Charter), the REIO clause remained the only unresolved issue within the Transit Protocol negotiations (Konoplyanik, 2009). Indeed, exemption of the EU from international transit norms indicated to Russia that the EU was mainly interested in the undisrupted transit of gas through Russia from Central Asia. The main issue raised in Moscow then became: why accept the regime, if it is in turn abandoned by the EU?

In December 2007, the Energy Charter Conference demanded to continue the negotiations at a multilateral level, although a political compromise between the EU and Russia was not fully founded. Moreover, during the Transit Protocol negotiations and consultations, Russia expressed concern regarding the aforementioned mediation procedures of Article 7 (7). Russia expressed concern about mediators' rights to control volumes and supplies during the 90 days of mediation (Belyi, 2009). The issue was about control of a strategic resource and therefore Article 7 (7) of the Treaty itself needed to be clarified.

It should be noted that from the very beginning, Moscow and Brussels had a different understanding of the use of the Protocol. For the Russians, the Protocol represented an opportunity to renegotiate the treaty, particularly the issue of transit tariffs and mediation mechanism of Article 7 (7). From the EU perspective, such a formula highlighted an important political position against the ECT itself as the Transit Protocol could not revise the ECT.

A divide between the EU and Russia over the Transit Protocol demonstrates a conflict in the interpretation of the norms of energy governance. For Russia, the Transit Protocol aims to secure the long-term supply chain (avoiding gas theft by transit countries, for example) and hence avoiding unnecessary competition. For the EU, the Transit Protocol aimed to bring new aspects of competition into the FSU area and therefore needed to provide flexibility in the gas supply chain. Likewise, the EU conception of the REIO was based on the principle that EU freedom of movement represents a more favourable system of governance than the ECT transit regime. Nevertheless, Russia considered the EU model mismatched supply obligations and access to capacities (Finon and Locatelli, 2007).

### 3.3 Transit Conflicts Impact on ECT Governability

A culmination of the controversy surrounding transit governance occurred during the gas transit crises between Russia on one hand, and Belarus and Ukraine on the other. Since the collapse of the USSR, Russia and the transit states (Belarus and Ukraine) established bilateral company-to-company relations, which allowed barter deals based on political accord. For instance, Russian Gazprom may pay a pre-established transit fee in natural gas in exchange for transit services made by a transit state. Up to 2005, Russia allowed its direct neighbors, Belarus, Moldova and the Ukraine, to enjoy lower tariffs through direct bilateral political agreement. Barter agreements have allowed Russia to avoid additional payments for gas transit. In response to the increased supply tariff, both

Belarus and Ukraine required a transit fee increase and used their "transit" position to strengthen their negotiating position with Russia.

In the case of Belarus, conflicts occurred in 2004 and then in 2007 and resulted in Gazprom getting a share of 50 per cent of Belarus' natural gas network company (Yafimava, 2011). Non-payment problems led again to a short-term gas supply cut during the summer of 2010 by Gazprom, which did not echo in Europe.

In the case of Ukraine, conflicts occurred several times and culminated in January 2009, when Russia decided to halt all deliveries through Ukraine and which represented up to 80 per cent of all its exports. One must consider the obscurity of these negotiations as well as the non-transparent situation around transit flows. Russia accused the Ukraine of taking more than the transit flow required. Ukraine, in turn, accused Russia of under-supplying gas (Pirani et al, 2009). Consequently, the situation was very similar to the crisis of January 2006. The difference, however, is that neither party made any further efforts to conclude an agreement. Consequently, a new gas crisis occurred, which caused much greater damage to the actors involved, as well as to the energy dependent European states compared to the 2006 crises (Mitrova et al, 2009).

An interesting question arose: could Russia use the ECT framework to resolve disputes with Belarus and Ukraine (Belyi and Klaus, 2007)? It could be argued that Russia actually could have used the transit dispute settlement of Article 7 (7) to its advantage. But an overall political understanding of the ECT did not allow Russian actors to use it to their advantage. Russia rejected the use of the ECT dispute settlement for transit mainly due to the perception of the Charter's pro-EU orientation.

In the context of REIO, Gazprom prefers to deal with the FSU partners bilaterally. Gazprom prefers to set a number of bilateral deals on transit with the transit states and avoid a capacity-supply mismatch, which can occur in the liberalized European energy markets (Finon and Locatelli, 2008). Therefore, Gazprom saw an opportunity in securing transit by increasing the so-called asset swaps: Gazprom opened parts of the upstream process to the European companies and in exchange received shares in the European downstream, hence securing access to the network. Nevertheless, the crisis with Belarus in 2010 demonstrated a vulnerability of controlling assets for energy security. Although Gazprom owned 50 per cent of the transit network, the gas supply was disrupted. Russia's approach of asset-exchange (Finon and Locatelli, 2008) does not remove the issues of access to capacities, of mismatch between supply and transit contracts and of cost-effectiveness of tariffs.

## 4 EU AND RUSSIAN POSITIONS: REINFORCING OR MARGINALIZING THE ECT?

Controversies surrounding the international energy governance further impacted on the respective positions of the EU and Russia regarding the ECT. There is a paradox in the fact that the idea of an international energy governance is backed in both Brussels and Moscow, in spite of the relatively marginal role of the ECT in their respective policy agendas. There is still a disagreement between the EU and Russia about the centrality of the ECT: Russia prefers to revise the existing treaty, whereas the EU hopes for its further expansion.

### 4.1 Russian Specific Position and "Conceptual Approach"

In the aftermath of the Russia-Ukraine gas transit crisis, Russia announced its withdrawal from the ECT"s provisional application. Indeed, prior to 2009, Russia applied the Treaty provisionally, which signifies that Signatories take their responsibilities after the Treaty before the completion of ratifications and the entry in force:

> Each signatory agrees to apply this Treaty provisionally pending its entry into force for such signatory in accordance with Article, to the extent that such provisional application is not inconsistent with its constitution, laws or regulations.

Nappert (2008) argues that effects of provisional applications are still unknown in international jurisprudence. The ECT jurisprudence opened the issue with *Kardassopoulos vs Georgia* case,[6] which declared that the provisional application between signature and ratification requires adherence to general treaty-based obligations. Article 45 (2) also states that a signatory can choose not to apply the treaty provisionally from the moment of signature. Australia, Iceland and Norway made a declaration, according to which the treaty is not applicable to them before ratification. Russia, by contrast, did not make a similar declaration and therefore is bound by the treaty (Nappert, 2008, 113).

The 2009 *Yukos* case,[7] which directly involved Russia, focused important attention to the obligations inherent in the provisional application. In essence, Yukos, a Russian oil company, was sold in 2004 after accusations of tax evasion. Subsequently, its largest subsidiary Yugoneftegansk was sold to state-owned Rosneft. In response, international shareholders initiated a case based on the discriminatory and retroactive tax reassessment of Yukos' tax payments (Waelde, 2008, 74). An important preliminary conclusion by the arbitral tribunal in this case was that Russia could be responsible under Article 45 of the treaty. It has been often argued that Russian withdrawal from the provisional application of the treaty in 2009 was largely conditioned by the arbitral decision on the *Yukos* case.[8] However, this would be unlikely because in 2004 Russia was still responsible under provisional application and therefore a withdrawal from the ECT in 2009 would not have made sense.

Russian provisional application came to an end in 2009 subsequent to Russia's formal announcement of non-ratification of the treaty. On 20 April 2009 Russia tabled an "alternative" to the ECT: the "Conceptual Approach to the New Legal Framework for Energy Cooperation" (Belyi and Nappert, 2009; Belyi et al, 2011). Broadly worded and in the form of a statement of principles at this stage, the "Conceptual Approach" includes many principles and practices which have previously been debated and adopted: sovereignty over natural resources, ensuring non-discriminatory access to markets, transparency, access to technologies, exchange of information, etc. Russia supported the idea of extending the ECT to other countries (including the US and producing countries) and covering a broader scope of energy sources (e.g., nuclear). Transit conflicts were given a more global dimension. This would create a governance of energy markets that would be broader and based more on political than legal considerations.

Several months after the proposal was made, Russia announced its non-ratification of the ECT itself. This constituted a clear political message that the ECT needed to be revised. At the same time, the effectiveness of the Russian proposal much depended on Ukraine's support of the new governance proposal. Importantly, Russia did not withdraw from the ECT in December 2009 despite domestic pressure to abandon the ECT.[9] It could be argued that the Russian "Conceptual Approach" aimed to create political grounds for a new round of Energy Charter discussions. A concrete result of the Russian moves towards revising the ECT consisted in a termination of the provisional application in early 2010.

Nevertheless, the Russian "Conceptual Approach" demonstrated a willingness to continue the multilateral process. Substantially, a declaration of non-ratification only repeated the earlier decision of the State Duma, back in 2001 (Konoplyanik, 2009, 100). In spite of the mixed messages on Russia's further participation in the ECT, Moscow did not withdraw its signature from the treaty. At the same time, a full withdrawal from the process may harm Russian investment and transit interests outside Russia in the longer run.

## 4.2 The EU's External Energy Policy and the Position Regarding the ECT

Albeit the EU external energy policy was not an evident step at the beginning of the process, it is now taking a certain shape. In 2011 and 2012 two important external energy policy (EEP) initiatives were issued by the European Commission and the European Parliament respectively (European Parliament, 2012). The European Parliament outlines the importance of the stability of long-term investment and of multilateral cooperation. Promotion of the internal market becomes the key point in the constitution of EEP. Furthermore, export of the internal market norms constitutes an important part of the EU's external policy. Communication issued by the European Parliament makes an explicit link to the Energy Charter as the global governance framework. The document reiterates the need to reinforce the Energy Charter Treaty by "extending the Energy Charter Treaty to countries which have not yet signed or ratified it".

The European Parliaments document remains silent on the pre-investment non-discrimination and on the involvement of the US into the multilateral framework. Although the EEP documents outline the importance of multilateral cooperation, the EU's influence remains limited in energy producing countries because of resource nationalism.

Moreover, the Energy Community Treaty of 2005 introduced a qualitatively new relationship between the EU and the abovementioned non-EU countries of Europe on energy trade. The Energy Community will follow the *acquis communautaires* (art. 5) related to the EU internal energy market as well as the European Community's competition norms (art 18). The impact of EU harmonization should also spread to the Contracting parties of the Energy Community Treaty.

Although the Energy Community Treaty does not explicitly contradict the multilateral Energy Charter Treaty, it gives a new character to the international energy governance (Prange-Ghstol, 2009). If the Energy Charter Treaty provides a framework for governance, including investment protection, the Energy Community Treaty aims at actual market transformation, which implies investors' commitments to competition. In turn, the transformation should be guided by the EU energy market norms. Thus, the multilateral character of the governance is replaced by a regional bloc governance with EU regional leadership. Thus, a hidden contradiction between a multilateral and a bloc governance might further affect the EU's evolving external energy policy conception.

## 5 CONCLUSION

The debate on energy security needs to be reshaped from the producer-consumer divide towards an institutional analysis of enhancing predictability of relations within international energy markets. A multilateral legal regime is a reflection of a demand for more predictable relations for both energy producers and consumers, as well as for states and investors. Predictability can be ensured by multilateral level governance.

At the same time, multilateralism supposes a number of shared values, which is still not the case with the ECT. Acceptance of norms has been hindered by opposing views on energy security. Russia's view evolved towards a security of demand combined with resource nationalism. The EU conditioned the international energy governance to the promotion of competition and apparently preferred a market transformation approach for the Energy Community Treaty. With the Energy Community Treaty, the EU promotes itself as a norm-creator, a position which further distances Russia from the framework. Hence, the ECT case study demonstrates an ever-evolving fragmentation of norms in the post-western hegemonic international political economy.

At the same time, it could be argued that a relative marginalization of the ECT did not lead to a rejection of an issue-specific multilateral regime in energy. The EU legal system of the Internal Market (especially, ownership unbundling and reciprocity clause) complicates the implementation of the ECT, but the treaty itself has so far not been questioned. The External Energy Policy conception reiterates the importance of the ECT. Despite its "alternative proposal", Russia did not withdraw from the treaty. Growing interdependence between producing and consuming countries, an increasing number of cross-border investments as well as possible mismatches between transport capacity and supply obligations led to a demand of predictability. On these grounds, we can note that at least four aspects demonstrate the relevance of the ECT-based energy governance: (1) the ECT provides the most overarching framework for protection of investments, including environmental ones; (2) the ECT represents a trend towards the MFN principle in the energy investment field, which could counterbalance the retaliatory reciprocity; (3) the ECT did not discredit itself during the transit crises as it was not used; and (4) the ECT remains the only currently existing realistic framework of legally binding relations involving the EU, Russia and the transit states.

International energy governance requires a certain level of acceptance by both producing and consuming states. In the meantime, the very understanding of energy security remains controversial and therefore hinders the applicability of the ECT.

## NOTES

1. The Energy Charter is the first overarching political and legal framework for an international energy governance, set as a political declaration in 1991 and then evolved to a Treaty in 1994. For a detailed legal analysis, see Walede (1996), Happ (2002), Salem Haighighi (2006), Blamberger and Waelde (2007), as well as the website of the Energy Charter Secretariat www.encharter.org (accessed February 2013).
2. Charter of Paris for a New Europe, 1990.
3. The concept of "seminar diplomacy" is defined by Adler (1997).
4. For details, see the website of the Energy Charter Secretariat: http://www.encharter.org/ (accessed February 2013).
5. Baker & McKenzie (2008), "Review of Amendments to the Law of the Russian Federation 'On Subsoil'", working paper, Moscow, April, pp. 1–2.
6. Case No. ARB/05/18, Ioannis Kardassopoulos (Greece) v. Georgia, ICSID.
7. Yukos Universal Ltd. (UK – Isle of Man) v. Russian Federation, Ad hoc UNCITRAL Arbitration Rules; arbitration administered by the Permanent Court of Arbitration (PCA) in The Hague.
8. This opinion was proposed by one of Yukos' arbitrators, E. Gaillard for *Financial Times*, 18 August 2009.
9. Konoplyanik (2010) has gathered quotations from Russian officials.

## BIBLIOGRAPHY

Adler, E. (1997), "Imagined (Security) Community", *Millennium*, 26(2), 275–277.

Belyi, A. (2003), "New Dimensions of Energy Security of the Enlarging EU and Their Impact on Relations with Russia", *Journal for European Integration*, 25(4), 351–369.

Belyi, A. (2009), "New regime for energy investment reciprocity", *Journal for World Energy Law and Business*, 2(2), 117–128.

Belyi, A. (2012), "The EU's Missed Role in International Transit Governance", *Journal for European Integration*, 34(3), 261–276.

Belyi, A. and U. Klaus (2007), "Dispute Resolution Mechanisms in Energy Transit – Missed opportunities for Gazprom or false hopes in Europe?", *Journal of Energy and Natural Resources Law*, 25(3), 205–224.

Belyi, A. and S. Nappert (2009), "New round of the Energy Charter: myth or reality", *Oil, Gas, Law Intelligence (OGEL)*, special issue Russia, 05-09.

Belyi A., S. Nappert and V. Poigoretsky (2011), "Modernizing the Energy Charter? Road map and Russian Draft Convention on energy security", *Journal of Energy and Natural Resource Law*, 29(3), 205–224.

Blamberger, Craig and Thomas Waelde (2007), "The Energy Charter Treaty", in Marta Roggenkamp, Catherine Regwell, Inigo Del Guayo and Anita Ronne (eds), *Energy Law in Europe*, Oxford University Press, pp. 145–195.

Collier, John and Vaughan Lowe (1999), *The Settlement of Disputes in International Law: Institutions and Procedures*, Oxford University Press.

Julia Dore (1996). "Negotiating the Energy Charter Treaty", in Thomas Waelde (ed.), *The Energy Charter Treaty*, Kluwer Law International.

Energy Charter Secretariat (1994), "Energy Charter Treaty", http://www.encharter.org/index.php?id=178 (accessed February 2013).

European Parliament (2012), "Resolution on Engaging in energy policy cooperation with partners beyond our borders: A strategic approach to secure, sustainable and competitive energy supply", http://www.europarl.europa.eu/sides/getDoc.do?pubRef=-//EP//TEXT%20TA%20P7-TA-2012-0238%200%20DOC%20XML%20V0//EN&language=EN (accessed February 2013).

Finon, Dominique and Catherine Locatelli (2007), *Russian and European Gas Interdependence. Can market forces balance out geopolitics?* LEPII, Cahiers de Recherche Grenoble.
Gilpin, Robert (1987), *The Political Economy of International Relations*, Princenton University Press.
Goldthau, A. (2012), "A Public Policy Perspective on Global Energy Security", *International Studies Perspectives*, 13(1): 65–84.
Happ, R. (2002), "Dispute Settlement under the Energy Charter Treaty", *German Yearbook of International Law*, 45, 331–362.
Hober, K. (2010), "Arbitrating Disputes under the *Energy* Charter Treaty", in K. Talus and P. Fratini, *EU Russia Energy Relations*, Brussels: Euroconfidentiel.
Keohane, R. (2001) "Governance in a Partially Globalized World", *American Political Science Review*, 95(1), 1–13.
Konoplyanik, A. (2009) "A common Russia-EU energy space: the new EU-Russia partnership agreement, acquis communautaires and the Energy Charter", *Journal of Energy and Natural Resources Law*, 27(2), 258–291.
Kuzemko, Caroline, Andrei Belyi, Andreas Goldthau and Michael Keating (2012), *Dynamics of Energy Governance in Europe and Russia*, Palgrave Macmillan.
Mitrova, T. (2009), "Natural gas in transition: systemic reform issues", in Simon Pirani (ed.), *Russian and CIS Gas Markets and their Impact on Europe*, Oxford University Press, pp. 13–54.
Mitrova, Tatiana, Simon Pirani and Jonathan Stern (2009), "Russia, the CIS and Europe: gas trade and transit", in Simon Pirani (ed.), *Russian and CIS Gas Markets and Their Impact on Europe*, Oxford Institute for Energy Studies.
Nappert, S. (2008), "Russia and the Energy Charter Treaty: The Unplumbed Depths of Provisional Application", *Oil, Gas, Energy Law International*, 3-08.
Pirani, Simon, Jonathan Stern and Katja Yafimava (2009), *The Russo-Ukrainian gas dispute of January 2009: a comprehensive assessment*, Oxford Institute for Energy Studies.
Prange-Gstohl, H. (2009), "Enlarging the EU's internal energy market: Why would third countries accept EU rule export?", *Energy Policy*, 37, 5296–5303.
President of Russian Federation (2009), "Conceptual Approach to the New Legal Framework for Energy Cooperation (Goals and Principles) ", Kremlin, http://archive.kremlin.ru/eng/text/docs/2009/04/215305.shtml (accessed February 2013).
Salem Haghighi, Sanam (2006), *Energy Security: the External Legal relations of the European Union with Energy producing Countries*, Florence: European University Institute.
Seliverstov, Sergei (2009), "Energy Security of Russia and the EU: Current Legal Problems", Paris: IFRI working paper.
Stickley, Denis (2007), *A Framework for Negotiating & Managing Gas Industry Contracts*, University of Dundee.
Talus, Kim (2011), *Vertical Natural Gas Transportation Capacity, Upstream Commodity Contracts and EU Competition Law*, Kluwer Law International.
Waelde, Thomas (1996), *The Energy Charter Treaty: An East-West Gateway for Investment and Trade*, Kluwer Law International.
Waelde, T. (2008), "Renegotiating acquired rights in the oil and gas industries: Industry and political cycles meet the rule of law", *Journal for World Energy Law and Business*, 1(1), 55–97.
Waltz, Kenneth (1979) *Theory of International Politics*, McGraw–Hill.
Willems A.R., J. Sul and Y. Benizri (2010), "Unbundling As a Defence Mechanism Against Russia: Is the EU Missing the Point? " in K. Talus and P. Fratini, *EU Russia Energy Relations*, Brussels: Euroconfidentiel.
Yafimava, Katja. (2011) *The Transit Dimension of EU Energy Security*, Oxford University Press.

# PART V

# ENERGY, THE ENVIRONMENT AND SECURITY

# 14. Governance dimensions of climate and energy security
*John Vogler and Hannes R. Stephan*

## 1 INTRODUCTION

For a long period 'security' was both a central, yet extraordinarily underdeveloped, concept in International Relations. Critical scholarly attention really dates from the pioneering work of Barry Buzan (1983). Since then varying security perspectives have proliferated. *Environmental* security became the subject of a long-running debate and the 1994 UN Human Development Report introduced the people-centred approach of *human* security (Dalby 2009). It is now commonplace not only to emphasize national border security, but also refer to food security, water security, and other 'sectoralized' security areas (Brauch et al. 2009). This expansive re-definition should alert us to the significance of the 'referent object' or, in other words, 'that which is to be secured'. In orthodox security studies, there is no doubt that the object of security policy remained the integrity of the state and its interests. There might be reference to people, but as Buzan (1983, p. 245) noted, there was always 'an unbreakable paradox' between state and individual security. In much recent security discussion notions of threat may have changed, as in the typical security triptych of 'terrorism, failed states and weapons of mass destruction', but the preservation of the state remains the essential object of policy. Energy security, often with overtones of control over contested scarce resources, is conventionally seen as a central component of the national interests of a state and not infrequently a *casus belli*. This is also true of the overwhelming bulk of environmental security discussions including those relating to the actual and possible conflict consequences of global climate change. However, the really radical move would be to shift the object of security from the state to human populations and then to the earth's climate upon which they depend.[1]

While there are several significant greenhouse gases (GHGs), current mitigation efforts concentrate on energy-related carbon emissions which, in 2005, accounted for around 61 percent of all GHG emissions (Baumert et al. 2005) and whose importance rises in line with increasing global energy consumption. In Europe, the figure is even higher at 80 percent (European

Commission 2007, p. 3). This physical link lies behind the evolution of the UN Framework Convention on Climate Change (UNFCCC) over the past decade. It largely explains why energy and climate change agendas have become increasingly intertwined. The particular characteristics of the Kyoto Protocol meant that, for an extensive period, most Parties were able to avoid this conjunction. Non-Annex I developing countries were not required to make any reduction in their fossil fuel-based emissions and the EU could sustain its climate 'leadership' without having to make significant cuts in energy use through the fortuitous use of the 1990 baseline in its burden-sharing agreement. The United States, which would under the Protocol have had to make real and economically damaging energy-related reductions, simply opted out; while others either failed to meet their obligations or were able to take advantage of carbon offsets. In the post-2012 discussions, which followed entry into force of the Protocol in 2005, the energy-climate connection became all too painfully clear and dominated the international discussions leading up to the 2009 Copenhagen Conference of the Parties (CoP). At the highest level, climate politics became international energy politics and could be portrayed as a competition to secure shares in a diminishing 'carbon space' or, perhaps, to ensure that the burdens of reductions in energy use should be borne by others. Energy security has habitually been associated with 'high politics' and it was noticeable that, in this regard, the climate CoP at Copenhagen departed markedly from other analogous 'low politics' environmental regimes.

The primary purpose of this chapter is to understand the energy-climate nexus within a security framework. We proceed by initially analysing both domains in their own terms. The energy security agenda is characterized by (geo)political and material (scarcity) constraints, and governance responses have largely been confined to the national arena. By contrast, climate change has long been subject to multilateral, UN-related governance processes. Explicit security lenses have been applied to the potential short- and long-term impacts of climate change. Associated policy responses can be broadly categorized as *reactive* or *preventive*. Finally, the third section provides a conceptual and institutional comparison between energy security and climate security agendas and considers the important question of 'synergies' between them, leading perhaps to the elusive 'win-win' solution under which a progressively de-carbonized economy might provide for really comprehensive security in terms of climate stability, sustainable energy and the avoidance of the more disruptive traditional threats associated with rapid climate alteration.

## 2  ENERGY SECURITY

The standard definition of energy security was forged amidst the oil crises of the 1970s and remains a conceptual cornerstone: 'access to secure, adequate, reliable, and affordable energy supplies' (Bordoff et al. 2009, p. 214). It should also be remembered that for producers, energy security means continued demand and market access. Energy security represents a broad, if rather vague, placeholder for a range of policy-making priorities. Admittedly, it does not adequately address other important aspects of energy governance such as carbon emissions or overall environmental sustainability. To keep these concerns separate nonetheless reflects the reality of policy-making where successful policy integration is rare, whilst parallel, competing tracks are still the norm. But even according to the orthodox definition, there have been plenty of reasons in recent years to highlight growing energy insecurity. The rise of major energy-consuming economies, such as China and India, has lowered overall confidence in 'secure' and 'affordable' energy supplies. Affordability may be compromised due to an increasing imbalance between the demand and supply of fossil fuels and especially the widespread recognition of 'the end of easy oil'. Secure access may be at risk because increasing scarcity implies greater international competition and encourages a move away from market allocation towards 'statist' forms of energy security.

Historically, realist theoretical assumptions have dominated thinking on energy security. Widespread recognition of the role of energy resources during the build-up and conduct of the Second World War ensured the status of energy as an issue belonging to the 'high' politics of national security. The role of energy as a 'strategic good' par excellence is not only related to its essential function in 'fuelling' military activities. Its price level and availability also play a fundamental role in a country's economic performance and socio-political stability (Lesage et al. 2010, p. 183). A realist interpretation of energy security was further reinforced by events in the 1970s when a trend towards the nationalization of energy supplies and the sporadic use of oil embargoes, orchestrated by the Organization of Petroleum Exporting Countries (OPEC), highlighted the dangers of energy dependence. Even today the privileged position of major energy-exporting countries still represents a constraint on the foreign policy agenda of major importers (Müller-Kraenner 2008, p. 27).

Market expansion and low energy prices from the 1980s until the mid-2000s encouraged the development of liberal approaches to energy security. Greater diversification of sources and a gradual shift to coal and natural gas all but eliminated the threat of an effective use of the 'oil weapon'. Well-functioning global markets for oil – and potentially for liquefied natural gas – have been increasingly promoted as effective mechanisms to provide cheaper energy inputs in an increasingly competitive, global economy and guard against both structural undersupply and short-term supply disruptions (Goldthau and Witte 2009). Realist notions of energy security, however, have not been superseded. On the contrary, Brazil, Russia, India, and China – the so-called BRIC states – are not just consuming increasing amounts of fossil fuels. They also employ the traditional, statist tools of energy security policy such as bilateral contracts and the promotion of national energy champions (Lesage et al. 2010, p. 27). China and India have struck numerous energy deals with oil- and gas-exporting countries from around the world, even if this has meant giving economic and military aid to 'pariah' states in Africa and Latin America (Müller-Kraenner 2008, p. 72). While this has served to raise rather than lower the availability of fossil fuels on global markets, it demonstrates that – given an uncertain future – no major power will rely exclusively on the market allocation of energy supplies.

When it comes to natural gas, a commodity still largely reliant on pipeline infrastructure and long-term supply contracts, overtly political considerations have remained dominant. The European Union, for example, has yet to produce a coherent energy policy or to perfect a 'real internal energy market' (European Commission 2007, p. 6). There are very significant differences in the energy mix and strategies of member states whose perspectives remain stubbornly national. Thus the Commission's principal approach has been to seek energy security through the perfection of a properly functioning, interconnected and transparent internal energy market. There has also been a largely unsuccessful attempt to extend EU liberalising regulatory practices to the EU's gas suppliers in its eastern neighbourhood. Failure was demonstrated in the twin Ukrainian gas crises of 2006 and 2009 which were only resolved through EU mediated political agreement between Russia and Ukraine.

Russia, having rejected the EU's invitation to subscribe to the Energy Charter Treaty, increasingly relies on its economic power derived from natural resources and energy services. It uses the mechanism of 'pipeline politics' to compensate for its loss of superpower status and to preserve its zone of influence, particularly in the Caspian region and Central and Eastern Europe (Müller-Kraenner 2008, pp. 47–56). The EU counterpart is the suggestion that security of supply can be achieved through diversification involving new pipelines circumventing Russian territory, Nabucco providing the best known example. Youngs (2009) suggests that the EU is in fact caught on the horns of a dilemma, between attempts to install market-based governance of energy supplies and an essentially realist approach to the geopolitics of pipelines. In the US, by contrast, new shale gas discoveries over the last few years have – for now – made the country virtually independent from imports. The situation is, of course, completely different for oil supplies, even though the US, if it was minded to incur the costs, could achieve a degree of autarchy in this sector too.

The uncertain future evoked by realist commentators is not merely concerned with 'above-ground', political-economic factors, but intimately bound up with the status of 'below-ground' energy reserves. While the momentous increase in energy prices during 2004–2008 may have been partly caused by the growing 'financialization' of energy markets and an upsurge in speculation (Bradshaw 2010, p. 276), there is now a strong chorus of voices pointing to underlying factors of supply and demand. Data problems caused by failure to report or intentional misreporting cannot conceal a general pattern of stagnant reserves (Owen et al. 2010). The possibility of a significant future shortfall in oil supplies is supported by a raft of additional arguments. First, significant additional demand will come from emerging economies, especially India and China, and may result in global energy demand growth of 36 percent by 2035, with demand for oil projected to grow by 15 percent (IEA 2010). Second, considerable investments will be needed to expand (or even maintain) supply because there will be growing reliance on non-conventional, more expensive oils from tar sands, enhanced oil recovery, or even coal liquefaction. Such investments, however, will be hindered by short-term price volatility.

Third, even those countries with the capacity to ramp up production of fossil fuels will struggle to increase exports. Many energy-rich countries – for example Saudi Arabia, Iran, Venezuela – are dominated by state-owned companies which frequently lack the capital or expertise to substantially increase production. Moreover, substantial energy subsidies have long been employed by these and other governments to reduce energy poverty and secure the consent of their populations. Expectations of cheap energy and a lack of interest in energy efficiency are now so entrenched in most energy-rich countries that a continued rise in energy demand, which could ultimately cancel out increased production, is entirely possible (Rubin 2009).

Critics of such projections highlight a decade-long history of erroneous predictions of scarcity. They argue that the supply of fossil fuels will be ensured by technological change, which can unlock previously unprofitable reserves, and higher prices triggering increased investment and exploration. At most, they acknowledge the potential for politically created supply crises through increasing resource nationalism and insufficient investment (Radetzki 2010). This riposte, however, is less forceful now than in the past. Even the traditionally conservative IEA has accepted the tenor of the end of 'easy oil' (Bradshaw 2010, p. 277) and conceded that crude oil production will never again reach its 'all-time peak' of 2006 (IEA 2010, p. 48). Because this entails a switch to non-conventional oils and a progressively lower 'energy return on investment', it is likely to contribute to rising oil prices.

Besides oil supplies, the general picture for fossil fuels is even more contested. Given recent technological advances in shale gas production and underground coal gasification, it remains very uncertain when tangible scarcities will materialize. With regard to oil, however, the significance of the 'peak oil' thesis is that both materially and politically induced supply shortages may well occur. The combined effect of uncertainty and price volatility cements the high status of energy security on governments' agendas because it suggests serious implications for economic development and heightened international competition for scarcer energy resources.

### 2.1 The Pursuit of Global Energy Security

Given the revitalized interest in energy issues, there is now a burgeoning literature both on *national* and on *global* energy governance, including the security dimension. In institutional terms, however, the idea of collective energy security has only made very limited progress over the last few decades. Continuing international discord is underpinned by the fundamentally divergent interests of fossil-fuel exporting and importing countries. For OPEC member states and other important exporters, energy security is primarily a question of stable and predictable demand from industrialized economies. The latter, on the other hand, created the International Energy Agency (IEA) in 1974 to coordinate their response to future oil crises and provide information and expertise for national energy policy. For importing countries, energy security hence equates to security of supply.

Although the fierce producer-consumer clashes of the 1970s are unlikely to return, the bifurcated structure of energy security policy has proved persistent. The establishment of the International Energy Forum (IEF) in 1991 was intended to signal a new era. It provides a basis for enhanced producer-consumer cooperation and already features initiatives on improving the transparency of oil and gas data with regard to production and investment levels. But a legacy of conflict and 'deep-rooted mutual suspicion' has so far stood in the way of major governance breakthroughs (Lesage et al. 2010, p. 62). The EU is a supporter of multilateralism but the approach of its member states to energy security is often to secure a network of bilateral deals with neighbours.

To bolster the case for global cooperation, commentators have underlined the high degree of interdependence (Yueh 2010) which typifies global energy relations. In theoretical terms, this condition has long applied in a globalising world. The drive towards an efficiently functioning global energy market (primarily for fossil fuels) is nothing less than an institutionalization of economic interdependence. But in practical terms, shared vulnerability was brought to the fore by surging energy prices during 2004–2008. From 2005 onwards, several G-8 meetings treated energy as a high priority and initiated a number of assessments and action plans by drawing on the IEA's expertise (Lesage et al. 2010, Ch. 7).

A minimal common ground between consumer and producer countries is the avoidance of extreme price volatility because it makes planning for the future exceedingly difficult. For example, the budgets of fossil fuel exporting countries were initially buoyed by rising revenues, then shrank suddenly when the financial crisis hit and prices collapsed. Given the nature of energy policy, however, it is unlikely that UN institutions will take the lead in this venture. Even though many UN organizations and programmes also pursue energy-related activities, the envisaged central organizational node, UN-Energy, is currently no more than an embryonic focal point. Therefore, in a 'business-as-usual' scenario of global energy governance, serious coordination will remain the preserve of 'coalitions of the willing', while broader multilateral processes are most likely to proceed through the UN climate change regime (UNFCCC) (Karlsson-Vinkhuyzen 2010, p. 193). As we argue in the final section of this chapter, this regulatory dynamic would likely increase the compatibility of energy security and climate change mitigation. To substantiate this point, however, we first turn towards the notion of climate security.

## 3 CLIMATE SECURITY

Climate security has its roots in the environmental security debate. The critical questions raised and empirical results first offered in the early 1990s are equally valid for today's discussions.[2] Unsurprisingly, in theoretical terms the precise meaning of climate security therefore remains contested. Understandings range from the adaptive capacity and resilience of societies in the face of extreme weather events to ambitious mitigation which reduces the risk of catastrophic consequences. Yet, in the realm of international climate governance, political consensus has developed around a precise number to distinguish 'manageable' from 'dangerous' climate change: this is the famous 2°C threshold. While the concept of *environmental* security has long been present in discussions about environmental governance, the related notion of *climate* security is a relative newcomer. This process of 'securitization' may be understood as a gradual and mainly discursive accomplishment from a constructivist perspective or as an inevitable and necessary development from a rationalist standpoint.

A constructivist approach would trace the rise of the climate security discourse over the past few years. Some have pinpointed the year 2002 as the point at which the political mainstream acknowledged potential security implications. According to Dupont (2008, p. 30), a report commissioned by the Pentagon (Schwartz and Randall 2003) helped trigger a learning process through which climate change 'metamorphosed from a boutique environmental concern to a first-order foreign-policy and national-security problem that is now being ranked alongside terrorism and the proliferation of weapons of mass destruction.' The following years did indeed witness a flurry of similar, if more sophisticated, assessments, most prominently a 2007 report by a US think tank (CNA Corporation), a 2008 EU report on 'Climate Change and International Security', and an explicit recognition by the 2010 US Quadrennial Defense Review and by the 2010 UK Strategic Defence and Security Review.

Although there are still some doubts about the extent and durability of the securitization process[3] (Scott 2008; Mobjörk et al. 2010), the debate on climate change and security in the UN Security Council (April 2007) may come to be seen as a genuine watershed. While most developing countries resisted the security framing and favoured environment/development discourses, small-island developing states (SIDS) – ranking among the most vulnerable nations – sided with industrialized countries to support an active role for the Council in climate change governance (Detraz and Betsill 2009, p. 312). In July 2011, Germany reintroduced the issue at the Council. There was agreement on the significance of climate change, but no consensus on the appropriate international forum for its discussion. China and Russia, supported by G-77 members argued that the Council was ill equipped to cover a topic that was the proper responsibility of the UNFCCC (MacFarquhar 2011). Despite the session ending without tangible results, these high-level diplomatic discussions have arguably ensured the presence of climate security considerations on the agendas of governments and international organizations.

A rationalist approach places greater emphasis on the expected impacts of climate change and the likely gamut of security responses they are likely to trigger. One of way of developing such predictive capacity is to construct scenarios based on the best available climate science. Another is to study historical instances in which climatic factors seem to have played a critical role, for example the decline or collapse of ancient civilizations (Dupont 2008, p. 31). Quantitative methodologies can equally be applied. Lewis (2009, p. 1199) thus cites a Chinese study concluding that 70–80 percent of 'peak war activity' in China's history took place during unusually cold or warm climatic periods which sharply reduced land productivity.

A combination of these analytical pathways holds insights for all major strands of security thinking. For proponents of national security priorities, climate change fits into the category of unconventional, destabilising 'threat multipliers' that could cause state failure, foment extremism, trigger migratory waves, and physically endanger military installations at home and abroad. This was the thrust of the 2008 paper developed, with special emphasis on climate change and the Arctic, by the EU's Javier Solana (European Council 2008). For advocates of human security approaches, climate change impacts pose grave challenges to the twin objectives of 'freedom from fear' and 'freedom from want' by undermining stable livelihoods and imposing significant and costly adjustments on frequently vulnerable communities.

Third, what marks out climate change from other non-conventional security threats is its disruptive effect on ecological, 'planetary' security. By adversely affecting the capacity of the atmosphere to render the 'ecosystem service' of providing a stable climatic system, rapid (and potentially abrupt) human-induced climate change may pose an existential threat to the biosphere, including human civilization itself. To use the terminology of the Copenhagen School, what is evident here is a shift in the referent object of security from the nation state, to the individual in society and finally to the planetary biosphere itself.

### 3.1 Reactive Climate Security

The argument about fundamental ecological security and climate change impacts has been implicit in the literature since the 1970s. But now that attention to the security implications of climate change is growing, the focus is often re-adjusted onto reactive policies, such as coping mechanisms and adaptation measures. This strategic shift in policy formulation is justified by two weighty arguments. First, the inertia of the climate system implies that the current concentration of $CO_2$ in the atmosphere is already sufficient to generate a significant degree of global warming over the coming decades. Second, the recognition of inevitable impacts is joined by pragmatic motivations. As David Keith (2009, p. 56) describes it, in contrast to globally coordinated mitigation measures, 'the self-interest of nations, firms and individuals will work to drive measures to ease adaptation to the changing climate since the benefits of adaptation can be captured locally where money is spent.'

Apart from positive results for human security, such benefits can equally be understood as the avoidance of violent conflict. Certainly, the causal connection between climate change and conflict remains hotly contested. Yet, a broad-based, if minimal, consensus has emerged around the proposition that violent conflict rests on numerous, complex socio-economic and political – as well as climatic – processes and that the latter may constitute a 'non-essential' causal factor (Mazo 2010, p. 40).

Among the expected consequences of climate change are sea-level rise, altered precipitation patterns, an increase in extreme weather events, melting glaciers, increasing burdens of infectious diseases, and the progressive acidification of the oceans (Dupont 2008, p. 32). When set against a number of separate trends, such as population growth, it is evident that climate change will contribute to increasing water and food insecurity around the world.[4] Whereas profound societal destabilization will not necessarily translate into inter-state warfare, it will harm the prospects for human well-being and may trigger unprecedented waves of migration, both within and across national territories. Some estimates suggest there may be 200 million environmental refugees by 2050 (Mazo 2010, p. 129). Furthermore, failing states could unwittingly 'export' insecurity well beyond their borders by becoming havens for international criminal or terrorist networks.

If security responses to such instabilities were designed by traditional military planners, one may expect the whole gamut of coping and containment tools to be applied. Individual states or alliances of states are likely to step up border security and the policing of major migratory routes (Rogers 2010). The beginnings of these trends can, for instance be discerned in the EU's Immigration, and Neighbourhood Policies. Active intervention in failed states, under the Common Security and Defence Policy, for the purpose of conflict prevention, conflict resolution or humanitarian assistance may also appear on the agenda alongside EU counter-terrorism efforts. However, many of these measures may also strengthen the widespread perception of an 'uncaring West'. This could bolster support for extremist groups[5] and perhaps provoke radical civic mobilization within developed countries themselves (Mabey 2008, p. 94).

On the other hand, reactive security responses need not be confined to military approaches. Following a human security perspective, there will also be increasing interest in emergency adaptation measures and, crucially, in 'pre-adaptation' strategies such as fostering resilience and 'climate-proofing' of critical infrastructures (Adger 2010; Mazo 2010, p. 102). Much of the climate aid for developing countries – projected to reach $100 billion annually by 2020 – is likely to be earmarked for this category of actions. Initial projects from late 2010, such as combating coastal erosion in Senegal or flood prevention in northern Pakistan, give an indication of what resilience and 'pre-adaptation' mean in practice.

## 3.2 Preventive Climate Security

While few would dispute the need for reactive security policy and adaptation measures, the central question is whether these will compete with the requirements of climate change mitigation. Given the slow progress of international climate regulation, adaptation funding – one of the few issues gathering widespread support – might well be employed for policies that further increase carbon emissions. On the other hand, the accompanying capacity-building may also serve to improve the effectiveness of mitigation policies (Mazo 2010, p. 132). Overall, a broad consensus exists that 'sustainable security' (Dalby 2009, p. 166) can only be achieved if both policy objectives are designed for compatibility. The governance arrangements for avoiding deforestation (REDD+), currently under discussion in the UNFCCC and affiliated fora, represent a test case for this integrated conception.

Regarding mitigation, ambitious reductions in greenhouse gas emissions are essential not only to curb the need for risky and expensive reactive security policies. Moreover, it is becoming increasingly clear that the process of climate change may not conform to the linear assumptions embodied by relatively conservative modelling exercises. Although the IPCC's Fourth Assessment Report states that climate change is very likely caused by human activity (IPCC 2007), detailed knowledge about the precise mechanisms of an enormously complex climatic system remains a work in progress. This is reflected by the broad ranges of possible temperature change given in the IPCC's scenarios. Rather than taking scientific uncertainty as a reason for hesitation, however, many commentators have pointed out that the probability functions of mainstream climate models could be too linear because the climate's sensitivity to GHGs might unexpectedly turn out to be much stronger. Recent developments – such as unprecedented reductions in mid-year Arctic sea ice in 2010 and 2012, sustained sea-level rises, and higher GHG emissions than projected – have bolstered these concerns (Mobjörk et al. 2010, p. 42ff.).

Unexpectedly rapid or strong climatic changes are not the only scientifically grounded scenarios that would likely have severe security consequences. It is equally possible that there are non-linear climatic dynamics scientists do not yet understand and which therefore cannot be integrated into their models. There are likely to be thresholds or 'tipping points' which could shift the global or, more likely, regional climate system into a new state. Mabey (2008, p. 22) presents a typology of such climatic events, distinguishing between 'high impact reversible events' (e.g. changing Asian monsoons, a weakening Gulf Stream), 'irreversible impacts' (e.g. melting glaciers, species extinction), and 'runaway climate change' whereby feedback loops – triggered by events such as melting permafrost soils releasing large quantities of methane – would push the climate system into an uncontrollable warming spiral.

Such 'high impact/low probability' scenarios are no mere figment of imagination, as the geological record shows that they have occurred in the distant past (Mabey 2008, p. 13; Mazo 2010, p. 29). These scenarios are now frequently recognized in the scientific and policy literatures. Innovative economic analysis has equally cast doubt on conservative 'median' damage functions commonly employed by mainstream 'gradualism'. Weitzman (2010, p. 24) thus proposes a 'fat-tailed' probability distribution of climate sensitivity to higher greenhouse gas concentrations. This implies a distinct chance (1 percent) of 10 or more degrees of global warming and suggests that ambitious mitigation targets would represent an insurance policy against catastrophic climate change.

Discourses of preventive security strike a similar note. The core argument here is that security analysts and military planners have been trained to rely on prudence and foresight which may lead them to consider worst-case scenarios rather than mere 'best guesses' (Dupont 2008; Mabey 2008; Rogers 2010). By implementing this form of assessment, analysts may come to recognize that reactive security responses cannot adequately deal with scenarios of extreme or abrupt climate change. First, every society has a limited adaptive capacity to profound perturbations. Second, as Mabey (2008, p. 13) puts it, 'while climate change raises many hard security problems, it [ultimately] has no hard security solutions.' In policy terms, both preventive security and risk-averse economic thinking point towards two major undertakings: a rapid transition towards an ultra-low carbon economy and enhanced international cooperation on climate governance.

## 4 GOVERNING ENERGY AND CLIMATE SECURITY

The first two parts of this chapter have come to different conclusions regarding the challenges of energy security and climate security. For the former, strong international governance mechanisms are desirable but difficult; for the latter, such advances are very challenging indeed, but ultimately indispensable. Energy security is largely subject to the vagaries of the market and the geo-political manoeuvres of major producers and consumers. Institutions such as the International Energy Agency, set up in 1974 by OECD countries, or agreements such as the EU-sponsored 1991 Energy Charter Treaty have not been able to fundamentally change this dynamic.

A range of existing governance mechanisms have extended their remit to cover climate change – typically viewed as a 'threat multiplier'. They provide numerous frameworks for collaborative international efforts to react to the effects of climate change. The climate issue has permeated the international institutional architecture from development organizations interested in adaptation to UN peacekeeping and the EU's Common Security and Defence Policy. Those who oppose a modification of the UN Security Council's activities prefer to keep the Climate Change Convention as the appropriate forum and to assert that climate security must be considered as a sustainable development issue. Despite recent growing concern with adaptation, the UNFCCC, therefore, remains the institutional location for efforts at preventive climate governance. Although hailed as a success, the Durban 2011 CoP extended the date for a comprehensive new agreement out to 2020. Given this delay and continuing uncertainty about the eventual agreement, many major economies have resolved to enact domestic climate and energy policies that pursue 'win-win' solutions, such as fuel-switching to low- and ultra-low-carbon sources, greater energy efficiency, demand reduction, and the development of cost-efficient carbon capture and storage (CCS) technology (Froggatt and Levi 2009).

### 4.1 'Synergistic' Climate and Energy Security Policy

The popularity of 'synergistic' approaches is reflected in policy developments in major economies around the world. For instance, emerging economies such as China and India continue to pursue traditional energy policy centred on diversification of energy sources, expansion of fossil-fuel-based energy generation, and the reduction of energy poverty. But they are also implementing 'win-win' energy-and-climate policies.

With Chinese oil imports predicted to rise from about half to well over 80 percent of domestic needs by 2030 (IEA 2007a), a target of reducing the energy intensity of its economy by 20 percent until 2010 has been raised to 40–45 percent by 2020. This goal is to be achieved primarily through energy efficiency measures, but is flanked by an ambitious programme of investment in low-carbon energy generation which might lead to renewable energy providing one third of total energy generation by 2020. India is projected to face an even greater degree of energy dependence, with up to 90 percent of oil and a rapidly growing share of natural gas and coal to be imported by 2030 (IEA 2007b). India's 2008 'National Action Plan on Climate Change' emphasizes significant future investments in solar energy and energy efficiency.

Even a high-income country such as the US has until now followed a similar policy pattern. US energy and climate policy has often depended on traditional notions of energy security emphasising domestic production of oil and gas. Yet, a temporary confluence with supporters of climate change mitigation brought about the 2007 'Energy Independence and Security Act' which yielded policies on biofuels,[6] energy efficiency and low-carbon energy generation (Bang 2010). However, there are some mid- and high-income countries which have committed to more ambitious and target-based action on climate mitigation. Mexico, for example, plans to reduce its GHG emissions by 30 percent below a business-as-usual scenario by 2020 and 50 percent from 2000 levels by 2050. Japan has pledged to reduce its emissions in the same period by 25 percent below 1990 levels, although this target will be difficult to achieve. And the EU proposed a 20–30 percent cut below 1990 levels, but made the upper figure conditional on stronger international reciprocity.[7] If current policies on renewable energy and energy efficiency are fully implemented, the EU might achieve a 30 percent reduction by 2030, but that still leaves a considerable gap to the long-term objective of cutting GHG emissions by 80 percent until 2050 (European Commission 2011).

By extrapolating from intra-European differences, one can try to deduce the main reasons behind the divergent ambitions of these two groups of countries. Marques et al. (2010) thus found that investment in renewable energy sources increased in line with an EU member state's dependence on energy imports. Unlike the US, most European states – as well as Japan and Mexico – do not currently have the option of expanding domestic production of fossil fuels in a cost-effective manner. The EU as a whole is projected to see its total import dependency increase from 82.6 percent for crude oil and 60.3 percent for natural gas (in 2007) to around 93 percent and over 80 percent by 2030 (Comolli 2010). In the wake of the 2005–2006 Russia-Ukraine dispute over natural gas deliveries, political momentum resulted in the 2008 'EU Climate and Energy Package'. The Commission has taken a synergistic view:

> Action on renewables and energy efficiency, besides tackling climate change, will contribute to security of energy supply and help limit the EU's growing dependence on imported energy. It could also create many high-quality jobs in Europe and maintain Europe's technological leadership in a rapidly growing global sector. (European Commission 2006, p. 10)

This is matched by calls for cooperation with other players US, China, India, Canada and Japan on energy efficiency and renewables, global market access and investment trends to achieve better results in multilateral fora such as the UN, the IEA and the G-8. 'If these countries reduce the use of fossil fuels, it will also be beneficial for Europe's energy security'

(European Commission 2006, pp. 16–17). 'Indeed energy must become a central part of all EU external relations; it is crucial to geopolitical security, economic stability, social development and international efforts to combat climate change' (European Commission 2007, p. 17). Existing momentum for climate policy was fuelled by newly salient energy security concerns which, in turn, were stoked by a more prolonged Russia-Ukraine gas dispute in January 2009 which caused severe gas shortages in several EU member states. An emerging European energy strategy focuses on investments in the diversification of import sources (e.g. alternative gas pipelines), the creation of a common internal energy market and exploiting untapped potential for energy efficiencies (European Commission 2010). Although significant potential for synergies remains, energy security considerations have by now largely replaced climate policy objectives as the main driver of regulatory evolution.

How do these various policies fare when compared with the overarching objectives of energy and climate security? The cautious energy-and-climate policy packages enacted by China, India, and the US score highly on affordability and reliability, while the latter economies benefit from enhanced security of access. Moreover, in the longer run, if predictions of rising energy prices come true, first-movers in energy efficiency and 'decarbonization' will reap substantial benefits: they will have already reduced their consumption of oil and thus improved their economic competitiveness.

In terms of preventive climate security, however, most policies are still inadequate. Although the US intends to cut GHG emissions by 17 percent between 2005 and 2020, the US Energy Information Administration (EIA) estimates that energy-related carbon emissions will grow by 0.2 percent annually until 2035 (EIA 2010, p. 128). In both China and India, domestic (carbon-heavy) coal will continue to play a dominant role in energy generation. EIA figures predict annual growth rates in energy-related carbon emissions of 2.7 percent (China) and 1.8 percent (India). By contrast, the EIA expects an annual reduction in energy-related carbon emissions in OECD-Europe and Japan by 0.2 percent and 0.6 percent respectively. Collectively, the climate and energy policies announced by major economies imply very limited progress on climate mitigation and in many other developing or emerging economies, the carbon intensity of power generation has in fact continued to rise (Tandon 2012). In November 2010, the UN Environment Programme calculated that targets and other pledges by major economies only amount to 60 percent of the GHG reductions needed to stay below the 2°C threshold.[8]

## 5 CONCLUSION: THE GOVERNANCE DILEMMA

According to the then EU Energy Commissioner, 'climate change and energy security are two sides of the same coin. The same remedies must be applied to both problems' (Piebalgs 2009, p. 2). There is certainly a conceptual overlap between energy and climate security. Not only are they both, to various degrees, concerned with the transition away from a carbon-heavy, fossil-fuel based global economy. They also have to confront fundamental scarcities: the scarcity of affordable and readily accessible fossil energy or the scarcity of atmospheric 'carbon space'. Both the energy and the climate challenge can therefore benefit from demand reduction as well as from supply-side measures which diminish both types of scarcity, such as low-carbon energy technologies. There is no denying the underlying attraction of this proposition and the way in which a 'synergistic' approach provides what must be the ultimate 'win-win' solution neatly addressing energy and climate security concerns through a move to a decarbonized economy.

The problem is, of course, making the political fit between climate and energy security. The orthodox vision of enhanced global energy security, grounded in economic globalization and increasing interdependence, still depends on a compromise between producers and consumers of fossil fuels. The sole likely benefit for climate governance would be reduced price volatility and hence greater predictability for alternative energy investments. Overall, strategies prioritising either national or global energy security are likely to result in incremental climate policy and a resort to *reactive* climate security. For such an approach the referent object will continue to be the state.

Furthermore, while the various benefits of energy security measures can be captured at the national or even regional- or EU-level and are not necessarily dependent upon international cooperation, the public good of climatic stability can only be attained by concerted efforts at the global level. This is because the global atmosphere can be regarded as having the characteristics of a commons. Climatic security defined in terms of stability is frequently understood as non-rival and non-excludable public good. It requires collective mitigation efforts amongst the largest emitters and mechanisms to ensure compliance and avoid 'free-riding'. The atmosphere also represents a finite 'common sink' for GHG emissions. Most current economic activities constitute a rival consumption of 'carbon space'. And strict international targets would partially enclose or 'privatize' this resource in order to limit ruinous over-consumption. However, the very notion of carbon space, which has become in recent years a key negotiating concept for some developing countries, illustrates the extent of the problem of arriving at an effective agreement beyond the first commitment period of the Kyoto Protocol.[9] The point here is that much of the world's limited carbon space has already been occupied by the industrialized countries and that justice demands that the remainder be used to realize the development objectives of the South. In the climate negotiations from Copenhagen (2009) to Durban (2011), the objective of the key players has been to avoid being trapped in an agreement that might imperil short-run national energy security.

The widespread preference for incremental policy reform signifies that national energy security will continue to be a 'far stronger policy driver' (Froggatt and Levi 2009, p. 1141) than climate security. Regardless of the progress made through synergistic measures, this 'gradualism' contains considerable structural bias. First, it permits the 'lock-in' of fossil- or biofuel-intensive infrastructures, which has important consequences for emission trajectories in rapidly industrialising countries. Second, it relies on domestic 'win-win' policy scenarios which – similar to local benefits derived from climate adaptation measures – favour outcomes consistent with *reactive* climate security. Furthermore, integrated global or regional markets for fossil fuels imply that national energy-and-climate policies result in 'carbon leakage' by lowering global/regional energy prices and stimulating energy demand elsewhere.

A different set of national and international policies would be required if governments were to pursue *preventive* climate security in earnest. At the national level, energy and climate policy would prioritize longer-term objectives – security of supply, greater foreign policy autonomy and ultra-low carbon emissions – without wholly ignoring short-term considerations of affordability, reliability, and political feasibility (Compston 2010). There are formidable difficulties here, where developed country governments need to overcome the incentives to operate on a short-term basis and to work with public opinion.

In terms of the latter there are mixed messages. For the European Union, '[a]n EU level solution to the climate problem has served as a convincing narrative to persuade EU citizens that there is a need to continue the process of European integration' (Adelle and Withana 2010, p. 331). There are also significant similarities in attitudes in the EU, the US, and other advanced countries. These include concern about climate change laced with some scepticism about the science and an unwillingness to make personally costly changes with the requirements of economic growth being placed above climate protection. While there is support for renewables in general, significant opposition exists to particular technologies such as wind farms (ibid., p. 327). One key difference which may help to explain divergence between EU and US policy and which is hardly merited by their respective situations is that 'energy security is considered a much more important factor in the US than in EU and is given greater priority than environmental protection by a significant number of Americans' (ibid., p. 329). Thus, 'for governments faced with tough policy choices, the public's reluctance to accept costly policy choices could limit the use and range of policy solutions in the transition to a low-carbon economy' (ibid., p. 328). Amongst the BASIC countries the demands facing policy-makers are bound to be more extreme, with energy priorities for development dominating other concerns and indeed being fundamental to the continuing legitimacy of governments.

There is also the question of the interaction between national and international climate and energy policies. Is it possible that international commitments could provide momentum for domestic policy reform? There is some evidence that the search for short-term national energy security does not always prevail and that international norms and commitments can have a significant effect, especially if the prestige and credibility of governments is engaged. The adoption of the Emissions Trading Scheme by the EU provides a case in point. This particular 'flexibility mechanism' had been opposed prior to Kyoto but became the bedrock of the Union's approach to climate, driven like the 'burden sharing' agreement before it by the requirements to maintain its leading position at the international level. The evidence is not as strong, but Chinese and Indian policy changes, involving the announcement of energy efficiency targets, were certainly stimulated by the need to generate a credible position in advance of the 2009 Copenhagen CoP. It remains the case, however, that more often than not energy security drives climate policy, although the result may not always be negative. In this regard the EU's 2006 gas crisis was one of the incentives to agree the 2008 'Climate and Energy Package' that provided the policy basis for the implementation of the Union's 20/30 percent emissions reduction commitment.

If global climatic stability became an actual policy priority, it would not only deliver important 'co-benefits' for global energy security, but would also provide a genuine basis for implementing a *preventive* climate security strategy. To dilute both domestic and international obstacles to ambitious climate policy, advanced industrialized countries would have to engineer an 'energy revolution' – through enormous investments, technology transfer, and capacity-building in developing countries. The necessary coalitions would have to be forged among major energy-consuming countries and not rely on older practices of producer-consumer conciliation (Mabey 2008, p. 68).

Energy and climate are thus not only materially intertwined, but also interdependent politically. Without increased availability of practical and affordable energy technologies to enable climate-friendly economic development, international climate governance will not progress substantially. What the public goods analysis makes clear is that the regulatory 'direction of travel' should still lead from climate to energy – rather than vice versa – because only this arrangement could ensure that long-term strategic foresight prevails over short-term pragmatism.

## NOTES

1. The substantial literature on environmental security was stimulated in particular by the ending of the Cold War. In general, it attempts to tease out the relationship between environmental degradation and conflict as in the extensive work of Homer-Dixon (1994). Deudney and Mathews (1999) explore some of problems of securitizing the environment.
2. For an excellent survey of different IR perspectives on climate security during this period, see Stripple (2002).
3. The reference here is of course to the Copenhagen School (Buzan et al. 1998). To some extent, this process of 'securitization' amplifies ideas proposed in the 1970s when natural resources and the environment were first recognized as security issues. In this sense, energy security and climate security, the core issues of this article, have merely been recast as critical components of security thinking in the twenty-first century.
4. The 2011 Climate Change Vulnerability Index by the British consultancy Maplecroft rates 16 countries as being at 'extreme risk', with Bangladesh, India and Madagascar among the top three.
5. Several southern diplomats have already described climate change as an 'act of aggression' and even extremist groups such as al-Qaeda have specifically referred to Western responsibility for climate change and the US refusal to sign the Kyoto Protocol (Scott 2008, p. 607; Mazo 2010, p. 129).
6. US biofuels such as corn-based ethanol, however, have been accused of producing higher GHG emissions than imported oil.
7. A significant proportion of Japanese and European emission reductions will likely be achieved by international offset procedures such as the Clean Development Mechanism.
8. See the UNEP Emissions Gap Report at www.unep.org/publications/ebooks/emissionsgapreport (accessed 21 February 2013). The outcomes of the 2010 Cancún and 2011 Durban climate change conferences have not altered the validity of these calculations.
9. 'Carbon space' is a special case of the more general concept of 'environmental utilization space' (Opschoor 1995). Calculated on a historical basis, it has become part of the negotiating position of India China and the BASIC group at international climate conferences (see Tata Institute of Social Sciences 2010). Calculations are made on a national basis, but it would be equally possible to arrive at a different outcome on the basis of individual carbon entitlements.

## BIBLIOGRAPHY

Adelle, Camilla and Srini Withana (2010), 'Public Perceptions of Climate Change and Energy Issues in the EU and the United States', in Sebastian Oberthür and Marc Pallemaerts (eds), *The New Climate Policies of the European Union: Internal Legislation and Climate Policy*, Brussels: VUB Press, pp. 309–336.

Adger, W. Neil (2010), 'Climate Change, Human Well-Being and Insecurity', *New Political Economy*, **15** (2), 275–292.
Bang, Guri (2010), 'Energy security and climate change concerns: Triggers for energy policy change in the United States?', *Energy Policy*, **38** (4), 1645–1653.
Baumert, Kevin A., Timothy Herzog and Jonathan Pershing (2005), *Navigating the Numbers: Greenhouse Gas Data and International Climate Policy*, World Resources Institute.
Bordoff, Jason, Manasi Deshpande and Pascal Noel (2009), 'Understanding the Interaction between Energy Security and Climate Change Policy', in Carlos Pascual and Jonathan Elkind (eds), *Energy Security: Economics, Politics, Strategies, and Implications*, Washington, DC: Brookings Institution Press, pp. 209–248.
Bradshaw, Michael J. (2010), 'Global energy dilemmas: a geographical perspective', *The Geographic Journal*, **176** (4), 275–290.
Brauch, Hans G., Úrsula Oswald Spring, John Grin, Czeslaw Mesjasz, Patricia Kameri-Mbote, Navnita C. Behera, Béchir Chourou and Heinz Krummenacher (eds) (2009), *Facing Global Environmental Change: Environmental, Human, Energy, Food, Health and Water Security Concepts*, Berlin: Springer.
Buzan, Barry (1983), *People States and Fear: The National Security Problem in International Relations*, Brighton: Wheatsheaf.
Buzan, Barry, Ole Waever and Jaap de Wilde (1998), *Security: A New Framework for Analysis*, Boulder, CO: Lynne Rienner.
Comolli, Virginia (2010), 'Chapter Eight: Energy Security', *Adelphi Papers*, **50** (414–415), 177–196.
Compston, Hugh (2010), 'The Politics of Climate Policy: Strategic Options for National Governments', *The Political Quarterly*, **81** (1), 107–115.
Dalby, Simon (2009), *Security and Environmental Change*, Cambridge: Polity.
Detraz, Nicole and Michele M. Betsill (2009), 'Climate Change and Environmental Security: For Whom the Discourse Shifts', *International Studies Perspectives*, **10** (3), 303–320.
Deudney, Daniel and Richard A. Matthews (eds) (1999), *Contested Grounds: Security and Conflict in the New Environmental Politics*, New York: SUNY Press.
Dupont, Alan (2008), 'The Strategic Implications of Climate Change', *Survival*, **50** (3), 29–54.
EIA (2010), *International Energy Outlook 2010*, Washington, DC: U.S. Department of Energy.
European Commission (2006), 'Green Paper: A European Strategy for Sustainable, Competitive and Secure Energy', Brussels, 8 March, COM(2006) 105 final.
European Commission (2007), 'An Energy Policy for Europe', Brussels, 10 January, COM(2007) 1 final.
European Commission (2010), 'Energy 2020: A strategy for competitive, sustainable, and secure energy', Brussels, 10 November, COM(2010) 639 final.
European Commission (2011), 'A Roadmap for moving to a competitive low carbon economy in 2050', Brussels, 8 March, COM(2011) 112 final.
European Council (2008), 'Report from the Commission and the Secretary-General/High Representative: Climate Change and International Security', Brussels, 3 March, 7249/08.
Froggatt, Antony and Michael A. Levi (2009), 'Climate and energy security policies and measures: synergies and conflicts', *International Affairs*, **85** (6), 1129–1141.
Goldthau, Andreas and Jan M. Witte (2009), 'Back to the future or forward to the past? Strengthening markets and rules for effective global energy governance', *International Affairs*, **85** (2), 373–390.
Homer-Dixon, Thomas F. (1994), 'Environmental Scarcities and Violent Conflict: Evidence from Cases', *International Security*, **19** (1), 5–40.
IEA (2007a), 'World Energy Outlook 2007: Fact Sheet – China', Paris: IEA, available at http://www.iea.org/papers/2007/fs_china.pdf (accessed 28 August 2012).
IEA (2007b), 'World Energy Outlook 2007: Fact Sheet – India', Paris: IEA, available at http://www.iea.org/papers/2007/fs_india.pdf (accessed 28 August 2012).
IEA (2010), 'World Energy Outlook 2010', Paris: OECD/IEA.

IPCC (2007), 'Climate Change 2007 Synthesis Report – Summary for Policy Makers', Geneva: IPCC, available at http://www.ipcc.ch/pdf/assessment-report/ar4/syr/ar4_syr_spm.pdf (accessed 28 August 2012).

Karlsson-Vinkhuyzen, Sylvia I. (2010), 'The United Nations and global energy governance: past challenges, future choices', *Global Change, Peace and Security*, **22** (2), 175–195.

Keith, David (2009), 'Dangerous Abundance', in Thomas Homer-Dixon and Nick Garrison (eds), *Carbon Shift: How the Twin crises of Oil Depletion and Climate Change Will Define the Future*, Toronto: Random House Canada, pp. 26–57.

Lesage, Dries, Thies van de Graaf and Kirsten Westphal (2010), *Global Energy Governance in a Multipolar World*, Farnham: Ashgate.

Lewis, Joanna I. (2009), 'Climate change and security: examining China's challenges in a warming world', *International Affairs*, **85** (6), 1195–1213.

Mabey, Nick (2008), 'Delivering Climate Security: International Security Responses to a Climate Changed World', *Whitehall Papers*, **69**.

MacFarquhar, Neil (2011), 'U.N. Deadlock on Addressing Climate Shift', *New York Times*, 20 July, available at http://www.nytimes.com/2011/07/21/world/21nations.html?_r=1 (accessed 27 August 2012).

Marques, António C., José A. Fuinhas and J. R. Pires Manso (2010), 'Motivations driving renewable energy in European countries: A panel data approach', *Energy Policy*, **38** (11), 6877–6885.

Mazo, Jeffrey (2010), *Climate Conflict: How Global Warming Threatens Security and What to Do About it*, London: Routledge.

Mobjörk, Malin, Mikael Eriksson and Henrik Carlsen (2010), 'On Connecting Climate Change with Security and Armed Conflict', Stockholm: Swedish Defence Research Agency (FOI), Report No. FOI-R-3021-SE.

Müller-Kraenner, Sascha (2008), *Energy Security: Re-Measuring the World*, London: Earthscan.

Opschoor, J. (1995), 'Ecospace and the fall and rise of throughput intensity', *Ecological Economics*, **15** (3), 137–140.

Owen, Nick A., Oliver R. Inderwildi and David A. King (2010), 'The status of conventional world oil reserves – Hype or cause for concern?', *Energy Policy*, **38** (8), 4743–4749.

Piebalgs, Andris (2009), 'EU Energy and Climate Policy', Speech at the 7th Doha Natural Gas Conference, 11 March, available at http://europa.eu/rapid/pressReleasesAction.do?reference=SPEECH/09/102&format=PDF&aged=1&language=EN&guiLanguage=en (accessed 27 August 2012).

Radetzki, Marian (2010), 'Peak oil and other threatening peaks: Chimeras without substance', *Energy Policy*, **38** (11), 6566–6569.

Rogers, Paul (2010), 'Climate Change and Security', Oxford Research Group, International Security Monthly Briefing – September 2010.

Rubin, Jeff (2009), *Why Your World is about to get a Whole Lot Smaller: Oil and the End of Globalization*, Toronto: Random House Canada.

Schwartz, Peter and Doug Randall (2003), 'An Abrupt Climate Change Scenario and Its Implications for United States National Security', report commissioned by the US Department of Defense.

Scott, Shirley V. (2008), 'Securitizing climate change: international legal implications and obstacles', *Cambridge Review of International Affairs*, **21** (4), 603–619.

Stripple, Johannes (2002), 'Climate Change as a Security Issue', in Edward A. Page and Michael Redclift (eds), *Human Security and the Environment: International Comparisons*, Cheltenham: Edward Elgar Publishing, pp. 105–127.

Tandon, Shaun (2012), 'Carbon efficiency failing to fight warming: study', *Agence France-Presse*, 28 August, available at http://www.google.com/hostednews/afp/article/ALeqM5gr3_nsyiZiz3L4JUmKplnM0tNDKQ (accessed 29 August 2012).

Tata Institute of Social Sciences (2010), 'Conference on Global Carbon Budgets and Equity in Climate Change', Mumbai, available at http://www.findthatpdf.com/download.php?i=7872348&t=hPDF (accessed 28 August 2012).

Weitzman, Martin L. (2010), 'GHG Targets as Insurance Against Catastrophic Climate Damages', Cambridge, MA: Harvard Project on International Climate Agreements, September, Discussion Paper 2010-42.

Youngs, Richard (2009), *Energy security: Europe's new foreign policy challenge*, Abingdon: Routledge.

Yueh, Linda (2010), 'An International Approach to Energy Security', *Global Policy*, **1** (2), 216–217.

# 15. Energy, climate change and conflict: securitization of migration, mitigation and geoengineering
*Jürgen Scheffran**

## INTRODUCTION: FROM ENERGY SECURITY TO CLIMATE CONFLICTS

In many of the world's conflicts energy resources have been an influential factor (Singer 2008; Singer and Scheffran 2004). This is partly due to the dual nature of energy: while energy use is important for human life and society, a precondition for social development and economic prosperity, it may also cause risk, destruction and death. Physical power from energy can be converted into political power, and physical force is a tool used in violent acts. Traditionally, energy security has been framed in terms of ensured access to energy resources to meet political and economic goals, while the lack of energy is perceived as a security threat. Security risks and conflicts can also result from the use and misuse of energy, its side-effects and unbalanced distribution. In turn, violent conflicts and social disruption impede access to energy resources. Each component of the energy system can become a target of attack or resistance by state and non-state actors, including dams, reactors and power grids. On the other hand, energy use as well as the prevention of risks and conflicts is a field for international cooperation and global security.

Energy security strongly depends on its geographical and geopolitical context. The components of the fossil-nuclear energy system – coal, oil and natural gas reserves, uranium mines as well as the connecting infrastructures, networks and transportation routes – have shaped the global conflict landscape in the past century (Yergin 1991, 2011). Geopolitical conflict lines may become sharper with growing energy demand, diminishing fossil fuel reserves, uneven distribution of energy resources and increasing North-South imbalances. Fossil energy use also affects the environment in many parts of the world. While most of these impacts are local or regional, global warming caused by greenhouse gas (GHG) emissions raises concerns about global security risks which are interconnected with regional and local conflict constellations.

Numerous studies have raised questions about the security risks and conflict potentials of climate change. Increasingly, strategies to counter and respond to climate change become subject to conflict themselves. When people migrate to evade the impacts of climate change, this could provoke security issues and conflicts in the regions or origin and destination. Mitigation and adaptation require substantial political and societal interventions that could provoke concerns and resistance (Scheffran and Cannaday 2013). While nuclear energy as a low-carbon technology poses a range of safety and security issues, with renewable energy geopolitical interests are shifting to the regional and local scale, giving more relevance to land use and the community level. In the emerging new energy landscapes conflict lines tend to be inner-societal but may reach national and international levels since they are interconnected through global markets, global warming and global governance. This may be even more the case when geoengineering techniques are developed to intentionally design a global climate system which could become a new field of international security and conflict (Brzoska et al. 2012).

Expanding these issues, the focus of this chapter is the potential security risks and conflict constellations of climate change as well as climate policies. This provides a basis for an assessment and comparison of human and societal responses to climate change, including migration, mitigation and geoengineering, which are analyzed in terms of narratives, empirical findings, critical perspectives and strategies. This integrated approach adds a new dimension to the discourses on the securitization of energy and climate change.

## CLIMATE CHANGE AS A SECURITY AND CONFLICT ISSUE

### Securitization of the Climate Change Discourse

Developed by the Copenhagen school (Wæver 1995, 1997; Buzan et al. 1998), the "securitization theory" offers a conceptual framework for the analysis of policy declarations ("speech acts") on the security implications of climate change. By declaring a development an existential threat, an actor turns this into a security issue that requires utmost priority and the use of extraordinary measures to respond to this threat. Securitization moves could be enacted by governmental representatives as well as by non-state actors. According to Buzan et al. (1998: 26) a successful securitization "has three components: existential threats, emergency action, and effects on interunit relations by breaking free of rules." To address the security challenges, extraordinary responses are considered, such as the use of violence or military and police actions, which lead to social interactions and conflicts, such as interstate war or civil war.

Critical assessments have pointed to several shortcomings of the securitization approach: its emphasis on "subjective" perceptions and "exceptional" measures, the focus on security experts and their often realist thinking, the Eurocentric focus of its research agenda (Bigo 2002; Methmann and Rothe 2012; Oels 2012). The relevance of securitization in the environmental field has been controversially debated. The conceptualization of environmental security emerged in the aftermath of the Cold War and was initially shaped by the discourse on environmental conflicts (Trombetta 2008, 2012). Several research projects assessed the linkages between environmental degradation, resource scarcity and violent conflict, based on qualitative regional case studies (Homer-Dixon 1991, 1994; Bächler and Spillmann 1996). In this period global warming remained largely an environmental issue while its security implications played only a marginal role, besides some political statements that used security-related language to express the dramatic consequences of climate change. Research efforts on the climate-security nexus were scattered until the end of the 1990s (e.g. Gleick 1989; Swart 1996; Scheffran 1997; Rahman 1999). In the past decade the issue gained ground by an increasing number of assessments (for an overview of the early literature see Brauch 2009).

The year 2007 became a turning point in the securitization of climate change. In its Fourth Assessment Report the Intergovernmental Panel on Climate Change (IPCC) addressed the risks of climate change, but not security and conflict issues. After its publication the public discourse on climate change intensified and increasingly focused on security dimensions (see Brauch 2009; Brzoska 2009, 2012; Scheffran and Battaglini 2011). Major agents in the securitization of climate change are those who project security threats or suggest actions to address them. A number of studies have investigated the potential security implications of climate change, involving researchers, consultants, think tanks, media, non-governmental organizations, national governments, as well as international organizations, institutions and regimes.

To give a few example cases, a US-based panel of experts has portrayed global warming as "one of the greatest national security challenges" that could breed new threats and conflicts (Campbell et al. 2007). A military think tank has characterized climate change as a "threat multiplier" that could heighten global tensions in fragile regions of the world (CNA 2007). The German Advisory Council on Global Environmental Change is concerned that without an effective climate policy the consequences "could well trigger national and international distributional conflicts and intensify problems already hard to manage such as state failure, the erosion of social order, and rising violence."(WBGU 2008) A recent study by a research unit of the German Federal Armed Forces points to the significant destabilization potential of climate change for states and societies, in particular when they have low problem solving capacity (resilience) to address climate impacts (Bundeswehr 2012).

On the European level, the EU High Representative and the European Commission suggested in 2008 that "climate change acts as a threat multiplier, worsening existing tensions in countries and regions which are already fragile and conflict-prone" (EU 2008). At the United Nations level, divergent perspectives became visible between developed and developing countries when the UN Security Council (UNSC) discussed the security risks of climate change in 2007 and 2011. Initiated by Great Britain, on April 17, 2007, the UNSC for the first time considered climate change as a security issue. On July 20, 2011 a statement of the UNSC – under German presidency – raised concerns regarding the threats posed by climate change to international peace and security (UNSC 2011). While a coalition of OECD countries and Pacific Small Island States stressed the need to address these threats in the UNSC, Russia, China and many G-77 states refused to a mandate of the UNSC for climate change (UNSC 2011). In 2009, the UN Secretary-General provided a report on the security implications of climate change (UNSG 2009), indicating that internationally climate change has reached highest levels.

**Security Risks, Pathways and Agents**

Assessments on the climate-security nexus differ regarding specific methods, interests, pathways and responses as well as affected regions and countries. They use similar narratives, storylines and speech acts that are summarized here in a generalized way. Usually they combine natural and social processes as drivers of security risks and conflicts. Global temperature rise has multiple physical consequences that alter natural processes and cycles, such as ice formation, ocean currents, soil conditions, precipitation and water distribution, ecological systems and biodiversity. Potential effects of these changes are extreme weather events (droughts, storms, floods), glacier melting and sea-level rise, biodiversity loss and spreading diseases. The environmental impacts affect various dimensions important for human livelihood, well-being and survival, including water resources, agriculture and forestry, human health and settlements, energy and economic systems. The impact of climate change on human beings and social systems is affected by their vulnerability which is a function of the "character, magnitude, and rate of climate change and variation to which a system is exposed, its sensitivity, and its adaptive capacity" (IPCC 2007: 21). Most vulnerable are poor communities in high-risk areas and developing countries which depend on agriculture and ecosystems services sensitive to climate stress. The stronger the climate exposure and the larger the affected region the more challenging it becomes for societies to absorb and adapt to the impacts. In highly vulnerable regions with low adaptive capacity, the impacts of climate change could turn into major security risks.

In estimating the security implications, the assessments use different security conceptions: national and international security at the level of nation states; global security on a planetary scale; environmental security for ecosystems and natural resources; human security for human life and wellbeing. It is widely argued that global warming acts as a "threat multiplier" that heightens global tensions in fragile regions of the world. Human responses to the security challenges of climate change and the resulting forms of social interaction have also raised security concerns. These include human displacement, the use of violence and military actions in conflict, as well as low-level forms of conflict such as protests, riots, rebellions, or extreme forms of violence such as genocide. Altogether it is concluded that the security risks and conflicts associated with climate change could undermine societal stability.

Going beyond generalized statements, climate change is supposed to affect security and conflict along multiple pathways. Among the major "conflict constellations" are the degradation of freshwater resources, the decline in food production, increasing storm and flood disasters, and environmentally-induced migration (WBGU 2008). Other pathways include sea-level rise, energy insecurity, deforestation, loss of biodiversity and fishery. Acting as a threat multiplier, climate change may interfere in complex ways with political, economic and social conflict factors, such as population growth, increased demand and unequal distribution of resources, or lacking political legitimacy of governments. For instance, due to water scarcity and soil degradation agricultural yields could drop, diminishing food supply and driving migration. Extreme weather events put the economic infrastructure at risk, including industrial sites and production facilities as well as networks for transportation and supply of goods.

For all these concerns there is a range of uncertainties leaving room for worst-case scenarios which are framing the discourse over potential climate impacts. Affecting societies in complex ways, there is a range of threatening narratives, including economic damage and risk to coastal cities and critical infrastructure; loss of territory, border disputes and environmentally-induced migration; resource conflicts and tension over energy supply. Some regional hot spots are seen as more fragile and vulnerable due to their geographic and socio-economic conditions and the lack of adaptation capabilities. As most vulnerable to climate change appear people living in poverty, fragile states with poor governance and conflict-prone societies where living conditions are already precarious. Exposed to these risks, "failing states" have inadequate management and problem solving capacities and cannot guarantee the core functions of government, including law, public order and the monopoly on the use of force. Although industrialized countries will be also exposed to climate stress, they are usually less susceptible and have better capacities to cope with the challenge.

Among the geographically-specific impacts are water scarcity, droughts and heat waves in Southern Europe, Northern Africa, the Sahel and the Middle East; floods and storms in Southern/Eastern Asia and the American Gulf region; glacier melting in the polar regions, the Himalaya and Andes mountains; sea-level rise on many of the world's coasts; biodiversity loss in various ecosystems, in particular in tropical forests, mountain regions and coral reefs (see WBGU 2008; Scheffran and Battaglini 2011, Scheffran et al. 2012a). In the worst-affected regions, climate change could spread to neighbouring states, e.g. through refugee flows, ethnic links, environmental resource flows or arms exports. Food insecurity in one country may increase competition for resources and force population to migrate into neighboring countries. Such spillover effects can expand the geographical extent of a crisis and intensify the erosion of social stability elsewhere.

A cycle of environmental degradation, economic decline, social unrest and political instability could destabilize the affected societies. When tipping points are reached, abrupt changes and cascading sequences in the climate system could have incalculable consequences. Examples are the potential loss of the Amazon rainforest, a shift in the Asian monsoon, the disintegration of the Greenland and West-Antarctic ice sheets, methane outburst from soils in Siberia or Canada, or the shutdown of the North atlantic circulation (Lenton et al. 2008). Worst-case scenarios may exceed adaptive capacities even in the most wealthy countries. The difficulty to protect against extreme weather events has been demonstrated by the 2003 heat wave in Europe which cost tens of thousands of human lives, and by Hurricane Katrina in the US in 2005 which caused record damages. Recent years have seen an increase in the frequency of such events, to mention the severe drought in the US Midwest or Hurricane Sandy on the US east coast in 2012.

## Empirical Findings on Climate Change and Violent Conflict

While policy-oriented statements make strong claims about the security implications of climate change, one question is how valid the underlying assumptions and perceptions are. Quantitative studies and data-bases provide empirical material together with statistical analysis to test hypotheses about relations between climatic variables (temperature, precipitation) and conflict-related variables (number of armed conflicts or casualties). The research literature reaches mixed evidence and differing conclusions on the causal relationship between climate change and violent conflict (for an overview see Gleditsch 2012; Scheffran et al. 2012b,d).

Studies over long historical periods tend to find a correlation between climate variability and armed conflict. One study concludes that cooler periods in pre-industrial Europe are more likely related to periods of violence than warmer phases (Tol and Wagner 2010). Similar results have been found for the Northern Hemisphere and Eastern China (Zhang et al. 2007, 2011). However, for recent periods empirical findings are more diverse. After the end of the Cold War the number of armed conflicts has declined, while temperature has increased. A significant linkage between temperature, precipitation and African civil wars has been found for the period 1981 to 2002 (Burke et al. 2009), a result that was challenged for alternative model specifications (Buhaug 2010). A statistical analysis of the El Niño Southern Oscillation (ENSO) for the period 1950–2004, concludes that the "probability of new civil conflicts arising throughout the tropics doubles during El Niño years relative to La Niña years" (Hsiang et al. 2011). Yet, key questions remain on the connection between climate change and the El Niño phenomenon. A more recent analysis of climate variability and conflict risk in East Africa for 1990–2009, involving geographical factors, finds mixed results about the impact of precipitation and temperature on violent conflict (O'Loughlin et al. 2012).

Explanations of the mixed empirical evidence point to the lack of adequate data. It is supposed that climate change affects armed conflicts between states less likely than low-level conflicts between societal groups which are harder to measure. In addition, future climate change may be much stronger than in the past for which data are available. To address and overcome methodological deficits, databases are being expanded, including low level violent events of non-state actors (such as riots). Geo-referencing of these events (as used in the mentioned study on East Africa) allows to represent spatio-temporal patterns of social interaction. In addition to large-N statistical methodologies researchers explore case studies, systematic assessments and modelling tools to analyze the complex pathways between climate change and violent conflict and consider human and societal responses. For some of the pathways the empirical findings are summarized in the following (Scheffran et al 2012d):

1. **Precipitation changes and variability:** strong deviations from average precipitation are related to violent conflict. In some regions (e.g. in Kenya) conflict on rain-fed agriculture or pastoralism is more likely in rainy than in dry seasons. More significant than environmental factors are political and economic marginalization.
2. **Freshwater resources and scarcity:** international river systems are more associated with low-level conflicts and diplomatic tensions than with full-scale wars. Violent conflict is outweighed by the number of international water agreements.
3. **Land and food:** climate change is likely to contribute to food insecurity, while food insecurity can contribute to conflict. In many countries, "food riots" were correlated with rising food prices between 2007 and 2011. Although frequently suggested, there is little empirical evidence yet of a strong link between climate change and food-related conflicts. Climate change plays an ambivalent role in conflicts among pastoralists and farmers.
4. **Weather extremes and natural disasters:** natural disasters could destabilize societies with weak economies, mixed political regimes and pre-existing conflicts. There are weak links to armed conflict, while disaster management is often a field of cooperation.
5. **Environmental migration:** there is a wide range of estimates on environmental or climate induced migrants and whether these will increase the likelihood of violent conflict. However, the empirical justification is weak. Recent studies suggest to consider migration as an adaptive response to climate change and develop the capabilities of migrant networks.

While empirical evidence has been mixed so far, with growing future climate change the various pathways to conflict could be intensified and combined.

**Critical Perspectives on Climate Securitization**

Contrary to many speech acts on climate-related security and conflict issues, the securitization of this policy field has met substantial criticism. In addition to the weak empirical evidence regarding the climate-conflict linkage, there have been no indications yet of "extraordinary measures" in response to climate change. While many countries mention climate change in their national security strategies as a potential threat, only a few governments treat it as an imminent threat or a possible source of violent conflict that justifies urgent actions in the security sector, with the exception of disaster preparedness (Brzoska 2012). Accordingly, some researchers conclude that the securitization of climate change has failed (Oels 2012; McDonald 2012).[1] It remains to be seen whether the securitization of climate change is a temporal trend limited to some Western countries or may increase in importance with growing climate challenges and spread to developing countries as well.

To what degree a growing climate stress will actually lead to violent conflicts depends on human and societal responses which are shaped by the conflict history, group identities, the organization and capacity of conflict parties, as well as the instrumentalization of resources for group interests and power structures. Conditions to reduce conflict potentials include political institutions and governance structures, resilient communities and sustainable livelihoods, cooperation and conflict regulation mechanisms. These aspects will be discussed in more detail for climate-induced migration which is one of the fields that has attracted a vigorous securitization discourse.

## SECURITIZATION OF CLIMATE-INDUCED MIGRATION

Migration is a possible response to climate change and has been identified as one of the potential conflict constellations in climate hot spots. In the following some of the issues in the securitization of climate-induced migration are highlighted.

### Pathways, Agents and Narratives of Securitization

In recent years, the link between climate change, migration and security has raised increasing attention in public statements and the research literature which identifies environmental migration as a potential security problem. Climate change has been described as a stress factor that increases migration pressure in climate hot spots (WBGU 2008; Warner et al. 2010; Gemenne 2011). The impacts of drought, water scarcity, food insecurity and extreme weather events on human livelihood are supposed to become driving factors for the displacement of people.

Environmental migration has multiple causes and multiple effects, depending on environmental as well as socio-political factors. People migrate for many reasons, including worsening conditions in their lives and incomes but also because they search for new opportunities. Regarding environmental migration, Jäger et al. (2009) distinguish between people who (a) move voluntarily from their residence primarily due to environmental reasons (environmental migrants); (b) are forced to leave their residence because of environmental risks (environmental displaced); or (c) are intentionally relocated or resettled due to a planned land use change (development displaced). In many cases it is difficult to separate the different drivers, thus environmental migrants will appear as economic migrants (e.g. farmers losing income) or as refugees of war (from environment-induced conflict). One of the main routes of environmental migration occurs from rural to urban areas, mostly within a country, in other cases as transnational migration.

The number of environmental migrants and displaced persons is expected to rise in response to climate change, as well as the indirect consequences such as economic decline and conflict. Where living conditions become unbearable, people are under pressure to leave their homes to seek refuge elsewhere. Environmental scientist Norman Myers estimated that the number of environmental migrants would rise to 150 million by 2050, up from 25 million in the mid-1990s (Myers 2002). Some studies present even higher estimates of the number of potential climate migrants up to several hundred million people who have to flee the impacts of climate change. This exceeds by far the number of registered refugees worldwide, according to the U.N. High Commissioner for Refugees. By the end of 2011, some 42.5 million people worldwide were considered as forcibly displaced due to conflict and persecution, including 15.2 million refugees and 26.4 million Internally Displaced Persons (IDPs) (UNHCR 2011).

A number of studies on the effects of climate change have described migration as a negative factor – a threat to human security and societal stability, and a generator of violent conflict, in particular in regions where migration is criminalized or inhibited by societies which fail to integrate migrants. Whether the risks prevail depends on several factors and conditions. A planned migration in response to a gradual environmental degradation is different from a sudden mass exodus after an extreme weather event which can undermine the stability of communities. Under adverse circumstances migration can increase the likelihood of conflict in transit and target regions when migrants compete with the resident population for scarce resources such as farmland, housing, water, employment, and basic social services. In certain cases immigrants are perceived to upset the "ethnic balance" in a region (Reuveny 2007). In countries without disaster management (e.g. weather warning systems or evacuation plans), extreme weather causes relatively greater damage and forces more people to migrate than in countries with these systems (WBGU 2008). Affected governments face enormous challenges if they have to handle sudden, unexpected and large-scale migration which can even overwhelm the management capacities of developed countries.

Most directly affected by climate change are the local communities as well as the migrant communities and networks, as far as human security is at stake. While in many cases the challenges are addressed at the area exposed by attempts for improving human living conditions, forced displacement can be seen as an extraordinary response to severe climate risks. In the latter case, affected communities, NGOs supporting them and those concerned about conflicting interests may use securitizing speech acts to make their case to protect or defend against "climate refugees" (e.g. for Small Island States). Regions that have been characterized as most vulnerable are coastal and riverine zones, hot and dry areas and regions whose economies depend on climate-sensitive resources. Although environmental migration will predominantly occur within national borders of developing countries, there is concern in industrialized regions about a substantial increase in external migratory pressure. Europe could see an increase in migration from sub-saharan Africa and the Arab world, and North America from the Caribbean, and Central and South America.

Treating the projected "climate refugees" as a security threat can be used to justify extraordinary measures in target regions that can drive not only the securitization, but also the militarization of climate-induced migration, including military and police action for border control (e.g. by the Frontex organization in Europe). Concerns about internal state security and international terrorism are driving the securitization of environmental migration. A new arsenal of technologies of political control and weapons has evolved that together with new military doctrines can be actively deployed against civilians in new public order roles, including negative human responses to climate change (Wright 2012). There is the risk that enhanced border control and crowd control initiatives merge with the massive funding for future security technology innovations provided in anti-terror and homeland defense activities.

**Critical Perspectives**

The securitization of the climate-migration discourse has received considerable attention but also criticism by scholars who question the underlying assumptions as well as the potential implications (for a collection of contributions see Scheffran et al. 2012a). One focus is the weak empirical evidence of some of the claims. There is a wide range of future estimates on the number of environmental and climate-driven migrants. Projections with huge numbers of climate refugees have been criticized as speculative and exaggerated, lacking justification and empirical foundations (Jakobeit and Methmann 2012). "Apocalyptic narratives" that forecast massive, abrupt and unavoidable flows of climate refugees are refused (Bettini 2012; Methmann and Rothe 2012). The maximalist quest for numbers faces conceptual problems of definition, causation, and prediction that stretch the predictive capacity of the social sciences beyond their limits. Empirical findings also reach no consensus whether environmental migration is a significant precursor for violent conflict (Barnett and Adger 2007; Reuveny 2007).

This is partly due to the complex and uncertain multiple causal chains between climate change, migration and conflict which preclude simple statements that isolate environmental factors from other migration drivers. New research approaches and integrated frameworks are needed to better understand the linkages between climate change, natural resources, human security, migration and conflict, based on case studies that allow for reconstructing the complex causal mechanisms (see WBGU 2008; Buhaug et al. 2008; Scheffran et al. 2012bd). Without a thorough understanding of the phenomenon, its prediction remains obscure.

Some scholars criticize the threat terminology regarding migrants who are struggling for their life (Oels 2011). Many non-governmental organizations emphasize that those who are vulnerable and exposed to climate risks are "victims", requiring international interventions and protection. Treating environmental migrants as a security threat could become a self-fulfilling prophecy when securitizing language justifies reactive strategies and precludes preventive approaches.

The maximalist approach has been questioned due to lacking awareness of different political frames and responses which often ignore the potentially severe consequences for the populations affected. A major concern is the instrumentalization of climate refugees in on-going political power games, which has been analysed within Foucault's "power-knowledge" framework (Oels 2012). Critical assessments challenge any military and security "solutions" to climate change (Wright 2012) and suggest to monitor and resist the securitization or militarization of the responses to climate change. It is seen as problematic if the well-funded security markets reorient their focus to environmental challenges and related social consequences, and address these problems with worst case scenarios and technical fixes to climate change.

One difficulty in this debate is that there is no commonly agreed definition of climate migration. Mixing up different meanings complicates constructive policies. If "regular" refugees are pushed into the same category as "environmental" refugees, this may bring about the demise of refugee rights. High numbers of "climate refugees" could spur anti-immigration discourses. An open debate may foster more emancipatory solutions to the problem of environmental migration such as an improved global social policy, local autonomy, and a more just migration policy.

**Strategies and Opportunities**

Instead of being a security threat migration can be an effective adaptation strategy to climate change, creating opportunities for the communities in both the regions of origin and destination as well as the migrants themselves. Strategies could be applied at international, individual, bilateral and national levels. The likelihood of conflict is affected by the functioning of local and national governments; and it is lower in countries where governments are well prepared for emergencies. It is important to analyse the current global governance of refugees, before reflecting on the political constraints that these proposals are likely to face.

The existing governance mechanisms are not sufficiently developed to deal with the challenge of climate-induced migration (Biermann and Boas 2012). The protection of climate refugees requires an effective system of multi-level governance, with a strong global framework providing vital support for, and coordination of, national and local efforts to develop a blueprint for a global governance architecture for the recognition, protection, and voluntary resettlement of climate refugees. To be successful, this regime needs to be tailored to the needs of climate refugees, and be appropriately financed and supported by the international community. Several governing principles were suggested: planned relocation and resettlement instead of temporary asylum; collective rights for local populations; international burden-sharing and international assistance for domestic measures; and financial support and compensation of climate refugees (Biermann and Boas 2012). The same authors suggest new legal instruments, a "Protocol on Recognition, Protection, and Resettlement of Climate Refugees" and a "Climate Refugee Protection and Resettlement Fund".

To overcome the threat-victim dichotomy of refugees, it is important to understand communities and migrants as active social agents that shape and create their livelihood under changing environmental conditions and find collective responses to the climate challenge. The poor and vulnerable are not just passive victims, but actively try to improve their livelihoods within their constraining living conditions. Throughout history, migration has been an adaptive response not only to poverty and social deprivation but also to environmental and climatic changes. Though migration was often associated with hardships, it also offers opportunities through the acquisition of new knowledge, income and other resources as well as the creation of social networks across regions, which can be used to increase resilience. Recent studies propose migration as an important adaptation measure to climate change (Foresight 2011), which could strengthen the resilience of affected communities. Adger et al. (2002) points to the significance of migration effects on social resilience and the natural environment in both sending and receiving areas.

An integrated view considers the ambiguity of the climate-migration nexus, emphasizes challenges and opportunities of migration for development and climate adaptation, and aims to improve the understanding under which conditions "distress migration" can turn into "migration as opportunity" (Scheffran et al. 2012c). Innovative approaches and institutional frameworks can make existing actions more efficient, open new action pathways and establish rules that strengthen social resilience. This includes co-development as an integrative concept that links strategies and networks between host and home countries of migrants (e.g. between the USA and Mexico or between Europe and Northern Africa). Cooperation across regions could support the less developed countries as a compensation for higher GHG emissions of the more developed countries (Oswald-Spring 2012). These compensation mechanisms could include the acceptance of a negotiated number of environmentally forced migrants. Preventive and cooperative migration practices focus on development, livelihood improvements and environmental services. This includes strong social and environmental policies, as well as jobs for migrants to meet the demand for labour (Oswald-Spring 2012). This would reduce the social vulnerability of migrants, to some extent diminish illegal and criminal activities, and improve the living conditions of immigrants in the destination country. In developing countries, environmentally-forced migration could be reduced by rural development policies, by preventive learning and early warning, and by appropriate adaptation and mitigation mechanisms.

## SECURITY AND CONFLICT ISSUES OF CLIMATE POLICIES

While the impacts of climate change have become subject to securitization, measures to prevent or adapt to climate change are also facing concerns about security and conflict implications (Tänzler et al. 2012). Examples are struggles on building dykes against floods, compensation of affected people, approaches to manage a disaster or resettlement of victims. Achieving a fair balance between those affected by climate change and those responsible for it is a continued field of conflict between North and South, rich and poor, state and non-state actors at local and global levels. Disputes on how to avoid climate change with mitigation strategies have been in the center of climate policies and could reach a new level with the transformation to a low-carbon society. While a wide range of mitigation strategies is possible, we will discuss potential security issues related to non-fossil energy sources as well as geoengineering approaches.

**Potential Interactions Between Nuclear and Climate Risks**

While public perception of nuclear power was initially shaped by its destructive potential in nuclear weapons and the East-West nuclear arms race, the civilian use of nuclear energy was associated with technical progress and almost unlimited energy supply. After a high growth in the mid-1970s, new nuclear reactor constructions declined, reaching low levels after the Chernobyl accident in 1986. India, China and other developing countries expand their nuclear power capacity to satisfy their growing energy demand. With emerging concerns about climate change, a "nuclear renaissance" has been proposed to expand nuclear power as a carbon-free source of electricity (Kessler 2012).

Since the 1970s nuclear power has been subject to domestic controversy. Due to protests and accidents public acceptance dropped in several countries which opted to phase out nuclear power, including Germany, Italy, Switzerland and Spain. Opponents have criticized nuclear power as a complex, unsafe, centralized and large-scale technology that is expensive, not commercially competitive and heavily subsidized (Acheson 2011). Most controversial have been the risks to human health and life. Radioactive materials are released and accumulated in the environment throughout the nuclear fuel chain, including uranium mining and fuel rod production, reactor operation and reprocessing, transportation and disposal. Nuclear facilities are vulnerable to accidents, natural disasters and extreme weather events, as demonstrated by the March 2011 earthquake and tsunami in Japan that destroyed the Fukushima Nuclear Power Station. Nuclear waste continues to rise and will remain a problem for many future generations, without responsible and affordable solutions in sight.

Most significant security issues are raised by the link between nuclear power and nuclear weapons. Despite safeguards attempts by the Non-Proliferation Treaty and the International Atomic Energy Agency, the number of nuclear weapon states has increased, becoming a driver of armed conflict and arms races, as demonstrated by the cases of Iraq and Iran. Further security concerns are the vulnerability of nuclear facilities against war or attacks by non-state actors, and the possibility that nuclear weapons or sensitive nuclear materials could fall into the hands of terrorists.

In the future, nuclear energy and climate change may interact in

334　*International handbook of energy security*

multiple ways. Nuclear power is a low-carbon energy source (though not carbon-free during the whole life-cycle) but it is unlikely that nuclear fuels will replace fossil fuels and prevent global warming due to high costs, limited and non-sustainable uranium resources, long planning cycles, inadequacy in combustion and as transportation fuel. The avoided risks of climate change need to be compared to nuclear risks, including nuclear proliferation as well as costly military responses such as missile defense and counter-proliferation. Although US-Russian nuclear arsenals have been significantly reduced more than 20,000 nuclear weapons still remain, enough to destroy the planet multiple times over. Recent scientific studies on nuclear winter suggest that even a limited regional nuclear exchange with a few hundred nuclear weapons would eject so much debris into the atmosphere that global temperatures could rapidly drop and food supply be severely affected for years (Toon et al. 2007).

In a confrontative environment with nuclear weapons it is more difficult to tackle the problem of climate change cooperatively. In turn, when climate change contributes to global insecurity and conflicts, this could create more incentives for states to rely on military force, including nuclear weapons. These scenarios suggest that a combination of nuclear and climate risks might create a volatile and unpredictable security environment (Scheffran 2011).

**Renewable Energy and Land Use Conflicts**

Growing concerns about energy security and climate change have increased the interest in energy from wind, hydro, solar, geothermal or bioenergy sources for meeting the society's electricity, heating and fuel needs. The proposed transition from fossil fuels to low-carbon renewable energy sources requires large areas of land and related infrastructure, creating new forms of "energy landscapes". On the one hand, renewables have environmental impacts and footprints which are usually local (Jacobson 2008) and may lead to conflict with other land uses. On the other hand, transnational cooperation in renewable energies contributes to technology-transfer, new economic options and development perspectives. A sustainable energy transition seeks to develop environmentally responsible and economically efficient land-use paths. Finding a balance between conflict and cooperation requires to address the specific characteristics of each renewable.

**Bioenergy**
Bioenergy has raised great expectations as a renewable, domestic and carbon-free energy source that creates jobs and income in rural areas of

developed and developing countries. In the past decade the biofuel industry has rapidly grown in a number of countries, driven by governmental mandates and subsidies. Due to low efficiency of the photosynthetic conversion of solar radiation into biomass, large areas of land are needed which increases competition with other uses of land, such as agricultural food production and nature protection. Throughout history, the extensive use of wood has contributed to deforestation and landscape changes. Burning of fire wood in homes causes severe health problems and deaths until today. Direct effects of bioenergy include local environmental impacts upon air, water and soil quality, the degradation of habitats and biodiversity. The agro-industrial expansion of bioenergy crops (rape seeds, oil palms, sugar cane, maize) affects food security, deforestation and marginalization of traditional agricultural systems and landscape degradation (Scheffran 2010; Scheffran and Summerfield 2009). Thus, a growing bioenergy use could increase food prices and land acquisitions which may contribute to food riots. Although in theory bioenergy is carbon neutral, there are carbon emissions during the whole lifecycle. While land clearing for bioenergy production releases carbon, biochar offers an opportunity for organic soil carbon sequestration. To improve the insufficient energy and carbon balance of the current generation of biofuels, more efficient bioenergy pathways are under development, including algae and second generation biofuels based on biotechnology. To constrain adverse impacts of bioenergy, sustainability criteria have been suggested and dialogues among stakeholders (Maestas 2012).

**Hydropower**
While in many industrialized countries the dam-building boom is over, some developing countries are expanding hydropower, especially in Latin America, East and Southeast Asia. Large hydro projects flood landscapes over hundreds of square kilometres, with severe impacts on ecological habitats, human livelihoods and cultural assets. Worldwide an estimated 40-80 million people were evacuated from their homes (Ponseti and Pujol 2012). In China alone more than 1 million people were displaced to build the Three Gorges Dam at the Yangtzekiang, the world's largest hydroelectric dam. A number of big dam projects provoked local resistance, for instance against the Narmada dam in India. In a few cases, international disputes emerged, e.g. between Hungary and Slovakia over the ecological and social impacts of the Gabcikovo dam on the river Danube. The GAP project of Turkey along the Euphrates River became a conflictive issue with the local Kurdish population and with Iraq as a downstream country. In response to rising protests, the World Commission on Dams involved civil society participants and released a

joint report in 2000 that defined criteria for the evaluation of big dams, favouring smaller projects.

**Wind power**
Due to the low energy density of the wind, wind power plants are spread over large areas and reach into high altitudes to increase efficiency. While some people directly benefit from additional income of wind power, others living in these areas complain about sound and shadows from rotors (Zoll 2001). Environmentalists are concerned about the impacts on birds and other animals or the effect of large off-shore wind power stations on marine ecosystems. Some complain about the transformation of landscapes. Occasionally there have been protests against large wind turbines and lawsuits to affect political decisions. While the impacts are subject to research, technical developments and economic gains have contributed to reducing the risk and building public support.

**Solar energy**
Although small-scale, decentralized use of solar energy is widely accepted among citizens, costs are still higher than alternative energy sources. Ecological and social interventions as well as conflict potentials may become more significant with large-scale industrial use and growing dependence on solar energy from other regions. The Desertec concept proposes to develop the solar energy potential in the world's deserts through cooperation between industrialized and developing countries which are connected by an electric power grid, e.g. across the Mediterranean. To avoid the conflict patterns in this region established for oil and natural gas, non-governmental organizations recommend criteria for economic viability, environmental sustainability, social acceptability and political stability that strengthen security, development opportunities, climate adaptation and capacity building for the local population (Klawitter and Schinke 2011).

Other renewable energy systems (geothermal, tidal or ocean wave energy) have specific impacts on the local environment and attract minor criticism but no major conflicts yet. Although the conflict potential of renewable energy sources cannot be neglected, it has to be compared to the substantial security risks and conflicts of fossil and nuclear energy sources. As long as renewable energy systems and their infrastructure are based on small-scale technologies, conflicts remain local. With the rapid and global expansion of space-intensive renewables to the same order as fossil energy, demands for investment, land and other critical resources could become an issue of conflict which involves a range of possible actions (demonstrations and protests, blockades and sabotage acts, police operations). How these evolve, depends on public acceptance and the effectiveness of regulatory mechanisms to balance benefits, costs and risks.

**Geoengineering the Climate: an Issue for Security Policy?**

Geoengineering is offered as a solution for reducing dangerous climate change by deliberate interventions into the climate system on a global scale. Two different types of measures have been suggested (for an overview see Royal Society 2009; Rickels et al. 2011).

- Carbon dioxide removal (CDR) reduces atmospheric carbon concentration as a cause of global warming. CDR techniques include carbon capture and sequestration in biomass, soil, underground or in the ocean through afforestation, biochar, ocean fertilization and enhanced weathering.
- Solar radiation management (SRM) diminishes the symptoms and effects of global warming by changing the earth's radiation balance. SRM technologies focus on the absorption of sunlight, e.g. by aerosol emissions into the higher atmosphere, similar to volcano eruptions; or by the reflection of sunlight, e.g. through surface brightening, cloud whitening or large reflectors in outer space.

Recent research has focused on the technical, environmental and economic dimensions of geoengineering, but increasingly includes political, psychological, ethical and social implications. There is still a lack of understanding for many critical issues, in particular the risk and conflict potential. Various arguments have been used: Geoengineering may pollute the environment, consume resources and investments, distract from climate mitigation policies, provoke security risks, public resistance, tensions, and conflicts. More specifically the following types of risks and conflicts of GE can be identified (Maas and Scheffran 2013; Scheffran and Cannaday 2013).

- Distribution of benefits, costs and risks: Geoengineering will not have a uniform global impact but can substantially differ regarding the local, regional, and international consequences which are specific to the respective environments. Some impacts are foreseeable, such as cooling or changing rainfall from aerosol emissions or change in ecosystems, vegetation and crop yields. Injecting sulphur into the stratosphere may have environmental impacts similar to pollution. Other consequences are unintended and unpredictable, e.g. when geoengineering triggers tipping elements in the climate system. Concerns about these risks and consequences provokes resistance and conflict between those who benefit from these measures and

those who are negatively affected and have inadequate coping mechanisms. Here regional differences can be quite significant.
- Competition for scarce resources: to realize geoengineering, comprehensive infrastructures and efforts are required which need resources (energy, raw materials, water, land), add pollution and change natural and social systems along the lifecycle. Compared to SRM, CDR requires a larger scale implementation to have a globally meaningful impact which needs substantial financial investments, resources and ecosystem services to have a significant effect. There are only limited places in which such activities are possible in technical, economic and political terms, leading to concerns of increasing pressure on resources and communities in affected regions. While SRM could be implemented by a few countries, a larger number of countries would need to engage in CDR to make a difference for the climate. CE measures compete with investments in other areas, in particular mitigation and adaptation strategies.
- Power games on climate control: during the Cold War the superpowers supported a small amount of research on weather control for offensive and defensive purposes (Fleming 2010). While climate control is more difficult to operationalize for military purposes, it contains significant potential dual use capabilities that could increase threat perceptions and worst-case scenarios. Countries may support research, testing and deployment of geoengineering as an instrument for power projection. When countries try to "optimize" their own climate, trans-regional effects create a potential for international disputes and security dilemmas. Although intended geoengineering may avoid some risks and conflicts of unintended climate change it could also contribute to the climate-conflict nexus. The question is whether the cure is worse than the cause.

Similar to other forms of environmental modification such as large-scale deforestation, building big dams along rivers, or genetic engineering, geoengineering measures will likely face resistance by local communities. One indicator is on-going protests against carbon capture and sequestration in Germany. Local actions could combine with international tensions between states, leading to complex multi-level interactions in the international system. Considering a hypothetical scenario, a series of droughts in the US Midwest could drive farmers to pressure their government to implement rapid SRM intervention which may provoke countermeasures by Russia or China. In a geoengineered world anyone applying these techniques could be blamed for weather-related disasters and associated damages elsewhere, whether justified or not.

Individual attempts to regulate global temperature could lead to security dilemmas from local to global levels and a range of competing actions, including local protests, diplomatic crises, violent actions against geoengineering components, terrorism, and armed conflict. Collective regulation of geoengineering requires an effective global governance system which needs to resolve a number of critical questions: Which climate is to be designed? Who is responsible for particular risks? Which consequences are judged as acceptable or inacceptable? How can the responsibility for prohibited actions be verified? How can compliance be enforced and conflicts be resolved?

Currently there are no legal mechanisms to regulate geoengineering although some agreements may be affected (for an overview see Rickels et al. 2011). If geoengineering activities are classified as pollution, they could fall under the respective environmental agreements, such as the Montreal Protocol for the Protection of the Ozone Layer, the Law of the Sea, and the London Convention and Protocol on the Prevention of Marine Pollution by Dumping of Wastes and Other Matter. The UN Framework Convention on Climate Change (UNFCCC) precludes dangerous anthropogenic interference with the climate system, which applies to dangerous climate change as well as dangerous geoengineering. However, the term "dangerous interference" has not been specified. Similarly, the 1978 Convention on the Prohibition of Military or Any Other Hostile Use of Environmental Modification Techniques (ENMOD) restrains the military or other hostile modification of the Earth's environment, but does not prohibit the peaceful use of geoengineering.

Activities for the regulation of geoengineering in the early stages include a moratorium on geoengineering activities recommended at the 2008 Convention on Biological Diversity (CDB) and a report by the British House of Commons that formulated the "Oxford Principles" for geoengineering research. These suggest to (i) regulate geoengineering as a public good, (ii) let the public take part in the decision-making process, (iii) disclose the results of research, (iv) have an independent impact assessment conducted, and (v) only begin with any implementation after a governance process is completed (Rayner et al. 2009).

## ENERGY TRANSFORMATION, DESECURITIZATION AND CONFLICT PREVENTION

Energy and climate change are increasingly subject to securitization. In the new emerging security and conflict landscapes, people are exposed to various risks of energy use, climate change and geoengineering. In this complex environment where everyone can contribute to the problems and be affected by it, traditional instruments of security policy, such as military buildup and interventions, will be ineffective. An integrated and coordinated global governance approach would combine measures in different regions and policy fields, including environmental policy, development policy and security policy, from local to global levels. Given the large uncertainties and complexities in the energy-climate-security nexus, anticipative and precautionary approaches are adequate that adapt to the knowledge about possible futures. Adaptive strategies constrain and adjust the path of GHG emissions towards a target domain of the climate system (e.g. a range of carbon concentration or global-temperature) to avoid dangerous areas. Continuously updated scientific information is essential to adapt future paths to sight, destination and unexpected events, as well as actions by other actors.

To achieve the proposed transformation of the fossil energy system towards a low-carbon society (WBGU 2011), energy pathways need to be systematically assessed and compared regarding benefits, costs, risks and conflict potentials. To build a secure, equitable and sustainable energy system that harvests the benefits while minimizing risks and conflicts, principles, criteria and standards for sustainable peace are to be implemented in legal norms, certification systems and monitoring processes. Mechanisms for desecuritization, mediation and conflict resolution can support this process, including stakeholder dialogues, participatory decisionmaking and arbitration. Then geographies of conflict on energy and climate change could be transformed into landscapes of cooperation on avoiding dangerous anthropogenic interference with the climate system.

## NOTES

\* Research for this study was funded in part by the German Science Foundation (DFG) through the Cluster of Excellence CliSAP (EXC177).
1. The argument that the securitization of the environment failed has already been suggested in Buzan et al. 1998, where they argue that the exceptional measures in the environmental sector are within the borders of normal politics. The argument is repeated in Buzan and Wæver (2009) where they suggest that climate change has determined a metapoliticization rather than a metasecuritization.

## REFERENCES

Acheson, Ray (ed.) (2011), *Costs, risks, and myths of nuclear power*, Women's International League for Peace and Freedom.
Adger, W.N, P.M. Kelly, A. Winkels, L.Q. Huy and C. Locke (2002), "Migration,

Remittances, Livelihood Trajectories, and Social Resilience", *Journal of the Human Environment*, 31(4), 358–366.
Bächler, Günter and Kurt R. Spillmann (eds) (1996), *Environmental Degradation as a Cause of War, Regional and Country Studies of Research Fellows*, Chur-Zürich: Rüegger.
Barnett, Jon and W. Neil Adger (2007), "Climate change, human security and violent conflict", *Political Geography*, 26, 639–655.
Bettini, Giovanni (2013), "Climate Barbarians at the Gate? A critique of apocalyptic narratives on 'climate refugees'", *Geoforum*, 45, March, 63–72.
Biermann, Frank and Ingrid Boas (2008), "Protecting Climate Refugees: The Case for a Global Protocol", *Environment*, 50(6), 8–16.
Biermann, F. and I.J.C. Boas (2012), "Climate change and human migration: Towards a global governance system to protect climate refugees", in Jürgen Scheffran et al. (eds), *Climate Change, Human Security and Violent Conflict. Challenges for Societal Stability?*, Berlin Heidelberg: Springer, pp. 291–300.
Bigo, Didier (2002), "Security and immigration: Toward a critique of the governmentality of unease", *Alternatives: Global, Local, Political*, 27(S2), 63–92.
Brauch, Hans Günter (2009), "Securitizing Global Environmental Change", in Hans Günter Brauch et al., *Facing Global Environmental Change*, Berlin/Heidelberg: Springer, pp. 65–104.
Brzoska, Michael (2009), "The securitization of climate change and the power of conceptions of security", *Security and Peace*, 27, 137–145.
Brzoska, Michael (2012), "Climate Change as a Driver of Security Policy", in Jürgen Scheffran, et al. (eds), *Climate Change, Human Security and Violent Conflict*, Berlin/Heidelberg: Springer, pp. 165–184.
Brzoska, Michael, P.Michael Link, Achim Maas, Jürgen Scheffran, (eds.) (2012), "Geoengineering: An Issue for Peace and Security Studies?", *Sicherheit & Frieden / Security & Peace*, Special Issue, **30** (4/2012).
Buhaug, Halvard (2010), "Climate not to blame for African civil wars", Proceedings of the National Academy of Sciences (PNAS), 107, 16477–16482.
Buhaug, Halvard, Nils Petter Gleditsch, Ole Magnus Theisen (2010), "Implications of Climate Change for Armed Conflict", in Robin Mearns and Andrew Norton (eds), *Social Dimension of Climate Change*, Washington: World Bank, pp. 75–101.
Bundeswehr (2012), *Klimafolgen im Kontext – Implikationen für Sicherheit und Stabilität im Nahen Osten und Nordafrika*, Berlin: Planungsamt der Bundeswehr.
Burke, M.B., E. Miguel, S. Satyanath, J.A. Dykema and D.B. Lobell (2009), "Warming increases the risk of civil war in Africa", *Proceedings of the National Academy of Sciences*, 106, 20670–20674.
Buzan, Barry and Ole Wæver (2009), "Macrosecuritization and security constellations", *Review of International Studies*, 35(2).
Buzan, Barry, Ole Wæver and Jaap de Wilde (1998), *Security. A New Framework for Analysis*, Boulder/London: Lynne Rienner.
Campbell, Kurt M. et al. (2007), *The Age of Consequences: The Foreign Policy and National Security Implications of Global Climate Change*, Washington, DC: Center for Strategic and International Studies, Center for a New American Security.
CNA (2007), "National Security and the Threat of Climate Change", CNA Corporation available at http://SecurityAndClimate.cna.org (accessed February 21, 2013).
EU (2008), "Climate Change and International Security", Paper from the High Representative and the European Commission to the European Council, Brussels, March 14, S113/08, available at www.consilium.europa.eu/ueDocs/cms_Data/docs/pressData/en/reports/99387.pdf (accessed April 30, 2009).
Feiveson, Harold A. (2009), "A skeptic's view of nuclear energy", *Daedalus*, 138(4), 60–70.
Fleming, Jim (2010), *Fixing the Sky*, New York: Columbia University Press.
Foresight (2011), "Migration and Global Environmental Change", Foresight Final Project Report, London.
Gemenne, Francois (2011), "Climate-induced population displacements in a 4°C world", *Philosophical Transactions of the Royal Society A*, 369, 182–195.

Gleditsch, Nils Petter (2012), "Special Issue on Climate Change and Conflict", *Journal of Peace Research*, 49, 163.
Gleick, Peter H. (1989), "The implications of global climatic changes for international security", *Climatic Change* 15(1/2), 309–325
Homer-Dixon, Thomas F. (1991), "On the Threshold: Environmental Changes as Causes of Acute Conflict", *International Security*, 16, 2 (Fall), 76–116.
Homer-Dixon, Thomas F. (1994), "Environmental Scarcities and Violent Conflict: Evidence From Cases", *International Security*, 19, 1 (Summer), 5–40.
Hsiang, Solomon, M., K.C. Meng and M.A. Cane (2011), "Civil conflicts are associated with the global climate", *Nature*, 476, 438–441.
IPCC (2007), "Climate Change 2007: Climate Change Impacts, Adaptation and Vulnerability", Working Group II, Intergovernmental Panel on Climate Change, Fourth Assessment Report.
Jacobson, Mark Z. (2008), "Review of solutions to global warming, air pollution, and energy security", *Energy & Environmental Science*, 2, 148–173.
Jäger, Jill, Johannes Frühmann, Sigrid Grünberger and Andras Vag (eds) (2009), "Environmental Change and Forced Migration Scenarios", Synthesis Report, ATLAS Innoglobe, Hungary.
Jakobeit, Cord and Chris Methmann (2012), "'Climate Refugees' as Dawning Catastrophe? A Critique of the Dominant Quest for Numbers", in Jürgen Scheffran et al. (eds), *Climate Change, Human Security and Violent Conflict: Challenges for Societal Stability*, Berlin: Springer, pp. 301–314.
Kessler, Günter (2012), *Sustainable and Safe Nuclear Fission Energy*, Berlin: Springer.
Klawitter, Jens and Boris Schinke (2011), *Desertec and Human Development at the Local Level in the MENA-Region*, Stuttgart: Germanwatch, Brot für die Welt.
Lenton, Tim M. et al. (2008), "Tipping elements in the Earth's climate system", Proceedings of the National Academy of Sciences (PNAS), 105(6), 1786–1793.
Maas, Achim and Jürgen Scheffran (2013), "Climate Conflicts 2.0? Climate Engineering as a Challenge for International Peace and Security", *Security and Peace*, Special Issue (30. Jg.) 4/2012, 193–200.
McDonald, Matt (2012), "The failed securitization of climate change in Australia", *Australian Journal of Political Science*, 47(4), 579–592.
Maestas, Aislinn (2012), "Major Milestone for Sustainable Biofuels", National Wildlife Federation, available at http://www.nwf.org/News-and-Magazines/Media-Center/News-by-Topic/Global-Warming/2012/02-10-12-Major-Milestone-for-Sustainable-Biofuels.aspx (accessed May 20, 2012).
Methmann, Chris and Delf Rothe (2012), "Politics for the day after tomorrow: The logic of apocalypse in global climate politics", *Security Dialogue*, 43(4), 323–344.
Myers, Norman (2002), "Environmental refugees: a growing phenomenon of the 21st century", *Philosophical Transactions of the Royal Society of London*, 357, 609–613.
Nordås, Ragnhild and Nils Petter Gleditsch (2007), "Climate change and conflict", *Political Geography*, 26, 627–638.
Oels, Angela (2011) "'Rendering climate change governable by risk: From probability to contingency", *Geoforum*, 45, March, 17–29.
Oels, Angela (2012), "From 'Securitization' of Climate Change to 'Climatization' of the Security Field: Comparing Three Theoretical Perspectives", in Jürgen Scheffran et al. (eds.) *Climate Change, Human Security and Violent Conflict*, Berlin: Springer, pp. 185–205.
O'Loughlin, John et al. (2012), "Climate variability and conflict risk in East Africa, 1990–2009", Proceedings of the National Academy of Sciences (PNAS), (early edition), available at www.pnas.org/cgi/doi/10.1073/pnas.1205130109 (accessed February 21, 2013).
Oswald Spring, Úrsula (2012), "Environmentally-Forced Migration in Rural Areas: Security Risks and Threats in Mexico", in Jürgen Scheffran et al. (eds), *Climate Change, Human Security and Violent Conflict*, Berlin/Heidelberg: Springer, pp. 315–350.
Ponseti, Marta and Jordi López-Pujol (2012), "The Three Gorges Dam Project in China",

*Revista HMiC*, IV, available at http://digital.csic.es/bitstream/10261/27902/1/Lopez-Pujol_01.pdf (accessed May 20, 2012).

Rahman, Atiq (1999) "Climate change and violent conflicts", In: Suliman, Mohamed (ed) *Ecology, Politics and Violent Conflict*. Zed Books, London, pp. 181–210

Rayner, S., C. Redgwell, J. Savulescu, N. Pidgeon and T. Kruger (2009), "Memorandum on draft principles for the conduct of geoengineering research", House of Commons Science and Technology Committee.

Reuveny, Rafael (2007), "Climate change-induced migration and violent conflict", *Political Geography*, 26, 656–673.

Rickels, Wilfried et al. (2011), "Large-Scale Intentional Interventions into the Climate System? Assessing the Climate Engineering Debate", Scoping report, German Federal Ministry of Education and Research, Kiel Earth Institute.

Rowe, J.W. (2009), "Nuclear power in a carbon-constrained world", *Daedalus*, 138(4), 81–90.

Royal Society (2009), "Geoengineering the climate: Science, governance and uncertainty", London, RS Policy document.

Scheffran, Jürgen (1997), "Conflict potential of energy-related environmental changes—the case of global warming" (in German), in: Bender, Wolfgang (ed) *Verantwortbare Energieversorgung für die Zukunft*, Darmstadt, pp 179–218.

Scheffran, Jürgen and Gale Summerfield (eds) (2009), "Sustainable Biofuels and Human Security", *Swords & Ploughshares*, XVII(2), Summer.

Scheffran, Jürgen (2010), "Criteria for a Sustainable Bioenergy Infrastructure and Lifecycle", in P.N. Mascia, J. Scheffran and J. Widholm (eds), *Plant Biotechnology for Sustainable Production of Energy and Co-products*, Berlin/Heidelberg: Springer Verlag, pp. 409–443.

Scheffran, Jürgen (2011), "Climate Change, Nuclear Risks and Nuclear Disarmament", Report for the World Future Council, May.

Scheffran, Jürgen and Antonella Battaglini (2011), "Climate and conflicts: the security risks of global warming", *Regional Environmental Change*, 11, S27-S39.

Scheffran, Jürgen, Michael Brzoska, Hans Günter Brauch, P. Michael Link and Janpeter Schilling (eds) (2012a), *Climate Change, Human Security and Violent Conflict*, Berlin/Heidelberg: Springer.

Scheffran, Jürgen, Michael Brzoska, Jasmin Kominek, P. Michael Link and Janpeter Schilling (2012b), "Climate change and violent conflict", *Science*, 336, 869–871.

Scheffran, Jürgen, Elina Marmer and Papa Sow (2012c), "Migration as a contribution to resilience and innovation in climate adaptation: Social networks and co-development in Northwest Africa", *Applied Geography*, 33, 119–127.

Scheffran, Jürgen, Michael Brzoska, Jasmin Kominek, P.Michael Link and Janpeter Schilling (2012d), "Disentangling the Climate-conflict Nexus: Empirical and Theoretical Assessment of Vulnerabilities and Pathways", *Review of European Studies*, 4(5) 1–13.

Scheffran, Jürgen and Thomas Cannaday (2013), "Resistance Against Climate Change Policies: The Conflict Potential of Non-fossil Energy Paths and Climate Engineering", in B. Balazs, C. Burnley, I. Comardicea, A. Maas and R. Roffey (eds), *Global Environmental Change: New Drivers for Resistance, Crime and Terrorism?* Nomos (forthcoming).

Schneider, Mycle, Antony Froggatt and Steve Thomas (2011), "World nuclear industry status report", *Bulletin of the Atomic Scientists*, 67(4) (2010/2011).

Singer, Clifford (2008), *Energy And International War: From Babylon to Baghdad and Beyond*, World Scientific.

Singer, Clifford and Jürgen Scheffran (2004), "Energy and Security – From Conflict to Cooperation", *INESAP Information Bulletin*, 24 (Dec), 65–70.

Socolow, Robert H. and Alex Glaser (2009), "Balancing risks: nuclear energy & climate change", *Daedalus*, 138(4), 31–44.

Swart, Rob (1996), "Security risks of global environmental changes", *Global Environmental Change*, Part A 6(3): 187–192.

Tänzler, Dennis, Achim Maas and Alexander Carius (2012), "Climate Change Adaptation and Peace", *Wileys Interdisciplinary Reviews: Climate Change*, 1(5), 741–750.

Tol, Richard and Sebastian Wagner (2010), "Climate change and violent conflict in Europe over the last millennium", *Climatic Change*, 9, 65–79.
Toon, O.B., R.P. Turco, A. Robock, C. Bardeen, L. Oman and G.L. Stenchikov (2007), "Atmospheric effects and societal consequences of regional scale nuclear conflicts and acts of individual nuclear terrorism", *Atmospheric Chemistry and Physics*, 7, 1973–2002.
Trombetta, Maria J. (2008), "Environmental security and climate change: analysing the discourse", *Cambridge Review of International Affairs*, 21(4), 585–602.
Trombetta, Maria J. (2012), "Climate Change and the Environmental Conflict Discourse", in Jürgen Scheffran et al. (eds), *Climate Change, Human Security and Violent Conflict*, Berlin: Springer.
UNHCR (2011), "A Year in Crisis – UNHCR Global Trends 2011", U.N. High Commissioner for Refugees, available at http://www.unhcr.org/4fd6f87f9.pdf.
UNSC (2011), "Maintenance of Peace and Security: Impact of Climate Change", Statement by the President of the UN Security Council on, S/PRST/1011/15, available at http://daccess-dds-ny.un.org/doc/UNDOC/GEN/N11/424/28/PDF/N1142428.pdf?OpenElement (accessed July 20, 2011).
UNSG (2009), "Climate change and its possible security implications", Report A/64/3, New York: UN Secretary-General.
Wæver, Ole (1995), "Securitization and Desecuritization", in R.D. Lipschutz (ed.), *On Security*, New York: Columbia University Press, pp. 46–86.
Wæver, Ole (1997), *Concepts of Security*, Copenhagen: Department of Political Science.
Warner, K., M. Hamza, A. Oliver-Smith, F. Renaud and A. Julca (2010), "Climate change, environmental degradation and migration", *Natural Hazards*, 55(3), 689–715.
WBGU (2008), "Climate Change as a Security Risk", Berlin: German Advisory Council on Global Change; available at www.wbgu.de (accessed February 21, 2013).
WBGU (2011), "World in Transition – A Social Contract for Sustainability", Berlin/Heidelberg: German Advisory Council on Global Change.
Wright, Steve (2012), "Policing Borders in a Time of Rapid Climate Change", in Jürgen Scheffran et al. (eds), *Climate Change, Human Security and Violent Conflict*, Berlin: Springer, pp. 351–370.
Yergin, Daniel (1991), *The Prize: The Epic Quest for Oil, Money, and Power*, New York: Simon & Schuster.
Yergin, Daniel (2011), *The Quest: Energy, Security, and the Remaking of the Modern World*, Penguin.
Zhang, D.D., J. Zhang, H.F. Lee and Y.Q. He (2007), "Climate Change and War Frequency in Eastern China over the Last Millennium", *Human Ecology*, 35, 403–414.
Zhang, D.D., H.F. Lee, C. Wang, B. Li, Q. Pei, J. Zhang and Y. An (2011), "The causality analysis of climate change and large-scale human crisis", Proceedings of the National Academy of Sciences (PNAS), 108, 17296–17301.
Zoll, Ralf (ed.) *Energiekonflikte. Problemübersicht und empirische Analysen zur Akzeptanz von Windkraftanlagen*, Münster: LIT, 2001, pp. 17–95.

# 16. Environmental implications of energy production
*Yolanda Lechón, Natalia Caldés and Pedro Linares*

## 1 INTRODUCTION

The assessment of the environmental implications of energy production is a wide and complex challenge. Many energy sources and technologies exist with very different environmental profiles and implications in the form of emissions to the environment, hazardous wastes, occupational risks, production of noise or visual intrusion or risks of accidents, etc. All of them impose risks on human beings, ecosystems and materials, producing damages that are *external* in the sense that they are not taken into account by the person or institution causing the effects. Such effects, known as externalities, are then not incorporated in the market price of the energy products they generate. The presence of externalities in the energy market is a market failure that results in the energy mix being inefficient from a social point of view.

When investment or operation decisions are made, e.g. about which power plant technology to use or where to locate a power plant, it would be beneficial for society to take environmental and health impacts into account and include the external effects into the decision making process. Quantifying these implications in a comparable manner would help energy policy makers in finding the energy mix which maximizes social welfare. In order to be able to assess and compare the external effects with each other and with private costs, it is convenient to express them in a common unit such as a monetary unit. Thus converting external effects into monetary units results in *external costs* that can be easily used in a cost-benefit analysis or be internalized through the appropriate environmental policy instruments.

In the first section of this chapter we will review the environmental implications of a wide variety of energy sources and technologies. In the case of fossil fuel technologies, as the previous chapter has examined the environmental impacts of fossil fuel extraction, this chapter will focus on those impacts produced during energy generation. However, in the case of renewable energies, the environmental implications of the energy generation stage of the fuel chain are in many cases negligible and most impacts are produced in either upstream or downstream stages. Therefore a discussion of these impacts will also be included.

In the second section of this chapter, the methodological options to quantify environmental implications and externalities of energy technologies will be briefly introduced, including the Life Cycle Assessment methodology and the ExternE methodology for externalities assessment. Finally, in the last section, some results of environmental impacts and externalities of different energy technologies will be presented.

## 2 ENVIRONMENTAL IMPLICATIONS OF DIFFERENT ENERGY SOURCES

### 2.1 Coal

Coal is the world's most abundant and widely distributed fossil fuel. Around 27 per cent of the world's total primary energy demand and 42 per cent of the world's electricity production is based on coal combustion (IEA, 2011). Several studies (MIT, 2007; IEA, 2011) conclude that coal will continue to have an important role meeting the world's energy needs in significant quantities.

The main use of coal nowadays is electricity generation although other uses include coking coal for steel manufacturing and industrial process heating whereas for electricity generation, pulverized coal fired is the most common installed technology nowadays (Bauer et al, 2008; IEA-ETSAP, 2010a). An average net thermal efficiency of 35–36 per cent is commonly assumed for large existing plants with subcritical steam burning relatively high quality coals.

Currently, supercritical pulverized coal (SCPC) power is the dominant option for new coal-fired power plants (IEA-ETSAP, 2010a). Super-critical pulverised coal (SCPC) power plants use supercritical steam as the process fluid to reach high temperatures and pressures and reach efficiencies up to 46 per cent. New ultrasupercritical (U-SCPC) power plants may even reach higher temperatures and pressure, with efficiency up to 50 per cent.

Integrated gasification combined cycles (IGCC) and Fluidized-bed combustion (FBC) are alternative coal-fired power technologies. IGCC can reach higher efficiencies varying from 39 per cent to 45 per cent while the primary driving force for the development of fluidized-bed combustion was the reduction in $SO_2$ and NOx emissions at the combustor (Bauer et al, 2008).

Pulverised coal-fired power plants produce a considerable amount of airborne emissions. Emissions of SCPC and IGCC power plants are quite smaller and shown in Figure 16.1. Coal combustion produces also trace elements such as metals, including nickel, mercury, arsenic, chromium and cadmium (CATF, 2001).

Coal-fired power plants use large quantities of water for producing steam and for cooling. Pollutants build up in the water used in the power plant boiler, cooling system, ash handling plants and flue gas desulphurization (FGD) plants (EC, 1995a; US-EPA, 2012). If the water used in the power plant is discharged to a lake or river, pollutants in the water can harm fish and plants. Further, if rain falls on coal stored in piles outside the power plant, the water that runs off these piles can flush heavy metals from the coal, such as arsenic and lead, into nearby bodies of water.

The burning of coal creates solid waste, called ash, which is composed primarily of metal oxides and alkali. On average, the ash content of coal is around 10 per cent (6–20 per cent). Solid waste is also created when air pollutants are removed from the stack gas. Much of this waste is deposited in landfills and abandoned mines, although some amounts are now being recycled into useful products, such as cement and building materials.

Finally, soil at coal-fired power plant sites can become contaminated with various pollutants from the coal and take a long time to recover, even after the power plant closes down.

## 2.2 Natural Gas

Natural gas is an extremely important source of energy that offers a number of environmental benefits over other sources of energy, particularly other fossil fuels. Natural gas is important in many sectors of the economy: for electricity generation, as an industrial heat source and chemical feedstock as well as for water and space heating in residential and commercial sectors. Moreover, natural gas is increasingly being used in the transportation sector.

Power plants use several methods to convert gas into electricity. The most common approach is to burn the gas in a combustion turbine to generate electricity in the so called open-cycle gas turbine (OCGT) plants, which offer a moderate electrical efficiency of between 35 per cent and 42 per cent at full load (IEA-ETSAP, 2010b). Another technology is to burn natural gas in a combustion turbine and use the hot combustion turbine exhaust to make steam to drive a steam turbine. This technology is called "combined cycle" (CCGT) and achieves a higher efficiency. This technology has been largely expanding in the last two decades (Bauer et al, 2008)

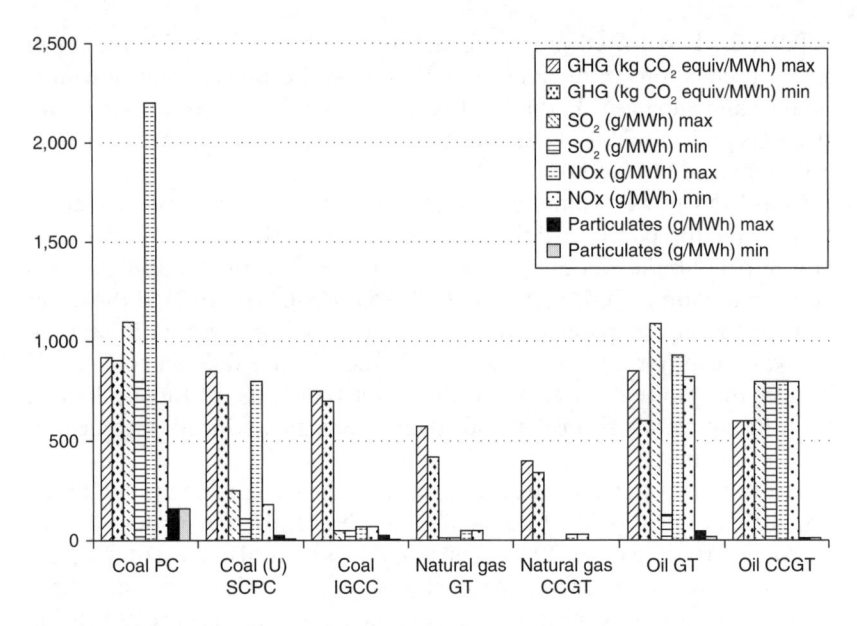

[Figure: bar chart showing GHG, SO₂, NOx, and Particulates emissions (max/min) for Coal PC, Coal (U) SCPC, Coal IGCC, Natural gas GT, Natural gas CCGT, Oil GT, and Oil CCGT]

Sources: EC, 1995a; IEA-ETSAP, 2010; EC, 1995b; US-EPA, 2011a; EC, 1995b; US-EPA, 2011a; IEA-ETSAP, 2010b.

Figure 16.1 *Airborne pollutant emissions from fossil fuel fired power plants*

and its net electrical efficiency ranges from 50–58 per cent (Bauer et al, 2008).

There is a wide variety of environmental impacts generated by the use of natural gas as a source of energy.

Composed primarily of methane, the main products of the combustion of natural gas are carbon dioxide and water vapour, very small amounts of sulphur dioxide and nitrogen oxides, virtually no ash or particulate matter, and low levels of carbon monoxide as well as other reactive hydrocarbons as shown in Figure 16.1.

Compared to the average air emissions from conventional coal-fired generation, natural gas produces half as much carbon dioxide, lower quantities of nitrogen oxides and almost no sulphur oxides at the power plant.

The burning of natural gas in combustion turbines requires very little water. However, natural gas-fired boiler and combined cycle systems do require water for cooling purposes. Pollutants and heat build up in the water used in natural gas boilers and combined cycle systems. When these

*Environmental implications of energy production* 349

pollutants and heat reach certain levels, the water is often discharged into lakes or rivers affecting water bodies' ecosystems.

The use of natural gas to create electricity does not produce substantial amounts of solid waste (EC, 1995b).

Other burdens from the operation of natural gas power plants include noise and visual intrusion (EC, 1995b). The construction of natural gas power plants, as any large facility, can destroy natural habitat for animals and plants. Possible land resource impacts include erosion, loss of soil productivity, and landslides.

## 2.3 Crude Oil

Crude oil is a vital source of energy for the world and will likely remain so for many decades to come (IEA, 2011). The vast majority of petroleum consumed in the world is used for transportation. The remainder is used to create the many oil-based products used in industry and our houses such as lubricants and plastics, to generate electricity and to heat our homes.

The previous chapter analysed the environmental implications of extracting crude oil, and the environmental consequences of the use of oil products are analysed in a subsequent chapter. This chapter will focus on the stages of the oil fuel cycle after oil extraction and up to the production of fuels and electricity.

### 2.3.1 Petroleum refining

The petroleum refining industry converts crude oil into refined products, including liquefied petroleum gas, gasoline, kerosene, aviation fuel, diesel fuel, fuel oils, lubricating oils, and feedstocks for the petrochemical industry employing a wide variety of processes.

Potential environmental impacts associated with petroleum refining include the following (IFC, 2007):

- Air emissions
- Waste water
- Hazardous materials
- Wastes
- Noise.

Air emissions can be exhaust gases – carbon dioxide ($CO_2$), nitrogen oxides (NOx) and carbon monoxide (CO) – from combustion of gas, fuel oil or diesel in turbines, boilers, compressors and other equipment for power and heat generation, venting and flaring emissions and fugitive emissions (hydrogen ($H_2$), methane ($CH_4$), volatile organic compounds(VOCs), polycyclic aromatic hydrocarbons (PAHs) and inorganic gases such as associated with leaks of hydrocarbon vapours from process equipment and evaporation of hydrocarbons from open areas (US-EPA, 2011b; IFC, 2007). Sulphur oxides (SOx) and hydrogen sulphide may also be emitted from boilers, heaters and other process equipment. Particulate emissions from refineries are associated with flue gas furnaces and catalyst based processes, handling of coke and incineration of sludge. Carbon dioxide is produced in significant amounts during petroleum refining from combustion processes, flares, hydrogen plants and catalyst regeneration (IFC, 2007).

Waste water from refineries includes "sour" water and alkaline waste water (IFC, 2007). Waste water is treated in treatment units before disposal. Oil refinery effluents contain many different chemicals at different concentrations including ammonia, sulphides, phenol and hydrocarbons. These effluents often have a toxic effect on the fauna, which is usually restricted to the area close to the outfall (Wake, 2005).

Petroleum refining facilities also use significant amounts of hazardous materials. Solid hazardous wastes from a refinery are spent catalysts, solvents, filters, used sweetening, spent aminas, activated carbon filters, sludge of several types, exhausted molecular sieves and exhausted alumina.

The main causes of noise in a refinery are the large rotation machines (compressors and turbines), pumps, air coolers and heaters.

### 2.3.2 Production of electricity from oil

Similarly to natural gas, oil is usually burnt in open cycle or combined cycle gas turbines to produce electricity with similar technological characteristics. Burning oil at power plants produces nitrogen oxides, sulphur dioxide, carbon dioxide, methane, and mercury compounds. The amount of sulphur dioxide and mercury compounds can vary greatly depending on the sulphur and mercury content of the oil that is burned. The average emissions are shown in Figure 16.1.

Oil-fired power plants use large quantities of water for steam production and cooling. When oil-fired power plants get the water from a lake or river, fish and other aquatic life can be killed, which affects those animals and people who depend on these aquatic resources. Moreover, power plants also release waste water – which contains pollutants and is generally hotter than the water in nearby lakes and streams – that can harm fish and plants.

Also, when oil is burned at power plants, residues that are not completely burned can accumulate, forming another source of solid waste which, if not properly disposed, can cause land contamination.

## 2.4 Nuclear Energy

Nuclear power is the use of sustained nuclear fission to generate heat and electricity. Nowadays, nuclear power plants provide around 13 per cent of the world's electricity (IEA, 2011).

Most of the world's reactors are light water reactors (LWR) that use the so called "once-through" open fuel cycle in which the spent nuclear fuel (SNF) from the LWR is sent to interim storage and eventually to waste disposal. Other options are a partly closed fuel cycle that is currently used in countries such as France, Germany and Japan in which the SNF is sent to a reprocessing facility where plutonium is recycled back to LWRs as mixed oxide (MOX) fresh fuel. Finally, there is a specific fast reactor fuel cycle in which the plutonium in SNF is separated and reused to produce fast reactor (FR) fuel. This strategy has been demonstrated but not yet commercially deployed. (MIT, 2011).

The nuclear fuel cycle entails all the stages from the mining and milling of uranium through the manufacture of the fuel, electricity generation in the reactor, transport and reprocessing of the spent fuel and the management of the associated waste in all these steps. The environmental consequences of all these stages are briefly discussed below.

Uranium is mined in either open-pit or underground mines. Then, the ore is processed at mills and the uranium is separated from the rock to get uranium oxide (yellow cake).

Environmental impacts associated with uranium mining are related to impacts on land and water due to waste water arising from mine drainage and from water use in drilling. Also occupational health impacts due to radon exposure are produced in this stage (EC, 1995c).

In the milling process, a high percentage of the radioactivity contained in the ore remains in the mill tailings (fine sands released by the milling plant which contain insoluble radium). Their associated environmental releases are the movement of the contaminated leachate to rivers or groundwater as well as the atmospheric dispersion of fine particles by wind.

From the mill, the yellow cake is converted into uranium hexafluoride ($UF_6$). To fabricate nuclear fuel, the $UF_6$ must be further enriched in the fissile isotope U-235. Enriched $UF_6$ is then transformed into uranium dioxide ($UO_2$) powder and pressed into pellets that are inserted into thin zircaloy or stainless steel tubes which are then sealed and assembled to form fuel assemblies.

The main potential hazard in the conversion stage arises from the toxicity of hydrogen fluorine and fluoride used in the production of $UF_6$. Gaseous and liquid releases of F- can be produced. Atmospheric and liquid releases of uranium isotopes are also produced in this stage (EC, 1995c).

The enrichment process uses a large amount of energy and atmospheric and liquid releases of uranium are produced. As the level of enrichment increases, the risk of criticality accidents also increases although this risk is considered to be very small (El-Hinnawi, 1978). The depleted uranium residue is stockpiled for possible future use. This material gradually produces nuclides such as 226Ra and 222Rn that can generate radiation hazard.

Fabrication of fuel elements is a non-hazardous process although some atmospheric and liquid releases of uranium isotopes and chemicals are produced. Manufacture of mixed oxide fuel is far more hazardous due to the higher toxicity of plutonium and its lower critical mass.

During normal operation of a nuclear reactor, radioactive fission and activation products are originated. Fuel elements retain most of the radioactive materials, but some are produced within or are diffused into the coolant which can later be removed by waste processing systems.

Nuclear power plants use large quantities of water for steam production and for cooling. As previously, some nuclear power plants obtain large quantities of water from a lake or river, and discharge waste heat to the cooling water which could affect fish and other aquatic life.

The solid waste from nuclear plants is mainly composed of spent fuel which is classified as high level waste and stored on site for some time and eventually taken to intermediate storage or reprocessing plants.

At the reprocessing plant the fuel is chemically dissolved and the residual fuel material – uranium and plutonium – can be recovered. Gaseous effluents from a reprocessing plant include gaseous fission products and also tritium, $^{14}$C and some Pu isotopes. Liquid effluents from a reprocessing plant are mainly composed of tritium, $^{14}$C and other minor radio nuclides. Intermediate and low level solid wastes produced by reprocessing plants are mainly structural elements of the fuel elements and can be contaminated with spent fuel.

High-level wastes from reprocessing plants comprise highly-radioactive fission products and some transuranic elements with long-lived radioactivity. They are later vitrified into borosilicate (Pyrex) glass, encapsulated into heavy stainless steel cylinders and stored for eventual disposal deep underground.

Long-term managed storage of the spent fuel preserves future options for its utilization at little relative cost (MIT, 2011). Managed storage can be done safely at operating reactor sites, centralized storage facilities, or geological repositories.

Public concern about reactor operation focuses on the possibility of accident occurrence leading to the release of a considerable amount of radioactivity to the surrounding environment. Various types of accidents are possible during the operation of a nuclear reactor and many safety devices are incorporated in the reactor design. In addition most power reactors are placed inside a containment building.

There have been three significant accidents in the 50-year history of civil nuclear power generation (WNA, 2012):

- Three Mile Island (USA 1979) where the reactor was severely damaged but radiation was contained and there were no adverse health or environmental consequences.
- Chernobyl (Ukraine 1986) where the destruction of the reactor by steam explosion and fire killed 31 people and had significant health and environmental consequences.
- Fukushima (Japan 2011) where three old reactors (together with a fourth) were damaged and the effects of loss of cooling due to a huge tsunami were inadequately contained. The fuel was badly damaged and there were significant off-site radiation releases.

The safety provisions in a nuclear plant include a series of physical barriers between the radioactive reactor core and the environment as well as the provision of multiple safety systems. These safety systems are "active" in the sense that they involve electrical or mechanical operation on command, but in some recent designs the passive or inherent features substitute for active systems. Such a design would have averted the Fukushima accident, where loss of electrical power resulted in a loss of the cooling function.

Another important aspect of the nuclear industry is the transport of radioactive materials. Spent fuel elements are transported in shielded cooled casks to prevent radiation exposure to transport workers. Accidents producing rupture of casks could have severe impacts on the environment. High level wastes are also transported to storage places in containers. Management, packaging, transport, and disposal of waste are strictly regulated and carefully controlled.

Plutonium-239, the isotope used in atom bombs, is produced in relatively high quantities in the nuclear fuel cycle. Environmental hazard of plutonium is related to the risk of the occurrence of a chain reaction if sufficient plutonium came together, with the subsequent emission and dispersion of plutonium to the environment.

## 2.5 Renewable Energies

### 2.5.1 Solar

Solar energy is the most abundant of all energy resources and there is a large family of different solar energy conversion technologies capable of meeting a variety of energy service demands.

Conversion of solar energy to heat (i.e., thermal conversion) is done by specialized techniques and devices such as optical coatings and mirrors. Generation of electricity can be achieved in two ways. Solar energy can be converted directly into electricity in a device called a photovoltaic (PV) cell. Solar thermal energy can also be used in a concentrating solar power (CSP) plant to produce high-temperature heat which is then converted to electricity via a heat engine and generator. Both approaches are currently in use. Furthermore, solar driven systems can deliver process heat and cooling, and other solar technologies are being developed that will deliver energy carriers such as hydrogen or hydrocarbon fuels – known as solar fuels. Large CSP plants may also prove effective for cogeneration to support water desalination (Arvizu et al, 2011; IEA-ETSAP, 2011).

Emissions associated with generating electricity from solar PV technologies are negligible because no fuels are combusted. PV systems do not generate any type of solid, liquid or gaseous byproducts when producing electricity. Also, they do not emit noise or use non-renewable resources during operation. However, two issues are of concern: the emission of pollutants and the use of energy during the full lifecycle of PV manufacturing, installation, operation and maintenance (O&M) and disposal; and the possibility of recycling the PV module materials when the systems are decommissioned.

Energy is required to manufacture and install solar PV components, and any fossil fuels used for this purpose will generate emissions. Most lifecycle GHG emission estimates range from about 30 to 80 g $CO_2$eq/kWh (Arvizu et al, 2011).

Some PV cells make use of scarce and rare materials like indium and tellurium. Large use of these cells would lead to resource depletion (IEA, 1998). Moreover, in its production lines, PV industry uses some toxic, explosive gases, GHGs, as well as corrosive liquids.

Photovoltaic systems do not require the use of any water to create electricity. However, in certain locations, periodic cleaning of the PV panels is required to maintain performance, resulting in non-negligible water requirements.

Construction and operation of PV systems can cause land use impacts on natural ecosystems. Main impacts are produced during construction activities and can be mitigated by minimizing earth movements and helping to reestablish the previous biodiversity (IEA, 1998).

Visual impacts of PV installations depend on the type of scheme. Centralized large schemes would have significant visual impacts while small scale roof mounted schemes would have much lower visual impact.

Decommissioning of PV systems may also cause some environmental impacts especially in the case of CdTe modules due to the high toxicity of cadmium. Recycling the material in PV modules is already economically viable, mainly for concentrated and large-scale applications (McDonald and Pearce, 2010; Arvizu et al, 2011).

For CSP plants, the environmental consequences vary depending on the technology. Life cycle GHG emissions estimates fall between 14 and 32 g $CO_2$eq/kWh (Arvizu et al, 2011). Main life cycle impacts are associated with the construction of the steel-intensive infrastructure for solar energy collection due to mineral and fossil resource consumption, as well as discharge of pollutants related to today's steel production technology (Lechón et al, 2009). However, most CSP solar field materials can be recycled and reused in new plants (SolarPACES, 2008).

The deployment of CSP can also cause some unintended environmental impacts. The main impacts are impacts on amenity and relate to the large area required for the technology. The main impacts identified are the following:

- Visual impacts: these impacts can be significant due to the large area occupied by the mirrors. Visual effects are most noticeable in tower CSP plants where very bright points appear in the rural landscape. An advantage is that CSP plants are often located in areas with limited amenity or aesthetic value (EASAC, 2011).
- Noise: noise will be generated from the steam generating plant (IEA, 1998).
- Ecological impacts due to land use: CSP plants have large land use requirements. To date, most sites are in arid or semiarid areas which have fragile soil and plant communities. Consequently, there could be an important risk of soil erosion and habitat loss due to CSP plants installations that could be significant in ecologically important areas (IEA, 1998). Mortalities caused to vertebrates are the main concern in respect of the local environmental impact of CSP. Direct mortalities take place by collision with top mirrors and buildings (the tower in particular), and heat shock or burning damage in the concentrated light beams plants (EASAC, 2011). Massive establishment of solar plants in an area may affect regional animal

or plant populations by cutting dispersion routes and partially isolating populations from each other (EASAC, 2011).
- If a plant is built on former agricultural land, available nutrients in the soil may facilitate growth of vegetation up to 1 m in height below and between solar collectors. The vegetation can dry up and contribute to fire risk (EASAC, 2011).
- Water resources impacts: CSP plants using conventional steam plants to generate electricity have a requirement for cooling water. This could pose a strain on water resources in arid areas (IEA, 1998). Because solar abundance and fresh water constraints often coincide geographically, the cumulative impacts of installing numerous CSP plants in a region raises policy concerns. Consequently, the trend is towards more fresh water efficient cooling technologies (Carter and Campbell, 2009). Water is also used for cleaning the mirrors to maintain their high reflectivity, although water use for cleaning is typically a factor of a hundred lower than that used for water cooling. For areas with high irradiation and available land close to the sea, using salt water for cooling could be an attractive option (EASAC, 2011).

Most of these impacts are local and are therefore highly affected by the siting of the technology. Consequently some of them can be minimized by a sensitive siting choice.

### 2.5.2 Geothermal
Geothermal resources consist of thermal energy from the Earth's interior stored in both rock and trapped steam or liquid water. Accessible geothermal energy from the Earth's interior can supply heat for direct use, can be used in combined heat and power applications as well as to generate electric energy.

Geothermal resources include basically low-enthalpy fields, which have long been used for direct heating applications and high quality high-enthalpy, which are used for power generation. In general, high-enthalpy geothermal fields are only available in areas with volcanic activity, whereas the rest of the fields are low- or medium-enthalpy resources (IEA-ETSAP, 2010c).

Currently, the basic types of geothermal power plants in use are steam condensing turbines and binary cycle units. Steam condensing turbines can be used in steam plants operating at sites with intermediate- and high-temperature resources ($\geq 150°C$). Binary-cycle plants are commonly installed to extract heat from low- and intermediate-temperature geothermal fluids (generally from 70 to 170°C), from hydrothermal- and EGS-type reservoirs.

## Environmental implications of energy production 357

Newer geothermal plants in New Zealand and Hawaii are combined-cycle or hybrid plants that provide relatively high electric efficiency (IEA-ETSAP, 2010c).

Emissions associated with generating electricity from geothermal technologies are negligible because no fuels are combusted. However, geothermal fluids contain minerals leached from the reservoir rock and variable quantities of gas, mainly $CO_2$ and a smaller amount of hydrogen sulphide (Goldstein et al, 2011).

$CO_2$ emissions from geothermal power plants range from 4 to 740 g $CO_2$/kWh with an average of 122 g $CO_2$/kWh (Goldstein et al, 2011). According to the published LCA studies of geothermal electricity, life cycle GHG emissions, including the emission from constructing, operating and decommissioning the power plants, are lower than 80 g$CO_2$equiv/kWh (Goldstein et al, 2011), value that compares very favorably to GHG emissions from fossil sources.

Apart from $CO_2$, geothermal fluids can, depending on the site, contain a variety of other minor gases, such as hydrogen sulphide ($H_2S$), hydrogen ($H_2$), methane ($CH_4$), ammonia ($NH_3$) and nitrogen ($N_2$). Mercury, arsenic, radon and boron may also be present. $H_2S$ is toxic, but rarely found in sufficient concentration to be harmful. $CH_4$ is also present in small concentrations.

Geothermal energy applications can have also some detrimental environmental impacts. Concerns are related to impacts on water use and quality. Geothermal power plants usually re-inject the hot water that they remove from the ground back into wells. However, a small amount of water used by geothermal plants in the process of creating electricity may evaporate and therefore not be returned to the ground. Also, for those geothermal plants that rely on hot, dry rocks for energy, water from local resources is needed to extract the energy from the dry rocks (US-EPA, 2012).

Geothermal power plants can possibly cause groundwater contamination when drilling wells and extracting hot water or steam. However, this type of contamination can be prevented with proper management techniques. In addition, geothermal power plants often re-inject used water back into the ground (through separate wells) instead of discharging the used water into surface waters. This prevents underground minerals or pollutants (boron and arsenic) from being introduced into surface waters (US-EPA, 2012).

Scrubbers reduce air emissions but produce a watery sludge high in sulphur and vanadium, a heavy metal that can be toxic in high concentrations. Additional sludge is generated when hydrothermal steam is condensed, causing the dissolved solids to precipitate out. This sludge is generally high in silica compounds, chlorides, arsenic, mercury, nickel, and other toxic heavy metals). Usually the best disposal method is to inject liquid wastes or redissolved solids back into a porous stratum of a geothermal well.

Moreover, drilling activities, production and reinjection of fluids can produce pressure and temperature changes in the fields that can lead to geo-mechanical stress that can in turn cause micro earthquakes, subsidence and steam eruptions.

Despite geothermal power plants typically requiring less land use than fossil fuel power plants, this is still an issue since new projects or power plants are often located within or adjacent to national parks or tourist areas.

### 2.5.3 Bioenergy

Bioenergy includes different sets of technologies for applications in various sectors. In 2009, biomass provided about 10 per cent of the annual global primary energy supply (IEA, 2011) from low efficiency traditional biomass use to high-efficiency modern bioenergy uses.

There is a wide variety of commercial bioenergy routes that starting with feedstocks – such as forest or energy crops or industrial, commercial or municipal waste streams and by-products – deliver electricity or heat, biogas and liquid biofuels. Some of these routes are already commercial and other are still in development stage (Chum et al, 2011; IEA Bioenergy, 2009).

As far as power generation is concerned, there are different technological possibilities: biomass can be used in medium-size power plants by way of mono-combustion as well as by way of co-combustion together with fossil fuels such as hard coal in large power plants. Another option is the gasification of biomass with subsequent combustion technologies with gas and steam turbines (IGCC) which allows efficient power and heat production (Gärtner, 2008).

Sustainably managed bioenergy can provide a number of beneficial effects (IEA Bioenergy, 2009; Royal Society, 2008) depending on the former use of the land where bioenergy is implanted, such as:

- enhanced biodiversity;
- soil carbon increases and improved soil productivity;
- reduced shallow landslides and local flash floods;
- reduced wind and water erosion;
- reduced sediment volume and nutrients transported into rivers; and
- improvements in growth and productivity and reduction of wildfire risk in forests.

Some options for a sustainable bioenergy production have been identified (Bringezu et al, 2009; Tilman et al, 2009; Ravindranath et al, 2009) and include:

- increasing yields and optimizing agricultural production (including double crops and mixed cropping systems);
- restoring formerly degraded land;
- use perennial crops;
- use sustainably harvested wood and forest residues;
- use of waste and production residues;
- cascading use of biomass.

The main environmental advantage of bioenergy is the fact that it is a part of the terrestrial carbon cycle. The $CO_2$ emitted due to bioenergy use was earlier sequestered from the atmosphere and will be sequestered again if the bioenergy system is managed sustainably. Thus, in a sustainable fuel cycle, there would be no net emissions of carbon dioxide, although some fossil-fuel inputs may be required for planting, harvesting, transporting, and processing biomass.

If biomass wastes such as crop residues or municipal solid wastes are used for energy, there should be few or no net greenhouse gas emissions. However, under some circumstances, GHG emissions from bioenergy can be very large. Drainage of peatlands or replacing tropical forests for growing energy crops can lead to big amounts of GHG emissions (Royal Society, 2008; Fargione et al, 2008; Page et al, 2011).

Diversion of crops or land into bioenergy can produce direct or indirect land use change (LUC or ILUC) mobilizing C stocks and emitting additional amounts of $CO_2$ that can in some cases offset the expected GHG emissions savings (De Cara et al, 2012; Laborde, 2011; EEA, 2011; Tilman et al, 2009; Melillo et al, 2009; Fargione et al, 2008).

Agricultural practices and specially the production and application use of fertilizers, give rise to important life cycle GHG and other pollutant emissions. $N_2O$ emissions arising after application of N fertilizers and from the N present in crops residues are of special importance. These emissions vary considerably with environmental and management conditions, including soil water content, temperature, texture, carbon availability, and therefore are difficult to assess.

A global greenhouse gas emissions policy that protects forests and encourages best practices for nitrogen fertilizer use can dramatically reduce direct and indirect GHG emissions associated with biofuels production (Melillo et al, 2009).

Like conventional agriculture and forestry systems, bioenergy can exacerbate soil and vegetation degradation associated with overexploitation of forests, too intensive crop and forest residue removal as well as water overuse. Water requirements of energy derived from biomass are about 70 to 400 times more than other energy carriers such as fossil fuels, wind and solar (Gerbens-Leenes at al, 2009). Therefore, large scale expansion of energy crop production could lead to a large increase in water use and in some countries exacerbating an already existing stressed water situation. Possible consequences of water overuse could be lowered groundwater levels, river depletion, and reduced downstream water availability (IEA Bioenergy, 2009). Several options to reduce these impacts exist including using less water demanding crops, increasing water use efficiency and managing water for multifunctionality (De Fraiture and Berndes. 2009). Bioenergy based on wastes and by-products of the food and forestry sector would also avoid these problems.

Feedstock cultivation can lead to leaching and emission of nutrients that increase eutrophication of aquatic ecosystems (Simpson, et al, 2009; Royal Society, 2008) and pesticide emissions to water bodies may also negatively impact aquatic life (Sala et al, 2009).

Increased biomass output for bioenergy can directly impact wild biodiversity through conversion of natural ecosystems into bioenergy plantations or through changed forest management (IEA Bioenergy, 2009; Royal Society, 2008). Habitat and biodiversity loss may also occur indirectly. Eikhout et al (2008) showed that, in most cases, the long-term positive effect of greenhouse gas reductions from biofuel production on biodiversity are not enough to compensate for biodiversity losses from land use change, at least not within a time frame of several decades. However, when crops are grown on degraded or abandoned land, such as previously deforested areas or degraded crop- and grasslands, the production of feedstocks for bioenergy could have positive impacts on biodiversity (Tilman et al, 2009).

Biodiversity impacts from the use of genetically modified energy crops could also arise (IEA Bioenergy, 2009; Firbank, 2008).

Several processes exist to convert feedstocks and raw materials into biofuels for use in transport. First-generation biofuels are produced through processes such as cold pressing/extraction, transesterification, hydrolysis and fermentation, and chemical synthesis, derived from sources such as starch, sugar, animal fats, and vegetable oil. Second-generation biofuels are produced through more advanced processes, including hydro treatment, advanced hydrolysis and fermentation as well as gasification and synthesis from ligno-cellulosic sources. Environmental burdens associated with these processes are related to the use of energy and chemicals.

Combustion of biomass and biomass-derived fuels to generate electricity and heat produces air pollution including carbon monoxide, nitrogen oxides, and particulates such as soot and ash. The amount of pollution emitted per unit of energy generated varies widely by technology. Compared to coal and oil stationary applications, sulphur dioxide ($SO_2$), toxic metals (cadmium, mercury and others) and nitrous oxide (NOx) emissions from bioenergy applications are mostly lower. Pollutant formation may occur due to incomplete combustion that can lead to high emissions of unburnt pollutants such as CO, soot, and PAH. Although improvements to reduce these emissions have been achieved by optimized furnace design, there is still a relevant potential of further optimization. Pollutants such as NOX and particles are formed as a result of fuel constituents. Air staging or fuel staging technologies can reduce NOx emissions by 50–80 per cent. As for other combustion technologies, if combined with secondary abatement measures such as selective catalytic reduction, reductions of up to 95 per cent can be achieved (Nussbaumer, 2003). The most serious problem is their particulate emissions, which must be controlled with special devices in place.

Raw municipal waste often contains toxic metals, chlorinated compounds, and plastics, which generate harmful emissions.

Carbon capture and storage (CCS) from fossil-fuelled power plant flue gases is being considered as a measure to reduce greenhouse gas emissions. In this context, CCS can also be applied to co-firing plants, which would enable the capture of carbon from biomass (biotic CCS), resulting in a net negative carbon emission or carbon sink associated with biomass combustion (IEA Bioenergy, 2009).

Biomass power plants require the use of water, because the boilers burning the biomass need water for steam production and for cooling. Whenever any type of power plant removes water from a lake or river, fish and other aquatic life can be killed, which then affects those animals and people that depend on these aquatic resources.

As is the case with fossil fuel power plants, biomass power plants have pollutant build-up in the water used in the boiler and cooling system. Pollutants in the water and the higher temperature of the water can harm fish and plants in the lake or river where the power plant water is discharged.

Biomass burning in boilers creates a solid waste called ash, which despite having extremely low levels of hazardous elements, must be properly disposed.

### 2.5.4 Wind

The use of wind energy requires that the kinetic energy of moving air be converted to mechanical and then electrical energy using wind turbines.

On shore wind turbines have been increasingly installed around the world providing nowadays 1 per cent of the world electricity (IEA, 2011). Offshore wind energy technology is less mature than onshore being the primary motivation to its development to provide access to additional and high quality wind resources.

Emissions associated with generating electricity from wind technology are negligible because no fuels are combusted. However, emissions are produced in other parts of the life cycle both upstream and downstream. The most significant are those associated with the processing and manufacturing of materials and components of the wind turbine (IEA, 1998; EWEA, 2009). The majority of lifecycle GHG emission estimates cluster between about 8 and 20 g $CO_2$eq/kWh (Wiser et al, 2011).

Wind power plants generally require the use of large areas of land, using space that might otherwise be used for other purposes. Special concern has been raised recently about wind farm installations in peat land (Nayak et al, 2010). Drainage of peat land produces the mobilization of the C stored in the soil and also other impacts such as erosion and mass movements.

Provided wind turbines are carefully located, away from scenic, tourism or highly populated areas, they tend to have low and localized impacts on the environment. However, sensitivity to some of these impacts may cause public opposition and make a project unacceptable (IEA, 1998).

Large wind farms pose aesthetic concerns. The dominant visual effect is the intrusion of the turbines and associated equipment. Wind turbines are highly visible structures often deployed in arrays of many machines on ridges or hilltops. Moreover, wind turbines and power plants have grown in size, making the turbines and related transmission infrastructure more visible (Wiser at al, 2011). Opposition to wind energy on the grounds of loss of visual amenity is one of the biggest problems in the wind industry (EC, 1995d) and concerns have risen for both on- and offshore wind energy (Ladenburg, 2009; Haggett, 2011).

Wind turbines that are improperly installed or landscaped may create soil erosion problems (US-EPA, 2012; NWCC, 1997). These problems can arise in deserts when the soil surface is disrupted during wind tower construction or in hilly environments when towers are located on ridgelines where soil-holding vegetation is removed. Access roads to wind turbine farms disrupt tracts of land and support equipment litters the surroundings.

Wind farms can also have noise impacts mostly due to the operating turbines (EC, 1995d). Possible impacts can be both audible and sub-audible. Sub-audible sound may cause health effects but little evidence of this effect has yet been found (Wiser at al, 2011). Regarding audible noise from turbines, it can be mechanical noise (produced by the gear boxes and generators) and aerodynamic noise created by the rotation of the blades. This audible noise can impact sleep patterns and well-being (Wiser at al, 2011). New blade designs are being used to reduce the amount of noise.

Bird and bat mortality has been an issue at some wind farms. The types of risks that may affect birds are mainly collision with turbines, habitat disturbance and habitat loss as well as interference in birds' movements between feeding, wintering, breeding and moulting habitats (EWEA, 2009). As offshore wind energy has increased, concerns have also been raised about seabirds (Wiser et al, 2011). The impact of wind power plants on bat populations is of particular concern, because bats are long-lived and have low reproduction rates (Wiser at al, 2011). Nevertheless, the number of bird fatalities at existing wind power plants appears to be orders of magnitude lower than other anthropogenic causes of bird deaths (Erickson et al, 2005).

Wind power plants can also cause habitat and ecosystem modification impacts on flora and fauna including avoidance of or displacement from an area, habitat destruction and reduced reproduction (EWEA, 2009).

The impacts of wind power plants on marine environment are also an issue for offshore wind farms. Constriction and operation of offshore wind farms can affect marine mammals mainly due to stress and increased diseases vulnerability caused by noise (EWEA, 2009). Offshore wind turbines foundations can also affect benthos-organisms living in the sediments at the sea bottom, through changes in the substrate and subsequent changes in the species composition of the benthos and the existing biodiversity in the area (EWEA, 2009; Köller et al 2006).

Impacts of wind energy on water are not very significant. In areas with little rainfall, wind farms may require the use of a small amount of water. If rainfall is not sufficient to keep the turbine blades clean, water is used to clean dirt and insects off the blades so that turbine performance is not reduced. Wind turbines do not discharge any water while producing electricity.

Finally, the use of scarce raw materials may also be an issue if we consider the use of neodynium in high performance magnets used in wind turbines generators (Köller et al, 2011).

## 2.5.5 Hydro

Hydropower is generated from water moving in the hydrological cycle from higher to lower elevations. Hydropower plants are often classified in three main categories according to operation and type of flow: Run-of-river (RoR), storage (reservoir) and pumped storage HPPs. A RoR HPP draws the energy for electricity production mainly from the available flow of the river. In hydropower projects with a reservoir, the generating stations are located at the dam or further downstream connected to the reservoir through tunnels or pipelines. In pumped storage plants water is pumped from a lower reservoir into an upper reservoir during off-peak hours and the flow is reversed to generate electricity during the daily peak load period.

Dams are constructed for several purposes and have several benefits for society. While the main one is water supply for domestic or industrial uses, other key purposes and benefits include: irrigation for agriculture, flood control, inland navigation and recreation (ICOLD, 1999).

Most of the impacts of the hydro fuel cycle are direct burdens on aquatic and terrestrial ecosystems.

IEA (1998) has identified several sensitive issues that need to be carefully assessed and managed to achieve sustainable hydropower projects. Some of these aspects are:

- Hydrological regimes: a hydropower project may modify a river's flow regime if the project includes a reservoir which significantly affects natural aquatic and terrestrial habitats in the river and along the shore.
- Reservoir: the creation of a reservoir transforms a terrestrial ecosystem into an aquatic one and a fast-flowing water course into an artificial lake generating some impacts such as erosion in the reservoir shorelines, changes in fish habitats and species compositions, etc.
- Water quality in the reservoirs: in the deeper anaerobic layers of large reservoirs many biochemical reactions may take place leading to a change in the nutrient balance of the reservoir and eutrophization.
- Increased sedimentation due to reduction in the sediment carrying capacity of the water body.
- Biological diversity effects due to the permanent loss of habitats, fluctuating water levels, downstream floods, introduction of exotic species and obstacles to fish migration.
- Public health impacts through increases in waterborne diseases.

Life cycle greenhouse gas emissions of hydropower plants are very variable depending on the technology and on the special characteristics of the location of the power plant. The main emissions of atmospheric pollutants associated with hydro schemes arise from the manufacturing and construction of the generation and transmission equipment including the dam (IEA, 1998). The majority of lifecycle GHG emission estimates for hydropower cluster between about 4 and 14 g $CO_2$eq/kWh, but under certain scenarios there is the potential for much larger quantities (more than 160 g/kWh) of GHG emissions (Kumar et al, 2011).

Hydropower creates no direct atmospheric pollutants or waste during operation, and GHG emissions associated with most lifecycle stages are minor (EC, 1995d). However, under certain conditions, $CO_2$ and methane ($CH_4$) emissions from reservoirs from the degradation of flooded vegetation might be substantial. Research suggests that emission levels in cold and temperate climates are generally low, and that elevated emissions may be observed in some tropical systems with persistent anoxia (UNESCO/IHA,2009). Decommissioning of the power plant can cause also important GHG emissions from the silt collected over the life of the plant.

There is evidence that the presence of massive bodies of water has influenced the geological stability in the local region around dams increasing the occurrence of earthquakes (IEA, 1998). The presence of large water bodies will affect the local climate, with higher humidity and fog formation in temperate climates.

Lastly, large dams can also have visual intrusion impacts (IEA, 1998) which are site specific.

### 2.5.6 Ocean energy

The ocean energy resource comes from six distinct sources:

- Waves
- Tidal
- Tidal currents
- Ocean currents
- Ocean Thermal Energy Conversion (OTEC)
- Salinity gradients (osmotic power).

Due to the scarce deployment and operation of ocean energy technologies there is little information regarding their anticipated environmental impacts.

Ocean energy does not directly emit $CO_2$ during operation. Life cycle emissions of GHG gases and other pollutants arise in the stages of construction, maintenance and decommissioning of the installation, including the emissions produced in the processes of manufacturing all the necessary components and in the extraction of all required raw materials. Lifecycle GHG emissions quantifications from wave and tidal energy systems are scarce but from the available studies (Sorensen and Naef, 2008; Lewis et al, 2011) they seem to be less than 23 g $CO_2$eq/kWh, with a median estimate of 8 g $CO_2$eq/kWh for wave energy.

A description of potential local environmental effects is given by Boehlert and Gill (2010) and Lewis et al (2011). Negative effects may include:

- reduction in visual amenity;
- loss of access to space for competing users;
- noise during construction;
- noise and vibration;
- electromagnetic fields;
- disruption to biota and habitats;
- water quality changes and possible pollution;
- water salinity and sediment movements in estuaries;
- moving parts may harm marine life;
- changes in the regional properties (temperature and nutrient characteristics) of seawater.

# 3 METHODS TO QUANTIFY ENVIRONMENTAL IMPACTS

### 3.1 Life Cycle Assessment

Life Cycle Assessment is a method for systematic analysis of environmental performance from a cradle to grave perspective. This analytic tool systematically describes and assesses all flows that enter into the studied systems from nature and all those flows that go out from the systems to nature, all over the life cycle.

The interest in LCA started in the 1990s and since then a strong development has occurred. The practice of LCA is regulated by the international standard ISO 14040 and 14044 (ISO, 2006a, 2006b), and there are several introductions (Guinée et al., 2002; JRC IES, 2010) and databases (Ecoinvent, 2007) available.

LCA is a robust and mature methodology although some aspects are still under development. A thorough review of the recent advances of the methodology can be found in Finnveden et al (2009).

A complete LCA study consists of four steps:

1. Definition of the goal and scope of the study;
2. Life cycle inventory (LCI phase) where the collection of all the environmental inflows and outflows takes place;
3. Life cycle impact assessment (LCIA) phase where the emissions and resource data collected in the former phase are translated into indica-

tors that reflect environmental and health pressures as well as resource scarcity; and
4. Interpretation of the results.

However, it is quite common that some LCAs only perform the inventory analysis delivering a list of emissions or only evaluate some of the impacts (like global warming impacts).

The purpose of LCIA is to provide additional information to help assess the results of the LCI in order to better understand their environmental significance. LCIA translates the inventory into potential impacts on the "areas of protection" that are: human health, natural environment and the man made environment (Udo de Haes et al, 2002). LCIA attempts to model any impact from the product system that can be expected to damage one or more areas of protection. In this sense, LCIA addresses toxic impacts from air pollution and also other impacts associated with emissions (global warming, stratospheric ozone depletion, acidification, photochemical ozone and smog formation) and waterborne effluents (eutrophization) as well as the environmental impacts of land and water use, noise, radiation and depletion of resources (Finnveden et al (2009).

### 3.2 Externalities Assessment

As explained in the introductory section of this chapter, all power generation technologies have some associated externalities – costs imposed on individuals or the community that are not paid for by the producer or consumer of electricity.

The most relevant project dealing with the determination of the externalities of energy is the European ExternE project (www.externe.info, accessed February 2013) which was launched in 1991 by the European Commission and the US Department of Energy. Since then, the European Commission has continuously supported this research field through several projects. The latest of these projects is the NEEDS Project (New Energy Externalities Development for Sustainability, www.needs-project.org/, accessed February 2013).

The ExternE methodology is widely accepted and considered as the world reference in the field by the scientific community. The quantification of the external costs is based on the "impact pathway" (IPA) methodology which was developed in the series of ExternE projects, and has been further improved in the NEEDS projects and other related projects like the EU CASES project (www.feem-project.net/cases, accessed February 2013). The impact pathway methodology aims at modelling the causal relationships from the emission of a pollutant to the impacts produced on various receptors through the transport and chemical conversion of this pollutant in the atmosphere.

The principal steps of an IPA can be grouped as follows (EC, 2005):

- Emission: specification of the relevant technologies and pollutants, e.g. kg of oxides of nitrogen (NOx) per GWh emitted by a power plant at a specific site).
- Dispersion: calculation of increased pollutant concentrations in all affected regions, e.g. incremental concentration of ozone, using models of atmospheric dispersion and chemistry for ozone formation due to NOx.
- Impact: calculation of the dose from the increased concentration, followed by calculation of impacts (damage in physical units) from this dose, using a dose response function, e.g. cases of asthma due to this increase in ozone.
- Cost: economic valuation of these impacts, e.g. multiplication by the cost of a case of asthma.

In practice, ExternE uses LCA in combination with IPA (impact pathway analysis) to get a complete assessment of external costs due to energy production, including impacts that occur upstream and downstream of the power plant itself.

Main receptors of the impacts are human health, crops, ecosystems and materials and welfare losses produced by these impacts are assessed using economic valuation methods.

Impacts categories, pollutants and effects considered in the ExternE methodology are summarized in Table 16.1.

Seven major types of damages have been considered in ExternE. The main categories are human health (mortality and morbidity effects), effects on crops and materials as well as global warming.

Global warming impacts assessment is subject to a very high degree of uncertainty. A good review of the literature on the economic impacts of climate change can be found in Tol (2009). Within NEEDS, the model FUND 3.0 was used to estimate the marginal external costs of GHG emissions (Anthoff, 2007). Results greatly differ depending on the assumptions regarding some very influencing parameters like discounting and equity weighting. Two sets of externals costs factors were used in NEEDS trying to reflect these uncertainty (Preiss and Friedrich, 2009).

Table 16.1  Impact categories, pollutants and effects considered in the ExternE methodology

| Impact category | Pollutant | Effects |
|---|---|---|
| Human health: Mortality | $PM_{10}$, $SO_2$, $NO_x$, $O_3$ | Reduction in life expectancy |
| | As, Cd, Cr, Ni | Cancer |
| | Accident risk | Fatality risk from traffic and workplace |
| Human health: Morbidity | $PM_{10}$, $O_3$, $SO_2$ | Respiratory hospital admissions |
| | $PM_{10}$, $O_3$ | Restricted activity days |
| | $PM_{10}$, CO | Congestive heart failure |
| | $PM_{10}$ | Cerebro-vascular hospital admissions |
| | | Cases of chronic bronchitis |
| | | Cases of chronic cough in children |
| | | Cough in asthmatics |
| | | Lower respiratory symptoms |
| | Pb | Neurotoxicidad |
| | $O_3$ | Asthma attacks |
| | | Symptom days |
| | Benzene, Benzo-[a]-pyrene 1,3-butadiene Diesel particles | Cancer risk (non-fatal) |
| | Noise | Myocardial infarction |
| | | Angina pectoris |
| | | Hypertension |
| | | Sleep disturbance |
| | Accident risk | Risk of injuries from traffic and workplace accidents |
| Building materials | $SO_2$ Acid deposition | Ageing of galvanized steel, limestone, mortar, sand-stone, paint, rendering, and zinc for utilitarian buildings |
| | Combustion particles | Soiling of buildings |
| Crops | $NO_x$, $SO_2$ | Yield change for wheat, barley, rye, oats, potato, sugar beet |
| | $O_3$ | Yield change for wheat, barley, rye, oats, potato, rice, tobacco, sunflower seed |
| | Acid deposition | Acid deposition Increased need for liming |

*Table 16.1* (continued)

| Impact category | Pollutant | Effects |
|---|---|---|
| Global warming | $CO_2$, $CH_4$, $N_2O$, N, S | Worldwide effects on mortality, morbidity, coastal impacts, agriculture, energy demand, and economic impacts due to temperature change and sea level rise |
| Ecosystems | Acid deposition Nitrogen deposition | Acidity and eutrophication (avoidance costs for reducing areas where critical loads are exceeded) |

*Source:* EC, 2005.

## 4 SOME RESULTS

In recent years, remarkable progress has taken place in analysing environmental impact and externalities of electricity generation thanks to several major projects (EC, 2005; NEEDS project www.needs-project.org/; www.feem-project.net/cases, accessed February 2013). LCA methodology has also been extensively used to assess mainly GHG emissions but also other emissions from the production and use of several biofuels of different origins compared to conventional fuels. A good review of published studies can be found in Menichetti and Otto (2009).

Some results on life cycle GHG emissions from several electricity generation technologies are shown in Figure 16.2. Results show that fossil technologies, especially coal and lignite fired power plants, produce the largest emissions. Results also show that non-combustion RE technologies and nuclear power cause comparatively minor emissions, only from upstream and downstream processes. Among these technologies solar PV power plants are the largest emitters.

Figures 16.3 and 16.4 show other pollutants' life cycle emissions of electricity generation technologies. Fossil fuel technologies and also biomass technologies produce the highest emissions of NOx and $SO_2$ followed by solar PV technologies. Oil and gas technologies produce the highest amounts of NMVOCs emissions, followed by biomass technologies. In the case of particulates, solar PV technologies show the worst results.

Results of GHG emissions from the production of biofuel – bioethanol

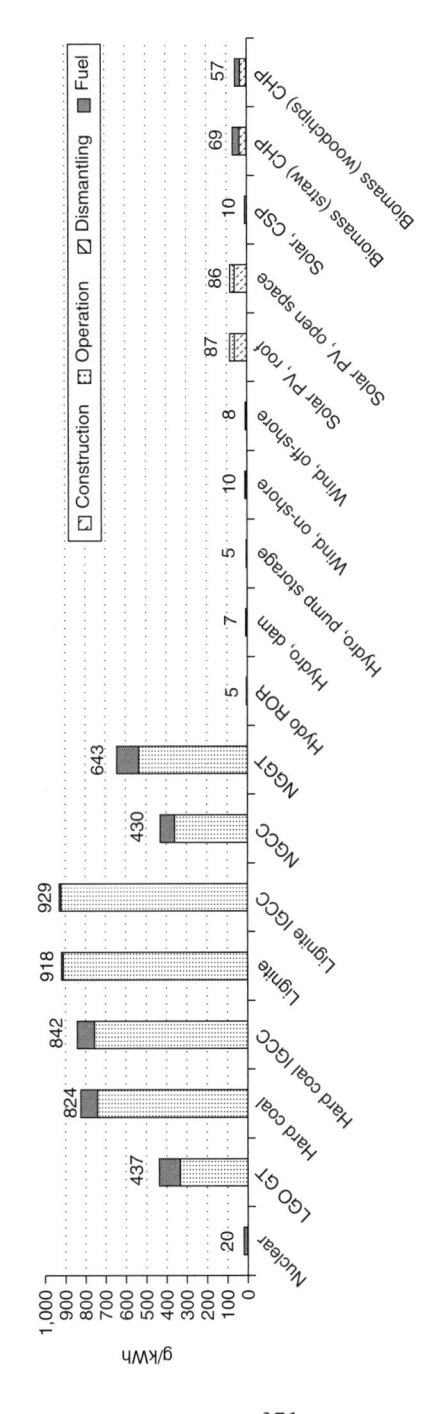

*Figure 16.2  Lifecycle GHG emissions of electricity generation technologies disaggregated by life cycle stage*

*Source:* CASES project (CASES, n.d.).

*Source:* CASES project.

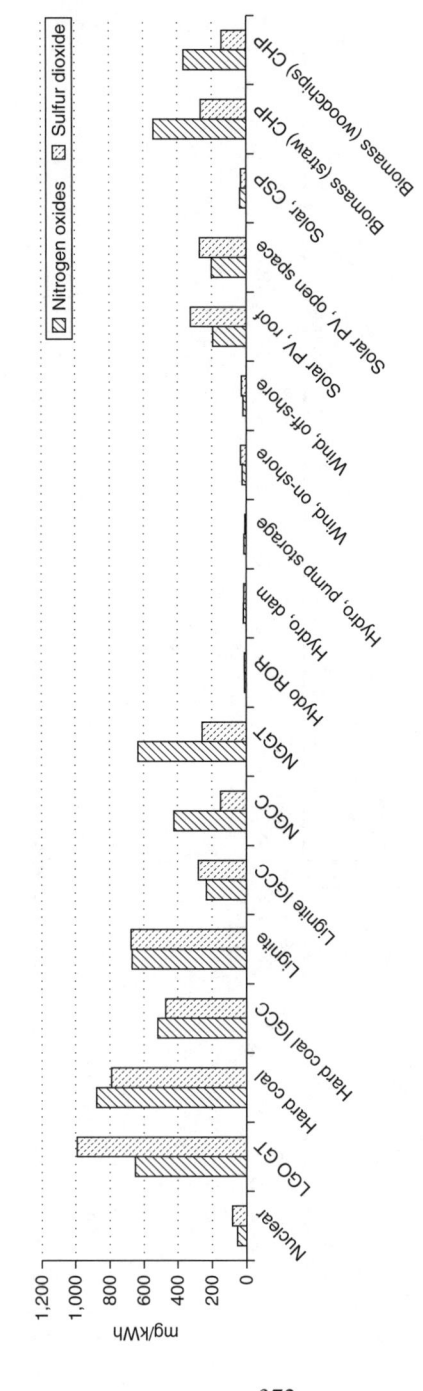

*Figure 16.3 Lifecycle NOx and SO₂ emissions of electricity generation technologies*

*Source:* CASES project.

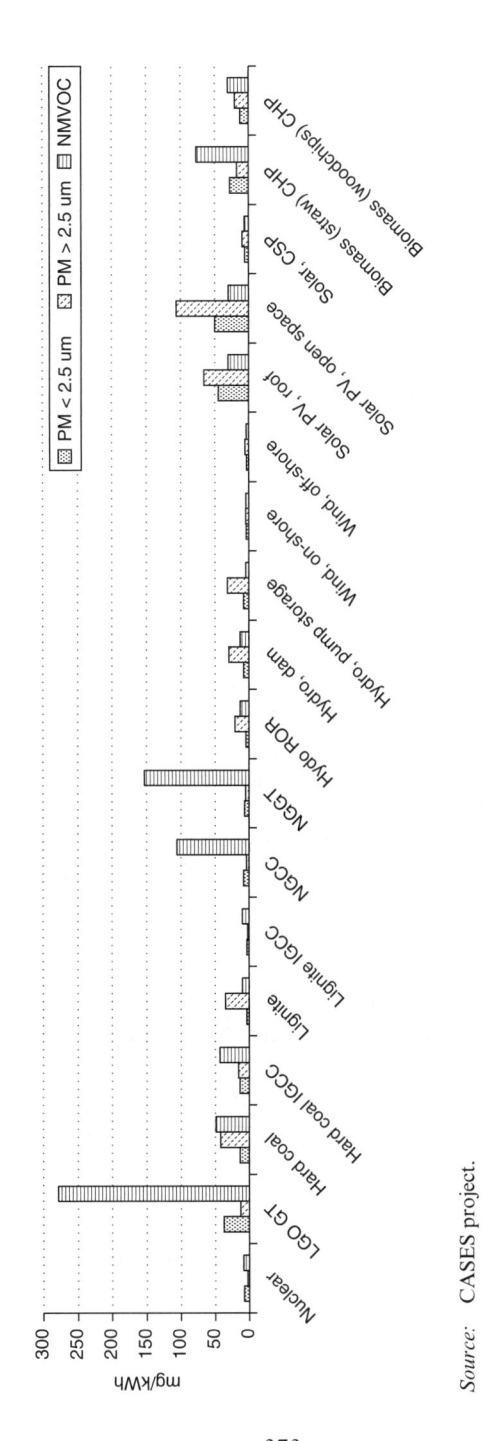

*Figure 16.4 Lifecycle particulate and NMVOC emissions of electricity generation technologies*

and biodiesel – obtained by different authors in the literature are shown in Figures 16.5 and 16.6. Dotted line on the graphs shows the reference emissions for fossil fuels. In general, results show GHG emissions savings for the vast majority of raw materials analysed, although some potentially important issues such as LUC or ILUC effects are not considered in the majority of the assessments. In the case of bioethanol production, sugar based raw materials are better than starch based ones with sugar cane showing the best results. In the case of biodiesel, the observed variability in the results is quite remarkable, especially in the case of palm oil biodiesel. Biodiesel from recycled oil followed by sunflower show the best results with a high emission saving potential. However, as mentioned before, the consideration of ILUC associated GHG emissions can have an important impact on these results (Laborde, 2011).

Results for damage costs per kWh obtained in the CASES project are shown in Figures 16.7 and 16.8 for various electricity generation technologies. As can be seen in those figures, results demonstrate that, compared to fossil fuel technologies – especially oil and coal fired generation – renewable energies have lower damages.

Fossil fuel technologies have external costs above 1.4 eurocent/kWh reaching more than 3 eurocent/kWh in the case of coal power plants. These costs are dominated by global warming impacts in the case of coal, lignite and natural gas and by health effects in the case of oil. Among fossil technologies, the ones with higher efficiencies have correspondingly less external costs per kWh.

Despite nuclear energy external costs appearing as very low in these calculations, it must be taken into account that they do not consider the effects of a possible nuclear accident or the effects on future environment and society of the possible accidental release of the nuclear waste that has been disposed of (Lecointe et al, 2007).

Among renewable technologies, biomass technologies show the highest external costs dominated by health and biodiversity effects while run of river hydro technologies show the lowest costs among all the technologies.

Among solar technologies, solar PV technologies have sensibly higher external costs than CSP and these costs are dominated by the health effects arising from the emissions originated by the energy requirements of the upstream processes related to the production of silicon and PV wafers (Frankl et al, 2006). Foreseen improvements in energy consumption and even better efficiencies for this technology would reduce the external costs significantly.

External costs of fossil fuel power plants are dominated by impacts produced during operation of the power plants while renewable energies have the most of their external costs associated with the construction stage of

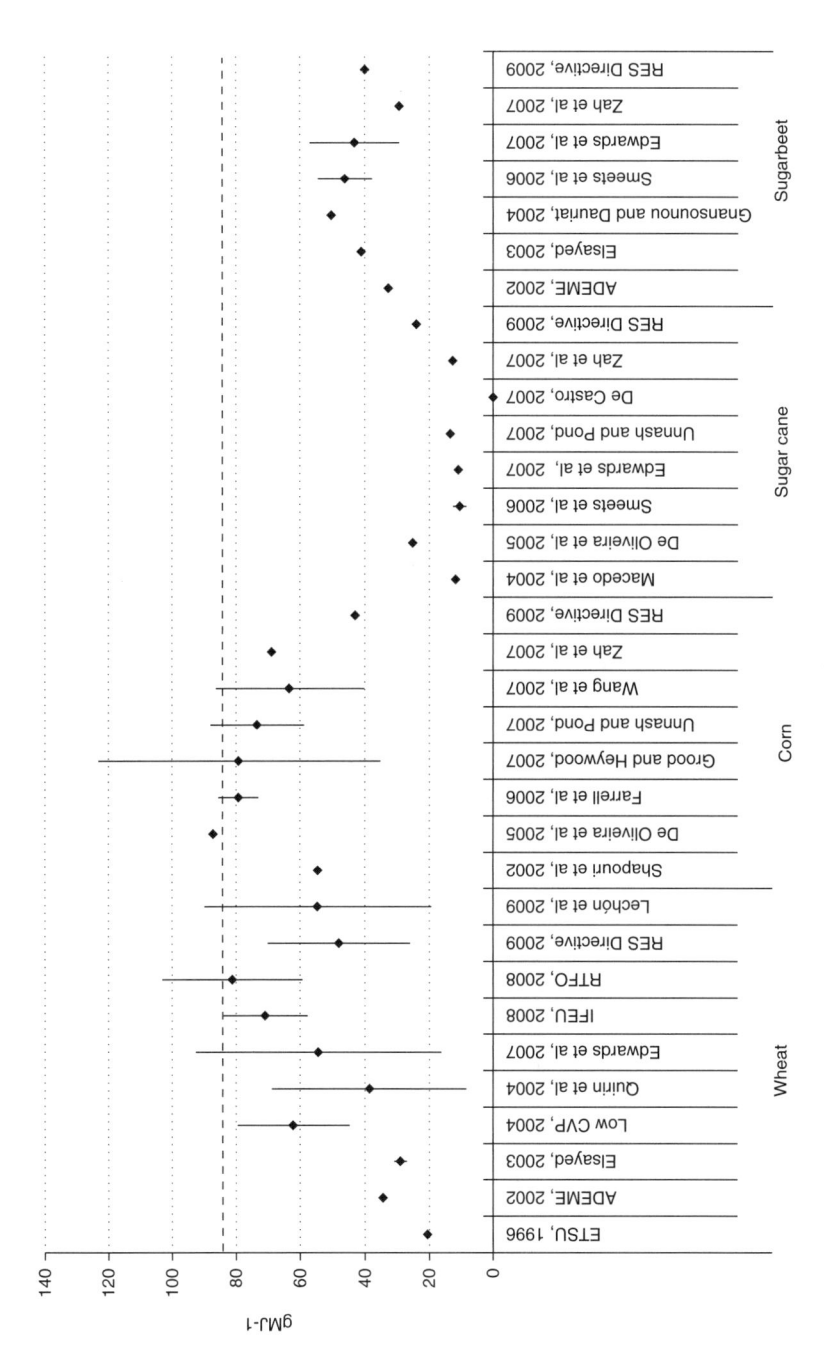

*Figure 16.5  Lifecycle GHG emissions of bioethanol production from different raw materials*

*Figure 16.6  Lifecycle GHG emissions of biodiesel production from different raw materials*

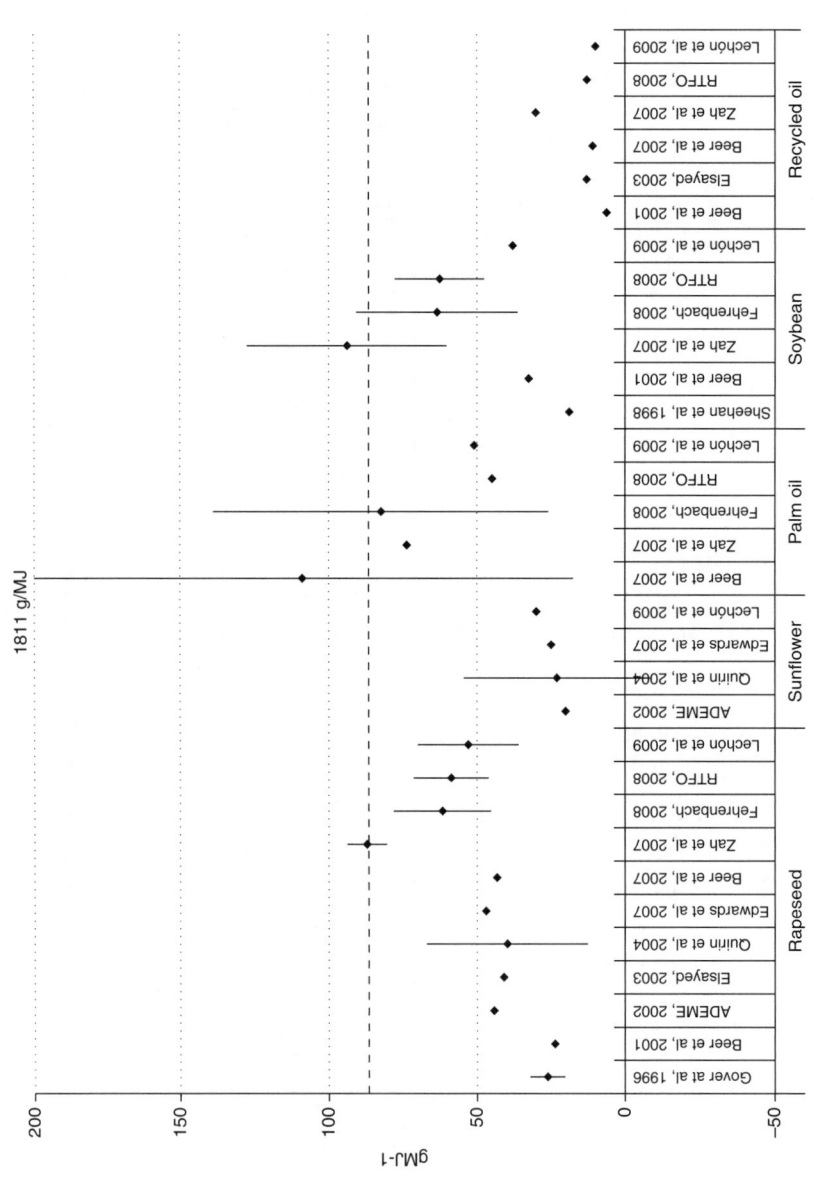

*Note:* The symbol ♦ represents the mean value of GHG emissions calculated by the authors in the references. Bars represent the highest and lowest values reported.

*Source:* CASES project.

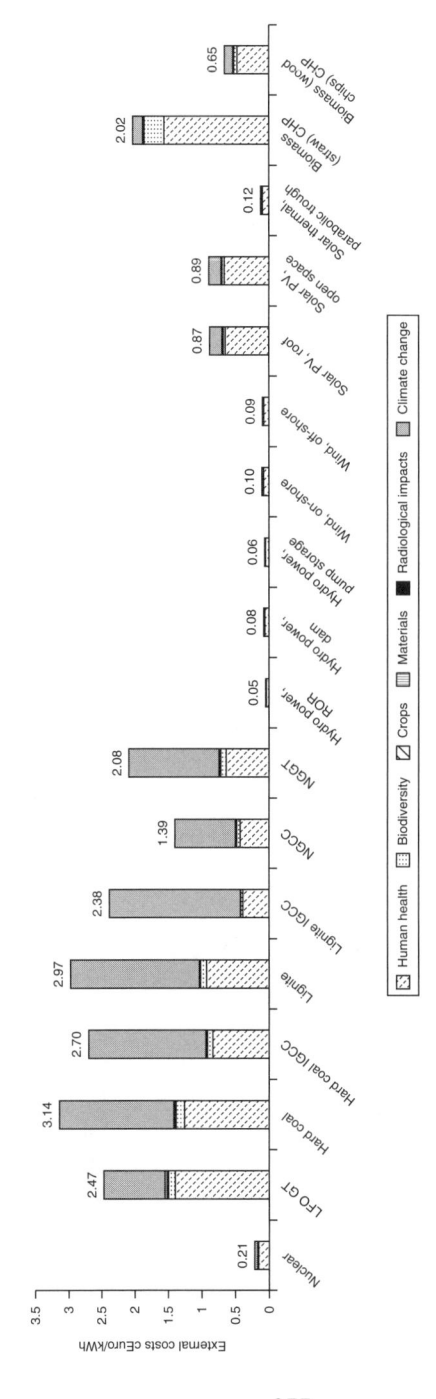

*Figure 16.7 External costs of electricity generation technologies disaggregated by impact category*

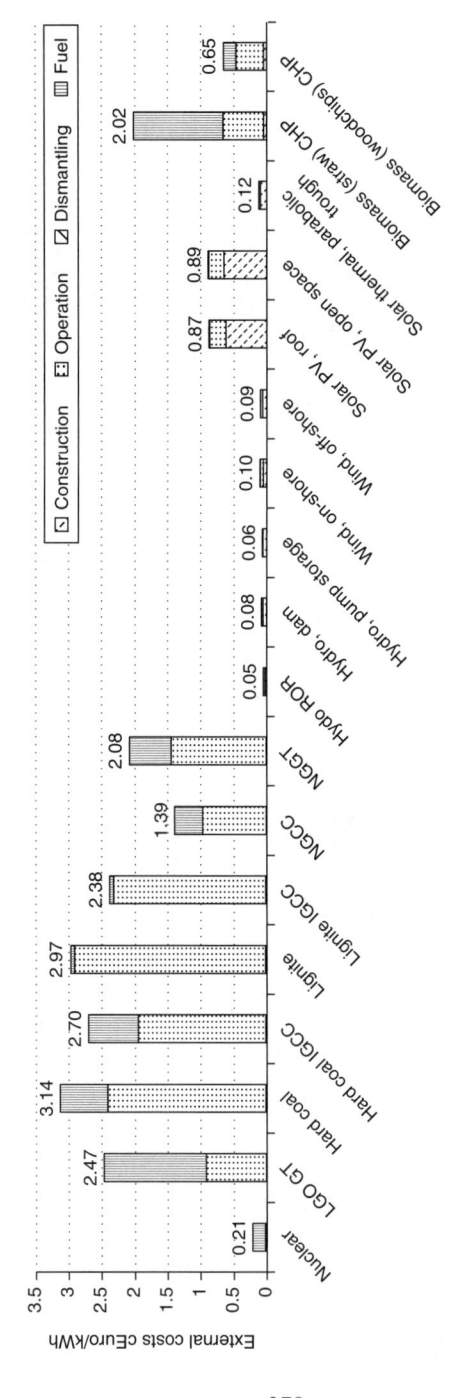

*Figure 16.8 External costs of electricity generation technologies disaggregated by life cycle stage*

*Source:* CASES project.

the fuel cycles. Fuel provision external costs are important in oil, coal and gas technologies but also in biomass power plants. The nuclear fuel cycle external costs are also dominated by fuel related activities.

## 5 CONCLUSIONS

This chapter has reviewed the environmental implications of a wide range of different energy production technologies showing results for different kinds of pollutant emissions, impacts as well as external costs calculations. One relevant conclusion emerging from this review is that renewable energies can play an important role in mitigating global warming emissions, a major concern in current environmental policy agendas. In terms of GHG mitigation potential, some renewable technologies seem to attain robust results in the reviewed studies, while others such as biomass derived electricity and biofuels show a more variable range of results. Of the latter technologies, some aspects of concern are related to the associated indirect effects produced by a large scale deployment. When mitigation of other impacts and pollutants are included in the picture, some renewable technologies show higher potential than others. Once more, biomass technologies but also solar PV technologies have a lesser potential to contribute to this mitigation.

When all of these effects are aggregated in a single indicator and are quantified in monetary units, the social welfare benefits of deploying renewable energy sources become clear. Consequently, if policy makers want to promote and pursue a sustainable energy system that maximizes social welfare, environmental externalities of all energy technologies must be taken into account. In order to do so, it is necessary to properly identify, quantify and later internalize the external costs in the price of energy through the various existing mechanisms.

## REFERENCES

ADEME, DIREM, Ecobilan, PriceWaterhouseCoopers (2002). "Bilans énénergétiques et gaz à effet de serre des filières de production de biocarburants." Rapport technicque.

Anthoff, D. (2007). "Report on marginal external costs inventory of greenhouse gas emissions." NEEDS Deliverable D5.2, RS1b.

Arvizu, D., P. Balaya, L. Cabeza, T. Hollands, A. Jäger-Waldau, M. Kondo, C. Konseibo, V. Meleshko, W. Stein, Y. Tamaura, H. Xu and R. Zilles (2011). "Direct Solar Energy." In O. Edenhofer, R. Pichs-Madruga, Y. Sokona, K. Seyboth, P. Matschoss, S. Kadner, T. Zwickel, P. Eickemeier, G. Hansen, S. Schlömer, C. von Stechow (eds), *IPCC Special Report on Renewable Energy Sources and Climate Change Mitigation*, Cambridge University Press, Cambridge, United Kingdom and New York, NY, USA.

Bauer, C., T. Heck, R. Dones, O. Mayer-Spohn and M. Blesl (2008). "Final report on technical data, costs, and life cycle inventories of advanced fossil power generation systems." Deliverable no 7.2 – RS 1a. NEEDS (New Energy Externalities Developments for Sustainability) Integrated project.

Baxter, L. and J. Koppejan (2004). "Biomass-coal Co-combustion: Opportunity for Affordable Renewable Energy." IEA Bioenergy task 32. Retrieved from: http://ieabcc.nl/publications/paper_cofiring.pdf (accessed August 2012).

Beer, T., T. Grant and P.K. Campbell (2007). "The greenhouse and air quality emissions of biodiesel blends in Australia." Report Number KS54C/1/F2.27. Report for Caltex Pty Ltd. CSIRO.

Beer, T., T. Grant, G. Morgan, J. Lapszewicz, P. Anyon, P. Nelson et al. (2001). "Comparison of transport fuels." Final report (EV45A/2/ F3C) to the Australian Greenhouse office on the Stage 2 study of Life Cycle Emissions Analysis if Alternative Fuels for Heavy Vehicles. CSIRO.

Boehlert, G.W. and A.B. Gill (2010). "Environmental and ecological effects of ocean renewable energy development: A current synthesis." *Oceanography*, 23: 68–81.

Bringezu, S., H. Schütz, M. O'Brien, L. Kauppi, R.W. Howarth and J. McNeely (2009). "Towards a sustainable production and use resources: Assessing Biofuels." UNEP.

Carter, N.T. and R.J. Campbell (2009). "Water Issues of Concentrating Solar Power (CSP) Electricity in the U.S. Southwest." Congressional Research Service. CSR Report for Congress.

CASES project (n.d.). "Cost assessment for Sustainable Energy Systems. Deliverables D.02.1 and D02.2". Retrieved from http://www.feem-project.net/cases/downloads_deliverables.php (accessed March 2013).

CATF, Clean Air Task Force (2001). "Cradle to Grave: The environmental impacts from coal". Retrieved from http://www.catf.us/resources/publications/files/Cradle_to_Grave.pdf (accessed February 2013).

Chum, H., A. Faaij, J. Moreira, G. Berndes, P. Dhamija, H. Dong, B. Gabrielle, A. Goss Eng, W. Lucht, M. Mapako, O. Masera Cerutti, T. McIntyre, T. Minowa and K. Pingoud (2011). "Bioenergy." In O. Edenhofer, R. Pichs-Madruga, Y. Sokona, K. Seyboth, P. Matschoss, S. Kadner, T. Zwickel, P. Eickemeier, G. Hansen, S. Schlomer, C. von Stechow (eds), *IPCC Special Report on Renewable Energy Sources and Climate Change Mitigation*, Cambridge University Press, Cambridge, United Kingdom and New York, NY, USA.

De Cara, S., A. Goussebaïle, R. Grateau, F. Levert, J. Quemener, B. Vermont, J.C. Bureau, B. Gabriell and A. Gohin (2012). "Revue critique des etudes evaluant éffet des changements d'affectation des sols sur les bilans environnementaux des biocarburants." ADEME. INRA.

De Castro, J.F.M. (2007). "Biofuels: an overview." Final Report. Prepared for DGIS/DMW/IB. Environmental Infrastructure and Impact Division, Environment and Water Department, Directorate-General for International Cooperation (DGIS), the Hague, Netherlands.

De Fraiture, C. And G. Berndes (2009). "Biofuels and water." In R.W. Howarth and S. Bringezu (eds), *Biofuels: Environmental Consequences and Interactions with Changing Land Use*. Proceedings of the Scientific Committee on Problems of the Environment (SCOPE) International Biofuels Project Rapid Assessment, 22-25 September 2008, Gummersbach Germany. Cornell University, Ithaca NY, USA. Retrieved from: http://cip.cornell.edu/biofuels/ (accessed February 2013), pp.139–153.

de Oliveira, M.E.D., B.E. Vaughan and E.J. Rykeil, Jr. (2005). "Ethanol as fuel: energy, carbon dioxide balances, and ecological footprint." *BioScience* 55(7): 593–602.

EASAC, European Academies Science Advisory Council (2011). "Concentrating solar power: its potential contribution to a sustainable energy future." Policy report 16. Retrieved from: www.easac.eu (accessed February 2013).

EC (1995a). "ExternE. Externalities of energy. Vol 3. Coal & Lignite." EUR 16522 EN.

EC (1995b). "ExternE. Externalities of energy. Vol 6. Oil & Gas." EUR 16523 EN.

EC (1995c). "ExternE. Externalities of energy. Vol 5. Nuclear energy." EUR 16524 EN.
EC (1995d). "ExternE. Externalities of energy. Vol 6. Wind & Hydro." EUR 16525 EN.
EC (2005). "ExternE Externalities of Energy. Methodology 2005 Update." EUR 21951. Peter Bickel and Rainer Friedrich (eds). Institut für Energiewirtschaft und Rationelle Energieanwendung, IER. Universität Stuttgart, Germany
Ecoinvent Database. Ecoinvent Centre (2007), Dübendorf, 2007. Retrieved from: http://www.ecoinvent.org (accessed February 2013).
Edwards R., J.F. Larivé, V. Mahieu and P. Rouveirolles (2007). "Well-To-Wheels analysis of future automotive fuels and power trains in the European context, v.2c." WTW Report 010307. JRC-IES/ EUCAR/ CONCAWE. Retrieved from: http://ies.jrc.ec.europa.eu/uploads/media/WTW_Report_010307.pdf (accessed February 2013).
EEA, European Environment Agency (2011). "Opinion of the EEA Scientific Committee on Greenhouse Gas Accounting in Relation to Bioenergy." Retrieved from http://www.eea.europa.eu/about-us/governance/scientific-committee/sc-opinions/opinions-on-scientific-issues/sc-opinion-on-greenhouse-gas/view (accessed March 2013).
Eickhout, B., G.J. van den Born, J. Notenboom, M. van Oorschot, J.P.M. Ros, D.P. van Vuuren and H.J. Westhoek (2008). "Local and global consequences of the EU renewable directive for biofuels. Testing the sustainability criteria." MNP Report 500143001.
El-Hinnawi, E. (1978). "Review of the Environmental Impact of Nuclear Energy." International Atomic. Energy Agency (IAEA) Bulletin 20, no. 2.
Elsayed, M.A., R. Matthews and N.D. Mortimer (2003). "Carbon and energy balances for a range of biofuels options." Project Number B/B6/00784/REP. Resources Research Unit Sheffield Hallam University.
Erickson, W.P., G.D. Johnson and D.P. Young Jr. (2005). "A Summary and Comparison of Bird Mortality from Anthropogenic Causes with an Emphasis on Collisions." General Technical Report, United States Forest Service, Washington, DC, USA.
EWEA (2009). "Wind Energy, the Facts." European Wind Energy Association (EWEA), Brussels, Belgium.
Fargione, J., J. Hill, D. Tilman, J. Polasky and P. Hawthorne (2008). "Land clearing and the biofuel carbon debt". *Science*, 319: 1235, DOI: 10.1126/science.1152747.
Farrell, A.E., R.J. Plevin, B.T. Turner, A.D. Jones, M. Ohare and D.M. Kammen (2006). "Ethanol can contribute to energy and environmental goals." *Science* 311: 506–508.
Fehrenbach H. 2008. GHG accounting methodology and default data according to the biomass sustainability ordinance (BSO). 2nd GBEP Task Force Meeting on GHG Methodologies. Washington DC: UN Foundation: 6–7.
Finnveden, G., Michael Z. Hauschild, Tomas Ekvall, Jeroen Guinée, Reinout Heijungs, Stefanie Hellweg Annette Koehler, David Pennington and Sangwon Suh.(2009). "Recent developments in Life Cycle Assessment." *Journal of Environmental Management* 91: 1–21.
Firbank, L.G. (2008). "Assessing the Ecological Impacts of Bioenergy projects." Bioenergy Research. 1: 12–19. DOI: 10.1007/s12155-007-9000-8.
Frankl, P., E. Menichetti, M. Raugei, S. Lombardelli and G. Prennushi (2005). "Final Report on Technical Data, Costs and Life Cycle Inventories of PV Applications." Ambiente Italia, Milan, Italy.
Gärtner, S. (2008). "Final report on technical data, costs and life cycle inventories of biomass CHP plants." Deliverable no137.2 – RS 1a. NEEDS (New Energy Externalities Developments for Sustainability) Integrated project.
Gerbens-Leenes W., A.Y. Hoekstra and T.van der Meer (2009). "The water footprint of bioenergy." *PNAS* 109(25): 10219–10223. Retrieved from: www.pnas.org/cgi/doi/10.1073/pnas.0812619106 (accessed February 2013).
Gnansounou, E. and A. Dauriat (2004). *Energy balance of bioethanol: a synthesis*, Lasen, Ecole Polytechnique Fédérale de Lausanne, Lausanne Switzerland.
Goldstein, B., G. Hiriart, R. Bertani, C. Bromley, L. Gutierrez-Negrin, E. Huenges, H. Muraoka, A. Ragnarsson, J. Tester and V. Zui (2011). "Geothermal Energy." In O. Edenhofer, R. Pichs-Madruga, Y. Sokona, K. Seyboth, P. Matschoss, S. Kadner, T. Zwickel, P. Eickemeier, G. Hansen, S. Schlomer, C. von Stechow (eds), *IPCC*

*Special Report on Renewable Energy Sources and Climate Change Mitigation*, Cambridge University Press, Cambridge, United Kingdom and New York, NY, USA.

Gover, M.P., S.A. Collings, G.T. Hitchcock, D.P. Moon and G.T. Wilkins (1996). "Alternative road transport fuels a preliminary life-cycle study for the UK." Vols. 1 & 2. A study co-funded by the Department of Trade and Industry and the Department of Transport. ETSU.

Groode, T.A. and J.B. Heywood (2007). *Ethanol: a look ahead*, MIT publication, Cambridge MA, USA.

Guinée, J.B., M. Gorrée, R. Heijungs, G. Huppes, R. Kleijn, A. de Koning, L. van Oers, A. Wegener Sleeswijk, S. Suh, H.A. Udo de Haes, H. de Bruijn, R. van Duin and M.A.J. Huijbregts (2002). *Handbook on life cycle assessment. Operational guide to the ISO standards. I: LCA in perspective. IIa: Guide. IIb: Operational annex. III: Scientific background.* Kluwer Academic Publishers, Dordrecht.

Haggett, C. (2011). "Understanding public responses to offshore wind power." *Energy Policy* 39: 503–510.

ICOLD (1999). "Benefits and concerns about dams." Retrieved from: http://www.swissdams.ch/Committee/Dossiers/BandC/Benefits_of_and_Concerns_about_Dams.pdf (accessed February 2013).

IEA Bioenergy (2009). "Bioenergy: A Sustainable and Reliable Energy Source." Main Report. IEA Bioenergy: ExCo:2009:06.

IEA- ETSAP, International Energy Agency Energy-Technology Systems Analysis Program (2010a). "Coal-fired power." Technology Brief E01, April 2010. Retrieved from: www.etsap.org (accessed August 2012).

IEA- ETSAP, International Energy Agency Energy-Technology Systems Analysis Program (2010b). "Gas-fired power." Technology Brief E02, April 2010. Retrieved from: www.etsap.org (accessed August 2012).

IEA- ETSAP, International Energy Agency Energy-Technology Systems Analysis Program (2010c). "Geothermal heat and power." Technology Brief E07 – May 2010. Retrieved from: www.etsap.org (accessed August 2012).

IEA-ETSAP, International Energy Agency Energy-Technology Systems Analysis Program (2011). "Photovoltaic solar power." Technology Brief E011 – February 2011. Retrieved from: www.etsap.org (accessed August 2012).

IEA, International Energy Agency (1998). "Benign Energy? The Environmental Implications of Renewables." OECD/IEA. Paris.

IEA, International Energy Agency (2010). "Technology Roadmap Concentrating Solar Power." OECD/IEA Paris France.

IEA, International Energy Agency (2011). "World energy outlook 2011." OECD/IEA Paris.

IFC, International Finance Corporation (2007). "Environmental, health and safety guidelines. Petroleum refining." Retrieved from: http://www1.ifc.org/wps/wcm/connect/52870d80488557e5be44fe6a6515bb18/Final%2B-%2BPetroleum%2BRefining.pdf?MOD=AJPERES&id=1323153091008 (accessed February 2013).

IPCC (2005). "IPCC Special Report on Carbon Dioxide Capture and Storage." B. Metz, O.Davidson, H. C. de Coninck, M. Loos and L. A. Meyer (eds). Prepared by Working Group III of the Intergovernmental Panel on Climate Change, Cambridge University Press, Cambridge, United Kingdom and New York, NY, USA.

ISO (2006a). "Environmental management – Life cycle assessment – Principles and framework." ISO 14040:2006. Retrieved from: http://www.iso.org/iso/home/store/catalogue_tc/catalogue_detail.htm?csnumber=37456 (accessed March 2013).

ISO (2006b). "Environmental management – Life cycle assessment – Requirements and guidelines." ISO 14044:2006. Retrieved from: http://www.iso.org/iso/home/store/catalogue_tc/catalogue_detail.htm?csnumber=38498 (accessed March 2013).

JRC IES (2010). "ILCD Handbook. General Guide for Life Cycle Assessment – Detailed Guidance." 1st edition. EUR 24708 EN. Luxembourg. Publications Office of the European Union.

Köller, J., J. Koppel and W. Peters (eds) (2006). *Offshore Wind Energy: Research on Environmental Impacts*, Springer, Berlin.
Kumar, A., T. Schei, A. Ahenkorah, R. Caceres Rodriguez, J.-M. Devernay, M. Freitas, D. Hall, A. Killingtveit and Z. Liu (2011). "Hydropower." In O. Edenhofer, R. Pichs-Madruga, Y. Sokona, K. Seyboth, P. Matschoss, S. Kadner, T. Zwickel, P. Eickemeier, G. Hansen, S. Schlomer, C. von Stechow (eds), *IPCC Special Report on Renewable Energy Sources and Climate Change Mitigation*, Cambridge University Press, Cambridge, United Kingdom and New York, NY, USA.
Laborde, D. (2011). "Assessing the Land use Change Consequences of European Biofuel Policies." Final report. IFPRI.
Ladenburg, J. (2009). "Stated public preferences for on land and offshore wind power generation – a review." *Wind Energy* 12: 171–181.
Lechón Y., C. de la Rúa and R. Sáez (2008). "Life Cycle Environmental Impacts of Electricity Production by Solarthermal Power Plants in Spain." *Journal of Solar Energy Engineering* 130: 021012-1.
Lechón,Y., H. Cabal, C. de la Rúa, N. Caldés, M. Santamaría and R. Sáez (2009). "Energy and greenhouse gas emission savings of biofuels in Spain's transport fuel. The adoption of the EU policy on biofuels." *Biomass and bioenergy* 33: 920–932.
Lecointe, C., D. Lecarpentier, V. Maupu, D. Le Boulch and R. Richard (2007). "Final Report on Technical Data, Costs and Life Cycle Inventories of Nuclear Power Plants." D14.2 – RS 1a, New Energy Externalities Developments for Sustainability (NEEDS), Rome.
Lewis, A., S. Estefen, J. Huckerby, W. Musial, T. Pontes and J. Torres-Martinez (2011). "Ocean Energy." In O. Edenhofer, R. Pichs-Madruga, Y. Sokona, K. Seyboth, P. Matschoss, S. Kadner, T. Zwickel, P. Eickemeier, G. Hansen, S. Schlomer, C. von Stechow (eds), *IPCC Special Report on Renewable Energy Sources and Climate Change Mitigation*, Cambridge University Press, Cambridge, United Kingdom and New York, NY, USA.
LOWCVP Low Carbon Vehicle Partnership (2004). "Well-to-wheel evaluation for production of ethanol from wheat." A report by the Low CVP Fuels Working Group, WTW Sub-Group. FWG-P-04-024.
Macedo, I., M.R. Lima, V. Leal and J.E. Azevedo Ramos da Silva (2004). "Assessment of greenhouse gas emissions in the production and use of fuel ethanol in Brazil." Government of the State of São Paulo, São Paulo, Brazil.
McDonald, N.C. and J.M. Pearce (2010). "Producer responsibility and recycling solar photovoltaic modules." *Energy Policy* 38: 7041–7047.
Menichetti E. and M. Otto (2009). "Energy balance and greenhouse gas emissions of biofuels from a life-cycle perspective." In R.W. Howarth and S. Bringezu (eds) *Biofuels: Environmental Consequences and Interactions with Changing Land Use*. Proceedings of the Scientific Committee on Problems of the Environment (SCOPE) International Biofuels Project Rapid Assessment, 22–25 September 2008, Gummersbach Germany. Cornell University, Ithaca NY, USA. Retrieved from: http://cip.cornell.edu/biofuels/ (accessed February 2013), pp. 81–109.
Melillo, J. M., J.M. Reilly, D. Kicklighter, A.C. Gurgel, T.W. Cronin, S. Paltsev, B.S. Felzer, X. Wang, A.P. Sokolov and C.A. Schlosser (2009). "Indirect Emissions from Biofuels: How Important?" *Science 4* 326(5958): 1397–1399. DOI: 10.1126/science.1180251.
MIT, Massachusetts Institute of Technology (2007). "The future of coal." An interdisciplinary MIT study. 2007 Massachusetts Institute of Technology.
MIT, Massachusetts Institute of Technology (2011). "The future of the nuclear fuel cycle." An interdisciplinary MIT study. 2011 Massachusetts Institute of Technology.
Nayak, D.R., D. Miller, A. Nolan, P. Smith and J.U. Smith (2010). "Calculating carbon budgets of wind farms on Scottish peatland." Mires and Peat 4: Art. 9. Retrieved from: http://www.mires-and-peat.net/map04/map_04_09.htm (accessed February 2013).
Nussbaumer, T. (2003). "Combustion and co-combustion of biomass: fundamentals, technologies and primary measures of emission reduction." *Energy and Fuels* 17: 1510–1521.

NWCC (1997). "Wind Energy Environmental Issues." Retrieved from: http://www.nationalwind.org/assets/archive/Issue_Paper_2.pdf (accessed February 2013).

Page, S.E., R. Morrison, C. Malins, A. Hooijer, J.O. Rieley and J. Jauhiainene (2011). "Review of peat surface greenhouse gas emissions from oil palm plantations in Southeast Asia." White paper Number 15. Indirect effects of biofuel production series. ICCT. Retrieved from www.theicct.org (accessed February 2013).

Preiss, P. and R. Friedrich (2009). "Report on the application of the tools for innovative energy technologies." NEEDS Technical paper no 2 7.2.– RS 1b

Quirin, M., S.O. Gärtner, M. Pehnt and G.A. Reinhardt (2004) "$CO_2$-neutrale Wege zukünftiger Mobilität durch Biokraftstoffe: Eine Bestandsaufnahme." Final report. By order of FVV, Frankfurt.

Ravindranath, N.H., R. Manuvie, J. Fargione, J.G. Canadell, G. Berndes, J. Woods, H. Watson and J. Sathaye (2009). "Greenhouse gas implications of land use and land conversion to biofuel crops." In R.W. Howarth and S. Bringezu (eds) *Biofuels: Environmental Consequences and Interactions with Changing Land Use*. Proceedings of the Scientific Committee on Problems of the Environment (SCOPE) International Biofuels Project Rapid Assessment, 22-25 September 2008, Gummersbach Germany. Cornell University, Ithaca NY, USA. REtrieved from: http://cip.cornell.edu/biofuels/ (accessed February 2013), pp. 111–125.

Royal Society (2008). "Sustainable Biofuels: Prospects and Challenges." Policy document 01/08,The Royal Society, London.

RTFO Department of Transport (2008). "Carbon and sustainability reporting within the renewable transport fuel obligation. Requirements and guidance." Draft Government recommendation to the RTFO Administrator. Retrieved from: http://www.arb.ca.gov/fuels/lcfs/lcfs_uk1.pdf (accessed March 2013).

Sala, O.E., D. Sax and H. Leslie (2009). "Biodiversity consequences of biofuel production." In R.W. Howarth and S. Bringezu (eds) *Biofuels: Environmental Consequences and Interactions with Changing Land Use*. Proceedings of the Scientific Committee on Problems of the Environment (SCOPE) International Biofuels Project Rapid Assessment, 22–25 September 2008, Gummersbach Germany. Cornell University, Ithaca NY, USA. Retrieved from: http://cip.cornell.edu/biofuels/ (accessed february 2013), pp. 127–137.

Shapouri. H., J. Duffield and M. Wang (2002). "The energy balance of corn ethanol: an update." Agricultural Economic Report No. 813. US Dept of Agriculture, Office of the Chief Economist/ Office of Energy Policy and New Uses, Washington DC.

Sheehan, J., V. Camobreco, J. Duffield, M. Graboski and H. Shapouri (1998). "Life cycle inventory of biodiesel and petroleum diesel for use in an urban bus." NREL/SR-580–24089 UC.US Department of Agriculture and US Department of Energy.

Simpson, T.W., L.A. Martinelli, A.N. Sharpley and R.W. Howarth (2009). "Impact of ethanol production on nutrient cycles and water quality: the United Staes and Brazil as case studies." In R.W. Howarth and S. Bringezu (eds) *Biofuels: Environmental Consequences and Interactions with Changing Land Use*. Proceedings of the Scientific Committee on Problems of the Environment (SCOPE) International Biofuels Project Rapid Assessment, 22–25 September 2008, Gummersbach Germany. Cornell University, Ithaca NY, USA. Retrived from: http://cip.cornell.edu/biofuels/ (accessed February 2013), pp. 153–167.

Smeets, E., M. Junginger, A Faaij, A. Walter and P. Dolzan (2006). "Sustainability of Brazilian bioethanol." University of Utrecht, Copernicus Institute, Utrecht, Netherlands.

SolarPACES (2008). "SolarPACES Annual Report 2007." International Energy Agency, Paris.

Sorensen, H.C. and S. Naef (2008). "Report on technical specification of reference technologies (wave and tidal power plant)." Final report on technical data, costs, and life cycle inventories of advanced fossil power generation systems. Deliverable no 16.1 – RS 1a. NEEDS (New Energy Externalities Developments for Sustainability) Integrated project.

Tilman, D., R. Socolow, J.A. Foley, J. Hill, E. Larson, L. Lynd, S. Pacala, J. Reilly, T. Searchinger,

C. Somerville and R. Williams. (2009). "Beneficial Biofuels – The Food, Energy, and Environment Trilemma." *Science* 325.

Tol, Richard S. J. (2009). "The Economic Effects of Climate Change." *Journal of Economic Perspectives* 23(2): 29–51.

Udo de Haes, H.A., G. Finndeven, M. Goekoop, M. Hauschild, E.G. Hertwich, P. Hofstetter, O. Jolliet, W. Klopffer, W. Krewitt, E. Lindeijer, R. Müller-Wenk, S. I. Olsen, D. W. Pennington, J. Potting and B. Steen (2002). "Life Cycle Impact Assessment: striving towards best practice." SETAC.

UNESCO/IHA (2009). "Measurement Specification Guidance for Evaluating the GHG Status of Man-Made Freshwater Reservoirs, Edition 1." IHA/GHG-WG/5. United Nations Educational, Scientific and Cultural Organization and the International Hydropower Association, London. Retrieved from: http://unesdoc.unesco.org/images/0018/001831/183167e.pdf (accessed February 2013).

Unnasch, S. and J. Pont (2007). "Full fuel cycle assessment: well to wheels energy inputs, emissions, and water impacts." Consultant Report prepared by TIAX LLC for California Energy Commission, Cupertino CA, USA.

US- EPA, United States Environmental Protection Agency (2011a). "Emissions Factors & AP 42, Compilation of Air Pollutant Emission Factors. AP 42, Fifth Edition Compilation of Air Pollutant Emission Factors, Volume 1: Stationary Point and Area Sources. Chapter 3 Stationary Internal Combustion Sources. 3.1 Stationary Gas Turbines." Retrieved from http://www.epa.gov/ttnchie1/ap42/ (accessed February 2013).

US- EPA, United States Environmental Protection Agency (2011b). "Emissions Factors & AP 42, Compilation of Air Pollutant Emission Factors. AP 42, Fifth Edition Compilation of Air Pollutant Emission Factors, Chapter 5: Petroleum Industry." Retrieved from: http://www.epa.gov/ttn/chief/ap42/ch05/index.html (accessed February 2013).

US-EPA, United States Environmental Protection Agency (2012). "How does electricity affect the environment?" Retrieved from: http://www.epa.gov/cleanenergy/energy-and-you/affect/index.html (accessed August 2012).

Viebahn, P., S. Kronshage, F. Trieb, and Y. Lechon (2008). "Final Report on Technical Data, Costs, and Life Cycle Inventories of Solar Thermal Power Plants." European Commission, Brussels, Belgium.

Wake, H. (2005). "Oil refineries: a review of their ecological impacts on the aquatic environment." *Estuarine, Coastal and Shelf Science* 62, 131–140.

Wang, M., M. We and H. Huo (2007). "Life cycle energy and greenhouse gas emission impacts of different corn ethanol plant types." *Environmental Research Letters* 2:024001

WCI, World Coal Institute (2005). "Clean Coal Technologies. Coal Power for Progress." Retrieved from www.wci-coal.com/uploads/ccts.pdf (accessed February 2013).

Wiser, R., Z. Yang, M. Hand, O. Hohmeyer, D. Infi eld, P. H. Jensen, V. Nikolaev, M. O'Malley, G. Sinden and A. Zervos (2011). "Wind Energy." In O. Edenhofer, R. Pichs-Madruga, Y. Sokona, K. Seyboth, P. Matschoss, S. Kadner, T. Zwickel, P. Eickemeier, G. Hansen, S. Schlomer, C. von Stechow (eds) *IPCC Special Report on Renewable Energy Sources and Climate Change Mitigation*, Cambridge University Press, Cambridge, United Kingdom and New York, NY, USA.

WNA (World Nuclear Association) (2012). "Safety of Nuclear Power Reactors." Retrieved from: http://www.world-nuclear.org/info/inf06.html (accessed August 2012).

Zah, R., H. Böni, M. Gauch, R. Hischier, M. Lehmann and P. Wäger (2007). "Ökobilanz von Energieprodukten: Ökologische Bewertung von Biotreibstoffen." EMPA, St. Gallen, Switzerland.

# 17. Washing away energy security: the vulnerability of energy infrastructure to environmental change
*Cleo Paskal*

## 1 INTRODUCTION[1]

Energy generation, extraction, refining, processing, distribution and delivery requires a complex, interlinked, costly and, sometimes, global infrastructure. However, much of that infrastructure lies in areas that are becoming increasingly physically unstable due to changes in the environment. Of particular concern are disruptions caused by, or exacerbated by, environmental change.

Climate change is only one (albeit critical) component of the larger challenge of direct, man-made environmental change. Humans often make direct and major alterations to the environment. Recently, for example, massive population increases have had a dramatic effect on the physical environment. At the turn of the 20th century, there were around 1.65 billion people on the planet. At the turn of the 21st, there were around 6 billion. The result is more groundwater pumped up potentially creating subsidence, more deforestation leading to erosion, more developments in flood plains putting infrastructure in vulnerable location, etc.

As we push the boundaries of the carrying capacity of the planet, a smaller degree of environmental variation has larger implications. As a result, climate change may significantly exacerbate existing challenges, but if there were no climate change, many of those challenges would still exist. For example, the massive disruption created by Hurricane Katrina in August 2005 was caused in large measure by problems with levee design and implementation on the part of the US Army Corps of Engineers, poor town planning, a failure of emergency services, and a breakdown in the chain of command. This naturally dynamic coastal region was going through a period of man-made environmental change, but much of that change was more direct than the sort caused by climate change. It included large-scale subsidence (in one area of New Orleans by about a metre in three decades) probably caused, at least in part, by the draining of wetlands, the extraction of groundwater and inappropriately designed waterways.

Katrina can be used to show how poor regulations, planning and emergency response can aggravate the environmental disasters that will almost certainly increase as a result of climate change, but one cannot say that the tragedy in New Orleans was caused by climate change alone. Similarly, the massive disruptions caused to the US Northeast during Sandy (2012) were exacerbated by major development in low lying regions known to flood. Curbing climate change without addressing other environmental vulnerabilities will not stop other 'Katrinas' and 'Sandys' (though it may keep the number from significantly accelerating).

A key node for security is energy supply. When energy is knocked out, there is a range of undesirable ancillary effects. So, within this matrix of environmental change, it is critical to understand the myriad vulnerabilities of energy infrastructure in this era of new variabilities.

When it comes to environmental change and energy infrastructure, there are two separate, but often interlinked, challenges. One is inherited, one is new. Both stem from the fact that energy infrastructure tends to have a long lifespan. The Hoover Dam in the Western US, was completed in 1935 and is still an important hydroelectric generator. China's Three Gorges Dam has an expected lifespan of at least 50 years. Nuclear power stations, from design through decommissioning, may be on the same site for 100 years. Additionally, constructions such as refineries, nuclear and coal power plants, and high voltage transmission lines can be perceived as undesirable to a community. As a result, when the time comes to build new installations, they are often erected in the same locations as the previous ones as the local population is already accustomed to the infrastructure. This means that sites chosen in the 2010s may still be in operation in 2110 and beyond.

The lifespan of existing energy infrastructure is well within the time frame predicted for potentially disruptive environmental change. When much of it was installed, the degree of change was not understood and so was not factored in to its design. This is an inherited challenge.

The new challenge involves upcoming investments. A substantial segment of energy infrastructure in North America and Western Europe is scheduled to be decommissioned in the coming decades due to revised environmental standards and general age-related retiring. And in much of the emerging world, major new builds are going in for the first time. Combined with stimulus packages in some countries, and development in others, it is likely that this is the beginning of an era of large-scale investments in new infrastructure. In some cases, there is now the necessary science to anticipate at least the minimum amount of environmental change that can be expected over the next century (well within the lifespan of most new investments). However, in too many cases, proposed new builds still do not incorporate likely effects of environmental change.

When planners talk about performing 'environmental impact assessments', almost invariably what is being assessed is how the construction would change the existing environment, not how a changing environment might affect the construction. While engineers and planners may perform a site inspection before designing an installation, they normally consider the parameters of that site a constant, not a variable. The general assumption is the coast won't move, river levels will remain constant, the ground won't subside and precipitation will be confined to historic norms. Most planners are not accustomed to, and often not trained for, incorporating new environmental variables into designs. An added problem is that while some change may be broadly predictable, there is likely to be wide variability in some areas, making precise projections impossible. The science is improving, but there are still many unknowns and a lack of fine graining. This in itself is sometimes used as a justification to avoid incorporating any change at all.

The result is that one could erect a multi-billion pound, high-tech, environmentally friendly installation in what will soon become a flood zone. Not only will the original investment be lost, the destruction of the property itself can cause new vulnerabilities.

It is no longer enough just to assess an installation's impact on the environment; one must also assess the impact of a changing environment on the installation. Then, as much as possible, the impact of that change must be integrated into planning and countered.

This chapter aims to add to that goal by identifying some of the most susceptible nodes in the global energy infrastructure and showing how they might be affected by moderate environmental change.

## 2 HYDROPOWER

The successful management of hydroelectric installations is contingent on being able to predict in advance the volume of water entering the system. Before construction, care is taken to assess river levels, hydrological cycles and precipitation patterns. Until recently those data were considered to be largely constants. For example, precipitation patterns might run on decadal cycles, but the cycles themselves were considered largely predictable and dams, turbines and reservoirs were designed accordingly. As the climate changes, what were constants are now variables. This causes problems for both primarily glacier-dependent and primarily rain-dependent power plants.

## 2.1 Glacier-Dependent Hydro Plants

Hydroelectric installations, such as some in the Himalayas, Alps and Andes that depend primarily on glacial melt, will likely face difficulties in managing widely varying flows both seasonally, and over the years. In Europe, mountain areas will likely see more flooding in the winter and spring, and drier summers. These fluctuations can disrupt hydroelectric power generation, erode infrastructure and damage valuable regional industries.

Many glaciers are currently retreating, producing more run-off than dams were designed for (SAARC Disaster Management Centre, 2008). In China, for example, virtually all glaciers are in retreat and the start of spring flow has advanced by nearly a month since records began (Stern 2005, 78). The Chinese Academy of Science estimates that by 2050 as much as 64 per cent of China's glaciers could have disappeared (UNEP 2005).

One immediate impact of that melt is flooding. An estimated 50 new lakes in Nepal, Bhutan and China have formed as a result of melting glaciers. Glacial lakes can be unstable and liable to burst their banks, as happened in Nepal in 1985 when one outburst washed away communities, and a hydroelectric installation (UNEP 2005). It is also possible that in areas that are already susceptible, the added geological stresses caused by the new lakes could be the 'last straw' that triggers an earthquake (Gupta Harsh 2002).

Eventually, once the glaciers reach a minimal extent, the flow may markedly decline, creating a new set of challenges, including a potential decline in hydroelectric production and increased competition with other sectors, including agriculture, for the water itself.

## 2.2 Rain-Dependent Hydro Plants

Hydro plants that depend primarily on predictable seasonal rains, such as many of those in India, will find it increasingly difficult to anticipate flow, potentially causing power generation declines, floods and irrigation problems.

Unexpected rainfall has already complicated the management of some of India's many dams (the country is one of the world's major builder of dams). In India, as in many other places in the world, dams often serve three purposes: flood control, irrigation and power generation. Most rain-dependent plants are designed to store water from the rainy season in order to be able to irrigate and generate power in the dry season. Those plans rely on predictable rain patterns. Some Indian dam managers are working on monsoon schedules that assume regular 35-year rainfall cycles. In 2008–2009 hydroelectricity generation in India saw an 8.42 per cent decline compared to 2007–2008. By the summer of 2009, the shortfall was 12 per cent. The loss was blamed on inadequate rainfall.[2]

The situation can be equally problematic when there is too much rain for the design of the installation. If the reservoir fills during the rainy season and then, due to changing precipitation patterns, the rain keeps falling well into what should be the dry season, the reservoir can back up and there will be a risk of inundating the villages upstream. If, in order to prevent that, the dam's floodgates are opened, the released water can add to the already swollen river and flood the cities downstream.

It was just such a downstream flooding that happened in August 2006 to Surat, an Indian city with a population of over 3 million people with a thriving economy as one of the world's largest diamond cutting centres. Unseasonably heavy rains overwhelmed dam management and led to the sudden release of water from an upstream dam. The resulting floodwaters covered around 90 per cent of the city and destroyed nearby villages. Over a hundred people are known to have died, hundreds more went missing, and disease spread as thousands of animals drowned and rotted in the waters. The financial cost was at least in the tens of million dollars and the loss of rare manuscripts from the city's academic institutions was incalculable (Mishra, 2006; Upadhyaya 2006; Thakkar 2006; Patel 2006).

### 2.3 Other Factors

These flow extremes, especially when combined with other environmental change factors such as deforestation, can cause erosion, subsidence, landslides and siltation, each of which can affect the efficacy and stability of hydroelectric power plants.

There are added political complications. Disputes between states, already concerned about power and water sharing, may only get worse as water supplies become even more erratic and hydroelectricity becomes less reliable. Additionally, Clean Development Mechanism financing and the push for low carbon power generation generally is resulting in a new era of dam building. A large percentage of all CDM projects are for hydroelectricity. As of June 2012, 1,531 hydro projects in China alone were registered or seeking registration with the CDM.[3] Some projects are well conceived. Others less so. It is critical that all new and existing plants be assessed to determine how environmental change over the lifespan of the dams will affect both their power generation viability as well as their structural integrity.

## 3  NUCLEAR POWER

Nuclear power generation may also face challenges with ensuring output and site security. Reactors usually require a large amount of water for cooling. As a result, they are generally positioned in areas that are susceptible to environmental change: they are normally either on the coast, making them increasingly vulnerable to sea level rise, extreme weather, storm surges, and tsunamis; or they are on rivers, lakes or reservoirs and are dependent on increasingly valuable, and variable, fresh water supplies.

Some installations have already been tested. There has been a degree of flooding at nuclear power plants in the US, France and India. In 1992, Hurricane Andrew caused extensive damage to the Turkey Point site in Florida. An earthquake, but mostly flooding, caused by the resulting tsunami, resulted in the meltdowns in Fukushima, Japan in 2011, the largest nuclear disaster since Chernobyl (1986).

In the UK, many of the existing coastal power stations are just a few metres above sea level. The Dungeness plant, in coastal Kent, is also built on an unstable geological formation. The site already needs constant management to stay protected. Many of these installations are aging, and there is momentum for new plants to be commissioned. However, as it is difficult to get communities to accept a nuclear power station in their region, in many cases the proposal is for the new plants to be located on the same sites as the old ones. The government has assured that builders would have to 'confirm that they can protect the site against flood-risk throughout the lifetime of the site, including the potential effects of climate change'(BBC 2008). It is, however, difficult to estimate both the lifetime of the site (those who built the installations that are there today did not factor in that new ones would be going in beside them, markedly extending the lifespan of the site), or the potential effects of climate change. For example, while sea level rise and storm surges may be increasingly well-understood, other disruptive factors, such as the possibility that changes in wave action could liquefy coastal sands, are not (Savonis et al. 2008). Coastal environments in particular are highly complex and unpredictable. Several times recently coastal nuclear power plants have had to power down as their intake valves were blocked by jellyfish, or 'jellyfish-like' animals. It happened in 2006 in Japan, forcing an emergency powering down (BBC 2006a); in 2011 in Scotland, where both reactors at Torness had to be shutdown (Miller 2011); and in 2012 in California, causing Diablo Canyon plant to cut production in one of its two reactors to 25 per cent of capacity (Chawkins 2012).

Riverside plants have different problems. In Europe, cooling for electrical power generation (including both nuclear and fossil fuel plants) accounts for around one-third of all water used. However, in some areas drought is reducing river, lake and reservoir levels at the same time as air and water temperatures are increasing. The effect on any form of power generation requiring a large amount of water (including coal powered plants) is likely to be substantial. The same heat conditions that peak demand will also increasingly result in some plants struggling to deliver.

The effects are already being seen. During Europe's record-breaking heat wave of 2003, temperatures reached more than 40 degrees Celsius across the continent. In France, 17 nuclear reactors had to power down or shut off, at a cost to Électricité de France of around €300 million (Kanter 2007). This was in spite of exemptions given allowing the plants to discharge water hotter than normally allowed into ecosystems, potentially disrupting other industries, such as fisheries.

The Hadley Centre predicts that, by 2040, heat waves like the one that seared Europe in 2003 would be 'commonplace'.[4] Additionally, as the average ambient temperature rises along with energy demand, the system may spend more time close to its limits and smaller events may result in shutdowns. The summer of 2006 was not as hot as 2003, but again France, Spain and Germany all had to power down nuclear plants because of heat and water problems (Jowit and Espinoza 2006). And in July 2009 France lost almost a third of its generation and had to buy power from the UK. A 2012 study published in *Nature Climate Change* projects:

> a summer average decrease in capacity of [thermoelectric, nuclear and fossil fueled] power plants of 6.3–19% in Europe and 4.4–16% in the United States depending on cooling system type and climate scenario for 2031–2060. In addition, probabilities of extreme (>90%) reductions in thermoelectric power production will on average increase by a factor of three. (van Vliet et al. 2012)

Given the high cost, long lifespan and potential for damage to nuclear power plants, it is essential that substantially more research be done on how they will interact with an increasingly volatile global environmental system.

## 4 OFF-SHORE/COASTAL PRODUCTION

As easier to access oil and gas sites are depleted, more difficult off-shore and coastal production may gain in importance. Off-shore and coastal oil and gas extraction is accomplished under a wide range of conditions, from the tropics to the tundra. The challenges vary depending on the location. In order to assess the variety of risks, case studies of the uncertainties in the Gulf Coast of the United States and the Arctic are instructive.

### 4.1 US Gulf Coast

Gulf of Mexico federal off-shore oil production accounts for 23 per cent of US crude oil production and federal off-shore natural gas production accounts for 7 per cent of total US dry production. The Gulf coast also contains over 40 per cent of US petroleum refining capacity and 30 per cent of US natural gas processing capacity (US Energy Information Administration, 2012).[5]

Climate change projections anticipate the US Gulf Coast will see increased flooding and extreme weather events. Storm activity has already affected supply. In the summer of 2005, Hurricane Katrina shut what amounted to around 19 per cent of the US's refining capacity, damaged 457 pipelines, and destroyed 113 platforms.[6] Oil and gas production dropped by more than half, causing a global spike in oil prices. Over the course of that year, close to three months of production was lost (Muller 2012). Much of the on-shore support infrastructure destroyed in 2005 was rebuilt in the same location, leaving it vulnerable to similar weather events.

In the summer of 2008, Hurricanes Gustav and Ike passed through the Gulf and destroyed 60 platforms. Interestingly, even before the hurricanes arrived, the economic effect was felt. What amounted to almost 10 per cent of US refining capacity, as well and much of off-shore Gulf production, was shut down in preparation for the hurricanes. The same thing happened again in 2012 when Isaac shut down most of the Gulf platforms and refining capacity (Morath 2012). This shows that even just the threat of extreme weather can affect supply and price. Climate change predictions imply that this sort of disruption is likely to become more common.

There are also other potential impacts. While most pipelines are buried, and so seemingly insulated from the effects of severe weather, there are nodes of exposed vulnerabilities, such as pumping stations and valves that can be affected. Also, it is uncertain how changes in water tables, soil structure, stability, erosion and subsidence might affect the pipelines.[7] Understanding how, or if, those factors may affect supply will require more research.

### 4.2 Other Low-Lying Coastal Facilities

Many of the world's largest oil and gas facilities (including Ras Tanura, Saudi Arabia; Jamnagar, India; Jurong Island Refinery, Singapore; Rotterdam Refinery; and major installations in the Niger Delta) are only slightly above sea level. This leaves them vulnerable to rising sea levels, storm surges, increasing storm activity, subsidence and changes in ground composition. If even one of these regions is affected, it could affect local security and global supply and markets.

### 4.2.1 Arctic

The US Geological Survey estimates that the Arctic might contain over one-fifth of all undiscovered oil and gas reserves.[8] One study postulated that Siberia could contain as much oil as the Middle East (Chalecki 2006). However, dreams of a resource bonanza in the north are premature. The environment is difficult and becoming increasingly unpredictable. Norway's northern Snohvit gas field cost 50 per cent more than the original budget and, in the fall of 2006, North Sea storms sank a 155-metre Swedish cargo ship and caused one oil rig to break away from its tow and be set adrift off the coast of Norway (BBC 2006b, IHT 2005). As one North Sea oil industry executive said: 'we've had our third "one-in-a-hundred-years" storm so far this year'.[9]

There are likely to be higher waves, increasing storm activity and more icebergs threatening off-shore rigs and complicating shipping (BBC 2006b, IHT 2005). Additionally, with warmer and wetter air freezing and thawing more often, ships, aircraft and infrastructure icing will become more common.[10] Also, many key elements of production, such as how to contain an oil spill in Arctic waters, are poorly researched. This is a highly complex operating environment. In 2012, Shell announced it was delaying its Alaskan Arctic drilling after a testing accident (Krauss 2012). All of this could result in high insurance costs, hampering exploration.

## 5 ENERGY PRODUCTION AND DISTRIBUTION IN COLD CLIMATES

An additional problem for off-shore Arctic energy extraction is that on-shore Arctic energy infrastructure is likely to suffer substantial damage. Coastal areas are already seeing more erosion and are weathering stronger storm activity. However the biggest problem may be the thaw of the permafrost.

Permafrost, essentially permanently frozen land, acts as a solid foundation for infrastructure in cold climates. It covers around 20 per cent of the planet's landmass, including large areas of Russia, parts of the Alps, Andes and Himalayas, and almost half of Canada. Many of these are energy production regions. These are also regions of energy transmission and distribution. The Trans-Alaska pipeline alone carries as much as 20 per cent of the US domestic oil supply.[11] As temperatures rise, the permafrost thaws. The ice trapped inside the frozen ground liquefies. If there is poor drainage, the water sits on the earth's surface and floods. If there is good drainage, the water runs off, potentially causing erosion and landslides.

Thawing permafrost has the potential to severely affect infrastructure in the north. Linear installations such as pipelines, electrical transmission line and railways are only as strong as their weakest point. If one section is destabilized, the entire supply can be disputed. Already in some cold climates, pipelines, roads, ports and airports are at risk of imminent structural damage and possible permanent loss. In Alaska, complete Arctic communities are being relocated. One of China's top permafrost experts who was involved in the multi-billion dollar, state-of-the-art Tibet railway, built in part on hundreds of kilometres of Himalayan permafrost, was quoted as saying 'Every day I think about whether the railway will have problems in the next ten to twenty years.' (Wolman 2006). While the railway is still in operation, sections of the foundation started sinking not long after it was opened (Mishra 2007). Often these problems can be engineered around, but they can add substantial cost and affect performance.

Construction and repair is also being affected. In cold climates, heavy equipment is often moved in the winter when the ground is most solid. With warming, that window is shortening. In some areas of Alaska, for example, the number of days per year heavy equipment can be driven on the tundra has halved.

Not only does environmental change create challenges for new cold climate resource extraction, existing installations that rely on iceroads, waste containment and pipelines built on thawing permafrost may need to be reassessed. In August 2006, a BP pipeline in Alaska corroded and broke. While not a direct result of environmental change, it gave an indication of the sort of vulnerabilities that may become more likely if thawing permafrost undermines pipelines. The line carried close to 2.6 per cent of the US daily supply and the closure created an immediate spike in oil prices and gas futures. The US government considered releasing emergency stockpiles and the Alaskan government faced a financial crunch (Pemberton 2006).

The stability of cold climate infrastructure is often overestimated. For example, with the retreat of Arctic sea ice a shipping route from Russian to Canada, through the Northwest Passage, has been proposed. Russia has offered to keep the Canadian section of the route open past the summer season with icebreaker convoys. The proposed Russian terminal is Murmansk. The proposed Canadian one is Churchill, Manitoba, on Hudson Bay. Shipping via Churchill can cut hundreds of kilometres off transit routes between Russian and the US Midwest. Under the plan, grain is the main proposed cargo, however fossil fuels could also join the route.

Some in Ottawa support the plan. However Churchill is only land-linked to the rest of Canada by rail (there is no road) and the railway, built in many places on permafrost, is already suffering from deteriorating tracks. There have already been derailments and at times in the summer the train can't travel faster than 10km an hour. This is an example of realities on-the-ground undermining economic and strategic analyses made in distant locations.

With environmental change, infrastructure problems in cold climates are likely to become more common. This section has looked primarily at permafrost, but there are many other challenges as well. For example, it used to be that in places such as parts of the northern United States and southeastern Canada, once winter came, it stayed below freezing until spring. Now, with rising temperatures, it warms and freezes, and warms and freezes. This opens up the possibility for more periods of freezing rain, and ice storms, like the one that cut the power off for millions in the winter of 1998.

It is going to take a major investment in permafrost and cold climate engineering research to finds way to rebuild Arctic and cold climate infrastructure in a manner that is viable over the long term.

## 6 OTHER CAUSES OF DISRUPTION

Any extreme weather event can extensively impair power delivery, and there are global predictions for an increase in those kinds of disasters. One UK government report, commissioned after the costly summer floods of 2007, found that potentially hundreds of UK substations are at risk of flooding (Shukman 2008). The wake-up call came that summer when a switching station near Gloucester, servicing around 500,000 homes and businesses, came within inches of being flooded.

Given the interconnected nature of global systems, and an increased reliance on just-in-time deliveries, unexpected failures are also appearing. During the Icelandic ash cloud incident of 2011, flights stopped to Scotland. Scotland's only refinery simultaneously refined multiple types of fuel. As jet fuel wasn't required due to the grounded flights, and there was little storage capacity as it had never been needed, the refinery came very close to having to shut down all refining because it couldn't off-load its jet fuel.

We are also seeing an increase in complex catastrophic failures, such as when Sandy hit the Northeastern US in 2012. Power stations shut down, critical installations flooded, powerlines came down, key personnel couldn't get to work, and more.

Extreme and unexpected events of all sorts are likely to become more common, jeopardizing energy security and, by extension, human and even potentially national security as well.

## 7 RENEWABLE ENERGY GENERATION

Every form of energy generation, including renewables, and every installation site chosen should be evaluated for stability in times of environmental change. For example, while solar plants may seem immune from disruption as long as the sun transits the sky, if they are built on flood plains, they can be rendered useless at the first heavy rain. Wind farms should assess if long-standing air currents may shift or if the hills they are often built on are likely to erode or suffer from landslides. Tidal generation should incorporate the effects of sea level rise, erosion, storm activity and so on. Just because an energy source is 'green' doesn't mean it is sustainable under environmental change conditions.

## 8 ENVIRONMENTAL CHANGE CAUSING ECONOMIC RECALCULATIONS

The clearest example of how energy supplies may be affected by a reevaluation of cost is how all of the above-mentioned disruptions (or even the potential for disruption) may affect insurance costs, potentially impairing the economic viability of certain investments.

Adding to the challenge is the fact that throughout our systems, real risk is being distorted, discounted and disguised. Potential market-based safety mechanisms, like insurance, are being undermined by politically expedient initiatives like the US's bankrupt National Flood Insurance Program, and 'investment friendly' initiatives such as caps on liability, both of which essentially offset the risk from the individual region or sector onto the population as a whole – and almost incidentally allow for critical pieces of energy infrastructure to be placed in highly vulnerable locations. For example, building a nuclear power plant in an earthquake and tsunami zone makes no sense from a physical reality perspective. However, that risk assessment was trumped by Japan's perceived domestic, economic and political imperatives. It was an illusory trade-off however, as Fukushima has done more to damage Japan's domestic, economic and political imperatives than choosing another location ever would have. Similarly, draining swampland in a hurricane zone like New Orleans to put up housing and critical infrastructure might make sense to developers and to the city officials that approve the plans for reasons of their own, but from a physical reality perspective, it is nonsensical. As was allowing BP to self-insure the Deepwater Horizon when insurance companies – some of the best in the world at assessing real physical risk – would not insure it. The problem is, those costs are adding up, and the public purse is increasingly light. As the environment changes, it will seem like we are building, and rebuilding, and re-rebuilding, sandcastles below the high tide mark. Something will have to give – either energy security, or the way risk is currently being assessed.

Other factors may change calculations as well. For example, if predictions of increasing water scarcity hold true, fresh, clean water may substantially increase in value. This would force a re-evaluation of the real cost of not only of hydro and freshwater-cooled nuclear installations, but of fossil fuel extraction and refining technique that contaminate water that could otherwise be used for drinking and irrigation.

Already China has abandoned or suspended the vast majority of its coal-to-liquid projects in part over concerns about water availability. Another potential area of concern is Canada's oil sands. The method of extraction used in the oil sands requires and contaminates large amounts of water. Currently Canada is perceived to have abundant fresh water, however the vast majority of that is non-renewable fossil water (once contaminated it is lost and not regenerated), and even that will be under increased stress in some regions as the climate shifts. Already there are concerns about water quality in some of the communities that share river systems with the oil sands. Apart from the domestic value of fresh water availability, ensuring a stable supply of fresh water for agriculture in Canada has wider implications. It is increasingly likely that, as other areas of the planet, such as Australia, become less fertile, Canadian agriculture's contribution to global supply will gain in relative importance, and so energy security may come in direct competition with food security.

## 9 GEOPOLITICAL FACTORS

Many of the potential disruptions mentioned could engender a political response. For example, in the case of Russian pipelines being undermined by thawing permafrost, if the engineering required for stabilization proves too costly, Russia might switch increasingly from pipelines to tankers, allowing them much more flexibility in delivery and subjecting the supply to greater politicization.

It is also possible, though quite controversial, that an increasingly parched US will look to Canada to supplement its water deficiencies. In some areas of the US, such as the agricultural belts and water-scare cities such as Las Vegas, water security might become more important than oil security. Other forms of energy may be found, but it is more difficult to find other forms of water. In that case, US energy security policy (which has been supporting the water-polluting Canadian oil sands) might come

*Vulnerability of energy infrastructure to environmental change* 399

into conflict with US water and food security policy (which would benefit by ensuring that an increasingly accessible supply to the north is not contaminated).

Another politics-of-water related problem that might affect energy supply could come about when dam building deprives one group, region or country of their expected supply of fresh water. It is possible that, should some become desperate enough as a result of increasing water scarcity, there could be attacks on the installations themselves. The goal would be to destroy the dam in order to secure water supply downstream.

## 10  CONCLUSION

There are concerns about both older installations not being designed for new conditions and new installations not integrating change into their planning. Either of those situations can result in marked decreases in energy output and risks to the installations themselves. That in turn, could affect energy prices and global security. Volatile energy prices have the potential to destabilize major economies.

Many of the challenges outlined above can be overcome with sufficient research, planning, engineering and financing. In some cases, it may even be possible to integrate change into planning in such a way that energy output increases with changes rather than decreases. For example, hydro installations in regions that are expecting higher rainfall could be designed to eventually take advantage of that excess flow, rather than be overwhelmed by it.

However, the reinforcement of global energy infrastructure is unlikely to happen overnight. It will take:

- an acknowledgement that the problems are real, integrated and wide ranging;
- the will to counter them;
- appropriate investment in research on potential impacts, assessment of real risks, and engineering and design solutions;
- implementation; and
- continuous reevaluation to ensure they are keeping up with environmental change predictions.

It is in the best interest of those concerned with energy security, such as national governments and the business community, especially the energy and insurance industries, to ensure this happens as quickly as possible. Until it does, it is to be expected that there will be increasingly frequent disruptions to energy supply, potentially in multiple locations and sectors at the same time. The economic, social and political costs are likely to be substantial.

At the same time, it may make sense to focus on building a more decentralized energy structure, preferably based on locally available renewables situated in secure locations. A degree of regional energy self-sufficiency could provide a better defence against the sort of large-scale outages that result when centralized power systems are compromised. This sort of regional, network-based system might also prove more flexible and adaptive, and therefore more able to cope with the increasing variability and unpredictability caused by environmental change.

Finally, it is worth remembering that energy infrastructure is often among the best funded, planned and maintained installations available. The challenges that even this well-supported sector will face are an indication of the vulnerability of other large sections of the critical infrastructure that support our economies, security and lives.

## NOTES

1. This chapter was updated and expanded from Cleo Paskal, 'The Vulnerability of Energy Infrastructure to Environmental Change', a joint publication of Chatham House and the Global Energy and Environment Strategic Ecosystem of the Department of Energy, July 2009.
2. 'Power generation growth plummets to 2.71% in FY'09', *Times of India*, 9 April 2009.
3. See http://www.internationalrivers.org/resources/spreadsheet-of-hydro-projects-in-the-cdm-project-pipeline-4039 (accessed 12 September 2012).
4. See http://www.metoffice.gov.uk/climatechange/science/explained/explained1.html.
5. 'Gulf of Mexico Fact Sheet', US Energy Information Administration, http://www.eia.gov/special/gulf_of_mexico/ (accessed 10 November 2012).
6. See http://www.mms.gov/ooc/press/2006/press0501.htm.
7. *Impacts of climate change and variability on transportation systems and infrastructure: Gulf Coast Study, phase 1*, 4–37 to 4–43.
8. See http://www.usgs.gov/newsroom/article.asp?ID=1980&from=rss_home.
9. Private conversation.
10. 'Naval operations in an ice-free Arctic Symposium', Office of Naval Research, Naval Ice Center, Oceanographer of the Navy and United States Arctic Research Commission, 2001, http://www.star.nesdis.noaa.gov/star/documents/2007IceSymp/FinalArcticReport_2001.pdf (accessed 5 March 2013).
11. 'Climate change in the Arctic and its implications for U.S. national security' (Chalecki 2007).

## REFERENCES

BBC (2006a), 'Nuclear plant struck by jellyfish', *BBC News*, 20 July 2006.
BBC (2006b), 'Nordic storm sink Swedish ship', *BBC News*, 1 November 2006.

BBC (2008), 'Flood risk "won't stop nuclear"', *BBC News*, 22 July 2008.
Chalecki, Beth (2007), 'Climate Change in the Arctic and its Implications for U.S. National Security', IDEAS, April 2007, http://ui04e.moit.tufts.edu/ierp/ideas/pdfs/issue2/Chalecki_Arctic.pdf (last accessed 5 March 2013).
Chawkins, Steve (2012), 'Diablo Canyon reactor gets unwelcome guests', *Los Angeles Times*, 26 April.
Gupta Harsh, K.(2002), 'A review of recent studies of triggered earthquakes by artificial water reservoirs with special emphasis on earthquakes in Koyna, India', *Earth-Science Reviews* 58: 279–310.
IHT, International Herald Tribune (2005), 'Arctic riches coming out of the cold', 10 October 2005.
Jowit, Juliette and Javier Espinoza (2006), 'Heatwave shuts down nuclear power plants', *The Observer*, 30 July.
Kanter, James (2007), 'Climate change puts nuclear energy in hot water', *The New York Times*, 20 May.
Krauss, Clifford (2012), 'Shell Delays Arctic Oil Drilling Until 2013', *The New York Times*, 17 September 2012.
Miller, David (2011), 'Jellyfish force Torness nuclear reactor shutdown', *BBC News*, 30 June 2011.
Mishra, Dinesh Kumar (2006), 'The unbearable lightness of big dams', *Hard News*, October 2006.
Mishra, Pankaj (2007), 'The train to Tibet', *The New Yorker*, 16 April 2007.
Morath, Eric (2012), 'Hurricane Isaac Likely Caused Some Ripples in Economy', Real Time Economics, *The Wall Street Journal*, 31 August 2012.
Muller, Grace (2012), 'Tropical Storm Lee May Put Long-Term Pressure on Oil Rigs', http://www.accuweather.com/en/weather-news/longterm-pressure-for-oil-rigs/54532 (accessed 8 November 2012).
Nautiyal, Monika (2006), 'Desert into sea', *Hard News*, October 2006.
Paskal, Cleo (2009), 'The Vulnerability of Energy Infrastructure to Environmental Change', a joint publication of Chatham House and the Global Energy and Environment Strategic Ecosystem of the Department of Energy, July 2009.
Patel, Ashok (2006), 'Modidom's watery grave', *Hard News*, October 2006.
Pemberton, Mary (2006), 'BP: Oil production may be closed months', *Associated Press*, 7 August 2006.
SAARC Disaster Management Centre (2008), 'South Asian Disaster Report 2007', New Delhi, 61–66.
Savonis, Michael J., Virginia R. Burkett, Joanne R. Potter et al (2008), 'Impacts of climate change and variability on transportation systems and infrastructure: Gulf Coast Study, phase 1', US Climate Change Science Program Synthesis and Assessment Product 4.7, US Department of Transport, March 2008, p. 4–38.
Shukman, David (2008), 'Flood risk fear over key UK sites', *BBC News*, 7 May 2008.
Stern, N. (2005) *Stern Review: The Economics of Climate Change*, HM Treasury, London.
Thakkar, Himnshu (2006), 'Damn it, this was designed!', *Hard News*, October 2006.
UNEP (2005) 'The fall of water', available at unep.org/PDF/himalreport.pdf.
Upadhyaya, Himanshu (2006), 'Cry me a river', *Hard News*, October 2006.
US Energy Information Administration (2012), 'Gulf of Mexico Fact Sheet', http://www.eia.gov/special/gulf_of_mexico/ (accessed 10 November 2012).
van Vliet, Michelle T.H. John R. Yearsley, Fulco Ludwig, Stefan Vögele, Dennis P. Lettenmaier and Pavel Kabat (2012) 'Vulnerability of US and European electricity supply to climate change', *Nature Climate Change* (2012), 3 June 2012.
Wolman, David (2006) 'Train to the Roof of the World', *Wired*, July 2006.

# 18. Paradoxes and harmony in the energy-climate governance nexus[1]
*Stéphane La Branche*

> Limits of survival are set by climate, those long drifts of change which a generation may fail to notice. And it is the extremes of climate which set the pattern. Lonely finite humans may observe climatic provinces, fluctuations of annual weather and occasionally may observe such things as 'this is a colder year than I've ever known'. Such things are sensible. But humans are seldom alerted to the shifting average through a great span of years. And it is precisely in this alerting that humans learn how to survive . . . They must learn climate.
> *Children of Dune*, Frank Herbert (Orion, 1976, 350)

## INTRODUCTION[2]

As early as 2008, climate change was seen by most international organisations as both the number one accelerator and amplifier of natural risks and the single most important obstacle to development efforts in the third world. This position has been reinforced since, driven home by new empirical data and studies from almost all disciplines. In the academic, public, economic and political spheres, climate change (CC) is raising deep issues about the economic structure of our society, debates and theorisation regarding post-carbon or non-carbon based production and consumption (or the green economy). Even a cursory look at this book shows how highly complex the interconnections between energy and climate governance can be. These links give some indications that CC is slowly inserting itself into *every aspect of our societies*.[3] Indeed, our argument is that climate governance is becoming a *meta*-governance, redefining, modifying and inserting itself into already existing ones.

But this movement toward a (climate) meta-governance encounters another primordial issue: the emerging global energy (in)security that is both due to 'classical' (due to peak oil and increased demand) and 'non classical' factors, such as restrictions imposed on carbon-based energy due to environmental efforts, especially CC, which creates an 'artificial' rarity. But, while climate governance and energy governance can indeed go in the same direction, they also can be at odds. From an analysis of institutional texts and reports and social science academic papers (national, European and international institutional reports, see bibliography) on the issue of climate and energy governance, I will explore the interactions between these two governances by focusing on areas of probable and already existing tensions. The exercise is not straight forward for several reasons: the transversality of the double energy-climate issue increases their complexity and both governances exist at different scales of analysis and in very diverse institutions. First, I need to develop that CC is, potentially at least, becoming a meta-governance aimed at managing climate as a *meta risk*.

## CLIMATE CHANGE: FROM META RISK TO. . .

In 2008, the International Panel on Climate Change (IPCC) received the Nobel prize for *Peace*, not science, because of the probable impacts of CC on our civilisation: changes in food production and water availability will lead to increased hardships for the poor and vulnerable, migrations, increased geopolitical tensions for resources and regional conflicts. Climate change has indeed emerged, for international organisations, as an international security issue. An American think tank working for the US Army has even declared CC as the number one security issue for the country in the 21st century, before terrorism (CNA 2007). CC has even helped redefine the security agenda setting and framing – especially those issues linked to the coming energy crisis . . . Furthermore, Wanneau (2011) showed how CC has even influenced security theory, forcing it to come out of its economic-military power analytical framework to broaden and take into account other non-classical issues. At the same time, from 2005, concerns for energy security and availability especially to the poor also increased. But while CC has really risen on the international political scene since the end of the 1990s, in most countries, it has emerged as a social and political reality since 2005 or so (depending on the country, the administration, or the region).

2005/6 was indeed a clear turning point, when climate and energy issues were explicitly and integrated in different international norms and declarations such as the UN's Millennium Development Goals (in 2005), the World Bank's development approach, water management and risks, as well as insurance policies, construction norms and urbanism. Note that in 2007, the UNDP's *Human Development Report* was subtitled: *fighting climate change*.

These issues thus started to be slowly redefined, reworked, reconceptualised, re-practised (and even re-taught in our teaching institutions) with climate and energy objectives in mind. While approaches to energy used to focus on energy supply, security and access, they now add its sustainability (and climate friendly), its role in health, education and access to clean water . . . But an even more fundamental change is that energy has ceased to be considered as a cheap, quasi infinite resource, without impacts on the ecosystem. Energy is now seen as an issue that needs to be tackled, planned for and managed. Energy as we know it and use it is no longer without environmental consequences, nor is it infinite nor cheap. In many texts, a link is made to the fact that the most vulnerable and poorest groups will be most adversely affected by CC, which will increase their vulnerability to natural hazards and decrease their capacity for development, for improving their quality of life and for adapting to CC impacts. Increased cost and decreased access to energy will mean a decreased capacity for coping with other difficulties and challenges, including the additional ones created by CC. All these issues, with their associated governance systems, are both affected by the 'objective' reality of CC (as evaluated by natural sciences) and by the social and 'subjective' efforts (as analysed by the social sciences) at dealing with it and managing it – i.e., governance. Indeed, the emergence of climate (meta) governance is less closely tied to its natural aspects than it is to its social, economic and political dimensions – which are even more complex – and which are intrinsically tied to energy.

Underlying these evolutions is the understanding among institutions that CC has become a meta-risk: while CC as such does not create new risks, it is seen as *the* risk above (and beyond) all others, in that it modifies and amplifies existing ones, such as floods, avalanche, droughts, waterfalls and storms and their consequences on food production and development efforts in the third world. For UNEP:

> climate change is every bit as alarming as any of the threats facing humanity, and probably more alarming than most, because – without drastic change – its impacts appear certain . . . What is at issue is not comfort, or lifestyle, but survival. Food security is at stake, climate refugees might hamper political security, and more uncomfortable changes will put humanity under strain. (UNEP 2008, 23)

In the 19990s water was declared by international institutions to be the second millennium's foremost issue. It is no surprise that it was replaced by climate at the turn of the century.

My reviews of studies on international, national and local climate governance indicate that the initial insertion of climate issues in natural risks and management has been spreading in the last 10 years to almost all other dimensions of human activity. Indeed, during the same period (2006–08), energy and climate issues started being integrated at the national and local levels into urban development planning (and theories) in some cities (it is now more widespread, with thousands of cities networking to fight and adapt to CC), construction goals and methods, agriculture, transports and to a much lesser degree, consumption (carbon footprint labels are a much more recent phenomenon in only some industrialised countries). Thus, CC first appeared as a meta-risk and then as the number one obstacle to development efforts and thus, very quickly, as a multi-stake, multi-issue governance problem at all levels – as an emerging meta-governance, perhaps? As such, it would be wrong to speak of an international climate regime since this notion is linked to a general governance mode linked to a specific issue. CC and its impacts are not issue-specific, they touch on every human activity – an anthropogenic issue.

The importance of the social, political and institutional dimensions of CC appears even more strongly in adaptation. Recent studies on adaptation to CC show that the most important factors playing a role in climate resilience (the capacity of a society to adapt to CC) are not natural factors (such as increased droughts or floods) but those social, institutional, political and economic factors that play a role in a society's capacity at preparing for and responding to climate-induced changes. This is leading to deep changes in how the developing world attempts to develop (influenced by the climate resilient and climate proof development approaches of the large international development institutions, with many national development agency following the lead). But the social factors playing a role in adaptation to CC are even less well understood than natural factors. It is really only in the last few years that the international scientific community has begun recognising the importance of social issues: the IPCC's 5th report (due in 2013) will focus on the economic and social dimensions of CC for both mitigation and adaptation. Hence, the emerging idea that CC is secondarily a hard science issue and increasingly, a social science one. The climate issue is having a direct effect on research and academic institutions, going beyond a simple effect on political speeches and having a real discursive effect in Foucault's sense of the term: becoming a truth, with its mechanisms and apparatus (Foucault 1980) and thus having effects on actual daily operations of institutions, companies, administrations and research.

## ... TO CLIMATE (META) GOVERNANCE?

The different examples given point to an emerging climate meta-governance. When I first started working on this issue, in 2008, it was in the framework of a prospectivist exercise, very rare in political science. I worked from weak signals, probable indications and signs of what might be emerging. In the last few years, these have all become strong signals;

the unlikely, more probable; the 'probable', more certain and emerging trends have become more mainstream. Now, examples abound: water, natural risks, insurance, urbanism, transports, biodiversity, security, housing, research, development, public policies, health, economic development, agriculture, cultural norms ... are all being re-defined by the climate issue either for mitigation or for adaptation. Even the research world has been affected by this movement: when one looks at the different calls for research in different fields, more often than not a CC dimension has to be integrated in the project and one's chance of getting funding increases if one adds a climate dimension. My different studies in my research programme on climate governance all suggest very strongly that CC is *emerging* as a new type of 'total' governance through three different means:

1. it is inserting itself in issue and sector-specific governance, as well as individual values and legal international, national and local norms;
2. it tends to redefine these types of governance, *from above*, as an overarching framework and;
3. it is *diffusing through* sectors in a transversal manner and from below (global and local civil societies).

For example, in the last few years, urban projects in most developed countries have had climate conditions attached to them. In construction, the introduction of energy and climate norms in France are leading to new needs in terms of labour skills (isolation, double-glazed windows . . .), new energy objectives to be reached, new architectural designs and approaches, new integrated urban planning, as well as efforts at coordinating different types of non carbon based energy sources (geothermal with wind and solar . . .). Indeed, the double climate-energy crisis has had profound effects on this sector.

Construction and urbanism actors have entered a new competitive field structured by new climate and energy efficiency norms. In order to achieve ambitious clean energy systems, buildings are now conceived as different units of a larger, energy-integrated, whole, at the neighbourhood level. This in turn requires a specific project management by city administrators and coordinating teams, who have to coordinate the energy related actors with the norms correctly. This is far from simple since different buildings may have different forms, styles, structure, materials, objectives, norms as well as different types of energy. This has led to a new role for architects who now need to go beyond drawing and start acting as orchestra leaders . . . To these different climate and energy norms and objectives, one also finds participatory norms and in many countries, minimal norms concerning social diversity (either economic or social). Thus, from an almost purely technical and economic issue, energy has gained in the last few years in complexity and now intersects politics, economics, technology, urbanism, networks, skills, labour, geostrategy, management, environmental, ethical, business models, energy security and autonomy . . .

The way climate governance is affecting water governance, to take another example, is multi-layered. Since 2005, I have not found a single paper on water governance without a climate component being present, notably its impacts on: water quality, quantity and access; floods and river flows; biodiversity and increasingly, on the impact of CC on water as an energy source. Indeed, much effort and hope is put on non-fossil fuels, and hydroelectric dams are at the centre of much attention, as the most promising source of renewable energy. But hydroelectric capacity will be directly affected by CC, through its impacts on variations in water quantity (either not enough or too much) as well as regularity. Indeed, in France, dams are used as equalisers of peaks in the national (nuclear) electricity demand but they may fail to fulfil this role if water becomes insufficient. Then water temperature has increased in average in most European countries' rivers in the last decade, but 28° is the limit at which a French nuclear plant must be shut down. In 2011, an energy crisis was very closely averted, when after two months of high temperature and little rainfall, it finally started raining, avoiding a shut down of one of the French nuclear power plants by 0.5°.

A last example is health governance. The 2003 European heat wave illustrated all the problems of emergency responses when this emergency is on a large scale and lasts too long: insufficient financial and human resources as well as management failures led to an ineffective response to the crisis and in taking care of the most vulnerable. Our current emergency responses system is entirely inadequate when faced with recurring catastrophes that become the norm: according to recent IPCC projections, the 2003 European heat wave will be a 'normal' year before 2050, two years out of three.[4] But the political and social difficulty is that an increase in *average* temperature not only translates into more emergency situations but also into permanent health problems since increased heat is associated with respiratory diseases, making already sensitive people (especially elderly) even more vulnerable to heart problems and so on . . . And this, in a society where the age pyramid is becoming increasingly heavier at the top.

Thus, climate as risk is also redefining the common perception of risk as such, from punctual and extreme events to recurrent, permanent pressure. In this sense too, it is a meta-risk that modifies our current understanding of adaptation to risk. From responding to *events*, we now need to learn to develop long-term self adjusting policies to constant but yet uncertain and future change, which includes threshold levels and tipping points which are yet not defined. Our governance system and culture is thus facing a real new challenge. To this we need to add that one society's vulnerability to risk is not another's. The literature on ecological inequalities shows that social, political and economic capacities play the key role (not exposure to natural hazards) in how much a crisis or a new stress will impact a society, including with regards to secure and diversified energy sources. But even wealthy societies have a limit: with all its wealth, knowledge, organisation capacity on dikes and sea water management, even the Netherlands will eventually reach a limit on its capacity to adapt to an increase in sea levels. Viewed globally, these different issues are linked to the phenomenon of path dependency, a notion underlying our argument.

## PATH DEPENDENCY

> Addiction is a terrible thing. It consumes and controls us, makes us deny important truths and blinds us to the consequences of our actions. Our society is in the grip of a dangerous greenhouse gas habit.[5]
> Ban Ki-Moon, UN General Secretary

Developed by Pierson (2000), the notion of path dependency puts forward the idea that any decisional process is affected by decisions taken at an initial stage. Some choices are more determinant than others, which causes a self-reinforcing process: the more one goes ahead on the path, the deeper it becomes and the more difficult it is to move out of it and the less there are political, social, cultural, economic and even conceptual alternatives. Indeed, the double climate-energy crisis may be first and foremost a crisis of our imaginative capacity at conceiving alternatives to the carbon path. Ingrained homogeneity limits efforts at heterogeneity; carbon energy dependency limits efforts at developing plural energy systems.

We are now touching the core of the tensions between climate and energy governance: our deep carbon dependency, inherited from the past, makes it very difficult to open a new path that will resolve the climate crisis. Indeed, this may well be the fundamental obstacle: is our carbon addiction so deep that only an in-depth social, political and economic transformation can take us out of it? Most texts seem to presuppose that a deviation suffices but the arguments they put forward do not allow such a conclusion, especially considering the depth of our present carbon path in all domains, including everyday behaviours and institutions.[6] In a special report, the World Bank warns:

> The environment ministry base provides core expertise and an established institutional home, but typically these are junior ministries. The linkage to the energy ministry engages a key sectoral connection (fossil fuel combustion lies at the root of climate change), but it also contains some risks: climate policy may be captured or overwhelmed by more established energy policy orientations and/or other sectoral connections may be neglected. (Meadowcroft 2009, 15)

Trying to understand our present energy transition efforts means understanding the multiple and complex contemporary inertias that influence efforts at moving to another energy system – which one exactly is still open for debate, and vision. Even if there were no climate crisis, and we considered only peak oil, we would have to reflect on these issues but the urgency would be less important (there is still enough coal and gas in the ground to last until the turn of the century, especially with unconventional sources). But the climate crisis amplifies and modifies the oil crisis, influencing how we adapt to it and how we attempt to manage the peak oil period. In turn, the oil crisis also affects the way we attempt to manage the climate crisis. The difficulty is that at times the two goals are compatible and at times they conflict with each other.

Considering the complexity of this question, it is surprising that there is very little consideration in international reports and texts for theories of transitions applied to energy governance, even while there are many operational definitions of energy governance in the texts reviewed. Such analysis of transition would have to include political, economic, social, cultural, institutional, behavioural and technological factors that would explain their effects on change: increasing or decreasing its speed, depth and quality or that could even block change. Most efforts and means devoted to both climate and energy governance proposed in most texts reviewed aim at rectifying the path dependency's trajectory. Few put forward that we would need to get out of it entirely. This seems to depend more on an institution's position regarding the severity of CC impacts than anything else ... and their deep beliefs regarding the capacity to govern such change. Indeed, the governance question here is primordial: part of the answer regarding whether we need to get out entirely or just divert the trajectory depends in part, not entirely, on how well climate and energy governance are coordinated with each other. In most texts, both climate and energy governance are conceptualised as mostly compatible with each other and when conflicts are found, energy is given more weight than climate in the political, economic and social arenas.

Still, CC has modified the way we conceive energy rarefaction. Traditionally, and to summarise to a great degree here, the oil peak is based on the amount of oil existing in the ground, plus accessibility and costs/prices linked to supply and demand, (as oil prices increases, previously 'unavailable' oil due to high extraction costs, becomes 'available'). But with climate governance, an induced and structured rarefaction has been emerging through limits imposed on carbon-based energy: carbon taxes, subsidies for non carbon energy, efforts at finding and developing alternative energy types, or even, individual agreement and efforts to modify behaviours to fight CC. Interestingly, this has a corollary but paradoxical effect: increasing the time period during which oil will be available . . . But other areas of tension need to be highlighted.

## ENERGY GOVERNANCE

Globally, in the texts reviewed, contemporary energy governance takes a threefold approach. The first objective is to ensure sufficient energy for our civilisation's needs – security of energy supply. But the oil crisis brings us to diversify energy sources and types – the second angle. Coal use is expanding at an incredible rate and so is natural gas, as well as research and explorations for non-conventional, hard to get carbon based energy sources, including in previously protected areas such as the Arctic. In these cases, we are clearly at odds with climate governance but not with decreasing energy dependency. The third angle, most recent, to energy governance concerns efforts to develop a non-carbon energy mix that includes wind power, geothermal, solar, nuclear and 'passive' forms of energy (notably isolation and consumption reduction – which raises the very difficult questions of values, comfort, behaviours, habits, rights to consume, freedom, liberalisation of energy markets, energy production by individuals and so on . . .).

Interestingly, most institutional reports do not offer clear *conceptual* definitions of climate and energy governance – besides the idea that they are part of a sustainable development framework. Yet, *operational* definitions are used. The IPCC defines energy governance as 'ensuring long-term security of energy supply at reasonable prices to support the domestic economy' (Sathaye et al. 2007, 719). Interestingly, this narrow definition does not include mitigation measures, the energy and the climate aspects not being linked to each other by the IPCC! But we do know what parts per million of $CO^2$ we should not surpass or the number of degrees we should strive for and the general means by which to do this. Also, several key notions and words have appeared which bridge the gap between energy and climate: *climate friendly development/investment/ business/technology; climate-compensated products, while climate neutrality* (UNEP 2008, 14), *climate resilient society, post-carbon society* or *non-carbon growth* are becoming more commonplace – in some circles at least. The energy question is omnipresent in the UNDP's 2008 Human Development Report, while a special section is entitled *Climate policy as human development* (UNDP 2008, 28). Indeed, the choice of words used in texts and the general goals offered (introducing a climate approach in all aspects of life, in every sector, in a transversal integrated way rather than a sectoral approach) all point to one direction: even though institutions do not phrase it in such a way, the global objective is indeed, to create a climate meta-governance.

The EU's Lisbon Strategy integrates both climate and energy dimensions and aims at reducing both consumption and energy dependency on foreign markets. Energy efficiency is both a means to face increasing oil prices and a predicted energy dependency of 70 per cent by 2030 (European Commission 2008a) for EU countries. Then, renewable energies are key to this effort since they also reduce GHG emissions. The 20/20/20 objective is to reach by 2020, a 20 per cent reduction in energy consumption, 20 per cent of renewables in the total energy package, with 10 per cent agrofuels for transports in each member state. It is estimated that about half the goal can be achieved with already existing measures. But this does not take into account social factors such as the social un/acceptability of energy reduction measures nor of non-carbon based energy diffusion. Indeed, the diffusion of locally or individually produced energy puts in the hands of individual citizens the capacity and the decision to implement and use these energies ... or not. Factors depend on price, return on investment, aesthetics, understanding, opportunity, ownership, values ... these all contribute to the emergence of a new energy consumer. From a passive individual simply paying his or her energy bills at the end of the month, citizens are now more and more involved in different energy choices, regarding sources, types, at what times of day and at what price, and through different technological means to control energy consumption. The same individual may even now produce energy.

As for climate governance, a transversal definition from the texts could be the sets of efforts in the short, middle and long terms to develop a non-carbon and climate-proof society through mitigation and adaptation measures (technologies, cultural change, economic levers, public policy, education, legislation ...) and whose aim is to cap the increase in temperature at $+2°$ (relative to industrial era) so as to reduce as much as possible social and political instability resulting from CC impacts while ensuring material development. These efforts are structured and slowed down by carbon based energy path dependency, whose associated problems are supposed to be resolved through the framework of sustainable development. To sum up, then, sustainable development aims at allowing for economic development in a participatory and ecological manner, while energy governance aims at ensuring economic activity, energy security and quality of life and finally; climate governance aims at stabilising greenhouse gases and climate change. While several single obstacles have been clearly identified, the general framework in which they operate, the carbon path dependency, is less well understood. Let us then explore further the energy pole of the climate and energy governance and sustainable governance trio.

## TENSIONS BETWEEN CLIMATE AND ENERGY GOVERNANCE

Climate governance is having an effect on energy governance but it is not, by a far cry, entirely setting the rules of the game. Energy needs and security are a pressing, immediate, necessity (or perceived as such) for our quality of life, our consumption/production and our general mode of living. But climate impacts are on the longer term and are yet uncertain. In this context, we see both an increase in the diversity of energy sources used and at the same time, a sharp increase in the use of coal, while shale gases are undergoing an explosive increase in production rates. For the time being, energy governance (with security of supply) remains the most important short-term obstacle to efficient climate governance and to the rise of climate governance as a meta-governance.

One of the most important differences between climate and energy governance is that the concept of energy security is an already well practiced and *politicised* notion. It is key to both decision makers and industrial actors. Necessary to industrial operations and profits, energy security is also inscribed in most nations' constitutions. Climate security still remains a climate expert notion: even while climate security is linked to CC's impacts on already existing risks, climate security is yet to be put on the political agenda of the world's nations – and still less inscribed in constitutions. But international organisations (see European Commission 2008a, UNDP 2008 or World Bank 2008) increasingly recognise climate security as a priority, especially through issues such as water stress, decreased agricultural production, droughts and violent natural disasters which will all contribute to population displacements. These institutions also make a link between climate and energy: the rarefaction of energy resources will combine with decreased agricultural production to increase *local* conflicts over resources as well as *international geopolitical* tensions to secure energy access by the major powers (Schwartz and Randall 2006), including China, the USA and India. Clearly, if CC has been undergoing a process of securitisation (Leboeuf and Broughton 2008), it is mostly within international organisations.

One notes clear differences in 'attitudes' toward both forms of governance: while climate governance is seen as *eventually*, leading in the middle and long terms, to new business and job opportunities, energy governance is more conceived in the short term and it aims at not losing jobs and to limit as much as possible the impact on the quality of life of citizens. Climate governance is presented in terms of potential future gains if one develops a good strategy, while energy governance is often termed as avoiding a potential crisis. Energy governance is a short term issue closer to political and economic temporal agendas; it is direct in terms of effects and costs, it is visible and easily perceivable by all stakeholders (who would want to experience a power shortage in the middle of winter?). But the fight and adaptation to CC are not yet part of national security agendas; they are long-term diffuse problems; they are not visible (only their indirect effects are quantifiable and they are not visible to the common eye); they are global (their local effects cannot yet be really seen in wealthy countries); their costs are still invisible and very indirect for most stakeholders. But politics tends to be short term and local and at best, for individual stakeholders, national. The UNDP puts it this way: 'The deeper problem is that the world lacks a clear, credible and long-term multilateral framework that charts a course for avoiding dangerous climate change – a course that spans the divide between political cycles and carbon cycles' (UNDP 2008, 5).

Most analysts see renewable energies (REn) as a way out of these difficulties because they reconcile most tensions between both forms of governance: they stimulate R&D, reduce GHG emissions and allow for local and decentralised energy production – thus reducing energy dependency on foreign markets, which has become a priority for many countries now and an integral part of energy policies in the US and the EU for example. In many ways, REn are seen as a mechanism to get out of the carbon path but there remains an essential condition linked to participatory democracy: that they are accepted by the populations. Indeed, REn create a new logic between individuals and energy production: the traditional passive energy consumer is now becoming a consumer with increasing choices for energy types and energy suppliers or even producers. More importantly, the consumer can now become self-sufficient in energy or even become an energy producer selling on the market, *given that such an individual consumer adopts REn*. That is an important issue since wind power continues to be refused by local communities, while adding insulation to already existing buildings is judged by owners to be too expensive and the reduction of energy consumption expected by reports is still not happening. In other words, the social un/acceptability of REn may have a direct impact

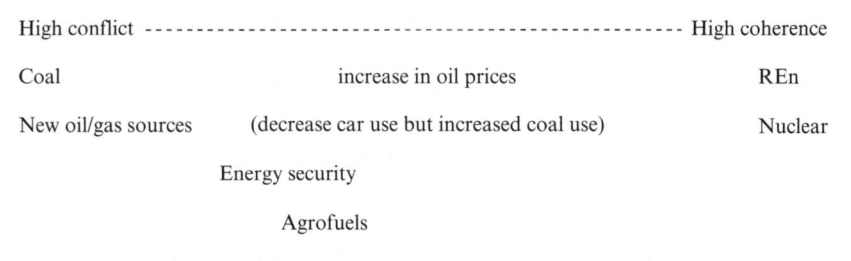

```
High conflict -------------------------------------------------- High coherence

Coal                        increase in oil prices                    REn

New oil/gas sources    (decrease car use but increased coal use)      Nuclear

                        Energy security

                           Agrofuels
```

*Figure 18.1   Climate governance versus energy governance*

on the speed and depth at which they are deployed and thus on the way out of carbon dependency.

The above modifies the argument that climate governance is becoming a meta-governance. But areas of conflict with energy governance block or slow down this movement. In many countries the emergence of energy as a national security issue in the last few years has led to new efforts at developing and finding non-carbon but also carbon based energy, such as coal, and previously hard to reach and expensive unconventional carbon-energy sources. At the same time, institutional reports abound with good examples of climate and energy governance practices. They insist on the fact that the best policies combine both energy and climate objectives but they do issue warnings that not all is post-carbon or carbon sober in a would-be climate governed world: for example, decoupling climate and energy is leading to serious detrimental policies, such as the development of coal energy in China or the USA (Gupta et al. 2007, 796). The diagrams below illustrate only a few other examples of the different obstacles on the road to a high sobriety, non-carbon society.

Note that agrofuels are not placed high on the coherence scale because forests are sometimes cut for their production. But there are high expectations for second and third generation (food leftovers for example and algae). What appears is that issues of conflict tend to show that in the short term, economic and national interests – and national security for several countries – are driven by energy, not climate.

Almost all texts recommend the diversification of energy sources (in terms of countries and types of energy) with an increase in REn; they expect a reinforcement of nuclear energy; they insist on the necessity to decrease energy consumption through technical efficiency and; push for (hope?) sobriety, i.e., a decrease in energy consumption through behavioural changes (daily behaviours such as cooking, heating and transport as well as consumption). But these measures sometimes conflict with democratic rights, expectations, beliefs and habits linked to consumption and the right to do so.

*Paradoxes and harmony in the energy-climate governance nexus* 415

High conflict ------------------------------------------------- High coherence

    Energy security                Rebound effect REn

                                  Nuclear

                                  agrofuels

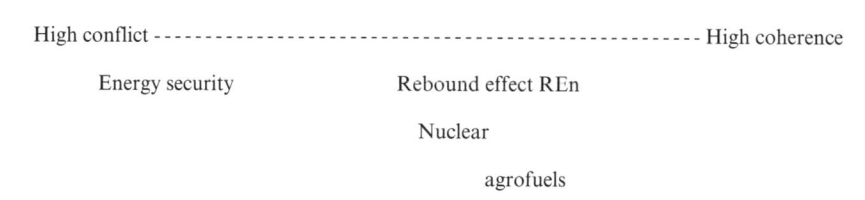

*Figure 18.2*   Climate/energy governance versus social acceptability

All texts agree that REn are a key element of the transition to a carbon sober society for both energy and climate reasons, but they are not sufficient to solve the energy issue as a whole. And in democratic wealthy countries, they are subjected to local acceptability and the capacity and willingness to change behaviours, or adopt REn where they are, or where they could be implemented. Their social acceptability depends on several factors linked to individual types of REn, but in general they include cost and time of return on investment (solar and isolation), aesthetics and wildlife (wind), historical protection, sense of freedom and autonomy (battery-fuelled cars), perception of security, comfort (Wan-Jung et al. 2010) and the rebound effect.

The rebound effect occurs when the expected gain in energy efficiency from a technological improvement is diminished or even cancelled by behaviour arising from the expectation of the gains in energy the technology is supposed to bring. The perception that the technology allows a decrease in energy consumption leads to behaviour that is even more energy hungry. For example, since fuel efficient vehicles make travel cheaper, consumers may choose to drive further and/or more often, thereby offsetting some of the energy savings achieved. This is termed the *direct* rebound effect. One may also have an indirect rebound effect, which occurs when 'any reductions in energy demand will translate into lower energy prices which encourage increased energy consumption ... The sum of direct and indirect rebound effects represents the *economy-wide* rebound effect' (Sorrell 2007, viii).

In the UK, a large government-funded programme aimed to improve energy efficiency with insulation and heating systems decreasing the cost of energy in the first year. But families increased the temperature of their houses (and energy consumption) so much that in the second year that the cost increased beyond what it had been before the improvements (European Environment Agency 2007, 271).

These examples highlight the complexity of the phenomenon of carbon-based energy path dependency due to its multi-factorial and multidimensional elements. In other words, for reasons of energy needs and path dependency, the way out of the fossil fuel path is arduous at best and it raises fundamental questions regarding our consumption and production as well as our political systems. The question of how quickly change will occur and how difficult it will be, depends only in part on technological progress, with economic, political and social factors playing a very important role, but the tensions between participatory democracy and energy governance are worth noting.

## CONCLUSION

The underlying transition question is: how do we stabilise GHG in a context where non-carbon based energies are insufficient to maintain even our present standards of living *made possible by fossil fuels* (this is *the* key point) and where these fuels are more than sufficient to drive us to a world ecological catastrophe? The dynamics of this evolution are complex, both harmonious and conflicting. Our initial hypothesis that climate governance is emerging as a meta-governance redefining other issues, including energy, holds but under certain conditions. It holds when climate and energy governance work harmoniously. But in cases where there is a conflict between both forms of governance, energy takes priority over climate, even if it is not entirely left aside. This is indeed not a binary, all or nothing, process: even while it may lose the short-term race, climate governance influences energy governance. Hence, even when a coal plant opens, there are efforts to implement carbon capture and sequestration technologies. Oil and coal extraction emit less GHG than before due to new techniques and technologies, while vehicle engines are also declining in emissions. But fossil fuels are still the rule and still get priority. While climate governance does seem to be emerging as a meta-governance, in the short term, the importance accorded to energy will give it priority when conflict arises between both, slowing its process down . . . perhaps until both become integrated in a climate compatible way.

The question then is: do we need to entirely get out of the carbon based economy and social path dependency or is a deviation of the path enough? According to Abbas (2011), who offers a macro economic analysis of climate governance, our modes of production and consumption are so intertwined with our carbon based energy structure that our efforts at resolving the climate crisis will likely *necessitate* an exit from the existing economic system – to which we can add the problem of urgency.

The questions raised here are important for the social sciences and policy makers, as well as for populations. Indeed, the first great human revolution, the Neolithic revolution, and then the sedentarisation of the human species – with the associated development of agriculture – happened 'on their own', without conscious volitions on our part. For the first time in human history, not only are we facing a really global and *total* crisis but we are also aware of this crisis and are attempting to create planned, conscious, solutions and apply them. Toward the creation of a total and global meta climate governance? Such an effort has never been attempted; we have no past experience in the matter and no lessons from history to draw on. Thus, indeed, the challenge is total and global in the sense that it involves all human activities. In this sense, CC is not only a phenomenon; it is an epiphenomenon (Olivier 2005), affecting the way we think and perceive the world globally. As such, climate has become increasingly a political issue in the large sense of the term, in terms of 'living together'.

## NOTES

1. This chapter presents the main argument from a multi-authored book: S. La Branche (ed.) (2011) *Le changement climatique. Du méta-risque à la méta-gouvernance*, Lavoisier.
2. My sincerest thanks to EDF (Electricité de France) who funded the initial research that led to the reflections presented here. 'La gouvernance climatique et énergétique face à la dépendance au sentier. Recension et Etude de prospectives'. 2008, EDF.
3. This has impacts on analyses as well since CC necessitates, arguably, more multidisciplinary approaches than any other environmental issue.
4. See among others, IPCC (2012). *Managing the risks of extreme events and disasters to advance climate change adaptation*, WMO UNEP.
5. In UNEP (2008), p. 6.
6. Young's recent studies fit rather well with this idea of path dependency, applied to institutions. To him, environmental efforts by an institution may well be blocked by its internal operations, its identity, its culture or its structure. This natural '*stickiness*' may require a change in identity in order for environmental goals to be reached (Young 2002a, 2002b).

## BIBLIOGRAPHY

Abbas, M. (2011), 'La lutte contre les changements climatiques vers une macro-transformation du capitalisme?', in S. La Branche, *Le changement climatique: du méta-risque à la méta gouvernance*, Lavoisier.

Ambrosi, P. and S. Hallegatte (2005), *Climate Change et enjeux de sécurité*, CIRED, pp. 15–20, available at: http://www.centre-cired.fr/spip.php?article464 (accessed February 2013).

Braithwaite, J. and R. Williams (2001), *Meta risk management and Tax system integrity – Centre for Tax System Integrity*, Research School of Social Sciences, Australian National University.

Chaumel, M. and S. La Branche (2008), 'Inégalités écologiques: vers quelle définition?', *Espaces, populations et sociétés*, 1, 101–110.

CNA Corporation (2007), *National security and the threat of climate change*, Washington.

European Commission (2005), 'L'efficacité énergétique – ou Comment consommer mieux avec moins', Livre vert de la Commission (22 June).
European Commission (2008a), 'Changement climatique et sécurité internationale', Brussels.
European Commission (2008b), 'Proposition de décision du parlement européen et du conseil relative à l'effort à fournir par les États membres pour réduire leurs émissions de gaz à effet de serre afin de respecter les engagements de la Communauté en matière de réduction de ces émissions jusqu'en 2020', Brussels.
European Environment Agency (2007), 'Europe's environment. The fourth assessment', European Environment Agency, Denmark, Copenhagen.
Foucault, M. (1980), *Power/Knowledge*, Harvester, Brighton.
Gleditsch, N.P. (2006), 'Changements environnementaux, sécurité et conflits', *Les Cahiers de la sécurité, Environnement, Climate Change et sécurité*, 63, 121–156.
Gupta, S., D.A. Tirpak, N. Burger, J. Gupta, N. Höhne, A.I. Boncheva, G.M. Kanoan, C. Kolstad, J.A. Kruger, A. Michaelowa, S. Murase, J. Pershing, T. Saijo and A. Sari (2007), 'Policies, Instruments and Co-operative Arrangements', in B. Metz, O.R. Davidson, P.R. Bosch, R. Dave and L.A. Meyer (eds), *Climate Change 2007. Mitigation. Contribution of Working Group III to the Fourth Assessment Report of the Intergovernmental Panel on Climate Change*, IPCC 4th Report, Cambridge University Press, Cambridge, United Kingdom and New York, NY, USA, Chapter 13, pp. 745–808.
Halsnæs, K., P. Shukla, D. Ahuja, G. Akumu, R. Beale, J. Edmonds, C. Gollier, A. Grübler, M. Ha Duong, A. Markandya, M. McFarland, E. Nikitina, T. Sugiyama, A. Villavicencio and J. Zou (2007), 'Framing issues', in B. Metz, O.R. Davidson, P.R. Bosch, R. Dave and L.A. Meyer (eds), *Climate Change 2007. Mitigation. Contribution of Working Group III to the Fourth Assessment Report of the Intergovernmental Panel on Climate Change*, IPCC 4th Report, Cambridge University Press, Cambridge, United Kingdom and New York, NY, USA, Chapter 2, pp. 117–168.
IPCC (2012), *Managing the risks of extreme events and disasters to advance climate change adaptation*, WMO UNEP.
Klein, R.J.T., S. Huq, F. Denton, T.E. Downing, R.G. Richels, J.B. Robinson and F.L. Toth (2007), 'Inter-relationships between adaptation and mitigation', in B. Metz, O.R. Davidson, P.R. Bosch, R. Dave and L.A. Meyer (eds), *Climate Change 2007. Mitigation. Contribution of Working Group III to the Fourth Assessment Report of the Intergovernmental Panel on Climate Change*, IPCC 4th Report, Cambridge University Press, Cambridge, United Kingdom and New York, NY, USA, Chapter 18, pp. 745–777.
La Branche, S. (2011), *Le changement climatique: du méta-risque à la méta gouvernance*, Lavoisier.
Leboeuf, A. and E. Broughton (2008), 'Securitization of Health and Environmental Issues: Process and Effects. A research outline', Paris: IFRI.
Mabey, N. (2008), 'Delivering Climate Security. International Security Responses to a Climate Change World', Colchester Essex, Routledge Journals, RUSI.
Meadowcroft, J. (2009), 'Climate Change Governance', Policy Research Working Paper 4941, World Bank.
Metz, B., O.R. Davidson, P.R. Bosch, R. Dave and L.A. Meyer (eds), *Climate Change 2007. Mitigation. Contribution of Working Group III to the Fourth Assessment Report of the Intergovernmental Panel on Climate Change*, IPCC 4th Report, Cambridge University Press, Cambridge, United Kingdom and New York, NY, USA, Chapter 2, pp. 117–168.
OECD (2008), 'OECD Environmental Outlook to 2030', Paris.
Olivier, L. (2005), *Le savoir vain. Relativisme et désespérance politique*, Liber.
Pierson, P. (2000), 'Increasing Returns, Path Dependency, and the Study of Politics', *The American Political Science Review*, 94(2): 251–267.
Sathaye, J., A. Najam, C. Cocklin, T. Heller, F. Lecocq, J. Llanes-Regueiro, J. Pan, G. Petschel-Held, S. Rayner, J. Robinson, R. Schaeffer, Y. Sokona, R. Swart and H. Winkler (2007), 'Sustainable Development and Mitigation' in B. Metz, O.R. Davidson, P.R. Bosch, R. Dave and L.A. Meyer (eds), *Climate Change 2007. Mitigation. Contribution*

*of Working Group III to the Fourth Assessment Report of the Intergovernmental Panel on Climate Change*, IPCC 4th Report, Cambridge University Press, Cambridge, United Kingdom and New York, NY, USA, Chapter 12, pp. 691–745.

Schwartz, P. and D. Randall (2006), *Rapport secret du Pentagone sur le Changement climatique*, Allia.

Sorrell, S. (2007), 'The Rebound Effect: an assessment of the evidence for economy-wide energy savings from improved energy efficiency', UK Energy Research Centre.

UNDP (2008), 'Human Development Report 2007/2008', Oxford University Press.

UNEP (1997), 'Protocole de Kyoto à la Convention cadre des Nations Unies sur les changements climatiques', Nairobi.

UNEP (2007), 'Assessment of Impacts and Adaptation to Climate Change Final Report of the AIACC Project', Nairobi.

UNEP (2008), 'CCCC. Kick the habit. A guide to climate neutrality', UNEP.

Wanneau, K. (2011), 'Sécuriser le changement climatique', in La Branche (2011).

Chou, Wan-Jung, Alistair Hunt, Anil Markandya, Andrea Bigano, Roberta Pierfederici Stephane La Branche (2010), 'Consumer Valuation of Energy Supply Security: an analysis of Survey results in three EU countries', Centre for European Policy Studies Policy Brief.

World Bank (2008), 'Towards a strategic framework on climate change and development for the World Bank Group', Concept and issues paper consultation draft, Washington.

Yohe, G.W., R.D. Lasco, Q.K. Ahmad, N.W. Arnell, S.J. Cohen, C. Hope, A.C. Janetos and R.T. Perez (2007), 'Perspectives on climate change and sustainability', in B. Metz, O.R. Davidson, P.R. Bosch, R. Dave and L.A. Meyer (eds), *Climate Change 2007. Mitigation. Contribution of Working Group III to the Fourth Assessment Report of the Intergovernmental Panel on Climate Change*, IPCC 4th Report, Cambridge University Press, Cambridge, United Kingdom and New York, NY, USA, Chapter 20, pp. 811–841.

Young, O.R. (2002a), *The Institutional Dimensions of Environmental Change: Fit, Interplay, and Scale*, Cambridge and Massachusetts: MIT Press.

Young, O.R. (2002b), 'Matching Institutions and Ecosystems: The Problem of Fit', *Gouvernance mondiale*, 2, Les séminaires de l'Iddri, available at: www.iddri.com/Publications/Collections/Idees-pour-le-debat/id_0202_young.pdf (accessed February 2013).

# PART VI

# ENERGY AND HUMAN SECURITY

# 19. Energy poverty: access, health and welfare
*Subhes C. Bhattacharyya*

## 1 INTRODUCTION

The United Nation's decision to declare 2012 as the 'International Year of Sustainable Energy for All' has once again caught global attention on sustainable energy in general and energy poverty in particular. Lack of access to clean or modern energy and an inability to use the desired energy when required tends to adversely influence the development prospects of the population and impose social burdens in terms of adverse health effects and welfare losses. Moreover, sustainability of energy provision and energy supply security implications of energy poverty have received limited attention in the past. While attempts are being made to extend basic energy services, the supply is often erratic and limited and does not ensure long-term needs, thereby causing concerns about a reliable supply. There is also the concern for future resource implications of enhanced energy access. This chapter reviews energy poverty, discusses the health and welfare implications of energy poverty and elaborates on the sustainability and energy security dimensions.

The chapter is organised as follows: the next section presents the definition of energy poverty and elaborates on the nature of the problem. The third section discusses the incidence of energy poverty, while section four presents the health and welfare implications of energy poverty. Finally, the fifth section considers the remedial measures and the last section presents the concluding remarks.

It needs to be highlighted here that this chapter does not cover the fuel poverty problem. Although fuel poverty is an important issue, particularly in some parts of Europe, the focus here is on the bigger issue of energy access.

## 2 ENERGY POVERTY

### 2.1 The Concept

Energy poverty is not a well-defined term. Pachauri et al. (2004) indicated that three types of measures are normally found in the literature:

1. Economic measures such as energy poverty line as used in the United Kingdom: This tries to find out the share of spending on energy by those below the national poverty line and compares this against a benchmark level (say 10 per cent). If a consumer spends more than this threshold level on energy, the consumer may be regarded as being below the fuel poverty line.
2. Engineering measures of minimum energy needs: this uses normative estimates of the basic energy needs of a household and anyone below this level is considered energy poor.
3. Measures based on access to energy services: this tries to find out whether consumers have physical access to the supply of energy, and access to markets for equipment.

   The issue arises because the term energy poverty draws parallel from the poverty literature where poverty is generally related to inadequate levels of income and consumption to fulfil the basic needs, which in turn, implies the deprivation of the basic minimum needs of a population. From this perspective, the energy poverty would mean ensuring a minimum quantity of energy to meet the essential needs of a population. Generally, either engineering estimates or normative values[1] are used to determine the essential needs but these estimates have their own issues as well due to inherent subjectivity.

Moreover, the needs might not remain unchanged over time and consequently the target itself can move, thereby creating the challenge of reaching a moving target. Further, 'the poor' is not a homogeneous category, and both endowments and entitlements can vary even within a country and across countries, implying that the needs are not homogeneous for the poor. In the case of energy this becomes important as the needs depend on geographical location, climatic conditions, resource endowments, etc.

Thus, Pachauri (2011) explains that reaching a consensus on the definition hinges on agreements on three elements: 1) consensus on services defining the basic needs basket, 2) a clear definition of the thresholds defining the basic needs, and 3) assessing the household expenditure on energy by different income class. Reaching an agreement on these elements is not easy.

A number of attempts in the literature try to capture the energy poverty through an indicator. The International Energy Agency has come up with an index, the Energy Development Index (EDI), along the lines of the Human Development Index. EDI is composed of the following four factors (IEA 2010):

*Table 19.1   Factor goalposts for 2010 EDI*

| Factor | Maximum | Minimum |
| --- | --- | --- |
| Per capita commercial energy consumption (toe) | 2.88% (Libya) | 0.03% (Eritrea) |
| Per capita electricity consumption in the residential sector (toe) | 0.08% (Venezuela) | 0.001% (Haiti) |
| Share of modern fuels in total residential sector energy use | 100% | 1.4% (Ethiopia) |
| Share of population with access to electricity | 100% | 11.1% (DR Congo) |

*Note:*   Toe = ton of oil equivalent.

*Source:*   IEA (2010).

- per capita commercial energy consumption;
- per capita electricity consumption in the residential sector;
- share of modern fuels in total residential sector energy use;
- share of population with access to electricity.

An index is created for each factor by considering the maximum value and minimum values observed in the developing world and determining how a particular country has performed. The following formula is used for this index:

$$\text{Factor index} = \frac{(\textit{Actual value} - \text{min}\textit{imum value})}{(\text{max}\textit{imum value} - \text{min}\textit{imum value})} \quad (19.1)$$

The goalposts (maximum and minimum values) are taken from the observed values within the sample of developing countries considered. For example, for calculating the factor goalposts for 2010 EDI, WEO (2010) used the values given in Table 19.1.

The simple average of four indicators gives the overall EDI. For any country, e.g. India, EDI can be calculated using Formula 1 and noting the goalposts as well as actual data for the country. For 2009, India's individual indicators are shown in Table 19.2.

Although this indicator provides a numerical value, it is not devoid of problems. It perpetuates the idea that a higher level of commercial energy or electricity consumption is synonymous to economic development. Accordingly, countries in the Middle East with high per capita energy use

Table 19.2  Example of EDI for India

| Factor | Indicator |
| --- | --- |
| Per capita commercial energy consumption | 0.140 |
| Per capita electricity consumption in the residential sector | 0.111 |
| Share of modern fuels in total residential sector energy use | 0.213 |
| Electrification rate | 0.62 |
| Average index | 0.272 |

*Source:* IEA (2010).

Table 19.3  Commonly used national and international indicators of energy poverty

| Scope | Indicator |
| --- | --- |
| International | Physical access to energy by households or population Energy Development Index |
| National | Physical access to energy at the village or community level |
|  | Minimum norms of energy needs for different uses or its variations |
|  | Share of energy expenses in the household budget |
|  | Share of energy expenses and annualised cost of end-use appliances in total household budget |
|  | Associate time costs |
|  | Health impacts |

*Source:* Pachauri and Spreng (2011).

rank better in this index. It also assumes that biomass energy use represents a symbol of under-development, which need not be the case, depending on how it is used. Finally, it does not pay any attention to sustainable energy supply or use.

Pachauri and Spreng (2011) provide a list of common national and international measures used in describing energy poverty (see Table 19.3). They also call for simple, measurable but meaningful indicators. More recently, Nussbaumer et al. (2012) have presented a multi-dimensional energy poverty index (MEPI) by capturing modern energy deprivation and the incidence and intensity of energy poverty. Yet, the sustainability issue has not been adequately captured in the indicators. Further work in this area is required.

## 2.2 Energy Poverty and Energy Security Linkage

Given that energy security aims for 'reliable and adequate supply of energy at reasonable prices' (Bielecki 2002), and because energy poverty is a manifestation of deprivation, the link between the two concepts becomes obvious. In a situation of energy poverty, not even the basic needs of energy are satisfied and there is a clear deficit in terms of reliability and adequacy of supply. Thus energy poverty also implies insecure energy supply. Further, the poor typically pay more for their energy needs and receive poorer quality of service due to inefficient technologies and poor infrastructure (Sovacool 2012), thereby making them worse off than other users.

The reliability and adequacy of supply also tends to have an urban bias, implying that urban consumers are traditionally given higher preference in terms of supply compared to the rural consumers. For example, electricity supply to rural India is often restricted to off-peak periods (10pm to 6am) when most household consumers do not need the supply. Rural electricity supply is often limited to a few evening hours in many countries and modern energies like kerosene or LPG reach the rural areas only infrequently. Lack of energy supply security thus reinforces the dependence on traditional sources of energy.

Dependence on traditional energy also leads to occupational hazards and human security issues. In most cases, the burden of traditional fuel wood collection falls disproportionately on women. Sovacool (2012) reports that on average an African woman carries 20 kg of fuel wood over 5 kilometres per day, which in turn inflicts injuries and other hazards including those of health and safety. It has been reported that women have faced assaults and violence while collecting fuel wood at times of civil unrest or war-like situations. The economic cost of such occupational hazards is barely estimated or considered in energy policy decisions.

Moreover, the remedial interventions to address energy poverty can also have energy security implications. For example, if the solution aims at meeting the basic needs for a limited period of time, then the supply is unlikely to be adequate and reliable and perhaps may not be affordable. This has emerged as the main complaint against many energy access programmes in recent times. Similarly, if the intervention envisages introduction of a non-renewable energy, it can impose supply security concerns depending on the country's present resource status. Thus energy poverty eradication joins the energy security challenge that cannot be ignored.

## 3 INCIDENCE OF ENERGY POVERTY

Energy demand in poor households normally arises from two major end-uses: lighting and cooking (including preparation of hot water).[2] Cooking energy demand is predominant in most cases and often accounts for about 90 per cent of the energy demand by the poor. Such a high share of cooking energy demand arises partly from the low energy efficiency and partly due to limited scope of other end-uses. As electricity is considered the appropriate form of energy for lighting, it is customary to associate the access to clean lighting to the level of electrification of a country. Access to clean cooking energies on the other hand can take different paths and therefore the access related information is generally presented for electricity and for cooking energies separately.[3] We maintain this distinction in this chapter.

### 3.1 Status of Electrification in Various Regions

The regional picture of electrification is presented in Table 19.4 (IEA 2011).[4] In 2009, more than 1.3 billion people (i.e. about 19 per cent of the global population) did not have access to electricity. Two regions stand out: South Asia, with 675 million (or 42 per cent of the population) without access comes first, while Sub-Saharan Africa comes second with a population of 587 million (or 40 per cent) of those without access to electricity. Outside these two regions, East Asia has 195 million without access to electricity (or about 13 per cent of those without access).

*Table 19.4  Level of electrification in various regions in 2009*

| Region | Population without electricity (Millions) | Electrification rate (%) Overall | Urban | Rural |
|---|---|---|---|---|
| North Africa | 2 | 99.0 | 99.6 | 98.4 |
| Sub-Saharan Africa | 585 | 30.5 | 59.9 | 14.2 |
| Africa | 587 | 41.8 | 66.8 | 25.0 |
| China and East Asia | 182 | 90.8 | 96.4 | 86.4 |
| South Asia | 493 | 68.5 | 89.5 | 59.9 |
| Developing Asia | 675 | 81.0 | 94.0 | 73.2 |
| Middle East | 21 | 89.0 | 98.5 | 73.6 |
| Latin America | 31 | 93.2 | 98.8 | 73.6 |
| Developing Countries | 1,314 | 74.7 | 90.6 | 63.2 |
| Global total | 1,317 | 80.5 | 93.7 | 68.0 |

*Source:* IEA (2011).

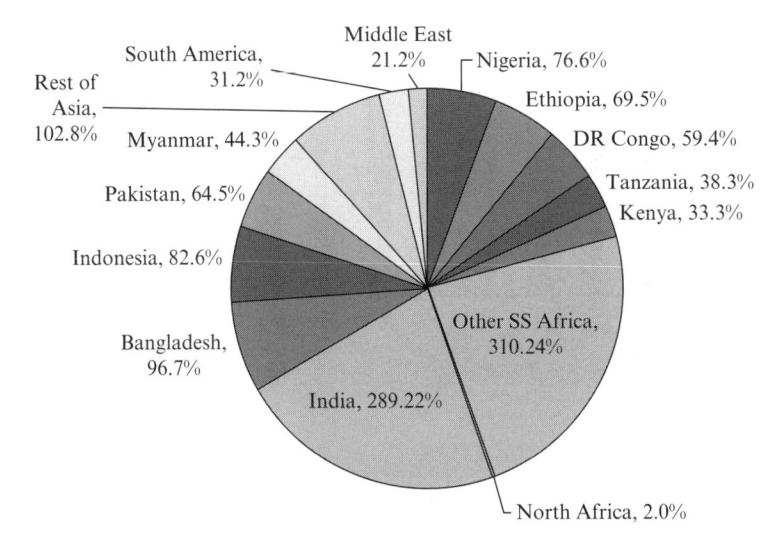

*Energy poverty: access, health and welfare* 429

*Note:* The first number in the label refers to population in million without electricity access.

*Source:* IEA (2011).

*Figure 19.1    Major concentration of population without access to electricity in 2009*

A closer look at the data shows that about 69 per cent of those lacking access to electricity reside in just 12 countries while the remaining 30 per cent is dispersed in all other countries (see Figures 19.1 and 19.2). The rural population in most of these countries lacks access, although in a few countries the urban population also lacks access. While the total number of people without access to electricity is high in South Asian countries, Sub-Saharan Africa fares worse in terms of rate of electricity access. In fact, out of 10 least electrified countries in the world, nine are from sub-Saharan Africa and Myanmar is the only country from Asia (see Figure 19.3).

Interestingly, the most populous country in the world, China, has achieved a very impressive record of providing electricity with only 8 million (or about 0.6 per cent of its population) without the facility. Similarly, a number of South East Asian countries such as Thailand, Malaysia, Vietnam and the Philippines have made impressive progress on the electrification front. In South America, Brazil, the most populous country of the region, has achieved an impressive record of about 2 per cent of its population without access to electricity, most of whom are located in the Amazon region.

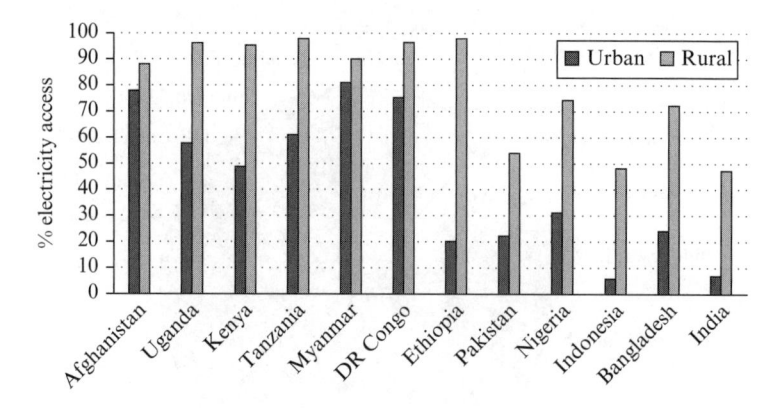

430   *International handbook of energy security*

*Source:* IEA (2009).

*Figure 19.2   Urban–rural electricity access disparity in major concentrations in 2009*

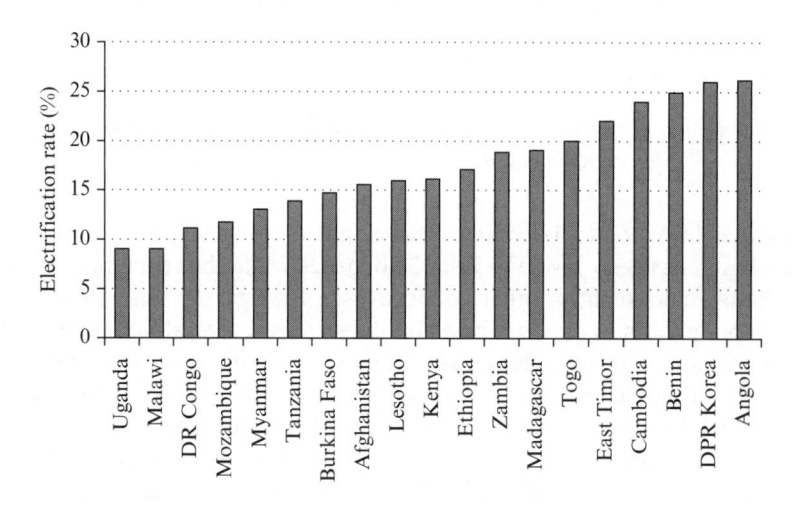

*Source:* IEA (2011).

*Figure 19.3   Ten least electrified countries in the world in 2009*

### 3.2   Status Of Cooking Energy Access

IEA (2011) provided some details about biomass use in the developing countries and estimated that about 2.7 billion people use biomass for cooking and heating purposes in these countries (see Figure 19.4).

*Energy poverty: access, health and welfare* 431

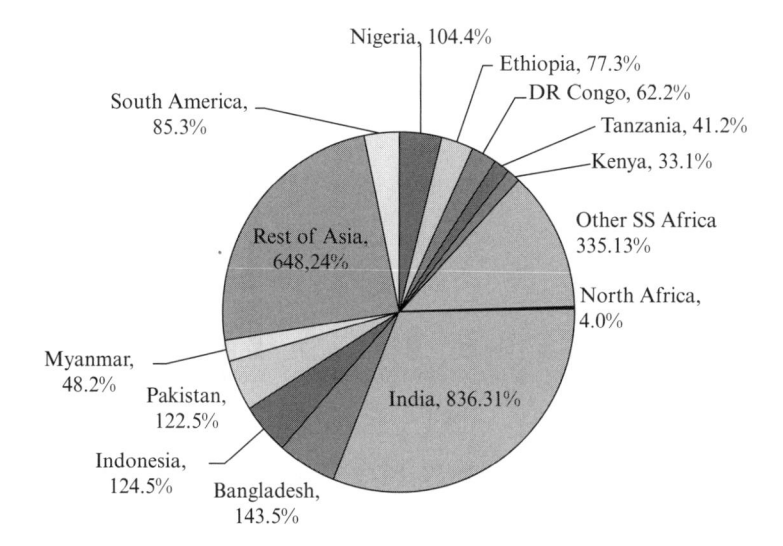

Nigeria, 104.4%
Ethiopia, 77.3%
DR Congo, 62.2%
Tanzania, 41.2%
Kenya, 33.1%
Other SS Africa, 335.13%
North Africa, 4.0%
India, 836.31%
Bangladesh, 143.5%
Indonesia, 124.5%
Pakistan, 122.5%
Myanmar, 48.2%
Rest of Asia, 648.24%
South America, 85.3%

*Note:* The first number in the label refers to population in million without clean cooking energy access.

*Source:* IEA (2011).

*Figure 19.4 Distribution of lack of cooking energy access in the world in 2009*

*Table 19.5 Reliance on biomass for cooking energy needs in 2009*

| Region | Total population % | Total Million | Rural Million | Urban Million |
| --- | --- | --- | --- | --- |
| Sub-Saharan Africa | 78 | 653 | 476 | 177 |
| Total Africa | 65 | 657 | 480 | 177 |
| India | 72 | 836 | 749 | 87 |
| China | 32 | 423 | 377 | 46 |
| Rest of Asia | 63 | 731 | 554 | 177 |
| Latin America | 19 | 85 | 61 | 24 |
| Total | 51 | 2662 | 2221 | 441 |

*Source:* IEA (2011).

At a disaggregated level, more than 80 per cent of the people lacking access to clean energies live in rural areas (see Table 19.5). Asia, with more than 72 per cent of those lacking access, has the largest share followed by Sub-Saharan Africa, but in contrast to electricity access where both the

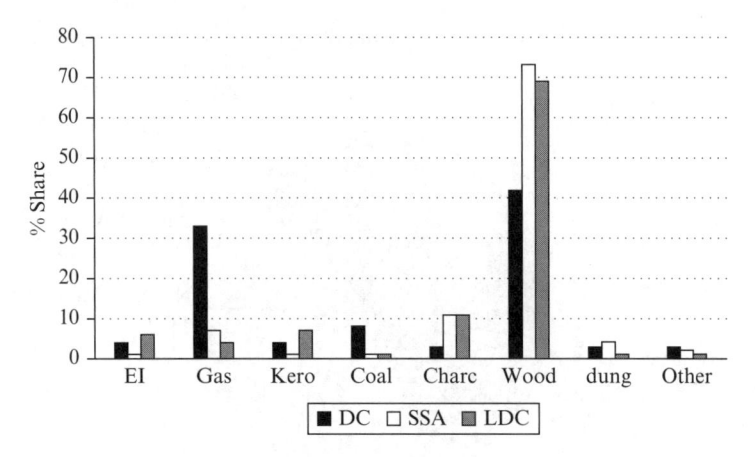

*Note:* DC: Developing countries, LDC: Least Developed Countries, SSA: Sub-Saharan African countries.

*Source:* UNDP-WHO (2009).

*Figure 19.5  Share of different cooking fuels in developing countries in 2007*

regions share similar sizes of population without access, here the picture is quite different.

The size of urban population lacking access to clean energies in both the regions is very similar but the rural population lacking access to clean cooking energies in Asia is 3.5 times more than that of Sub-Saharan Africa. India has the single largest concentration of people lacking clean cooking energy access in the world.

Although a range of fuels is used for cooking, according to the UNDP-WHO (2009) study, about 2.6 billion people rely on traditional energies and 400 million rely on coal. There is significant regional variation in terms of fuel use (see Figure 19.5), but the rural population is generally more reliant on solid cooking fuels, including traditional energies. Moreover, the use of improved cooking stoves is limited to only 30 per cent of those relying on solid fuels but the dependence on traditional stoves is predominant in the least developed countries (LDCs) and Sub-Saharan Africa (UNDP-WHO, 2009).

### 3.3 Future Outlook

But more importantly, forecasts by IEA (2011) suggest that almost 1 billion people will still lack access to electricity in 2030 while 2.7 billion

*Table 19.6   Expected number of people without electricity access in 2030*

| Region | In 2030 |  |  |
|---|---|---|---|
|  | Urban | Rural | % of population |
| Sub-Saharan Africa | 107 | 538 | 49 |
| India | 9 | 145 | 10 |
| China | 0 | 0 |  |
| Rest of Asia | 40 | 181 | 16 |
| Latin America | 2 | 8 | 2 |
| Middle East | 0 | 5 | 2 |
| Total of Developing world | 157 | 879 | 16 |

*Source:*   IEA (2011).

*Table 19.7   Outlook for biomass use for cooking in 2030 (million)*

| Region | By 2030 |  |  |
|---|---|---|---|
|  | Urban | Rural | % |
| Sub-Saharan Africa | 270 | 638 | 67 |
| India | 59 | 719 | 53 |
| China | 25 | 236 | 19 |
| Rest of Asia | 114 | 576 | 52 |
| Latin America | 17 | 57 | 14 |
| Total Developing world | 485 | 2230 | 43 |

*Source:*   IEA (2011).

people will not have access to clean cooking energies. Although the forecast assumes a significant level of investment ($13 billion per year on average), increases in the population in developing countries of South Asia and Sub-Saharan Africa will mean that electricity access will remain a problem. According to IEA (2011) 356 million in South Asia and 645 million in Sub-Saharan Africa will still live without electricity access (see Table 19.6).

In terms of access to clean cooking energies, the situation will be even worse. IEA (2011) suggests that 485 million urban population and 2.2 billion rural population will still continue with traditional energy even by 2030 unless more specific interventions are made (see Table 19.7). The size of population relying on traditional energies in Sub-Saharan Africa will increase to 900 million by 2030 and one-third of the global population without clean cooking energy will reside there. This represents a deterioration compared to the present situation. The situation in Asia improves marginally but still about two-thirds of the population without clean cooking energy will be found there by 2030. India will continue to remain the country with the highest concentration of population without clean cooking energy. Clearly, the future does not appear to be very promising and serious thoughts in terms of policy analysis and implementations will be required to address these issues.

However, governments of most of the countries are now aware of the problem and a large number of countries have set energy access targets. According to UNDP-WHO (2009), almost one-half of all the developing countries have set targets for electricity access, with Sub-Saharan countries emerging as leaders in setting targets. In the case of clean cooking energies, only a few countries have set targets but Sub-Saharan countries are appearing to be more proactive here as well. The regional distribution of countries with targets for electricity access and clean cooking energies is presented in Figure 19.6.

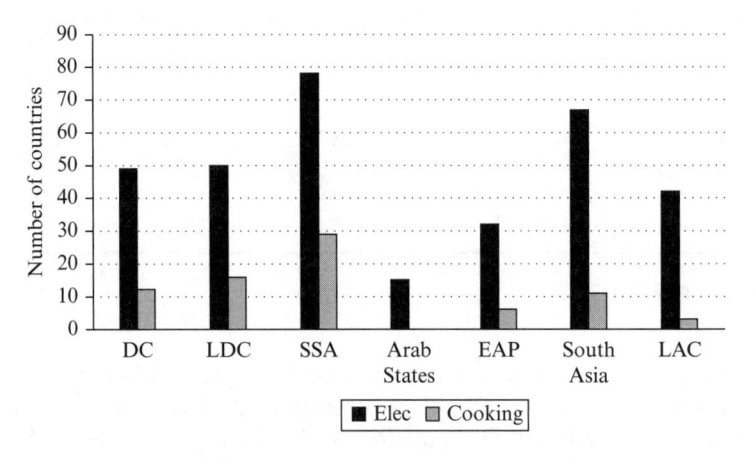

*Note:* DC: Developing Countries, LDC: Least Developed Countries, SSA: Sub-Saharan Africa, EAP: East Asia and the Pacific, LAC: Latin American Countries.[5]

*Source:* UNDP-WHO (2009).

*Figure 19.6 Regional distribution of countries with targets for electricity and clean cooking energy access*

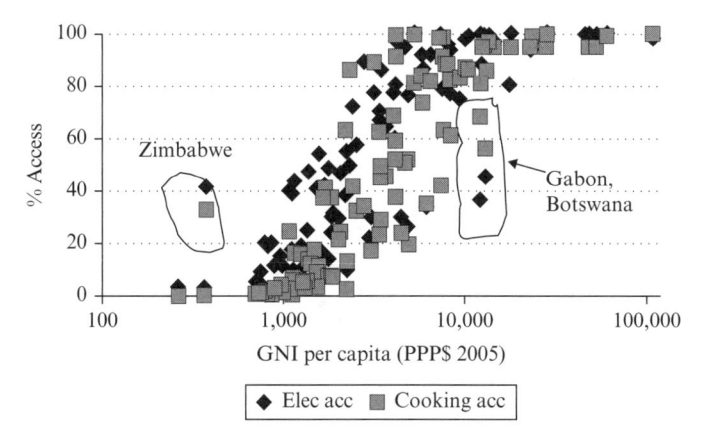

*Figure 19.7   Energy access improves with per capita income*[6]

## 4   ENERGY POVERTY AND WELFARE ISSUES

### 4.1   Energy Poverty: Development Linkage[7]

As is generally expected, higher levels of energy access are normally associated with a higher income level but a rapid improvement in access level occurs within an income band bounded by a lower threshold income level of about $1,000 per person in PPP terms 2005 and an upper saturation level of about $15,000 per person in PPP terms (see Figure 19.7). Those below the lower threshold clearly lack access to clean energy, while everyone above the upper threshold has access to clean energy services. However, the scatter plot shows a significant level of dispersion within the upper and lower thresholds, implying that some countries are able to reach better energy access at low income levels while some with high income have failed to deliver energy access to their population. Clearly, income does not automatically ensure high level of energy access of a country and there are other drivers that play an important role. However, a detailed analysis of the causes, drivers and lessons from the successful/ unsuccessful cases is beyond the scope of this chapter and is an area of further research.

The Human Development Index of a country, on the contrary, bears a better correlation with energy access than income. Using the HDI data for 2011 and energy access data presented in UNDP-WHO (2009), a few indicators linking HDI and economic development are presented in

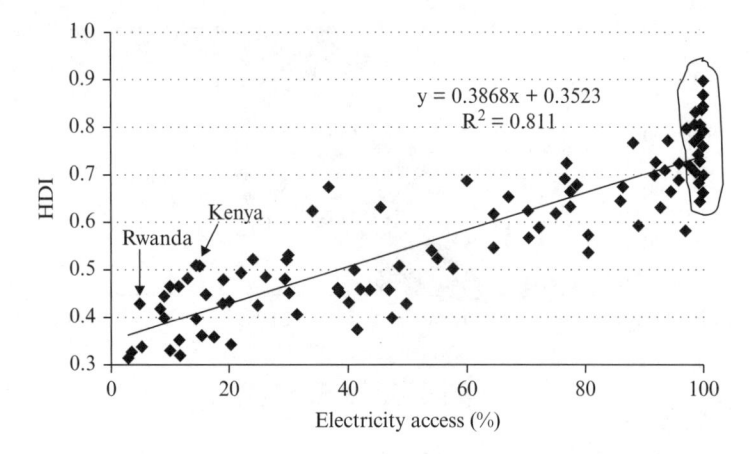

436   *International handbook of energy security*

*Source:* HDI data for 2011 and UNDP-WHO (2009) for electricity and cooking energy access.

*Figure 19.8   HDI and electricity access*

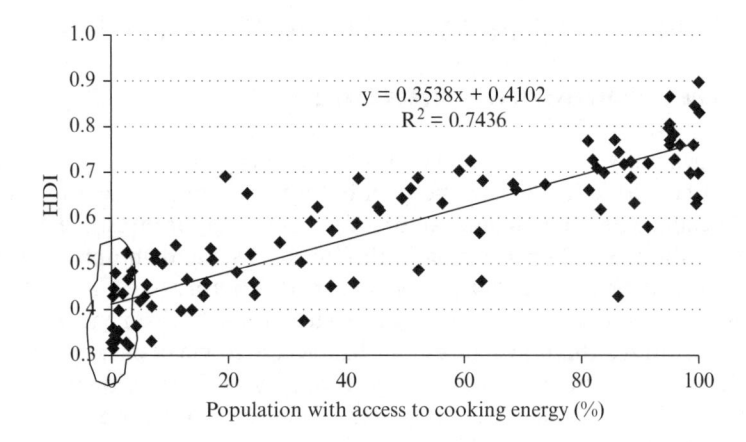

*Source:* HDI data for 2011 and UNDP-WHO (2009) for electricity and cooking energy access.

*Figure 19.9   HDI and cooking energy access*

Figures 19.8 to 19.11. Figure 19.8 shows that better HDI scores are generally associated with higher levels of electricity access, while Figure 19.9 shows that the HDI also is positively correlated with access to cooking energy. Similarly, the life expectancy and mean schooling years are also positively correlated to clean cooking energy access and electricity access

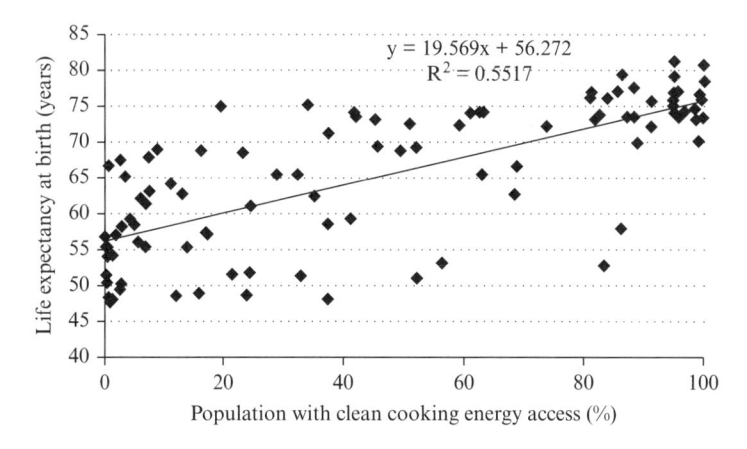

*Energy poverty: access, health and welfare* 437

*Source:* HDI data for 2011 and UNDP-WHO (2009) for electricity and cooking energy access.

*Figure 19.10 Life expectancy at birth and cooking energy access*

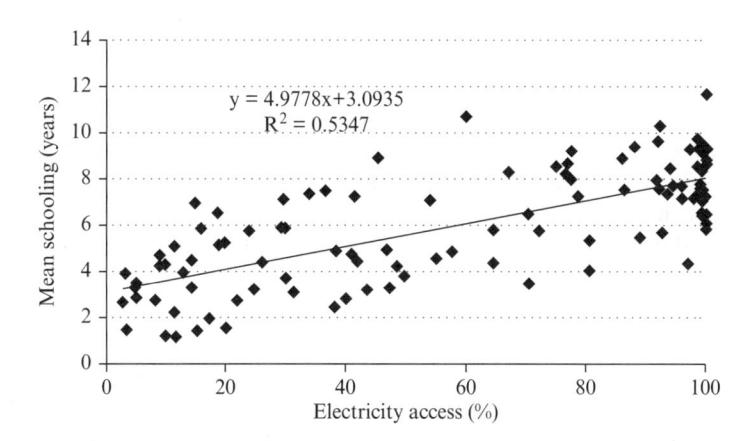

*Source:* HDI data for 2011 and UNDP-WHO (2009) for electricity and cooking energy access.

*Figure 19.11 Mean schooling years against electricity access*

(Figures 19.10 and 19.11), although the goodness of fit of a linear relationship is less strong than the previous two cases.

AS HDI focuses on three equally-weighted components (namely life expectancy at birth, mean schooling years and GNI per capita), and because energy is one of the many drivers (and thus has an indirect influence) behind the performance of these components, the influence of energy on HDI is not always very straightforward.

### 4.2 Health Impacts of Energy Poverty

The health problem associated with energy poverty arises mainly due to indoor air pollution caused by burning solid fuels. Poor households rely on primitive technologies and solid fuels and family members, often women and children, are exposed to excessive smoke from incomplete combustion of these fuels. As the smoke contains substances like carbon monoxide and small particulate matter, that are toxic to human health, severe health impacts result. Polski and Ly (2012) report that while the acceptable concentration of PM10 is 50 micro-gram/m3, average concentration of 600 micro-grams/m3 and high levels of 2000 micro-grams/m3 are not uncommon.

Indoor air pollution is responsible for acute respiratory infections, chronic obstructive pulmonary disease and diseases like asthma, lung cancer, low weight at birth, cataract, tuberculosis, etc. (WHO 2007). The average level of exposure to smoke from fuels is three hours per day, which is equivalent to smoking two packets of cigarettes daily (Bruce et al. 2000). Such exposure levels cause respiratory infections in children and are the major cause of child mortality. Similarly, solid fuel use is responsible for airflow restriction in the pulmonary system (found mostly in women above 30), which is not fully reversible and lung cancer is also a result of smoke from solid fuel combustion.

According to UNDP-WHO (2009), 1.94 million deaths per year can be attributed to solid fuel use in developing countries. The least developed countries and Sub-Saharan Africa suffer most severely in terms of death rate per million population. Globally, one-third of the deaths from pneumonia, COPD and lung cancer are attributable to solid fuel use but in the LDC and Sub-Saharan Africa, this share increases to one-half (UNDP-WHO, 2009). A broader measure, disability adjusted life years (DALY), that captures the premature death and the years lived with a disease, reveals even more concentrated effects of solid fuel use in these regions. Out of 40 million DALYs attributable worldwide to solid fuel use, 45 per cent occur in LDCs and 44 per cent in Sub-Saharan Africa (UNDP-WHO, 2009). According to Sovacool (2012), the cost to the national health system ranges between $212 billion to $1.1 trillion. Using the projected biomass fuel use trend WHO estimates that indoor air pollution would lead to 1.5 million premature deaths per year by 2030. This would pose significantly higher health risks compared with other infectious diseases like malaria and tuberculosis.

The health problem related to energy poverty has a gender bias as well. Women are generally more affected than men due to higher exposure to smoke. Children are also critically vulnerable due to their immature metabolic system pathways. 44 per cent of the deaths due to solid fuel use occur in children and 60 per cent of the adult deaths involve women (UNDP-WHO, 2009) in developing countries. The human development index bears a close association with disease burden: lower HDI is associated with high disease burden and vice-versa. The Sub-Saharan African countries generally register the highest disease burden and lowest HDI.

As the population in Asia and Africa increases, the overall health impact will be aggravated unless remedial measures are taken. This will inflict significant economic loss to these countries and cannot be allowed to happen. Consequently, remedial measures need to be considered.

## 5 OPTIONS FOR REMEDIAL INTERVENTIONS

A number of intervention options for reducing the health impacts of solid fuel use can be found in the literature. These can be categorised into three groups: 1) changing the source of pollution, 2) improving the living environment and 3) modifying the user behaviour (Isihak et al., 2012). Figure 19.12 provides the details of such options.

Generally, the policy focus has centred on promoting cleaner petroleum fuels (such as LPG or kerosene) and clean cooking stoves. However, as the poor often cannot afford the cost of petroleum fuels and the associated devices, such an option requires a subsidised supply of the fuel and a mechanism to ensure access to the required appliances. This issue then ties well with the energy supply security concerns due to increased vulnerability of the countries to international price fluctuations and consequent economic hardships. Such a policy also becomes vulnerable due to its poor long-term prospects, arising from tight budget conditions of governments and increasing demand for funds for other uses. Further, low demand for alternative fuels reduces the business incentive for commercial suppliers and in the event of supply constraint; the supply to the poor is greatly affected, thereby forcing them to return to the old, traditional systems. Thus the energy poverty issue rejoins the supply security problem that has an equity dimension.

The technology intervention option through improved cooking appliances has seen a mixed fortune so far. According to UNDP-WHO (2009), only a third of the biomass-using population in the world is using improved cooking stoves and about two-thirds of those using them live in China and another 20 per cent in other Asia-Pacific countries. But Sub-Saharan

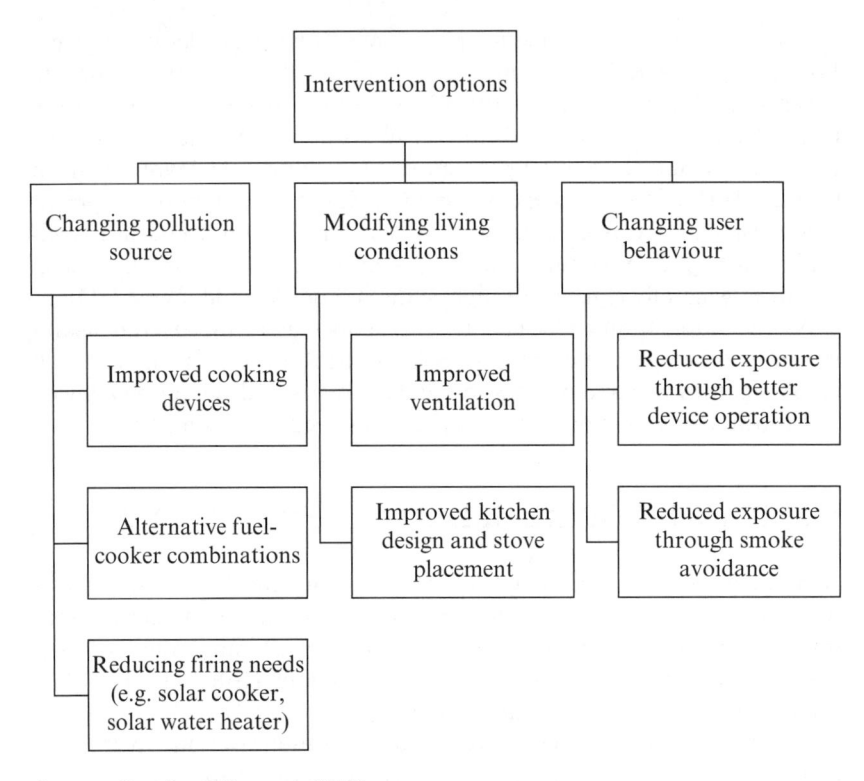

```
                    ┌─────────────────────┐
                    │ Intervention options │
                    └─────────────────────┘
          ┌──────────────────┼──────────────────┐
┌─────────────────┐  ┌─────────────────┐  ┌─────────────────┐
│ Changing pollution│  │ Modifying living │  │ Changing user   │
│      source      │  │   conditions    │  │   behaviour     │
└─────────────────┘  └─────────────────┘  └─────────────────┘
```

Figure 19.12  *Intervention options for reducing health impacts of solid fuel use*

Tree:
- Changing pollution source
  - Improved cooking devices
  - Alternative fuel-cooker combinations
  - Reducing firing needs (e.g. solar cooker, solar water heater)
- Modifying living conditions
  - Improved ventilation
  - Improved kitchen design and stove placement
- Changing user behaviour
  - Reduced exposure through better device operation
  - Reduced exposure through smoke avoidance

*Source:* Based on Salisu et al. (2012).

Africa, where 80 per cent of the solid-fuel using population lives, accounts for only 4 per cent of the improved cooking stove-using population. This implies that the region with most needs has not benefited much from this intervention. Foell et al. (2011) argue that, despite the gravity of the problem, global attention on clean cooking and heating energies has been relatively low compared to that for electrification. This is evident from the IEA (2011) estimate of investments in energy access. IEA (2011) estimated that $9.1 billion was invested in providing access to energy in 2009 – of which only $70 million went to provide advanced biomass cooking-stoves (benefitting 7 million people) and the rest was used in providing access to electricity to 20 million people. Electricity accounts for a minor share of rural households' energy needs and electricity is unlikely to be competitive with traditional firewood (or biomass-based fuels) used by the poor for cooking purposes. As traditional solid biomass fuel imposes little private monetary cost burden, the poor will prefer to use this when they cannot afford modern fuels, although these fuels impose heavy social costs. Thus a rebalancing of priorities and significantly higher investment in clean cooking energy supply is urgently required.

There is a further window of opportunity in this respect. As solid fuel is a major source of greenhouse gas emission, any intervention to reduce the pollution also brings climate change benefits. Therefore, this potential co-benefit opens up opportunities for accessing carbon finance for mitigating interventions. There are initiatives in this respect but more needs to be done to face the challenge.

## 6 CONCLUSIONS

The issue of energy poverty and consequent health impacts have significant economic and social implications. The poor in many developing countries are forced to rely on such dirty options due to their economic conditions, and the continued reliance on such fuels pose greater public health risks due to population increase. Energy poverty has a significant gender bias and inflicts disproportionate social inequity through health impacts, occupational hazards and reduced human capital development potential. There is also significant loss of social welfare and ultimately this reduces the prospects for economic development of these countries. Although interventions have been attempted in the past, the level of investment and the rate of success are not encouraging. Greater efforts are required to face the challenge. Fortunately, the climate co-benefits of the interventions can be exploited to access carbon financing for such interventions. More efforts will need to be directed to the poorer countries so that a change becomes visible.

## NOTES

1. For example, the Indian Planning Commission used to rely on such norms.
2. In some climatic conditions heating may also be an important source of energy demand.
3. It is important to mention that the quality of data on this subject is relatively poor, although major efforts are being made by the international organisations to improve the situation. The data quality is affected, among others, by the distributed and dispersed nature of the population being considered, lack of any administrative arrangements for systematic records on traditional energies, definitional issues, limited availability of comprehensive surveys, and poor communication and infrastructure facilities. This aspect needs to be kept in mind with respect to any analysis on the subject.
4. See also UNDP-WHO (2009) for a detailed review of energy access.

5. The details of the above classification are indicated in UNDP-WHO (2009), Annex 2, Table 13.
6. The horizontal axis is presented in logarithmic scale to capture the wide range of income variation across countries.
7. This section is largely based on Bhattacharyya (2012).

# REFERENCES

Bhattacharyya, S.C. (2012), 'Energy access programmes and sustainable development: A critical review and analysis', *Energy for Sustainable Development*, 16(3):260–71.
Bielecki, J. (2002), 'Energy security: Is the wolf at the door?', *The Quarterly Review of Economics and Finance*, 42:235–50.
Bruce, N., R. Perez-Padill and R. Albalak (2000), 'Indoor air pollution in developing countries: a major environmental and public health challenge', *Bulletin of the World Health Organization*, 78(9):1078–1092.
Foell, W., S. Pachauri, D. Spreng and H. Zerriffi (2011), 'Household cooking fuels and technologies in developing economies', *Energy Policy*, 39(12):3479–86.
IEA (2010), 'World Energy Outlook 2010', International Energy Agency, Paris.
IEA (2011), 'Energy for all: Financing access for the poor', Special early excerpt of the World Energy Outlook 2011, International Energy Agency, Paris.
Isihak, S., U. Akpan and M. Adeleye (2012), 'Interventions for mitigating indoor air pollution in Nigeria: a cost-benefit analysis', *International Journal of Energy Sector Management*, 6(3):417–29.
Nussbaumer, P., M. Bazilian and V. Modi (2012), 'Measuring energy poverty: Focusing on what matters', *Renewable and Sustainable Energy Reviews*, 16(1):231–43.
Pachauri, S. (2011), 'Reaching an international consensus on defining modern energy access', *Current Opinion in Environmental Sustainability*, 3(4):235–40.
Pachauri, S., and D. Spreng (2011), 'Measuring and monitoring energy poverty', *Energy Policy*, doi:10.1016/j.enpol.2011.07.008.
Pachauri, S., A. Mueller, K. Kemmler and D. Spreng (2004), 'On measuring energy poverty in Indian households', *World Development*, 32(12):2083–2104.
Polsky, D. and C. Ly (2012), 'The health consequences of indoor air pollution: a review of the solutions and challenges', White Paper, University of Pennsylvania, USA.
Sovacool, B. (2012), 'The Political economy of energy poverty: A review of key challenges', *Energy for Sustainable Development*, 16(3):272–82.
UNDP-WHO (2009), 'The energy access situation in developing countries: A review focusing on the Least-developed countries and Sub-Saharan Africa', United Nations Development Programme, New York.
WHO (2007), 'Indoor air pollution: National Burden of Disease Estimates', World Health Organisation, Geneva, available at http://www.who.int/indoorair/publications/indoor_air_national_burden_estimate_revised.pdf (accessed 25 February, 2013).

# 20. Ethical dimensions of renewable energy
## Hugh Dyer

Our sense of insecurity with respect to energy is enough to warrant consideration of how relative energy security might be obtained, yet the most obvious source of insecurity is our collective failure to plan adequately for inevitable changes, which will be forced upon us sooner or later. On the assumption that justice and equity must underwrite the feasibility of any energy strategies, we need an ethical framework for energy which includes as a central concern the lack of human security in respect of the allocation of limited resources. For the sake of our common humanity, and for posterity, there appears to be a clear moral imperative for pursuing renewable sources of energy. At the same time, political and economic trade-offs suggest this has not yet been taken seriously.

It seems clear enough that the pursuit of any meaningful energy security policy will require anticipation of future post-carbon scenarios. This requires a perspective on 'the age of petroleum' as only a recent and relatively short-run phenomenon in the long run of human energy supply (up to the late 19th century provided by biomass and animate labour, and now in the 21st century increasingly by renewables). The alternatives to fossil fuels clearly exist, though it 'will take a new industrial revolution' (Scheer 2002) or an 'energy revolution' (Geller 2002). At the same time there is certainly evidence of growth of electricity, heat, and fuel production capacities from renewable energy sources, including solar, wind, biofuels, hydropower, and geothermal, etc, which will be discussed below in an ethical context. Heinberg notes that the 21st century ushered in an era of declines, in a number of crucial parameters: global oil, natural gas and coal extraction; yearly grain harvests; climate stability; population; economic growth; fresh water; minerals and ores, such as copper and platinum.

> To adapt to this profoundly different world, we must begin now to make radical changes to our attitudes, behaviors and expectations . . . the cultural, psychological and practical changes we will have to make as nature rapidly dictates our new limits (Heinberg 2007).

Thus moral issues arise as the idea of a post-petroleum economy gains new currency as a security issue. Some years ago, intergovernmental bureaucracies (e.g. UNFAO 1982, 'Planning for the post-petroleum economy') were addressing what now seems a novel and urgent issue, perhaps because the sense of urgency has re-emerged in the confluence of energy and climate concerns. Both producers and consumers of energy have already taken some steps to reflect concern with energy insecurity, by experimenting with different practices (recycling, improving efficiency, slowly introducing new technologies, attempting to manage the energy situation collectively, etc), and yet a remaining element of denial is reflected in the slow pace of change. Even market actors who might otherwise be thought to hold neutral views on energy sources *per se* are becoming agitated about the potential economic costs of delayed action on energy policy: institutional investors are now pressuring governments to act clearly and promptly (Fogarty 2012). It seems clear that maintaining current assumptions about economic growth while addressing climate change will at the very least require prompt application of new technologies and a regulatory and fiscal environment to support them (Sachs 2008). This suggests a fairly radical shift of practices, and it remains to be seen whether currently familiar assumptions about economic growth and energy consumption will survive. Given the range of issues implicated in a discussion of renewable energy, there are various links here to discussion in other chapters in this volume on Energy Security Policy and Democracy Chapter 2); Resource Conflicts (Chapter 4); Energy, Climate Change and Conflict (Chapter 15); Energy Poverty: Access, Health and Welfare (Chapter 19); Low Carbon Economy and Development in Africa (Chapter 21); Centralized vs. Decentralized Energy Systems (Chapter 22); and Human Security and Energy Security (Chapter 23).

Clearly energy is central to our lives. Macfarlane (2007) calls it 'the issue of the 21st century'. Kimmins (2001, 31) notes that 'any consideration of the ethical aspects of these efforts will, therefore, involve an analysis of energy'. Interestingly, he speaks of a 'universal vision' in respect of energy ethics: all potential solutions to individual energy questions involve a social cost, an ethical dilemma and an impact on the way other problems are resolved. Thus, they can only be looked at within a broader consideration of the functioning of the world system of which energy is but one intimately woven component (Kimmins 2001, 35). This is at odds with the typically narrow national perspectives of state governments where energy is seen as fundamental to a way of life and our national security. Kimmins also captures the intergeneration and future-oriented requirements for approaching energy policy in saying that 'many ethical issues arise as a result of unequal access to energy and of the environmental repercussions' and this requires 'that we consider the consequences for future generations of satisfying the energy needs of the present', while also pointing clearly to the long-term requirement for renewables:

The only question is how rapidly we should move to such sources and what mix should be used in various parts of the world over time' (Kimmins 2001, 37–38).

As an indication of how the ethical issues are set in an political-economic context, comments from a regional conference suggest that 'ethics of energy' required that people 'have access to affordable and reliable supply of energy for their basic needs' (UNESCO 2007, 3). Yet how such energy is supplied remains an open question – for example, nuclear energy is on the list of alternatives. As Shea notes:

> tightness in the market has re-ignited the debate over alternative energy supplies such as biofuels or solar power not to mention a renewed interest in nuclear power (Shea 2006).

We should note that a negative experience in Japan and reaction to it in Germany has tempered enthusiasm in the latter. As we will see below biofuels and other renewable can also present difficulties, and if some support for the nuclear option comes from unlikely quarters (James Lovelock, of Gaia fame) it remains controversial due to significant ecological issues, whatever its short-term appeal as a panacea for addressing low-carbon energy security. The UNESCO ethics report also raised doubts over privileging human interests, and advocated 'harmony with nature' (UNESCO 2007, 3). It is noted that earlier negotiations indicated 'objections of certain countries to the development of a potentially binding commitment on environmental ethics that might relate to economic issues' (UNESCO 2007, 4). It was questioned 'whether we could really depoliticize choices about energy', and indeed it seems strange to 'depoliticise' an issue as central to the political-economy as energy, except from a purely technocratic perspective. Certainly the human-environment relationship is facing considerable challenges, which renewable energy may help to address, and the challenge is in large measure an ethical one (Crist and Rinker 2010). So if the political domain extends to include our ethical concerns, then it is not surprising that as 'fossil fuel supplies were dwindling and climate change was accepted as a reality, clean renewable energies, like wind energy, geothermal, wave, tidal, hydropower, and photovoltaic were the way of the future' (UNESCO 2007, 5).

Not surprisingly, the energy mix in any country 'depended on the existing governance and the international sourcing or supply chain', and there is:

> already a large population in lesser developed countries who did not have good access to conventional technology such as electricity and fossil fuels (UNESCO 2007, 8–9).

This perspective appropriately challenges the notion that energy security is an issue of the future – it is clearly an issue for many of us now, and will soon be an issue for all of us. Providing such energy security through renewable sources is a challenge, though the ethical requirements involve an ecological context as much as an economic one. Already renewable sources of energy account for a fifth of electricity generation, and that will increase to a third in the next decades, but overall energy consumption continues to rise and it isn't yet clear that renewable sources can meet the demand. Baer et al. identify the basic dilemma in noting that 'there is no road to development, however conceived, that does not greatly improve access to energy services', and yet there is 'not enough "environmental space" for the still-poor to develop', hence the emergency situation requires 'a wholesale reinvention of the global energy infrastructure on the basis of low-emission technologies' (Baer et al. 2007, 23–26). To the extent that this dilemma is now recognised in political debates, and to some extent in policy initiatives, there is already evidence of an ethical turn and its structural implications. This is the clearest indication that the ethics of renewable energy are tied closely to the pursuit of energy security.

## ANTHROPOCENTRIC AND ECO-CENTRIC MOTIVATIONS

A key distinction underlies the difficulty of squaring the distributional demands of economic and ecological perspectives on ethics. Anthropocentric (human-centred) and eco-centric (ecologically centered) perspectives represent a significant schism in both political theory and moral philosophy. This gives rise to rather different perspectives in ethics and politics. The ethics of renewable energy draws on both perspectives, even if they do not sit comfortably together. One (familiar) perspective is instrumental in its approach to non-human nature, and the significance of renewable energy, understood in terms of human interests in isolation. The other is more holistic, encompassing both human and non-human interests within a wider ecological context. A possible connection between the two perspectives is that human interests may be best (or only) met through adopting the more holistic understanding of our situation. Thus one view is apparently 'realistic' in political terms and consequentialist (ends-oriented) in ethical terms, the other apparently more 'idealistic' in political terms and deontological (means-oriented) in ethical terms. In fact there is more common ground in both practices and pragmatic thought than this distinction suggests, but the schism in traditions of thought and practice is obvious enough. A social-constructivist perspective on the ethics of renewable energy would point to common ground in ethics and politics, where both involve negotiation of sorts – we should note that morality is normative ethics: ethical positions supported by social consensus or norms. This sociological perspective involves communities of people creating or 'constructing' their own forms of knowledge, amongst themselves in collaboration, thus developing a culture of shared meanings – in this case about energy sources and uses. This seems a quintessentially political and pragmatic process, if sometimes dressed in the apparently high-minded clothing of ethics to secure the claims being made in what are otherwise political exchanges. As always, our political exchanges about interests are also laden with values. In ethical terms, an anthropocentric view follows the political logic of distributional issues but does so only with reference to equity among humans, where an eco-centric approach will also take into account the balance in the ecosystem of which humans are a part. For renewables, ecosystem balance suggests the lighter touch of relatively passive energy systems (such as solar). However, such a holistic perspective also requires considering the whole supply chain of components for renewable systems (such as energy intensive aluminium) such that 'green products' are produced by a 'green process'. Being consistent and coherent is a challenge for both ethics and politics. If there are limited opportunities to engage ethically with 'others' and the 'market place' for empathy is troubled by imperfect knowledge, then individuals may have difficulty matching supply and demand for ethical commitment as much as for sustainable energy.

# NATURE AND RESOURCES: PEOPLE, STOCKS, AND FLOWS

Since our interests (and values) are not always or only static, delimited by place and time, the significance of our ethical stance on renewable energy colours our political negotiations. The resulting decisions or actions will have implications beyond a single national society, and beyond a single generation of any society. So our attitude to non-human nature requires the longer, wider view that ethical perspectives are generally better at expressing than political perspectives constrained by bounded sets of interests or electoral horizons. The question of which humans we're talking about is of course a central feature of both ethics and politics at the global (and often local) level. The context for renewable energy is the set of people (here and there, present and future) that depend on the same natural resources. From an anthropocentric viewpoint, nature exists only to provide for our needs, with needs defined in terms of our economic practices and normally confused with our wants and expectations. The natural resources that supply these needs may be viewed as either stock or flow resources, with stocks being finite and flows infinite (unless disrupted). If nature is provident, it also has its limits – and we're getting closer to them all the time. Importantly, we are not all in the same situation, with some feeling the limits more immediately as others buy time. We might view some stocks as being flows, if they renew themselves within human timescales (such as crops or forests), whereas most stocks will start to run out at some point (peak oil). Renewable energy finds itself in this mix of circumstances, such that if a source of energy qualifies as renewable in some sense it at least shouldn't run out, but its availability in time and place may still vary greatly. Although it might be possible to provide abundant energy, for some time, for some people (for some cost to the ecological balance – of the climate in particular), it won't be possible to provide so much for all, for ever. Thus if renewable energy practices are not deployed in the right place at the right time, sustainable availability of energy is in question, and it is precisely such distributional issues which make renewable paths to energy security both ethical and political. Thus the moral context for energy is a matter of distributional justice as between the 'haves' and the 'have-nots'. If uneven development raises familiar distributional issues across populations, in terms of intergenerational equity the 'have-nots' may be future generations. Energy security, importantly, is generally presented as an issue of increasing supply to meet increasing demand. The distributional issues are typically buried under the priority of national economic growth; if it is already difficult to achieve equity in national distribution, how much more so trans-nationally in a globalised economy. So distributional aspects of energy supply are thus determined as much by the structure of 'state sovereignty', and national authority structures of varying degrees of accountability, as by markets. Barnett notes that energy security is just 'the use of national power to secure supplies of affordable energy' in support of economic growth (Barnett 2001, 35), where 'economic and energy security takes priority over environmental security' (Barnett 2001, 76). This is how energy concerns may be forced into a conventional security framework, and why renewable energy in particular is more challenging. However, the long chains of energy production, supply, and demand are not conducive to national energy independence. So, the ethical implications of such structural features may adjust as both states and markets are encompassed in wider social trends, including the increasing use of renewable energy.

## RENEWABLE SHARE OF ENERGY PRODUCTION: A REALITY CHECK

In terms of electricity generation energy sources, renewables comprise 20 per cent and rising, with nuclear at 12 per cent, leaving the bulk of electricity produced by fossil fuels which are heavily subsidised. The World Energy Outlook (2012) up to 2035 suggests that renewables growth in the OECD will come 'mainly from wind (47%), bioenergy (16%), solar PV (15%) and hydro (11%)'. In non-OECD countries hydro will account 'for 42% of the increase in renewables, but wind (25%), bioenergy (16%), and solar PV (10%) also play an important role'. Investment in generation capacity will be more than 60 per cent renewable, 'principally wind (22%), hydro (16%), and solar PV (13%)'. 'A steady increase in hydropower and the rapid expansion of wind and solar power has cemented the position of renewables as an indispensable part of the global energy mix.' What such encouraging developments may hide is a fundamental (and familiar) ethical issue: over a billion of our fellow humans have no access to electricity, renewable or not, and sub-Saharan Africa in particular reflects the great inequality of energy access. The World Energy Outlook also indicates the increasing use of limited water resources (ironically a potential renewable energy source itself) in biofuel crop production, as well as the production and distribution of fossil fuel energy (World Energy Outlook 2012). In such future scenarios, running up to 2050 (Shell 2008, 36), the anticipation is that renewable energy will gain significance due to public pressure, local initiatives and industrial concerns with predictability.

## CONTEXTS: DIFFERENT TYPES OF RENEWABLE ENERGY AND THEIR ETHICAL IMPLICATIONS

The political-economic circumstances are ripe for renewable energy development, and the political will to support them is emerging, but the ethical implications are typically more complicated than single policies or one-size-fits-all solutions may suppose. While each type of renewable energy has somewhat different implications, it is likely that all types will be needed in some context. Furthermore, they all have in common the prospect of meeting ethical demands whether human-centred or ecological; and it is perfectly possible and pragmatic for them to do both. If the global capitalist system has something to answer for in both ethical and ecological terms, the central actors in that system inhabit the same planet as everyone else (and so will their descendents). They also know which side their bread is buttered on, and which way the wind is blowing; like populist politicians, market players will respond to clear signals. If the signals are not clear however, then political-economic alignments and agreed regulatory frameworks are less likely, and this is why achieving an ethical appreciation of renewables is vital to their successful implementation. The Investor Network on Climate Risk (comprising 100 institutional investors with assets of more than $11 trillion) has called on governments to take action on serious climate dangers and to increase clean energy investment, due to purely financial risks' (Fogarty 2012). This is one of many indications that actors of all kinds (not just governments) are seeking a clearer view of the future energy scenario in which renewables are a dominant feature. Even where government action is either possible, expected or necessary, there is variation: while the US federal government is hesitant, the government of California (it's most significant state) is investing heavily in green energy (Carroll 2012).

Similarly, cities are taking the lead over national governments in many cases. A brief examination of a range of renewable sources and technologies will illustrate the issues, concentrating on biofuels as an illustration, but also pointing to the ethical implications of other kinds of renewable energy.

**Biofuels**

Biofuels include plant material such as wood, crops, and their waste or residue, which are potentially carbon-neutral in so far as the growing plants fix carbon at the same rate that carbon is released in their combustion. Biofuels refers to the conversion of biomass into liquid fuels for transportation, and is related to other terms like biopower (burning biomass directly, or converting it into gaseous fuel or oil for electricity generation) and bioproducts (conversion of biomass into chemicals to make products that would otherwise be made from petroleum). If biofuels are an important part of our past, they are also a key part of our future: 'traditional biomass's share has declined slightly, while modern renewable energy's share has risen' (REN21 2012).

Biofuels emerged as a potential 'silver bullet', solving at once the problem of energy security and security from climate change, but were quickly mired in controversies arising from the energy-food-water nexus. This controversy around biofuels as an alternative energy source has an obvious moral dimension. Not surprisingly, as with other renewables, the morality and the practicality of biofuels are closely connected. It seems clear that initial enthusiasm for biofuels has been tempered by unintended (or unrevealed) consequences, including the net energy/environment benefits, and the impact on food crops with knock-on effects on the price and availability of food. The EU therefore reconsidered, though didn't suspend, its targets in this area: 'Both the EU Environment Commissioner and Defra's own chief scientist today went on record to say that current plans to vastly increase the amount of fuels such as bioethanol and biodiesel might need to be reconsidered' (Greenpeace UK 2008). US policy remains tied to domestic political commitments, particularly on ethanol production in parts of the country with electoral influence (Congress approved a five-fold increase in use of biofuels in 2007), but is also facing opposition:

> ... a reaction is building against policies in the United States and Europe to promote ethanol and similar fuels, with political leaders from poor countries contending that these fuels are driving up food prices and starving poor people. Biofuels are fast becoming a new flash point in global diplomacy, putting pressure on Western politicians to reconsider their policies (New York Times 2008).

A low rate of Energy Returned on Energy Invested (EROEI), heavy use of water, and fossil fuels needed in ethanol production all created a complex food and agriculture problem. There was little consideration to how ecological and sustainable biofuels might be, given soil degradation and fertiliser requirements:

> on the face of it, growing biofuels to support the car habit is a suicidal prospect ... What is the morality in light of the growing numbers of mouths to feed? (Energy Bulletin 2008).

The political (and ethical) stakes involved are quite high, with the EU Commission having to reject claims that production of biofuels is a 'crime against humanity' (Agence France Press 2008), even with the relative proportion of biofuels in the energy mix being so far limited.

Policies such as the European Renewable Energy Directive, through which European Union (EU) states committed in 2008 to source 10 per cent of their transport energy needs from renewable fuels by 2020, have 'backfired badly'... If a biofuels technology meets all the proposed environmental and ethical standards, then there is a 'duty' to develop it (Gilbert 2011).

Second generation biofuels, as with other advanced renewables, may avoid some of the apparent difficulties. Over the long term biofuels will likely continue to be part of the renewable energy mix, but the problems illustrate the moral issues surrounding energy security, particularly with respect to distributional justice. If there are still potential advantages to be gained from biofuels, these may not be equally distributed: currently, we are only on the receiving end of the negative effects of the global biofuels trend, instead of the positive – the country is being forced to absorb the inflationary high food prices, without getting the benefit of direct investments in rural areas and the associated job creation. It would seem that the key to making sense of these suggestions is for policymakers to re-evaluate biofuels through the prism of rural and industrial development rather than simply employing the somewhat populist food/fuel framework (Creamer 2008).

Clearly biofuels can serve a range of purposes from substituting petroleum fuels to encouraging agriculture and rural development, but this seems to water down the energy security strategy often implied and completely undermines it if the net use of energy doesn't actually reduce petroleum dependency. There are economic motives here, and even old-fashioned energy efficiency ('negawatts') could be significant for energy and climate alike, but the 'rebound effect' of increased access and lower prices for fuel leading to greater consumption could cancel 26–37 per cent of any gains (Economist 2008), but the ecological and moral motives seem somewhat distant, and the political coherence of energy security is thus limited. A focus on a narrow area of policy concern without a wider perspective is likely to lead to problems, both political and ethical. While the 'configuration and context of business at the global level' is transforming and there is 'a growing need for sustainability coupled with growth' there are '*some tensions, however, between the imperatives for developing renewable and biofuel resources, and the imperatives for advertising and promoting high energy consumption luxury products*' (UNESCO 2007, 17, emphasis added). Some tensions, no doubt. As the promise of advanced biofuels is explored, the problems of conventional biofuels may become irrelevant over time (VLAB 2010). Nevertheless, as biofuels seem most obviously linked to both human-centred and eco-centred development, ethical issues will remain close at hand, and consequently ethical principles for biofuels policy-making are being established (Buyx and Tait 2011, 633).

**Hydro**

Hydro (water) power includes a range of renewable energy production, including large-scale hydro-power dams, small-scale low head hydro turbines, micro-hydro, and related pumped storage plants which use water to store potential electricity. Hydro, utilising an existing flow resource, has the potential to produce unlimited zero-carbon energy. At the same time, it may involve expensive infrastructure and its potential is so far less fully realised in the developing world than in the industrial or industrialising world – at present China is the leading hydro power producer. However, the consequences of large-scale hydro dams for the ecosystem, and for local and upstream/downstream human populations (who may not even benefit from the energy) means that large-scale hydro may not be viewed as renewable from the wider perspective on sustainability. This immediately raises the ethical issue of 'inequity in distribution of impacts among different social groups':

> Diversion of the river resulted in loss of water sports (for high-income groups both local and remote), loss of historical monuments (for remote high-income groups) and recreation losses (for local poor). Removal of forest cover leads to loss of non-timber products (for local poor) and carbon storage (for remote high- and low-income groups). Loss of home garden productivity was borne by local poor groups. Benefit of the project, generation of 145 GWh annually, was a gain for the grid connected groups. (Gunawardena 2010, 726)

Issues of population displacement, land use, and the complex water-energy relationship create tensions, as noted in the IEA report 'World Energy Outlook 2012'. So for example, 'large hydro' is excluded from measurements of renewable contribution:

> ... total investment in renewables excluding large hydro last year increased 17% to a record $257 billion, a six-fold increase on the 2004 figure and 94% higher than the total in 2007, the year before the world financial crisis, says the UNEP report, *Global Trends in Renewable Energy Investment 2012* (REN21 2012).

Such a substantial contribution to energy investment and production may raise as many issues as it addresses, when viewed from an ethical perspective, but as with other renewables the potential advantages of hydropower must not be dismissed simply because political and economic failings have left ethical issues in their wake. In particular smaller-scale hydro power is both decentralised, providing electricity locally as well as potential surplus for a grid, and can have a low impact on the environment.

**Wind**

Wind, like micro-hydro, can produce energy on a small local scale which is likely to be more environmentally friendly overall, as well as offering great potential as a mainstream source of grid energy as is currently demonstrated by wind farm development in many parts of the world. Large-scale wind farms or plants carry with them both the economic advantages of scale, and could compete with other significant energy infrastructures, but equally with the controversies of any large development. On shore wind farms, or even single wind turbines, can spark off typical objections on the grounds of visual and other impacts on local environments which resonate in local and national politics (the NIMBY problem – 'not in my backyard'). Off shore wind farms, being to some extent out of sight and thus out of mind (for most of the population) are therefore less often objected to on the grounds of intrusions into settled environments. However, even when environmental amenity arguments don't play strongly in the balance of concerns, concerns about the relative costs and benefits (on or off shore) find their way into public debate. If, as for most renewables, there are initial costs and a need for regulatory and financial support, this must surely be weighed against the long-term costs of climate change and energy insecurity.

While there are ambitious government targets to increase the share of renewable energy in many countries, it is increasingly recognised that social acceptance may be a constraining factor in achieving this target. This is particularly apparent in the case of wind energy, which has become a subject of contested debates in several countries largely due to its visual impact on landscapes (Wüstenhagen et al. 2007, 2683).

Wind farm development illustrates the significance of perceived fairness in accepting renewable energy, such that the eco-centric concerns with environmental justice are balanced with anthropocentric concern with procedural justice.

Decisions concerning the siting of infrastructure developments or the use of natural resources have the potential to damage a community's social well-being if the outcomes are perceived to be unfair. Justice is accepted as central to the good functioning of society with fairness being an expectation in day-to-day interactions. Outcomes that are perceived to be unfair can result in protests, damaged relationships and divided communities, particularly when decisions are made which benefit some sections of the community at the perceived expense of others (Gross 2007, 2727).

**Geothermal**

Geothermal energy, derived from underground sources, seems to avoid some of the political difficulties of visual impact. For local (residential) heating energy can be drawn from shallow ground with few visible signs of the source. Larger-scale use of geothermal energy draws on either hot water or hot rock for heating, or deploying the heat (directly or indirectly) to drive steam turbines for electricity generation. This might also attract objections on the grounds of visual impact, but not more so (perhaps less so) than other forms of electricity generation. As with other renewable sources, the absence of unpleasant by-product or pollution should carry public opinion, even if the more serious concerns of global climate change do not.

## Oceans

The oceans, covering most of the planet's surface, are an important factor in renewable energy provision as indeed they are in all aspects of our relationship with the environment (oceans are also a significant carbon sink, for example). Like geothermal energy, ocean thermal can provide a ready source of heating for conversion. Unlike geological sources, and more like hydro energy, oceans can also provide a mechanical source of energy in tides and waves. Barrage or dam systems can convert tidal energy into electricity via turbines. Waves can power floats or oscillating water column systems to drive hydraulic pumps. As these are typically on or near shorelines, the visual impact issue arises, as does cost – though the same moral imperatives should bear on the political and economic calculus.

## Solar

Solar energy technologies can provide heat, light, hot water, electricity, and even cooling. There is a considerable global industry in photovoltaic (PV) systems which produce electricity directly from sunlight, and solar seems an obvious solution to both energy and environmental issues. Solar PV, though advancing steadily towards a mainstream source of renewable energy, is dogged by political manipulations of 'feed-in tariffs' which determine the cost and benefit of linking distributed solar PV energy production to the electrical grid. Ideally renewables would not need such support (and may not need it in time), but like all new technologies the start-up costs are considerable, and in the case of energy supply there are also infrastructural issues which clearly implicate the public interest in diverse energy supplies.

The link between ethics and renewable energy is short and clear: we do not have the right to destroy the conditions for life on earth by continued use of climate-damaging energy sources; we must replace them with clean, safe, abundant and geopolitically-benign renewable energy: the mechanism proven to deliver the fastest, lowest-cost renewable energy deployment is the feed-in tariff (FIT), often referred to in the US as the advanced renewable tariff (ART) (Mendonca 2007).

Another common and relatively uncomplicated and inexpensive technology provides immediate domestic hot water using solar energy through heat conversion, and is a common sight on rooftops in warm climates. Of course like other heat sources, it can be used to drive steam turbines for electricity generation, and large-scale desert solar electric plants have been developed. But as a low-tech option solar energy is equally attractive, and this endless resource can be used by simply allowing the sun to provide passive heating and lighting through intelligent design (rather than keeping both out); and there are somewhat more advanced applications of such technology for industrial and commercial heating and cooling. If the sun seems such an obvious source of energy, solar power, like wind power, faces frequent objection on the grounds of intermittent availability. However, since it is normally the resulting electricity supply which is expected to be available on demand around the clock (day and night, rain or shine, calm or storm), the issue easily transfers itself to the mix of energy sources contributing to the grid and its storage capacity – for example, one aspect of hydro power is its use as a means of storage by absorbing any excess electricity to pump water up to a reservoir, and easing the peaks of grid demand through opening up the flow to turbines as required.

**Other Options: Hydrogen, Nuclear, Waste**

Much has been made of the potential for hydrogen to replace fossil fuels, as it is portable and produces water rather than carbon as a by-product. Hydrogen fuel cells are being tested for transportation applications, operating in place of a battery to provide electricity, and might also be used for heat and electricity for buildings. The source of hydrogen and its production, however, may detract from this apparent panacea. Hydrogen is extracted by heating hydrocarbons – mostly natural gas, as well as methane or gasoline – though it can also be produced from water by electrolysis (requiring electrical energy up-front of course). Interestingly, some algae and bacteria can produce hydrogen utilising solar energy. So while hydrogen may not provide a ready renewable energy solution, its portability could make it an important energy carrier, and a potential energy storage medium for renewable sources like wind and solar.

Nuclear power, if carbon neutral in terms of its direct emissions, is not renewable in wider ecological terms because of long-term (very long-term, given half-life of fuels) waste management issues, and short-term ecological risks, including those directly relating to human health. Kristin Shrader-Frechette (2011) has argued extensively about the miscalculations of energy planning that make nuclear power seem feasible or even sustainable, and advocates renewable as the only option for the future. It remains that if climate change is not addressed promptly (which seems likely) then nuclear power may be the least worst option in the medium term. In this context, the Nuclear Energy Agency argues that the 'critical path structure' should include 'concurrent risk, economic, and environmental impact analyses . . . for all technologies and proposed actions for the transition to a post-petroleum economy' (Nuclear Energy Agency

2004, 37). However, while nuclear power remains under consideration, and hydrogen technology emerges as a potential portable fuel (though electricity intensive in production), there are many more positive solutions to the challenge.

Utilising our own existing *waste*, both domestic and industrial, is another source of energy. While recycling waste in the energy context can only be a good thing, it depends on generation of the waste in the first place. An ecological approach would reduce waste overall, and thus also reduce that available for energy production.

## INFRASTRUCTURES: ON AND/OR OFF GRID?

Much of what passes for insurmountable difficulty in renewable energy technologies rests on our current expectations and traditional practices. Both may have to change somewhat, if we are facing a dual crisis of energy security and climate change, and exacerbating ethical shortcomings. Current energy infrastructures were designed for and are still dominated by fossil fuels, though could be supplied by renewables. Modern energy infrastructures offer a range of possibilities, from large-scale smart grids to small-scale smart meters, which contribute to efficiency (if not immediately encouraging simple sufficiency). If centralised energy systems make sense in dense population areas, distributed energy resources have such great advantages that they may dominate the future. For many of us, traditional and distributed energy, or none at all, is the norm. That in itself raises the central ethical issue in relation to energy, which is its distribution across the human population and not simply across the economy. In these more deprived circumstances, even modest (and thus ecologically sustainable) forms of energy supply greatly improve life chances (refrigeration for medicines) and life experiences (lighting for reading). Very basic wind and solar energy technology may provide for household needs. Small local grid systems may provide community-level energy from a shared renewable source. Such distributed and decentralised energy supply may aid social and economic development with relative independence from centralised supply grids and urban centres. Development may end up involving a combination of both, since a variety of small, modular power-generating technologies can improve the operation of electricity delivery systems. Portable energy storage also supports the many aspects of increased mobility in the modern world – not least in the area of information and communications. If many small sources of energy are better suited to a smarter way of life, then 'energy harvesting' from human motion, or the scavenging and recovery of residual energy, may prove to be the wave of the future. For the moment, it seems our ravenous appetite for large volumes of both material and energy consumption can only be satisfied by large-scale grid systems and energy intensive transportation.

## CONCLUSIONS

UNESCO discussions suggested some realism about the application of ethical principles on renewable energy: 'although a normative declaration would be nice, it was not feasible in the current political environment', even though this suggests the need for a 'global eco-ethics' (UNESCO 2007, 7–8). However it is precisely the point that energy security dilemmas are seen as a challenge. In the case of renewable energy the challenges are both human and ecological, since meeting ecological expectations in some respects may not guarantee full human participation, which in itself is an ecological issue.

Conventional energy technologies and deployment approaches cannot be relied upon to eliminate energy poverty in Africa. Innovations in energy access are necessary (Agbemabiese et al. 2012).

The ostensible aims of governments (wealth, health, security) look different from an ecological perspective, which may lead to a change in political and economic practices, as well as to our ethical consciousness. Minteer (2005) suggests that such reconsiderations of the public interest is a necessary aspect of linking ethical questions to policy debates.

> Ethically, for reasons of environmental justice and towards meeting the goals of sustainable development, further research into alternative energy sources is an imperative for all nations capable of such research and development. A failure to increase research into alternative energy sources is unethical, in that it not only favors the current status quo in relation to energy infrastructure, and among other reasons, it increases the burden of costs for future generations in the both the clean-up of higher levels of atmospheric carbon as well as the inevitable costs to further innovate. However it is neither wise, economical, nor just to proceed developing certain sectors of alternative energy technologies without a proper analysis of the ethical dimensions of various proposed strategies. (Schienke 2007)

This reflects Hayward's (1998) argument that environmental values are supported by enlightened human interests, and furthermore that this link must exist to promote ecological goods, and that consequently there are serious implications in fully integrating environmental issues into our disciplinary concerns. A key aspect of the political barriers to ethical practices is the tendency to leave everything to a market economy, allowing current short-term calculations to determine the value of energy sources. It may require some regulatory intervention, or public and energy-consumer action to shift the balance of interests in this regard.

While it is true that markets are often excellently self-organising, with many concomitant benefits, fossil fuels are not sufficiently required to address the damage that they cause to life on earth, *and* they continue to attract large subsidies – in itself a market distortion. Ernst Ulrich von Weizsacker, founder of Germany's Wuppertal Institute, referred to this as 'telling the ecological truth' (Mendonca 2007).

Hargrove (1989) makes an argument for anthropocentric, aesthetic sources of modern environmental concern by identifying attitudes that constrain ('idealism', 'property rights') and support (scientific and aesthetic ideals) our ecological perspectives. However, even if such approaches make renewable contributions to energy security easier to sell politically, it remains that an ecological ethic more directly reflects the limits we face. Renewable energy carries with it the prospect of ethical as well as ecological sustainability.

# REFERENCES

Agbemabiese, Lawrence, Jabavu Nkomo and Youba Sokona (2012) 'Enabling innovations in energy access: An African perspective', *Energy Policy* 47, Supplement 1, June 2012, pp. 38–47.

Agence France Press (2008) 'EU defends biofuel goals amid food crises', 14 April 2008, http://afp.google.com/article/ALeqM5gp1nkJeC-IhlYkVtsvPfp3u7mOWQ (accessed 25 November 2012).

Baer, Paul, Tom Athanasiou and Sivan Kartha (2007) 'The Right to Development in a Climate Constrained World: The Greenhouse Development Rights Framework', Berlin: Heinrich Böll Foundation, Christian Aid, EcoEquity and the Stockholm Environment Institute, http://www.ecoequity.org/docs/TheGDRsFramework.pdf (accessed 25 November 2012).

Barnett, Jon (2001) *The Meaning of Environmental Security: Ecological Politics and Policy in the New Security Era*, New York: Zed Books.

Buyx, Alena M. and Joyce Tait (2011) 'Biofuels: ethics and policy-making', *Biofuels, Bioproducts & Biorefining* 5, pp. 631–639.

Carroll, Rory (2012) 'As U.S. hesitates, California pours billions into green energy', *Reuters*, 14 November, http://www.reuters.com/article/2012/11/14/us-clean-energy-california-idUSBRE8AD0F720121114 (accessed 25 November 2012).

Creamer, T. (2008) 'Is the SA biofuels debate properly framed?', http://www.polity.org.za/article.php?a_id=129266 (accessed 28 March 2008).

Crist, Eileen and H. Bruce Rinker (eds) (2010) *Gaia in Turmoil: Climate Change, Biodepletion, and Earth Ethics in an Age of Crisis*, Cambridge, MA: The MIT Press.

Economist, The (2008) 'The elusive negawatt', 10 May 2008, pp. 93–95.

Energy Bulletin (2008), 16 April, http://www.energybulletin.net/stories/2008-04-16/food-agriculture-apr-16 (accessed 25 November 2012).

Fogarty, David (2012) 'Global investors call for action on serious climate danger', *Reuters*, 20 November, http://www.reuters.com/article/2012/11/20/us-climate-investors-idUSBRE8AJ0AB20121120 (accessed 25 November 2012).

Geller, Howard (2002) *Energy Revolution: Policies for a Sustainable Future*, Washington DC: Island Press.

Gilbert, Natasha (2011) 'Biofuels need enforceable ethical standards', *Nature*, News 12 April 2011, http://www.nature.com/news/2011/110412/full/news.2011.230.html (accessed 25 November 2012).

Greenpeace UK (2008) 'Senior EU and Defra figures agree: we were too hasty on biofuel targets', posted 14 January, http://www.greenpeace.org.uk/blog/climate/we-were-too-hasty-on-biofuel-targets-20080114 (accessed 25 November 2012).

Gross, Catherine (2007) 'Community perspectives of wind energy in Australia: The application of a justice and community fairness framework to increase social acceptance', *Energy Policy* 35:5, May, pp. 2727–2736.

Gunawardena, U.A.D. Prasanthi (2010) 'Inequalities and externalities of power sector: A case of Broadlands hydropower project in Sri Lanka', *Energy Policy* 38:2, February, pp. 726–734.

Hargrove, Eugene C. (1989) *Foundations of Environmental Ethics*, Englewood Cliffs, NJ: Prentice Hall.

Hayward, Tim (1998) *Political Theory and Ecological Values*, New York: St Martin's Press.

Heinberg, Richard (2007) *Peak Everything: Waking Up to the Century Of Declines*, Gabriola, BC: New Society Publishers.

Heinberg, Richard (2004) *Powerdown: Options and Actions for a Post-Carbon World*, Gabriola, BC: New Society Publishers.

Kimmins, James Peter (2001) *The Ethics of Energy: A Framework for Action*, UNESCO.

Macfarlane, Allison M. (2007) 'Energy: The issue of the 21(st) century', *Elements* 3:3, June; 165–170.

Mendonca, Miguel (2007) 'Energy, Ethics and Feed-in Tariffs', http://www.renewableenergyworld.com/rea/news/article/2007/04/energy-ethics-and-feed-in-tariffs-48310 (accessed 22 November 2012).

Minteer, Ben A. (2005) 'Environmental Philosophy and the Public Interest: A Pragmatic Reconciliation', *Environmental Values* 14:37–60.

New York Times (2008) 'Fuel Choices, Food Crises and Finger-Pointing' http://www.nytimes.com/2008/04/15/business/worldbusiness/15food.html (accessed 25 November 2012).

Nuclear Energy Agency (2004) 'Nuclear Production of Hydrogen', proceedings of the Second Information Exchange Meeting, Argonne, Illinois, USA, 2–3 October 2003 (OECD).

REN21 (2012) 'Renewables 2012 Global Status Report', http://www.ren21.net/Portals/97/documents/GSR/GSR2012_Press%20Release%20short_ENGLISH.pdf (accessed 25 November 2012).

Sachs, Jeffrey D. (2008) 'Technological Keys to Climate Protection: Dramatic, immediate commitment to nurturing new technologies is essential to averting disastrous global warming', *Scientific American* March.

Scheer, Hermann (2002) *The Solar Economy*, London: Earthscan.

Schienke, Erich William (2007) 'An Initial Ethical Evaluation of Alternative Energy Strategies: Introduction', 11 July, http://rockblogs.psu.edu/climate/2007/07/an-initial-ethical-evaluation-of-alternative-energy-strategies-introduction.html (accessed 25 November 2012).

Shea, Jamie (2006) 'Energy security: NATO's potential role', *NATO Review* 2006:3 Autumn, http://www.nato.int/docu/review/2006/issue3/english/special1.html (accessed 25 November 2012).

Shell International BV (2008) 'Shell energy scenarios to 2050', http://s08.static-shell.com/content/dam/shell/static/future-energy/downloads/shell-scenarios/shell-energy-scenarios2050.pdf (accessed 23 November 2012).

Shrader-Frechette, K.S. (1988) 'Planning for Changing Energy Conditions' *Energy Policy Studies*, 4, pp. 101–137.

Shrader-Frechette, K.S. (1984) 'Ethics and Energy' in Tom Regan (ed.), *Earthbound, New Introductory Essays in Environmental Ethics*, New York: Random House, pp. 107–146.

Shrader-Frechette, Kristin (2011), *What Will Work: Fighting Climate Change with Renewable Energy, Not Nuclear Power*, New York: Oxford University Press.

UNESCO (2007) 'Report of the launch conference for the project Ethics of Energy Technologies in Asia and the Pacific', UNESCO Bangkok in collaboration with the Ministry of Science and Technology and the Ministry of Energy, Thailand, Imperial Tara Hotel, Bangkok, 26–28 September 2007, prepared by Dr Darryl Macer and Anna Engwerda-Smith, http://www.unescobkk.org/fileadmin/user_upload/shs/Energyethics/EnergyEthics2007Report.pdf (accessed 25 November 2012).

UN Food and Agriculture Organisation (1982) 'Planning for the post-petroleum economy', *Ceres: The FAO Review on Agriculture and Development*, 15(4), July, pp. 26–32.

VLAB, Stanford-MIT Venture Lab (2010) 'Biofuels 2.0: Sustainable Startups – from Garage to Gargantuan', February, http://www.vlab.org/article.html?aid=306 (accessed 22 November 2012).

World Energy Outlook (2012) http://www.worldenergyoutlook.org/publications/weo-2012/ (accessed 12 November 2012).

Wüstenhagen, Rolf, Maarten Wolsink and Mary Jean Bürer (2007), 'Social acceptance of renewable energy innovation: An introduction to the concept', *Energy Policy* 35:5, May, pp. 2683–2691.

# 21. Low carbon development and energy security in Africa
*Chukwumerije Okereke\* and Tariya Yusuf*

## 1 INTRODUCTION

Energy poverty is without a doubt one of the most critical development challenges facing African countries today. Out of a population of about a billion, over 547 million Africans do not have access to electricity but depend on biomass for their basic energy needs (IEA, 2011). Wide scale energy provision is therefore a vital requirement for achieving the economic growth and development aspirations of African countries. This is more so the case in Sub-Saharan Africa which has the worst poverty in the world (EIA, 2011).

An equally important challenge facing Africa and the rest of the world is how to deal with the problem of climate change which is caused mostly by carbon emissions implicated in energy production and consumption. Climate change, through its impact on drought, desertification, health and extreme weather events, will exacerbate energy poverty in Africa and lead to the further impoverishment of millions (IPCC, 2007a). In fact, worsening energy security problems is one the most critical ways in which climate change is affecting, and will continue to affect Africa (IPCC, 2011).

It is clear therefore that the central development dilemma facing African countries and their development partners today is how to address the problem of climate change while at the same time pursuing the quest for rapid economic growth and universal energy access (IEA, 2011; Sokona et al., 2012). The answer to this dilemma lies, to a large degree, in the concept of low carbon development. Africa has plenty of cheap renewable resources that can be harnessed to achieve the triple wins of combating energy poverty, mitigating climate change and building a low carbon economy. A low carbon development path for Africa carries the promise of multiple co-benefits including wider systematic economic resilience, improvement in health condition, resource conservation, energy security, reduced foreign exchange need and reductions in budget deficit (Bowen and Frankhauser, 2011; Diog and Adow, 2011; Khennas, 2012). Indeed, the current situation in many African countries where high cost and carbon intensive energy is imported through distant transcontinental routes is incongruous with reason and basic economics of good development. Pursuing low carbon development options has the potential to facilitate economic integration and market access among African countries while also helping the continent to contribute its quota in helping the global community address the problem of climate change. That said, harnessing cheap renewable energy resources in Africa will require appropriate capital, national regulatory and investment policy frameworks which may not necessarily be easy to achieve. It will also require an equitable distribution of the cost of transition to a global low carbon future, a significant aspect of which will have to come from substantial North-South financial and technology transfer.

## 2  ENERGY POVERTY IN AFRICA

One of the most significant issues affecting development in Africa is poor access to modern energy services. Energy access is critical for meeting many human needs including lighting, heating, cooking and communication. It also plays a central role in driving productive enterprise such as agriculture, transport and industrial activities. The link between energy consumption and economic growth is fairly straightforward and well established (Abanda et al., 2012; IEA, 2011; Kanagawa and Nakata, 2007). For Africa a key feature of the economic landscape is extremely poor access to and low consumption of modern energy. This is despite the fact that Africa has abundant natural resources and is a major contributor to the world's primary energy oil production. Africa has about 9 per cent of the proven world's oil reserve and accounts for about 12 per cent of total world oil production (IEA, 2011). The continent has about 7 per cent and 6 per cent of the world's total gas and coal reserves respectively (BP, 2012). There is also abundant natural gas, hydro, solar, biofuel, geothermal and nuclear energy resources.

Despite these endowments, Africa with its 14 per cent of global population accounts for just about 3 per cent of the world's primary energy consumption (BP, 2012). Africa has the lowest electrification rate of all the world at 26 per cent of households, with as many as 547 million people without access to electricity (EIA, 2011). Accordingly, a vast proportion of people in Africa depend on traditional biomass fuels from woods, agricultural residue and dung for heating and cooking needs. According to IEA reports, more than 80 per cent of Sub Saharan African households amounting to 653 million people use biomass for cooking. This has had devastating consequences for people and the environment. In 2009, more than 1.45 million African lives were lost to household pollution caused by inefficient biomass cooking stoves. Fewer people died from malaria (IEA, 2011).

On current trends less than half of African countries will reach universal access to electricity even by 2050. Generation capacity in Africa at 39 MW per million population is about one-tenth of the levels found in other low income regions of the world. Per capita electricity consumption in Africa (excluding South Africa) averages only 124 kilowatt-hours a year, barely one per cent of the consumption typical in high income countries (EIA, 2011). This is hardly enough to power one light bulb per person for six hours a day. The number of people without electricity is either static or increasing because population growth is outstripping the pace at which households are being connected. In other words, the annual rate of new connections in Africa (less than 1 per cent) is not keeping pace with new household formation (1.9 per cent).

Within Africa there is wide variation across the sub regions. In the recent past, North African countries have made significant advances in dealing with the challenge of energy access leaving the problem mostly for sub-Saharan Africa. For example, two out of three Sub Saharan African (SSA) households live without electricity. In stark contrast, 99 per cent of North African households have electricity supply (IPCC, 2011). Only 14 per cent of rural Sub Saharan African (SSA) households are linked to the grid. In comparison, 74 per cent of rural households in Latin America are connected to the national grid (Khennas, 2012).

In Sub-Saharan Africa, East Africa has the lowest consumption of modern energy services per capita and this is in spite of the growth of national economies witnessed in the last decade or so. For example, per capita consumption of electricity in Tanzania is alarmingly low at 65 kWh. This represents only about 2.4 per cent of world consumption of 2,751 kWh/per capita (World Bank, 2007).The picture is even grimmer for Uganda. Here, only 10 per cent of the population has access to electricity and the per capita electricity consumption stands at 44 kWh. Kenya's per capita electricity consumption is comparatively better. It is estimated to be 128 kWh (UNEP, 2012).

But the picture is not much different even for large countries like Nigeria which is widely regarded as one of the two super economies in Africa – the other being South Africa. It is estimated that well over 50 per cent of Nigerian's 152 million population does not have access to electricity (Eleri et al., 2011; Okoro and Chikuni, 2007). The country, which requires a minimum of 10,000 MW of electricity to meet her energy demands currently has a production capacity of just about 3,000 MW. Even though the installed capacity of electricity is much greater than 3,000 MW, infrastructure utilization is historically very poor and power supply has been epileptic as a result of lack of maintenance and unscheduled outages (Oseni, 2012). Investment needed in the power sector by Nigeria to generate anything near the quantity needed is estimated at USD20 billion (Eleri et al., 2010; Okafor, 2008; Okoro and Chikuni, 2007).

While the figures on the number of households connected to the grid provide a very good indication of the extent of the problem, they do not tell the whole story. In reality, a high percentage of those that are connected to the grid do not have regular access to electricity due to high frequency of blackouts and unstable power supply. The World Bank estimates that SSA households experienced between 91 and 105 days of blackouts in 2007. There are several instances where communities that have been connected to the national grid actually never get to enjoy electric power supply for more than three months in a year. Frequent outages and load shedding very much characterize the experience of a vast majority of populations and businesses in most sub-Saharan African countries (Akinbami et al., 2001).

High-cost and unreliable energy services in Africa are a significant drag on economic growth and competitiveness in the region. Every year, African households and business spend upwards of USD17 billion on fuel based lighting that is often of poor quality and hazardous (World Bank, 2009). The World Bank further estimates that the economic value of power outages noted above amounts to as much as 2 per cent of GDP for countries affected. This figure according to the International Energy Agency (see Figure 21.1 below) will continue to grow unless new policies and programmes to increase access are implemented.

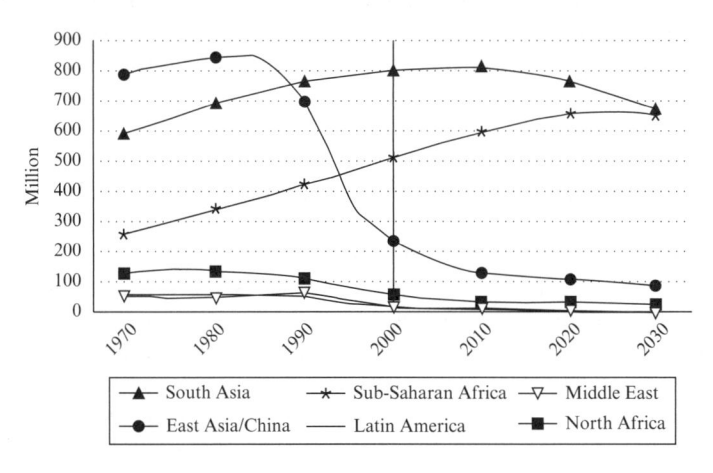

*Source:* World Bank (2009).

*Figure 21.1  Number of people without electricity (actual and projected) by region under current policies*

Beyond low electrification, energy poverty challenges in Africa extend to inefficient and perilous forms of domestic energy for cooking attributable to a lack of modern fuels and clean cookers. Africa currently has the highest energy intensity in the world. It uses far more energy for every dollar of gross domestic product (GDP) than any other region (IEA, 2011). Africa's inefficient energy system is characterized by energy that is imported, expensive, environmentally unsustainable and dependent on coal, oil, wood fuels and natural gas (IPCC, 2011). Massive dependence on imported fossil fuels consumes a high portion of Africa's export earnings. Even Nigeria, the region's largest exporter of crude oil, has to import refined fuels. Fluctuation of oil and gas prices further complicate the task of delivering a secure energy supply in the region. Sub Saharan African countries spent 14 per cent of their GDP on fuel imports in 2000 (EIA, 2011). The focus of delivering centralized conventional electricity through thermal power from oil, gas and coal or from large-scale hydropower has not effectively delivered either energy access for poor people or the rate of economic growth that sub-Saharan African countries aim for (Anozie et al., 2007).

It is generally accepted that for Africa to achieve the millennium development goals (MDG) embodied in sustained economic growth leading to poverty reduction, improved standard of living, adequate and reliable energy services have to be made available. There is a close association between reducing the need for poorer households in developing countries to use biomass for cooking and heating and reaching the MDGs on universal primary education, promoting sex equality and the empowerment of women and reducing under-five child mortality. A rapidly changing climate will however make this task more difficult and further exacerbate energy poverty in Africa (Ebohon, 1996).

## 3   CLIMATE CHANGE AND ENERGY POVERTY

The threat of climate change to humankind and to the planet as a whole has gradually become more evident. Africa is the continent most vulnerable to climate change impacts and the least prepared to deal with its effects (IPCC, 2007b). In fact, Africa is already experiencing severe negative impacts of climate change on its people, environment and economy (IPCC, 2007b). These include among others, prolonged periods of droughts, surface mean temperature increase, reduced agricultural yield, erratic precipitation patterns, flooding, ecosystem collapse, malnutrition, spread of tropical diseases and deaths. One of the key ways in which climate change is impacting and will impact Africa is in the worsening problem of energy poverty (IPCC, 2011).

Climate change and energy poverty has an intimate and complex relationship especially in the context of developing regions like Africa. First, it is well known that energy provision is a critical requirement for achieving economic growth and development. Hence, African countries would need to vastly increase their energy generation and consumption in order to reduce poverty, build climate adaptive capacity and achieve their development aspirations (Bowen and Frankhauser, 2011; Sokona et al., 2012). At the same time, it is well known that the vast proportion (up to 82 per cent) of anthropogenic carbon emissions come from energy related activities, including electricity generation, transport, building, and industry (IPCC, 2007b). The paradox then is that the effort to achieve development, which is crucially needed to escape energy poverty and adapt to climate change, could in turn exacerbate the problem of climate change leading to poverty (Okereke and Schroder, 2009). Second and related, the global aspiration to combat climate change has serious implication for available options for energy provisioning in Africa and other developing countries (Ouedraogo, 2012).

One important emerging issue in this regard is the reluctance of many policy makers, international aid agencies and environmentalists to consider the full range of energy options to meet the energy needs of the poor on the grounds of the need for reducing greenhouse gas (GHG) in the atmosphere (Sanchez, 2011). A famous case was the controversy that was generated in 2010 when the World Bank sought to lend Eskom in South Africa about USD3.75 billion to finance its coal fired power plant investment project at Medupi. Then (and as of now) the contention by many NGOs was that the money would have been better used to finance investment in more expensive but less carbon intensive renewable energy projects. Sanchez (2011) has also noted that in some cases this imperative to achieve development in the context of reducing global GHG emissions may result in the use of uncompetitive options to pursue development objectives. One example was where a development agency insisted on using a renewable energy system to pump underground water for drinking and farming even when it would have been cheaper to use small diesel engines (Sanchez, 2011). Yet, another scenario which has recently gained currency in literature and public policy relates to situations where efforts to conserve tropical forests in the context of climate change can deny local people access to valuable forest products including wood fuel (Okereke and Dooley, 2010).

Thirdly, climate change, through erratic rainfall, flooding and drought can have a direct impact on energy infrastructure with a serious impact on generating capacity. This has already been witnessed in various countries in Africa where a drastic decrease in precipitation rates has resulted in severe drought affecting hydropower generation. For example Kenya and Ghana, both of which currently rely heavily on large hydropower dams, have experienced significant power shortages in recent years due to unusually long droughts. Droughts generally lead to massive load shedding and decreased electricity supply with the result of huge economic disruptions and losses (UNEP, 2012). Excessive flooding on the other hand contributes to a rapid build-up of silt in hydropower dams, affecting the amount of water available for electricity generation.

For developing countries in Africa, the recent food, commodities, and oil price shocks – all of which are to some degree a resulting impact of climate change – are already having severe implications for energy access particularly among the poorest. The World Bank estimated that the high food, oil, and other commodity prices since January 2007 have reduced the gross domestic product of Africa by 3 to 10 per cent. The terms-of-trade effects of the combined food and energy price increases are in excess of 10 per cent of GDP in more than 15 developing countries, where the room for manoeuvre on the macroeconomic front is limited. With millions of Africans living on the margin between subsistence and starvation, high food and fuel prices may represent a threat to their survival and further heighten energy poverty (UNIDO, 2007). At the same time, poverty and high fuel prices can cause people to engage in high profile deforestation which might in turn exacerbate climate change.

It is important to point out the international justice implication of the above. Historically and currently, Africa's contribution to climate change is very low compared to other world regions (IPCC, 2007). This implies that the changing climate in Africa and associated energy poverty consequences is very much an issue of global justice and equity (Okereke and Schroeder, 2009; Okereke, 2010). Emissions of carbon dioxide in Africa represent only a small fraction (3.6 per cent) of the total CO2 emissions per year worldwide. In sharp contrast, the contributions from Europe, Latin America and the US are 14 per cent, 20 per cent and 14 per cent respectively. Looking forward, Höhne and Blok (2005) calculate that by 2050 the OECD will be responsible for about 41.7 per cent of global average surface temperature increase due to fossil CO2 while Africa and Latin America combined would be responsible for just 17.05 per cent. So both in present and future terms, it is the African poor without access to modern energies and who have not shared in the benefits of wealth created from the intensive use of energy in the last century that will be the most affected by the impacts of climate change due to greenhouse gas emissions.

In any case, it is clear from the above discussion that nearly all policy measures adopted in order to pursue GHG emission reductions will have implications for the developing countries' economies in general and energy provisioning in particular. Specifically, investments aimed at reducing GHG emissions in Africa may result in reduced expenditure or investments in the energy sector of various African economies (UNEP, 2012). As stated, an urgent challenge for Africa in the light of climate change is how to achieve growth in the context of growing international mitigation policies and the impact these are having on their economies. While this challenge may appear overwhelming at first, there are indications that Africa could turn climate change into opportunities by pursuing climate resilient low carbon growth strategies. Such strategies will not only enable the continent to manage the development risks associated with global climate mitigation efforts, it will also help them to build a diversified and energy secure and robust local national economies.

## 4 LOW CARBON DEVELOPMENT IN AFRICA

Low carbon development has recently gained currency in academic and policy circles. Although fuzzy and poorly defined, the concept has nevertheless captured public imagination as a possible means of reconciling the need for economic development and climate mitigation. Low carbon development is an imperative for all countries, developed and developing alike, if the global aspiration to combat climate change is to be achieved. Yet, it is widely recognized that the concept has a particular relevance to developing and fast emerging countries where wide scale development initiatives are either needed or already taking place (UNEP, 2012). A cardinal feature of the low development paradigm is the notion of decoupling economic growth from carbon emission (Mulugetta and Urban, 2010). In practice, this entails embracing low carbon designs, structures and industrial activities as central parts of development plans. It also requires technological intervention to enhance the energy efficiency of key sectors, implements and practices (Bowen and Frankhauser, 2011).

Given the abundance of renewable energy sources in Africa, it has been suggested that the effort needed to shift production towards cleaner sources may be less than can be expected in some other world regions (Abanda et al., 2012). Africa has significant hydro power potential which is thought to be in the region of 40,000MW (Kalitsi, 2003; Diog and Adow, 2011). The Democratic Republic of Congo (DRC) and Ethiopia together account for more than 60 per cent of Africa's hydropower potential. Currently, though, only about 20 per cent of this energy source is being utilized. Africa has an abundant reserve of natural gas mostly concentrated in Algeria, Egypt, Libya and Nigeria. The BP statistical Review (BP 2012) estimates that total proven reserve in Africa is 513.2 trillion cubic feet which amounts to about 7 per cent of total world reserve. Only a fraction of this is currently being utilized and a vast portion especially in Nigeria is being flared. It is in fact estimated that gas flaring accounts for about 12 per cent of Nigerian's annual greenhouse gas emission (Anozie et al., 2007).

Estimated geothermal resources on the African continent are around 14GW (EIA, 2011). Of this, only 0.6 per cent has been exploited. Currently, the only countries using geothermal for electricity in Sub-Saharan Africa are Kenya (127MW) and Ethiopia (7MW) (Diog and Adow, 2011). It is thought that some of the most promising undeveloped rift systems are the East African Rift in Mozambique, and in Uganda. Further research is required to explore possible geothermal potentials in Tanzania, Eritrea and Zambia (Sokona et al., 2012). In 2007, Africa had about 476MW of installed wind energy generating capacity; a significantly low proportion of the estimated Sub-Saharan Africa-wide capacity (93,000MW) (Okoro and Chikuni, 2007). Countries like Djibouti, Eritrea, Mauritania and Madagascar among others experience strong wind speed which can be readily converted into useful energy. The Sub-Saharan African countries are well exposed to sunlight with some of the highest solar intensities in the world (Wolde-Rufael, 2009). Northern and southern Africa, particularly the Sahara and Kalahari deserts, have particularly promising conditions for concentrated solar plants for large-scale power production. Kenya has made good strides towards the utilization of solar with more than 30,000 very small solar panels, each producing 12 to 30 watts sold annually. It is the world leader in the number of solar power systems installed per capita (Abanda et al., 2012). However, to date, only South Africa is generating appreciable solar thermal power (0.5MW) in the Sub-Saharan region. Africa has vast land mass which can be farmed to produce biofuel in a sustainable manner (Akinbami et al., 2001; Amigun et al., 2011) Countries with suitable land include South Africa Angola, Zambia and Mozambique among others.

In general, with adequate economic, technological and governance infrastructure a vast portion of African energy could be produced from clean and low carbon sources. This of course is not to suggest that the exploitation of fossil fuel will have to be completely abandoned. Nor is it the case that all the energy sources mentioned above do not have social and environmental impacts. It is general knowledge that in most cases, achieving energy security requires that countries strive to maintain a balanced mix of energy sources in their portfolio (Sokona et al., 2012).

On the consumption side, low carbon development would require significant changes in values especially with respect to consumption behaviour and patterns. For Africa then, the goals of low carbon development should be to: (a) achieve energy security by significantly enhancing access and reducing reliance on imported fuels; (b) contribute to tackling climate change by avoiding high profile emission in the path to economic development; (c) achieve diversified, equitable and climate resilient economic growth; and (e) realize effective climate adaptation. It is apparent that achieving low carbon development would have far reaching implications for governance and institutional design. For example the internalization of environmental costs of growth would require appropriate pricing of goods and services through a range of policy and economic measures targeting both the production and consumption sides. These would include national and sub national strategies, and master plans, targets, taxes, subsidies, infrastructure and public awareness campaigns.

## 5   BENEFIT OF LOW CARBON DEVELOPMENT FOR AFRICA

The concept and practice of low carbon development in Africa implicate a number of opportunities and threats. Some of these will resonate with other developing countries and some will be specific to Africa with its unique set of resource-base, technical capacity, financial situation. One threat of low carbon development to Africa would be if the approach impedes economic growth by requiring African countries to bear a disproportionate financial burden relative to business as usual development approach (discussed further below). Another would be if commitment to low carbon development results in unnecessary intrusion into the domestic policy processes by foreign actors. That said, there are evidently several reasons while it is in the interest of African governments to embrace and pursue low carbon development. Below, we sketch some of the pertinent benefits of low carbon development for Africa. In doing so, we bear in mind that Africa is a very diverse continent and that a number of points discussed here may not be equally applicable to all the countries within the continent.

The first obvious potential benefit of low carbon development is that the approach provides an opportunity for African countries to build more resilient and diversified economies. Many African countries can make their economies more resilient by moving away from conventional single source energy generation towards a more diversified energy portfolio. Countries that get their energy from a mix of sources would naturally be far less vulnerable than those that rely on single sources. Specifically, low carbon development has the potential to help African countries reduce the economic vulnerability associated with dependence on oil (Okereke and Tyldesley, 2011). Almost one-third of African countries (15) are landlocked with no access to the ocean or seas. Many of these are oil importing countries with no proven oil reserve. Examples of such landlocked countries include Botswana, Burkina Faso, Burundi, Central African Republic, Chad, Ethiopia, Malawi, Mali, Niger, Rwanda and Uganda. In the absence of directly accessible oceans and seas, these countries transport crude and refined oil by road over very long distances. Oil is not only an expensive commodity, but one that is uniquely prone to dramatic price spikes. The dependence of these countries on imported diesel and heavy fuel oil means that their economies are very vulnerable to highly fluctuating oil prices. Furthermore, it also implies that the country's electricity generation is firmly tied to insecure oil sources. Oil price spikes directly affect GDP, and in oil-dependent countries the effect can be quite high. On average, it is estimated that every 10 per cent increase in the oil price results in a global drop in GDP of around 0.2 per cent (Owen et al., 2010). This figure is much higher for many African oil importing countries.

Economic resilience and security can also be enhanced through a range of other sectoral measures and policies widely associated with the concept of green growth such as low carbon development cities and transport. In urban design, for example, countries that opt for dense and compact cities with ample allowances for buses and cycle routes are likely to have both economic and social advantages over those that favour extensive sprawls with very little and expensive integration. Low carbon houses can save energy and help reduce demand on national grid. The main source through which African countries contribute to climate change is currently through land use and deforestation (IPCC, 2007a). Low carbon development would have to incorporate effective forestry and land use management (Bowen and Frankhauser, 2011). This should help to protect the natural environment and reduce the vulnerability to flash flooding and destruction caused by wide fire.

Shifting away from inefficient biomass energy to modern sources would also help eradicate poverty and increase economic growth and development. Cleaner energy sources can enhance the adaptive capacity of households to climate change with several important co-benefits. For example using clean cooking stoves and solar lighting will reduced the need to walk large distances to collect firewood. Accordingly, women and children who are usually responsible for these tasks will have more time in the day to engage in productive activities such as trading, household duties or educational activities. Increasing the amount of time for such tasks can support gender equality in the home, and the wider socio-economic benefits associated with empowering marginalized groups such as improved access to community decision-making processes (Diog and Adow, 2011).

Renewable energy can be particularly suitable for rural and remote areas where, transmission and distribution of energy generated from fossil fuels can be difficult and expensive. Producing renewable energy locally can offer a viable alternative. In such situations, renewable energy can also contribute to education, by providing electricity to schools.

In addition to reducing wider economic vulnerability related to oil price fluctuations, the purist of low carbon development can specifically increase African countries' energy security. Many African countries currently depend on very high cost imported electricity to meet their power needs (World Bank, 2011). Paradoxically, in most cases these countries have abundant renewable sources that could easily be harnessed to meet internal energy demands. In fact with good planning and targeted investment, these countries could eventually be net energy exporters. However, current reliance on externally sourced energy not only implies that scare financial resources are spent on the importation of electricity; it also results in huge insecurity in energy supply. Energy insecurity is further exacerbated by the fact that supplies in some cases come from countries and regions that are politically unstable. Currently, Morocco is the largest energy importer in northern Africa. Morocco produces small volumes of oil and natural gas from the Essaouira Basin and small amounts of natural gas from the Gharb Basin (UNIDO, 2007). However, over 90 per cent of its energy resources come from external sources. Much of these imports are transcontinental from Spain via cables laid beneath the sea and across the Strait of Gibraltar. It is estimated that the country's total yearly costs for energy imports range from USD1 to 1.5 billion. With the increase in oil prices in 2005 the cost of import rose to approximately USD2 billion resulting in a huge budget deficit for the country (UNEP, 2012).

Another high energy importing county is Togo which imports as much as 80 per cent of its electricity. Out of a total consumption of 726 GWh in 2006, Togo imported 505 GWh mainly from Ghana with additional imports coming from Nigeria and Ivory Coast. Zimbabwe imports up to 400MW of electricity from neighbouring countries including Zambia and the Democratic Republic of Congo. In 2008 the country experienced wide power cuts because an accumulated debt of about USD100 million prevented it from importing larger amounts of electricity. Similarly, also in 2008 Botswana and Namibia, both of which imports over 50 per cent of its electricity from South Africa, were hard hit when an internal energy crisis in South Africa forced Eskom to ration its internal supplies and drastically reduce the amount exported to neighbouring countries. Drained by five principal rivers including the Zambezi, Mozambique is richly endowed with considerable hydropower potential. This has been estimated at 12,500 MW, with a corresponding annual energy generation potential of 60,000

Gwh per year. However, so little of these resources have been exploited. Hence, up to 70 per cent of the country still depends on inefficient and unsustainable biomass sources for energy (World Energy Outlook, 2012).

In general, energy sources such as river basins in Africa are under-utilized, with only 20 per cent of the total potential of hydropower plants under use (Kalitsi, 2003; World Bank, 2011). Mozambique, Lesotho and Swaziland, all of which have abundant renewable resources, continue to rely on South African Eskom for significant amount of their energy needs. Other heavy electricity importing African countries include Egypt, Niger, Namibia, Tanzania and the Republic of Benin.

It has to be acknowledged that there are instances when it is cheaper for a country to import rather than produce its own electricity. Furthermore, some have argued that regional power 'pooling' schemes provide the benefit of scale and as such could help solve African's energy challenges (Khennas, 2012). However, cheap importation and power pooling should not be seen as substitutes for developing in-country energy sources, especially when this can be done from abundant renewable and sustainable sources. Currently though, far too many African countries are either importing from very insecure sources or relying on thermally generated electricity to meet their energy needs.

However, the high price of oil means that electricity generated using oil-based generators is very costly and must be subsidized by the government in order to make it accessible to consumers. Even so, many African countries still pay exorbitant prices for their electricity. This is unsustainable especially in the context of very limited resources, high budget deficit and spiralling external debt. The purist of low carbon development, if well planned, could help lower the cost of energy in many African countries. On 1 July 2008 Namibia and Botswana increased their electricity tariffs by 18.6 per cent and 20.4 per cent respectively and warned that prices will continue to rise. These increases had to do with the need to offset the high cost of electricity importation from South Africa. In 2005 the government of Rwanda spent around 8.4 billion RwF (USD13,356,000) on fuel for electricity generation which produced 55.2 GWh. This is in comparison to 1.5 billion RwF (USD2,385,000) spent on 86 GWh of imported electricity from regional hydropower stations. At the same time electricity prices in Rwanda rose from 82 RWF (USD0.13) in 2005 to 112 RWF (USD1.78) in 2006 as the government struggled to cover the high cost of thermal generation and importation (MINIFRA GoR, 2008).

A further incentive to move away from a fossil fuel-based economy is the effect that the high level of imports has upon African countries' trade deficit. Many African countries that have huge budget deficits spend large amounts of their foreign exchange on oil importation. The current situation, where the import of oil-based fuel for electricity generation and transport continually saps scarce foreign reserves, is not sustainable for long-term economic development. It has been suggested that in 2008 Ethiopia spent up to 96 per cent of foreign earnings on oil import. With the extraordinary rise in oil prices within this period, it is no wonder that inflation soared to as high as 39.4 per cent. Given the very low purchasing power of the African population, high electricity costs are far from ideal. In addition, high electricity costs discourage industries and businesses wishing to set up in the continent. Utilizing domestic renewable energy resources would allow cheaper electricity generation which would enable greater electricity access.

Lastly, low carbon development will help Africa to prevent what is commonly known as carbon lock-in. Carbon lock-in refers to a situation where a high carbon infrastructure in a country inhibits drive and options to pursue alternative energy sources. Many African countries are currently investing or making serious plans to commit large sums of money to develop their energy infrastructure. With the rapid economic growth experienced in the last 10 years massive investments are also being made or planned in other development infrastructure such as roads, railways, airports, and cities. In general, these infrastructures have an average life span of about 40 years and buildings can last much longer than this. The implication is that decisions made today about the types of power plants, roads, railways and buildings funded or constructed will have carbon and energy implications for up to 40 or 50 years at least from now. In other words countries can be 'locked-in' to a high carbon or energy pathway by making wrong or short-sighted decisions on which development plans they adopt. By thinking through future energy and carbon implications of decisions and adopting low carbon alternatives, African countries could avoid high carbon lock-in while continuing to develop in a sustainable fashion.

The World Bank and OECD estimates that a total of USD40.8 billion a year in investments is needed for Africa's power sector, with USD26.72 billion for capital expenditure and USD14.08 billion for operations and maintenance (World Bank, 2009). For the IEA Africa needs about USD344 billion to create additional electricity capacity, upgrade installed equipment and extend transmission and distribution networks to households and factories. Countries that are already spending large sums of monies in their energy sectors include South Africa, Nigeria, Botswana, and Namibia among others.

The government of South Africa estimates that just keeping up with growing demand from industries and the population will require doubling its generating capacity by 2025 at a cost of USD171 billion. Of this, the government plans to spend up to USD45 billion by 2013. Between 2000 and 2007 the Nigerian government under President Olusegun Obasanjo spent about USD16 billion to revamp its power sector. With precious little achieved, the present government has recently pledged to spend USD5.7 billion over the next four years on the power sector. Botswana is spending USD28 billion to construct an integrated coal mine and power station in Mmamabula that could generate 4,800 megawatts (MW) for about 40 years. Zambia is estimated to need USD billion to raise power output to meet its expanding demand and Rwanda has pledged to invest up to USD 4.74 billion in its energy sector between now and 2017.

A quick look at energy development strategy of many African countries however indicates that for the most part these countries are investing or planning to invest their money on conventional hydrocarbon-based generating facilities rather than on cutting edge low carbon technologies. This is, in a sense, very understandable as these countries are concerned about the high cost of low carbon technologies especially given the difficulty of raising finance from the public and private sector. Perhaps the most recent high-profile example is the 4,800MW Medupi coal-fired power plant that is being constructed by the South African government with financial assistance from the World Bank. When the plans by the South African government to borrow about USD3.75 billion came to light, many green NGOs argued that the money should have been better spent towards generating from renewable sources such as wind and solar. However, South Africa argued that investing in renewable sources will cost much more and generate far less electricity than a coal power plant. Given the urgent need to increase supply to support rising demand and economic growth, the country felt it had little option than to commit to building the high carbon power plant.

## 6 CHALLENGES TO LOW CARBON DEVELOPMENT AND ENERGY SECURITY IN AFRICA

Achieving low carbon development and energy security in Africa will not be easy. The pursuit of low carbon growth poses a challenge for countries all over the world and there are reasons to believe that these challenges may be particularly acute for African countries. One of the most obvious and frequently cited obstacles that may hinder Africa from harnessing the advantages of green growth is lack of finance. Although technological advances that help lower their costs are being made, clean energy in most cases still costs far more than its conventional alternatives. For example, figures from the 2012 report of the Energy Information Administration (EIA) of the US Department of Energy (DOE) suggest that the average capital cost of energy from conventional coal, solar thermal and off-shore wind are USD65.8 million MW/h, USD204.7 million MW/h and USD300.6 million MW/h. Even when one factors in the operation and maintenance, fuel costs and transmission investment, all of which are in most cases lower for clean energy, the average total system costs of renewable technologies are still mostly higher than the conventional sources. Generating from renewable sources may also have benefits such as those related to health. However, their costs remain mostly 'front-loaded relative to their benefits' (Bowen and Frankhauser, 2011, 149).

There are of course a number of renewable options such as low head hydro and biomass that are economically competitive especially as a means of supply to sparsely populated rural communities, but even these may require technical assistance and the correction of market failures which may be difficult for many African countries to deliver. Moreover, even when there are strong long-term economic and environmental arguments for investing in clean energy, the notion might persist that in taking a low carbon route, Africans are paying a price imposed on them by foreign governments whose main interest is global climate mitigation. These are some of the main reasons why the provision of adequate and predictable finance and technical assistance by developed countries is absolutely essential in encouraging African countries to embrace low carbon growth paths.

Unfortunately, it is well known that current financial and technical support from the developed countries is far too low compared to what is needed. The Global Environmental Facility established in 1991 serves as the main operating entity for the international climate regime and has the longest track record on environmental funding. However, it received just over USD1 billion during its fourth replenishment period (2006–2010). Similarly the other two funding sources within the convention, the Least Developed Countries Fund (LDCF) and the Special Climate Change Fund (SCCF) (both also administered by GEF), have only disbursed USD108 million and USD80 million respectively since their inception in 2002 (Sokona et al., 2012). These figures represent a tiny fraction of what is needed to help developing countries adequately invest in low carbon development and energy security. To compound matters, African countries, mainly because of poor capacity and their position in the global economic structure, have not been the favourites in attracting multilateral and private sector climate finance. Fast-start finance agreed at Copenhagen in 2010 was just USD10 billion per year up until the end of this year; with a long-term goal of mobilizing USD100 billion per year by 2020. And while this may not be enough, it is doubtful that the global North will commit the amount of money required to meet this ambition.

In addition to the general lack of climate funding, there is also an indication that availability of finance is not linked to what is actually needed or to where it is most urgently needed (Diog and Adow, 2011). Rather, climate finance delivered has reflected the political preferences of developed countries, and the 'low-hanging fruits' opportunities offered by the higher emitting, middle income countries, particularly India and China (Diog and Adow, 2011). It is therefore imperative that adequate and predictable funding is available in order to incentivize low carbon development in Africa. A good illustration of the problem of lack of finance can be found in Rwanda. The Government of Rwanda has ambitious plans to develop its energy sector as a wider plan for achieving low carbon and climate resilient development. Up to 90 per cent of Rwandans have no access to electricity. Of the 20 per cent living in urban areas only about 25 per cent are connected to the national grid. And at about 44 kWh per capita per year Rwandan electricity consumption per capita is among the lowest in the world. The Energy Sector National Policy and Strategy set out the aim of installing a total of 1,000MW electricity generation capacity by 2017 (up from 85MW at present). This expansion of electricity generation capacity is planned to come from four main sources; geothermal (310MW), hydropower (300MW), methane (300MW) and peat (100MW). It is planned that a rapid national grid expansion programme will accompany this increase in generation capacity. The grid will be extended by 2,100 km (700 km of HV and 1,400 of MV), increasing the number of connections to 1,200,000 up from 175,000 today (Safari, 2010). The aim of this extension is to enable 50 per cent of the population access to the national grid by 2017. In addition, by 2017 the GoR plans to ensure that all health centres, local administration offices and all schools in the country have access to electricity, either off or on grid (Eggoh et al., 2011). The increase in access to electricity is intended to provide alternatives to traditional sources of energy, hopefully reducing the dependence upon biomass and limiting risks of deforestation. The target is to reduce the use of biomass from 86 per cent of primary energy use today to 65 per cent in 2017. This will also be accompanied by efficiency measures such as improved cooking stoves. Alternatives such as biogas will also be introduced.

The big challenge however is that the energy generation and access plan is estimated to cost around USD5 billion. To put this in context, the entire budget for Rwanda including both recurrent and domestic spending for the fiscal year 2009/10 was USD1 billion. And of this figure, donor support accounted for about 41 per cent. One can immediately see the difficulty faced by the country in overcoming the problem of energy access which is fundamental to economic growth.

While lack of international climate finance is a major factor limiting low carbon energy secure future for Africa, it is important to stress that this is only one side of the coin. The other is widespread lack of technical capacity and good governance in Africa. There is an abject paucity of technical skills needed to design and implement conventional development projects, let alone cutting edge low emission growth plans (Okereke and Tyldesley, 2011). The majority of African countries have critical capacity gaps in all the key phases involved in the low carbon development delivery chain from conception, through design and planning to implementation. Many governmental ministries have just one or two experts who have to draft or vet project proposals, study sophisticated engineering designs, conduct rigorous economic analysis, negotiate complex legal contracts and undertake the other several highly technical tasks associated with policy development and implementation.

Closely related to, and perhaps the primary reason for poverty, is the problem of poor governance. Decades of poor governance in Africa have resulted in underinvestment in education, human capital, and research and technology development. Similarly, there is widespread underinvestment in fundamental development infrastructure upon which to leverage green growth policies. Lack of infrastructure, poor institutions and widespread corruption provide the platform for the dominance of the energy market by monopolies which are often controlled by a few elites. These monopolies are usually not interested in widening energy access but have been known to actively block the market entry of green energy providers. The result is lack of private capital, pervasive market failure, chaotic regulatory environment and usually high costs for investment in clean energy. To achieve energy security and low carbon development, then, African countries will need to undergo radical governance reforms aimed *inter alia* at minimizing corruption, increasing technical and human capacity, correcting market failure and boosting investment in infrastructure and technology development.

## 7 CONCLUSION

The concept of low carbon development offers plenty of prospects for Africa to grow its economy, achieve energy security while contributing its own quota in the global effort to fight climate change. Given the critical importance of modern energy to wellbeing and to economic development in general, achieving universal energy access should definitely be the priority of African countries. However, there is no reason why Africa must follow the development path towed by the West with its negative impact on environment and humankind. Rather, emphasis should be on harnessing the abundant renewable natural resources present all across the continent. However achieving low carbon development would require the massive upscale of climate and clean energy finance, large-scale investment in technology and human capacity as well as radical governance reforms. Africa has the right to expect significant financial assistance from the international community to offset the additional cost associated with low carbon development. It currently makes an insignificant contribution to the global carbon pool while bearing the brunt of much of the negative impact of climate change. Indeed, the World Bank (2010) calculates that scaling up access to electricity access in Africa would add only a small fraction of projected global emissions from 1.5 per cent of global annual energy related CO2 emissions today to 2–3 per cent of global emissions by 2050. Provisions of much needed basic energy services to the poor would therefore contribute only 1 per cent to global CO2 emissions (World Bank, 2011). The poor deserves basic energy services like everyone else. Moreover, they have made little contribution to climate change. Global justice and equity is therefore at the heart of the debate about climate change, energy security and climate mitigation.

## NOTE

\* Chukwumerije Okereke wishes to acknowledge research funding from The Leverhulme Trust, Grant Reference ECF/7/SRF/2010/0624.

## REFERENCES

Abanda, F., A. Ngombe, R. Keivani and J.H.M. Tah (2012), 'The link between renewable energy production and gross domestic product in Africa: A comparative study between 1980 and 2008', *Renewable and Sustainable Energy Reviews*, **16**, 2147–2253.

Akinbami, J.F.K, M.O. Iloori, T.O. Oyebusi, I.O. Akinwumi and O. Adeoti (2001), 'Biogas energy use in Nigeria: current status, future prospects and policy implications', *Renewable and Sustainable Energy Reviews*, **5**, 97–112.

Amigun, B., J.K. Musango and W. Stafford (2011), 'Biofuels and Sustainability in Africa', *Renewable and Sustainable Energy Reviews*, **15**, 1360–1372.

Anozie, A., A.R. Bakare, J.A. Sonibare and T.O. Oyebisi (2007), 'Evaluation of cooking energy cost, efficiency, impact on air pollution and policy in Nigeria', *Energy*, **32**, 1283–1290.

Bowen, A and S. Frankhauser (2011), 'Low-Carbon development for the least developed countries', *World Economics*, **12** (1), 145–162.

BP 2012, Statistical Review of World Energy, available at http://www.bp.com (accessed 16 November 2012).
Diog, A. and M. Adow (2011), 'Low carbon Africa: Leap frogging to a greener future', available at http://www.christainaid.org.uk/images/low carbon/Africa.pdf (accessed 15 November 2012).
Ebohon, O.J. (1996), 'Energy, economic growth and causality in developing countries', *Energy policy*, **24** (5), 447–453.
Eggoh, J.C., C. Bangake and C. Rault (2011), 'Energy consumption and economic growth revisited in African Countries', *Energy Policy*, **39**, 7408–7421.
Eleri, E., O. Ugwu and P. Onuvae (2011), 'Low Carbon Africa: Leap frogging to Green future: Nigeria', available at http://www.christainaid.org.uk/resources/policy/climate/low-carbon-africa (accessed 17 November 2012).
EIA, Energy Information Administration (2011), 'International Energy Statistics' available at http://www.eia.gov/ipdbproject/IEDIndex3.cfm (accessed 20 November 2012).
Höhne, N. and K. Blok (2005), 'Calculating historical contributions to climate change: discussing the "Brazilian proposal"', *Climatic Change*, **71**, 141–173.
IPCC (2007a), 'Climate Change 2007 – Impacts, Adaptation and Vulnerability',
Contribution of Working Group II to the Fourth Assessment Report of the IPCC, Cambridge: Cambridge University Press.
IPCC (2007b), 'Climate Change 2007 – Mitigation of Climate Change',
Contribution of Working Group III to the Fourth Assessment Report of the IPCC, Cambridge: Cambridge University Press.
IPCC (2011), 'Renewable energy sources and climate change mitigation. Summary for policy makers', Special Report of the IPCC, Cambridge: Cambridge University Press.
IEA, International Energy Agency (2010), 'World Energy Outlook', available at http://www.iea.org/publications/freepublications/publication/name,27324,en.html (accessed 3 March 2013).
IEA, International Energy Agency (2011), 'Statistics and Balances', available at http://www.iea.org/stats/index.asp (accessed 20 November 2012).
Kalitsi, E.A. (2003), 'Problems and Prospects for hydropower development in Africa', available at www.un.org/esa/sust/dev/sdissues/energy/op/nepadkalitsi.pdf (accessed 20 November 2012).
Kanagawa, M. and T. Nakata (2007), 'Analysis of the energy access improvement and its socioeconomic impacts in rural areas of developing countries', *Ecological Economics*, **62**.
Khennas, S. (2012), 'Understanding the political economy and key drivers of energy access in addressing national energy access and policies: African Perspective, *Energy Policy*, **47**, 21–26.
MININFRA GoR (2008), 'National Energy Policy and National Energy Strategy 2008–2012', Kigali.
Mulugetta, Y and F. Urban (2010), 'Deliberating on low carbon development', *Energy Policy*, **38**, 7546–7549.
Okafor, E. (2008), 'Development crisis of power supply and implications for Industrial Sector in Nigeria', *Stud Tribes Tribals*, **6** (2), 83–92.
Okereke, C. (2010), 'Climate justice and the international regime', *WIREs Interdisciplinary Review*, **1**, May/June, 462–474.
Okereke, C. and K. Dooley (2010), 'Principles of justice in proposals and policy approaches to avoided deforestation: Towards a post-Copenhagen climate agreement', *Global Environmental Change*, **20**, 82–95.
Okereke, C. and H. Schroeder (2009), 'How can the objectives of justice, development and climate change mitigation be reconciled in the treatment of developing countries in a post-Kyoto settlement?', *Climate and Development*, **1**, 10–15.
Okereke, Chukwumerije and Sally Tyldesley (2011). 'Low Carbon Africa: Rwanda', in Alice Diog and Mohamed Adow (ed.), 'Low carbon Africa: Leap frogging to a greener future', available at http://www.christainaid.org.uk/images/low carbon/Africa.pdf (accessed 15 November 2012).

Okoro, O.I. and E. Chikuni (2007), 'Power Sector reforms in Nigeria: Opportunities and Challenges', *Journal of Energy in Southern Africa*, **18** (3), 52–57.

Oseni, M. (2012), 'Households' access to electricity and energy consumption pattern in Nigeria', *Renewable and Sustainable Energy Reviews*, **16**, 990–995.

Ouedraogo, N. (2012), 'Energy Security, Global Climate change and Poverty. Interrelated Challenges of energy poverty, energy security and climate change mitigation and adaptation in Africa', available at http://www.iaeu2012.it (accessed 18 November 2012).

Owen, N.A., O.R. Inderwildi and D.A. King (2010), 'The Status of Conventional World Oil Reserves – Hype or Cause for Concern?', *Energy Policy*, **38** (8), 4743–4749.

Safari, B. (2010), 'A review of Energy in Rwanda', *Renewable and Sustainable Energy Reviews*, **14**, 524–529.

Sanchez, T. (2011), 'Climate Change and Energy Poverty in Africa', Africa Energy Yearbook 2011, available at http:// www.practicalaction.org (accessed 12 November 2012).

Sokona, Y., Y. Mulugetta and H. Gujba (2012), 'Widening energy access in Africa: Towards Energy Transition', *Energy Policy*, **47**, 3–10.

UNIDO (2007), 'Powering Industrial Growth-the challenge of Energy Security issues for Africa', available at http//www.ofid.org (accessed 17 November 2012).

UNEP (2012), 'Financing renewable energy in developing country. A study and survey', UNEP, February.

Wolde- Rufael, Y. (2009), 'Energy consumption and Economic growth: The experience of African countries revisited', *Energy Economics*, **31**, 217–224.

World Bank (2009), 'Africa Energy Poverty.G8 Energy Ministers Meeting', available at http://www.g8energy2009.it (accessed 15 November 2012).

World Bank (2011), 'Energy in Africa. An Overview', September, available at http://web.worldbank.org (accessed 5 March 2013).

World Energy Outlook (2011), 'Executive Summary', available at http://www.iea.org/publications (accessed 12 November 2012).

World Energy Outlook (2012), 'Early Release Overview', available at http://www.iea.org/publications (accessed 12 November 2012).

# 22. The road not taken, round II: centralized vs. distributed energy strategies and human security
*Ronnie D. Lipschutz and Dustin Mulvaney*

## INTRODUCTION

Thirty-five years ago, in the midst of the energy crisis of the 1970s, Amory Lovins published an article in *Foreign Affairs* warning against the United States' plans to increase its reliance on nuclear power as a path toward reducing dependence on imported oil (Lovins 1976). Not only would a large nuclear program – an order of magnitude more nuclear plants are in operation in the U.S. today – raise significant safety and security risks, argued Lovins, preventing the theft of weapons grade nuclear materials would require a highly-centralized and authoritarian infrastructure, with corrosive effects on American democracy and individual freedom (see also Ayres 1975). A decade later, Langdon Winner (1986) echoed these sentiments and argued that technological artifacts such as nuclear reactors have important implications for social order because of their entanglements with questions of nuclear weapons proliferation. Both Lovins and Winner were essentially arguing that socio-technical systems, such as those that make possible electrification of human society, have social tendencies independent of human intention, and that differing technological paths would have different kinds of social impacts and, by extension, effects on human well-being and security.

Lovins also argued that, by contrast, solar energy technologies did not rely on either risky or far-away energy sources and were more amenable to decentralized modes of energy procurement and generation. These features, he claimed, would promote energy autonomy, democratization and even reduce the political and economic power of electric utilities and energy corporations. Solar energy is freely available to anyone able to capture it *on site*, making it suitable for decentralized, small-scale, distributed power generation (DG).[1] Photovoltaic (PV) technologies, in particular, enable people to electrify homes and businesses with limited reliance on or even no connection to the electricity distribution grid, fostering a sense of energy autonomy and freedom, and enhancing well-being and security. Already in the early 1970s, *The Whole Earth Catalogue* (http://www.wholeearth.com/index.php, accessed February 22, 2013), Stuart Brand's counterculture publication offering "designs for living," included plans and advertisements promising PV systems that could increase the energy self-reliance of communities and individuals (Turner 2006), and the country as a whole.

In the intervening decades, solar energy and PVs have become well-developed and mature technologies, proving cost-effective in a wide range of settings yet largely failing to fulfill the promises made by Lovins, *Whole Earth* and many others. Even today, as the cost of rooftop PV systems has enabled numerous middle-class consumers and companies to solarize their homes and businesses (Cardwell 2012), distributed generation (DG) still provides only a small fraction of the country's electricity needs. As far as reducing fossil fuel dependence, renewables have, so far, proved unable to substitute for much oil, which is not much used in the power sector but dominates in transportation (a shortcoming shared with nuclear power). Nuclear power is unlikely ever to achieve the lofty goals set in the 1970s, while coal and natural gas, two of the world's mainstays for generating electricity, remain remarkably cheap. Most of the world has found it economically and politically difficult to shift away from fossil fuels even in light of growing risks from climate change. What has gone wrong? Or, in the immortal words of China's Prime Minister, Chou En-Lai, when asked for his assessment of the French Revolution, replied, "It is too soon to tell" (McGregor 2011). Perhaps the solar revolution has not yet arrived, even if it is around the corner.

What is arguably different today are the rising risks and impacts of continued reliance on fossil fuels and nuclear energy, exacerbated by impending climate change. The Fukushima disaster of 2011 demonstrated not so much the hazards of nuclear power as the enormous social and economic vulnerabilities of heavy reliance on technologies that sometimes suffer catastrophic failures. Numerous predictions that global fossil fuel consumption will continue to increase during the coming decades, notwithstanding the threats and risks of climate change, suggest that concerted action to transform energy systems is unlikely to happen. Continued reliance on an aging power distribution infrastructure, evident in large-scale regional power blackouts, raise questions about greater concentration and centralization of power generation strategies. Finally, repeated warnings that cyberhackers, whether civilian or military, could gain control over or disrupt the power sector, with catastrophic consequences for the country's infrastructure, should not be ignored. All of these shortcomings are of concern where human welfare and well-being are concerned; the effects of extended power outages and broken distribution systems become only too evident in the aftermath of destructive hurricanes, tornadoes, floods, tsunamis and even human error.

None of this is to say that renewables are wholly safe and risk-free, or could wholly eliminate threats to human security. A full life-cycle analysis of various solar, wind and biofuel technologies reveals a host of associated problems, such as land use change, toxic materials, damage from mining of rare earth elements and other minerals, threats to species diversity and even waste disposal issues. If deployed in a centralized fashion, renewable power is as vulnerable to regional blackouts as any other fuel source. Vaclav Smil (2010) points out that renewables require more land due to lower power densities, which means that extensive deployment will invite more opportunities for land use conflicts with other species and communities. PV is ideal for DG and can be sited atop the human footprint, and can fulfill many regions' energy needs. A full tally of the ecological and social costs of renewable energy sources indicates that they offer greater benefits with fewer externalities than fossil and nuclear sources, especially if climate change is taken into account. More to the point, and somewhat reminiscent of Lovins's arguments, if deployed in a distributed, decentralized fashion, in the form of electricity micro-grids, there could well be a number of positive and socially-beneficial consequences that would not only reduce overall environmental costs but also enhance human security. An important caveat: these effects would not result from any inherent qualities of the technology but, rather, deliberate and intentional policy and design.

In this chapter, we undertake a comparative assessment of the impacts of centralized and decentralized energy sources and strategies, their inherent and constructed political and social qualities, and the environmental, health, safety and security implications of the two alternatives, especially as they affect "human security." In this instance, the term *human security* refers primarily to societies' and individuals' health and well-being, rather than national or military security. The two concepts are not mutually exclusive, but the latter has been plumbed in great depth for at least 40 years, while the former has not. We begin the chapter with a brief discussion of "human security" and what is encompassed by the concept. Although national security has been challenged as the basis for public and foreign policy since the 1980s (e.g., Ullman 1983; Lipschutz 1995), as we shall explain, the changing organization and structure of markets, in particular, has driven a shift towards greater focus on individual security in the context of broad practice.

We then turn to a discussion of the relative costs and benefits of centralized and decentralized energy strategies, with particular attention to the vulnerability of large-scale power plants and grids. As we shall see, while first order analyses focus on reliability, resilience and vulnerability, it is also important to consider second order environmental and third order life-cycle impacts. Such assessments are not quite as straightforward as they might seem, as illustrated by the parallel case of the miniaturization and decentralization of computing technologies, which have highly-sophisticated, low-cost devices available to many of the world's people. But they have also resulted in complex and potentially vulnerable networks as well as growing volumes of front- and back-end toxic wastes, with concomitant health risks. By contrast, large, mainframe computers and communication networks might not offer the freedom and creativity available from desktops, laptops and tablets, but they do concentrate many of the risks and hazards associated with the latter.

In the third part of the chapter, we turn to a more detailed description and inventory of centralized and decentralized *renewable* power production (the risks and hazards of fossil and nuclear fuels are relatively well-known, and we address them here only briefly). We also offer a few case studies of both, in order to illustrate comparative costs and benefits, especially from the human security perspective. To characterize the tensions and problems arising from large-scale centralized energy production systems, we examine the California desert, where several renewable energy projects have generated considerable controversy and opposition (Ivanpah solar, Genesis solar, and Ocotillo wind). We contrast these with decentralized projects and project proposals to integrate solar energy projects into the footprint of existing human infrastructure. Note, however, that the latter are, for the most part, conceptual and in the design phase, and so the full panoply of potential issues has not been inventoried from actual experience.

In the concluding part of the chapter, we summarize some of the risks and vulnerabilities to which both types of systems are subject, ranging from economic and political to social and environmental, and suggest areas of further research that could help to clarify the comparative costs and benefits of the two strategies. We also return to the issue of human security, and discuss how the two approaches might affect it. We should note that, although there are good reasons to value the individual freedoms that might result from less reliance on centralized power systems, most of the world's people lack the capital to access decentralized technologies, and over 1.5 billion people live in energy poverty, having no access whatsoever to electricity. The security benefits to the latter of *any* generating technology would be substantial.

# DEFINING HUMAN SECURITY

Although the concept of "human security" has only entered the political lexicon over the past couple of decades, it is not a new idea. Human security encompasses not only health, welfare, and well-being, but also environmental and structural conditions that might affect social order and diminish people's quality of life. As such, people are "secure" when the circumstances of their lives provide access to basic necessities, such as food, water, energy, housing, healthcare, education, employment and income as well as protection against depredations and oppression by others, whether neighbors, governments or other countries. To this, we might also add those freedoms, liberties and opportunities that offer people at least the possibility of a full life and the right to decide how they will be governed, and by whom (see, e.g., Sen 1999; Newman 2010). By most accounts, there are as many as two billion people – and possibly more – who are not secure and lack one of these critical necessities (Lipschutz and Romano, 2012).

In this chapter, our particular concerns are with the ways in which the systems and mechanisms that provide energy – what might be called the "energy assemblage" – can affect human security. We have enumerated some of these effects above, but offer here a more detailed, albeit incomplete, list.

- Lack of adequate energy supplies, especially in intemperate climates, can have a direct impact on individual and community health; in those circumstances in which local fuel materials are in decline or depleted, the very project of accumulating them may have far-ranging social and environmental consequences.
- Situations in which supplies of potential energy are enclosed and access restricted to people – as, for example, in the case of dams and reservoirs – can result not only in water scarcity but also in denial of access to food, land, and other basic needs.
- The high costs of liquid fuels, such as kerosene, drives people to rely on wood, charcoal, dung and other biofuels, all of which can lead to deforestation and have deleterious health effects, especially if burned in poorly-ventilated spaces.
- The processes of extracting fossil fuels are very dirty and where poorly-regulated, as in the Niger Delta of Nigeria or the mountains of West Virginia, can undermine local living conditions and environments, not to mention general health and well-being.
- The manufacture of semiconductor devices, including solar PV cells, can expose both workers and environments to toxic materials released into the factory environment, water sources, arable land and the atmosphere, problems exacerbated with the rise in contract manufacturing.

- The operation of large-scale power plants continues to contribute to air pollution and, where coal is being burned, to large quantities of coal fly ash, even though, in some parts of the world, these have been greatly reduced over the past 60 years.
- While nuclear power plants have a fairly good operating and safety record – with a few notable exceptions – the safe and effective disposal of nuclear wastes, especially in the United States, has yet to be successfully accomplished, and there is a great deal of uncertainty about the long-term safety of storing spent nuclear fuel rods in "swimming pools" at reactor sites.
- Energy distribution systems are subject to accidents that can have deadly consequences, such as when oil and gasoline pipelines are sabotaged, or improperly monitored, and explode with sometimes considerable loss of life; high voltage transmission lines are much safer, although there continue to be concerns about the effects of exposure to them.
- Improper disposal of decommissioned energy systems and components, such as batteries, PV semiconductors and nuclear plant parts, can have both short- and long-term health effects, if not property contained.
- Excessive reliance on fossil fuels increasingly appears to be driving global climate change in highly-undesirable directions, which will have a broad range of negative effects on human security.

The ways that the energy assemblage affects and undermines human security are very unevenly distributed, leading to frequent and severe cases of environmental injustice and racism. Those living in wealthy countries tend to experience fewer of these inequalities, and they often have recourse and income to respond to the majority of them. By contrast, the world's billions of poor people are much more vulnerable to the uncertainties of energy supply, risks and hazards. This is not to argue that renewable DG is, somehow, a panacea for the lack of security faced by those billions; there are many other sources of insecurity facing them that energy cannot address. From our perspective, a critical question for human security is how to provision electricity. How will choices that affect the degree of centralization versus decentralization shape the effectiveness of energy delivery (volume, time, reliability) and cost (health, social, environmental and economic)? As suggested above, a full understanding of the comparative costs, risks, hazards and vulnerabilities of energy alternatives, and their implications for human security, is far beyond our capacity or the space available in this chapter, but we endeavor to address the question as comprehensively as possible.

# ROADS NOT YET TAKEN

Certain aspects of warnings issued by Lovins and others during the 1970s remain salient today, and are being recapitulated in contemporary debates around renewable energy, especially solar and wind. Lovins called for soft energy paths with soft social impacts, arguing that they would reduce inequality, social conflict, vulnerability, democracy, and freedom. New discourses of clean energy focus primarily on climate change impacts, sometimes at the expense of social impacts. Whereas, 40 years ago, it was the "energy crisis" and nuclear safety that provided the context for analysis and public policy, most of the emphasis on energy transitions today is in the context of climate change and political instability, which has refigured politicians', scientists' and engineers' preferences in terms of the technological composition of future energy infrastructures. The new emphasis on "low-carbon" or "clean," rather than simply "renewable," energy sources has broadened the scope of energy scenarios on offer to include natural gas and nuclear for their lower carbon emissions profiles. Indeed, this emphasis has led even the coal industry to tout the benefits of "clean coal" and warn darkly of growing reliance on natural gas from fracking.[2]

The transition to new sources of energy has also been framed in terms of technological and distribution scales, on the assumption that meeting the challenge of global climate change requires a planet-wide and rapid response. The lack of progress in international climate change negotiations has raised widespread doubt that such a global energy strategy will ever be forthcoming. A further complication is that new energy technologies typically require 50 years to be fully integrated into supply and demand systems, while electrical infrastructure operates on a 40-year replacement cycle (one that has been lengthened for a growing number of power plants). The potential for "tipping points" and catastrophic climate change (Lenton et al. 2008) has underscored the importance of getting new energy sources to scale as rapidly as possible, which has led most countries to focus on large-scale, centralized renewable energy projects with short-term planning horizons and hastened approval processes. What is not so clear is where the large volumes of capital required to finance a new power infrastructure will come from, whether renewable or not. The International Atomic Energy Agency (Birol 2004), for example, has predicted a need, between 2001 and 2030, for about $350 billion per year to meet global electricity demand. While the net cost of a wholly renewable electricity assemblage might be greater, the price of individual DG projects would be far smaller than the $1–5 billion required for many large power plants.[3] Moreover, while larger projects may have lower levelized costs of electricity, the question is prices for whom? Renewables may not yet be able to compete in wholesale electricity markets delivered through centralized systems, but DG is already at grid parity for many electricity consumers who must buy electricity at retail rates (Denholm et al. 2009a). To take advantage of this, however, still requires considerable upfront capital.

What, then, does a DG system look like? The basic principle underlying DG is that any particular location that experiences a sufficient inward flow of solar or wind energy – complemented, perhaps, by heat pumps, micro-hydro, solar hot water heaters and biogas generators – can provide a significant fraction of that location's electricity (and energy) needs. These vary, of course, from basic cooking and light all the way to large-screen TVs, pool heaters and electric cars – which is why it is easier to meet the supply criteria in poor societies, where electricity demand is relatively low. In wealthier communities, it is often less costly to reduce electricity demand through efficiency and conservation than to install capacity to meet peak usage. To date, the most common deployment of DG systems is to supply a single house or building; it is more cost-effective and efficient in the long-run to design and size DG for a group of consumers, whether constituted by a neighborhood, a cluster of businesses or some other configuration. Not only does this make available greater rooftop area for PV panels, it also affords greater opportunities for internal load-leveling, behavioral change and user involvement in system operation as well as offering some flexibility during grid failures.

At the same time, however, renewable energy sources are intermittent. The sun does not shine 24 hours a day nor does the wind blow at a constant velocity. Sometimes, a DG system generates more than is demanded; at other times, less. There are various ways to address this mismatch. At the present time, most DG systems in the industrialized world are hooked into regional power grids, which rely on them to effectively reduce power demand. Critical to effective DG is energy storage, whether through batteries, off peak hot water heating, fly-wheels, pumped-hydro storage, or compressed air (Denholm et al. 2010). Load leveling, which redistributes tasks from high demand to low demand periods, can also reduce storage and backup requirements. What remains to be seen is how DG systems and larger utility grids will be integrated (Shirmohammadi 2010; Delucchi and Jacobson 2011) as well as the role that consumers will play in operating such systems (Fischer 2006).

While the technical capabilities to deliver decentralized or distributed renewable generation continue to improve and, in some instances, are more cost-effective than conventional sources, tensions with centralized renewables and their operators are growing in prominence (Marnay and Bailey 2004). For technical, financial and management reasons, electric utilities, power generators and even governments prefer to pursue the traditional unidirectional model of power delivery: large mega- and giga-watt generators with extensive and interconnected distribution grids ultimately feeding electricity to end users. Where DG is being embraced by public policy, popular imagination and a broad range of small-scale development projects, investor-owned utilities in the United States are coming to regard DG as highly-problematic, for several reasons. First, as a result of deregulation, electric utilities are less and less involved in the business of generation and more and more in purchasing power from independent companies operating large plants, on the theory that this introduces competition to the power market. Assuming that they would provide their surplus power to the grid, DG would simply multiply the number of *small* independent generators with which a utility would have to deal, greatly complicating both business and technical aspects of managing the system. Second, a plethora of DG systems linked into a larger grid could introduce load instabilities due to changing demand and excess power when it might not be needed elsewhere. Third, and perhaps of greatest concern to utilities, is their complaint that customers with PV systems expect to be paid for the surplus power they put into the grid but not have to pay for the capital and maintenance costs of transmission and the distribution infrastructure.[4]

Resistance to DG is therefore considerable, and utilities appear to prefer centralized renewable generation. Large organizations tend to develop "standard operating procedures" tied into particular forms of technology and experience; changing those practices impose significant transaction costs and involve steep learning curves. It is much easier to keep doing what one does reasonably well – although, as American auto companies have learned, this can be a path to perdition – rather than adapting to new technologies, rules and practices. This suggests that centralized systems are subject to entrenchment, owing to the substantial periodic investments required to maintain them. It also points to another reason that utilities and power generators have shied away from DG.

As we noted earlier, centralized power generation is not without costs and risks, including environmental and other impacts and implications for human security and safety. Decentralized energy is not a panacea for all risks and hazards, and there is no predetermined relation between technology and democracy, as Langdon Winner (1986) has warned. But there are good reasons to pay attention to the comparative risks and benefits of energy system design, scale, and organization and their implications for human security and well-being. Typically, such costs – which economists call "externalities" – only become apparent after a great deal of time, energy and money has been spent. For renewable DG, the short-term on-site risks and hazards are a good deal smaller than for centralized power plants, yet those will be much more widely-distributed. More homeowners are liable to fall off their roofs than technicians off power plant catwalks (but there will be fewer catastrophic accidents). The risks and hazards experienced by the DG owners and operators are voluntarily incurred rather than unilaterally imposed, as is the case with air pollutants, fracking and nuclear power. On a life-cycle basis, the greatest risks and hazards from renewable DG are to be found in materials mining and processing, manufacturing and, ultimately, disposal (Fthenakis et al. 2008). Unfortunately, externalities in renewable energy supply chains tend to be "out of sight, out of mind," and efforts to regulate them are still quite limited (McDonald and Pearce 2010). Redistribution of the health and safety impacts of electricity generation from the wealthy to the poor is hardly a new practice, but that is no reason to continue it (SVTC 2009).

Where DG could well have its greatest impacts is in those parts of the world now lacking reliable power supplies or any electricity at all. It is estimated that close to 1.5 billion people lack access to electricity, while many more are connected to power systems that operate only sporadically. Even small amounts of power – as little as 100 watts and one kilowatt-hour/day – can make enormous differences in the well-being and quality of life of the world's poor, while a few kilowatt-hours daily can be used to procure water, electrify maternity wards, and refrigerate vaccines. Given the high capital costs of building new centralized power plants and distribution systems, paying for the fuel and collecting tariffs from the very poor, locally-deployed solar, wind and biogas systems, with ordinary lead-acid batteries, can provide a very modest level of power at relatively low upfront cost (Narula et al. 2012). DG can also reduce electricity theft and the very real risks and health hazards that arise from stealing power from the centralized grid.

## THE COSTS AND BENEFITS OF LARGE-SCALE CENTRALIZED RENEWABLES

To recap the essence of Amory Lovins's argument, centralized energy delivery and distribution systems pose a host of safety and political problems that could be avoided through a transition to decentralized, on-site renewable sources of energy. Lovins did not consider comparative life-cycle issues, nor did he offer any more than a deductive argument regarding the relationship between decentralization and democracy. Today there is increasing attention to the fragility of the global petroleum system, the emergence of climate change as a political issue, the ever-rising capital costs of centralized power plants, the water and carbon pollution problems arising from extracting unconventional fossil fuels such as shale and tar sands, and the complexity and vulnerability to disruptions of critical data transmission networks.

At the same time, the costs of electricity from smaller-scale renewable technologies continues to decline and their popularity continues to grow. Major changes have taken place in the organization of power production and distribution, especially as a result of deregulation and independent generation, but repeated attempts to incentivize a large-scale transition to DG have met with only limited success. Why? It might be useful to compare the power and telecommunication sectors to see how the combination of technology, social innovation and cultivation of demand have led to near-revolutionary changes. To be sure, the degree of miniaturization achieved through micro- and nanotechnology are, as yet, hardly applicable to the power sector, especially given the low power densities of renewable sources. However, there are useful parallels.

In the case of telecommunications, we see how the American state, in particular, played a significant role in jumpstarting technological innovation, through the defense sector. Not only was the early development and demand for semiconductors funded and stimulated by government demand, the Internet also was born out of a government-sponsored project to develop a secure and invulnerable communication system and to further rapid contacts among research scientists. The original microelectronics companies – Hewlett-Packard, IBM, Xerox and others now forgotten – were government contractors before they were commercial providers of technology. While they have not proven as agile in the market as later startups – especially Apple – they did provide an important organizational bridge between state and market demand. Ultimately, the decreasing size and rising capacities of computers and other electronic devices, stimulated further research and consumer demand, bringing the devices into the reach of even the very poor. The growing availability of various forms of narrow and broadband communication have stimulated network growth and transmission speeds, again leading to further innovation across a broad range of technologies and practices. This is not to argue that the result has been wholly salutary; such complex systems lend themselves to both productive and destructive activities (Noble 2011).

In the case of DG, if various PV and wind generators represent the technology – both of which have grown out of initial developments in the military sector – and national and transnational electricity grids represent the network – albeit not quite so unitary as the Internet – it is the social organization of electricity markets that have yet to undergo significant transformation. The Internet was never run as a public utility, subject to the kinds of government regulation that entails, and so it has not had to throw off that institutional form.[5] By contrast, deregulation in the power sector has not sought to break up large utilities, whose commitments to large-scale generation remain steadfast. On the Internet, anyone can hang out a shingle and try to sell goods and services; so far, small-scale generators are not permitted to do so (how this might come about is a public policy question we cannot as yet answer). We could imagine arrangements in which DG plants were permitted to contract with other DG plants nearby in some kind of power exchange, although this would require cooperation of the owner-operator of the distribution system.

One result of this centralizing tendency in renewable generation is a rush to stake out solar- and wind-intensive sites, with strong encouragement and incentives from government. According to Dian Grueneich (2012), 29 states, two territories and the District of Columbia have developed Renewable Portfolio Standards, while the DSIRE database (2012)[6] indicates that every American state and several territories offer financial incentives to renewables. The future for centralized renewables looks bright, albeit with visible and growing smudges. As of mid-2012, across the United States some 31.5 GW of solar (SEIA 2012) and 60 GW of wind (AWEA 2012) generating capacity, with a footprint of some 8,350 square miles[7] (Jacobson and Delucchi 2011) are in operation or in the planning and construction stage. Already, a number of shortcomings and problems are becoming evident, generating opposition and stoking what some have called a "Green Civil War" (*New York Times* Editors, 2012). Controversial large-scale solar energy and wind projects are impacting sensitive ecological habitats and literally running over endangered species as well as widespread public concerns (Solar Done Right 2012). The imaginary of those who promote large-scale solar is best captured by the "Solar Grand Plan," whose supporters argue that public lands in the desert southwest would be well suited for solar electricity generation (Zweibel, et al. 2008). Indeed, the U.S government is setting aside tens of millions of acres of public lands in the Southwest for such projects (Fears 2012), in the expectation that dispatchable solar and wind projects will be developed by private firms to deliver electricity to utilities. What do such projects look like, and what are their environmental and social impacts? Here, we offer three cases.

**Ivanpah Valley and BrightSource**

The Ivanpah Valley is situated about 25 miles south of Las Vegas in the California desert, just outside of the California-Nevada border town of Primm, Nevada. The valley is vast and gently sloping, 30 miles across from north of Primm to the headwaters of the New York Mountains to the south. The valley's higher elevation flatlands, such as near Cima, California (population: 21), contain the densest population of Joshua trees (*Yucca brevifolia*) in the world. Creosote (*Larrea tridentate*) chaparral dominates the lower elevation flatlands such as those at the proposed project site (a clonal colony of creosote nearby is considered by some to be the oldest living organism on Earth). The Ivanpah solar project site is adjacent to the Mojave National Reserve and known to desert ecologists as an important habitat for the threatened desert tortoise (*Gopherus agassizii*) among several other important reptile, mammal, and plant species.

BrightSource is a solar energy company whose founder, Arnold Goldman, received the World Economic Forum's Technology Pioneer Award for his work on solar power tower technology. He was the designer of 10 solar electricity generating systems (SEGS) built from 1984 to 1990 by Luz Industries, called SEGS I through IX (only nine were built), that still operate in the California desert today, providing over 350 MW of power. As early as 2006, BrightSource filed a petition to acquire about 8,000 acres of public lands in the Ivanpah Valley. Upon hearing of the project, most local ecology and public lands groups opposed the project, yet a number of national environmental groups endorsed the project because of its potential to mitigate GHGs from electricity generation. This became an important proposal, as it pitted local groups against the national organizations. For example, the desert chapter of the Sierra Club vehemently opposed BrightSource's Ivanpah project, while the national chapters chastised the local groups for lacking a commitment to efforts to slow climate change.

While numerous ecological impacts were noted during the siting hearings, the major impact identified was to the desert tortoise. The United States spends more money protecting the desert tortoise than on the grizzly bear, bald eagle, and gray wolf combined. The project would require the company to obtain an incidental take permit, implying that the project will harm up to a specified number of the threatened tortoises. To estimate the number of take permits required to construct the facility, consultants initially surveyed the project and suggested that the project would impact 12 to 24 individual tortoises. Yet, when preparations for construction were initiated, more than 36 tortoises were discovered living on less than one-third of the proposed project area, which triggered a re-consulation with permitting agency (the US Fish and Wildlife Service) as stipulated in the permit. By the time the entire project site had been cleared, more than 150 desert tortoises had been found (BLM 2011), consistent with desert activists and ecologists' claims that the Ivanpah Valley should be designated as an Area of Critical Environmental Concern, a designation that would offer greater protection for this sub-population of desert tortoise.

To lessen the impacts on endangered species of this particular $2.2 billion dollar project, the company has spent over $56 million on various mitigations (BrightSource Energy 2012). In this instance, a centralized renewable system has experienced a double whammy: it has been made economically vulnerable and rendered a population of desert tortoise less resilient to the impacts of climate change. Scientists suggest that the Mojave Desert will be impacted by climate change more than any place south of the Arctic. For species to adapt, they may have to migrate in response to changes in heat stress and precipitation. Large-scale projects such as these not only pose physical threats to animals, they also present obstacles to species migrations in the future. Some have argued that such ecosystems must be sacrificed in the name of climate change mitigation. Others contend that decentralized systems of energy procurement and generation would obviate these impacts.

**Genesis Solar Energy and Ford Dry Lake**

In 2009 a subsidiary of Next Era Energy Resources proposed building a solar energy trough system in the Colorado Desert, close to the Colorado River boundary between Arizona and California. A number of issues similar to those raised at Ivanpah appeared at this site, although the area is also of significance from a cultural heritage perspective. Giant intaglios, figures tens of meters across in so-called desert pavement – small pebbles firmly settled atop desert soils – are found throughout the area. Native Americans have occupied this region for thousands of years and it is believed to have been an important area for settlement by the ancestors of the Aztec, Mayan, and Incan civilizations, as humans migrated across the Bering Strait and toward Central and South America. Native American elders earlier warned that the solar plant was sited near a desert watering hole along an ancient trail connecting the Colorado River to the Pacific Ocean. Not long after construction began on this multi-billion dollar plant, a Native American burial ground was found on the site. Native American tribes justly claim that there was no prior consultation with them on the site, a requirement for public lands through a special arrangement with the U.S. government.

The project has also had critical ecological impacts on an undistributed desert ecosystem, the site of several resident Mojave kit fox populations, another threated species and the smallest canine in the world. To comply with the rules established prior to construction, the developer was required to evict the kit foxes, which they did by spraying coyote urine into their dens. Shortly after construction commenced, however, seven dead foxes were found on the site, killed by distemper, a disease previously never detected in the kit fox population (Sahagun 2012). Presumably, the attempt to drive away the animals played a role in their contracting the disease.

Centralized power systems are also vulnerable to localized weather events (so are DG systems, but the effects on total generating capacity is likely to be much less). Although deserts are dry by definition, they are also subject to flash flooding. Activists had warned argued that siting the project in a desert alluvial fan would make it vulnerable to flooding. In July 2012, a flash flood did several million dollars worth of damage to the project site.

**Ocotillo Express Wind Energy Project**

Large-scale wind farms already have a significant presence in the California Desert, especially near the San Gorgonio and Tehachapi passes that link the northern and southern parts of the state. The Ocotillo Express project is a wind farm south of Palm Springs, California, in close proximity to the Coyote Mountains Wilderness Area, the Jacumba Mountains Wilderness Area, and Anza Borrego State Park. Pattern Energy proposed to raise 112 wind turbines across 10,000 acres of undisturbed desert ecosystem, connected to the Sunrise PowerLink, a recently-approved and controversial high-voltage transmission line that will carry energy over the 117 miles from the Colorado Desert to metropolitan San Diego. Sunrise Powerlink has been justified as providing San Diego access to the abundant renewable energy resources around the Imperial Valley region of California.

Wind farm footprints differ from those of solar plants, as turbine spacing allows for multiple uses include ranching, farming, and open space conservation. A major ecological consideration, however, is that turbine blades often strike birds and leave a pressure wave in their wake causing significant mortality in bats, of which 12 species are found in the area. The Ocotillo Express region is home to golden eagles and other raptors such as burrowing owls. Because wind turbines are known to pose particular hazards to golden eagles, the developer must obtain an incidental take permit for "taking" (killing) a number of the birds over the life of the project. Under federal law, it is illegal to intentionally kill eagles (or even possess eagle feathers and parts), and recent legislation prohibits the Fish and Wildlife Service from issuing taking permits for eagles. Several California desert tribes, including the Quechan, Kumeyaay and Cocopah Nations oppose the project because the landscape is a sacred part of their cultural heritage and close to a recently-discovered prehistoric cremation site (which has subsequently been withdrawn from the project area).

All three projects were specially designated as "fast-track" projects for their potential to quickly add renewable energy to the grid. As various ecological and cultural resource conflicts gain legal and political traction, however, it is less evident that such efforts are the most expeditious way to deploy renewables. Activists argue that the wind resources to make the project financially viable are actually not available, since the turbines would be sited on the leeward side of the coast range (developers have supplied data to the contrary). Unlike solar power plants, there is less flexibility in siting wind farms. Ridge tops are windiest but also most visible, so that wind farms tend to be particularly controversial on aesthetic grounds. Distributed alternatives for wind power are fewer but research suggests that larger wind farms with multiple turbines face more resistance than single turbines (Devine-Wright 2004). Micro-wind turbines have sometimes been touted as an alternative to large machines, but some argue that these are inconsequential energy suppliers except in low-demand, remote applications (Mackay 2009).

**Not All Large Projects Have Significant Impacts**

Not every large-scale solar and wind project has such negative ecological impacts, and not every project has generated public opposition. Denmark has gone quite far in deploying on- and off-shore wind farms, based on very large turbines (which can be easier for some birds to navigate – though the turbine tips move much more swiftly). There are serious concerns about the impacts of these wind generators on birds and the ideal sites for turbines are usually also important flyways. Where objections have been raised, it is primarily about impacts on the "viewscape," as in the case of proposed windfarms off the Massachusetts coast and islands (ironically opposed by Robert Kennedy Jr. who lambasted environmentalists for opposing BrightSource, the case described above). But there is no reason why large-scale projects in ecologically-sensitive areas are the only way to go. There are numerous degraded landscapes throughout the desert southwest – abandoned mines, agricultural lands, etc. – where large-scale projects might be more appropriate. For example, the Aqua Caliente project will provide 290 MW of electric power on 2,400 acres of former agricultural lands, where no endangered species or cultural resources are present. But activists engaged in these issues emphasize the need to develop distributed energy resources first. It has been estimated, for example, that the technical potential for rooftop photovoltaic generation in California exceeds 70 Gigawatts

(GW) (CEC 2007), compared to a peak summer demand of less than 60 GW (CEC 2011).

**Tackling Energy Poverty, Improving Human Security**

The three projects described above are all located in the Global North, where electricity supplies are plentiful, if sometimes costly, and consumers suffer few, if any, power shortages. The same cannot be said across vast swath of the Global South, where urban and industrial demand for energy is rising rapidly. The common response is to build large-scale plants, whether hydroelectric, coal or nuclear. We will not reiterate the risks and hazards of these energy sources, except to note that, quite often, those most affected by the negative externalities of nearby energy production facilities often realize the fewest benefits from them. For the most part, energy, especially electricity, is directed to those areas where there is at least some semblance of a distribution system and consumers with some potential to pay for the energy. Europe has made plans to build large-scale solar generating plants across North Africa, sending most, if not all, of the power across the Mediterranean. The host countries will realize royalties from use of their solar resources, but their people, and especially the poor, are unlikely to gain much access to that electricity.

By contrast, DG has already made an impact across the Global South, especially in areas remote from power grids, enhancing human security while obviating many of the externalities associated with centralized systems. PV electricity has reached grid parity in areas where there is currently no electricity infrastructure, with such efforts dating back to the 1970s, when a humanitarian named Father Verspieren sought to electrify remote parts of African with PV (Perlin 1999). Today, organizations such as 'Engineers without Borders' are making great strides in installing PV, biogas, and micro-hydro systems in remote villages, providing light to hospital maternity wards and refrigeration to keep vaccines cool. There are major challenges in using DG PV for rural electrification including bringing down the cost of delivering these systems to remote parts of the world, as well as ensuring that local communities are adequately trained to operate, maintain and repair systems. Nonetheless, there are also major successes. In *Chasing the Sun*, for example, Neville Williams (2005) recounts efforts to electrify and light places mired in energy poverty. His Solar Electric Light Fund has helped assist anti-poverty NGOs to raise the standard of living for thousands of people who, in the absence of any adequate electricity infrastructure, previously relied on wood and charcoal for cooking, and kerosene for lighting.

## THE COSTS AND BENEFITS OF RENEWABLE DISTRIBUTED GENERATION

A key political principle in operation around the world today is *subsidiarity*, the proposition that policies and practices should be implemented by the least centralized authority. Historically, the inefficiencies of utility competition in the same market as well as economies of scale drove the consolidation and enlargement of energy-producing companies. Moreover, as air and water pollution came to be of growing concern, regulation of a few large point sources was much more straightforward than many small ones. But such pollution disappears with renewable DG. Moreover, while economies of scale operate on the production side of renewable technologies, it is not relevant on the supply side, where much of the retail cost of electricity, in particular, is attributable to distribution, operation and maintenance of the utility grid. If such grids remain in existence, for example, in order to provide backup power to DG during periods of low renewable generation, the costs of maintaining the grid and backup generators will have to be paid, most probably through some kind of tax or tariff on DG systems.

Economies of scale are not wholly absent from DG, however; the particular design and sizing of a DG system will make a difference in terms of reliability, flexibility and vulnerability and, by extension, for human security. The relatively high cost of single unit systems, such as rooftop PVs supplying a single house, the intermittent nature of wind and solar energy and the problem of energy storage for high load and low energy flux periods today make DG feasible only for consumers with adequate access to capital. One consequence is that those most in need of low-cost, reliable energy – the poor and elderly – can least afford renewables.

One potential approach to these shortcomings are renewable energy "microgrids," combined power and heating systems that serve, as noted earlier, bounded neighborhoods and districts. Such a microgrid might, for example, include rooftop PV and solar water heaters, wind microturbines, heat pumps providing district heating, cooling and hot water, energy storage in the form of batteries, water tanks, compressed air and electric vehicles, possibly a biogas digester, and various monitors and controllers. Of course such major overhauls in energy supply technology must be complemented by demand management, including conservation improvements in buildings, replacement of inefficient appliances, real-time energy use information, incentives for load-shifting as feasible, timed devices on washers and dryers, and so on. If necessary, a microgrid can be hooked into the larger utility grid, pumping surplus power into it and drawing power from it when local supply drops below demand.

One important impact of climate change is higher temperatures in urban areas, from the urban heat island effect. The urban heat island effect is caused largely by human-made developments that emit thermal radiation at night, preventing cities from cooling the way that suburban and rural areas do. This presents an opportunity for DG to be deployed to make cities more resilient to the impacts of climate change. DG PV can be used to cover parking lots and buildings, thereby lessening the quantity of solar energy absorbed by pavement and other thermal masses that are primary constituents of the heat island effect. These systems would also reduce the losses from transmitted electricity, and allow for better integration of peak shaving strategies such as PV-powered air conditioning for large buildings.

A number of ancillary questions arise with such DG microgrids. Who will manage and maintain them? This could be done either by a committee of users, some of whom would need to be knowledgeable about or trained in system mechanics and requirements. Alternatively, management companies providing such services – a potential new source of green employment – could be hired for the task. How will microgrids be financed? Utilities might find it in their interest to pay for microgrids, billing users for electricity consumed. Or, local power cooperatives may develop and choose to withdraw from existing energy distribution systems. Currently, some solar companies will finance solar PV systems upfront, either retaining ownership or through a long-term loan, in return for payments for electricity generated and saved. Another approach involves bank financing of cooperative ownership shares: on the island of Samsø in Denmark, shares in wind turbines were financed through 10–20-year loans repaid through the monthly revenue from the sale of wind-generated electricity. What risks and hazards are associated with microgrids? Clearly, no technological system is failsafe and foolproof. People fall off roofs; birds try to fly through wind turbines; blackouts don't disappear; toxics are not eliminated from either the front or back ends of renewable energy life cycles. Overall, however, the net benefits of DG as opposed to centralized renewables, especially in terms of human security, appear to outweigh the costs, especially for the poor.

**Which Road This Time?**

Energy supply systems and companies are notoriously slow to change: there are no longer Seven Sisters in petroleum, but their descendents continue to operate globally and with relative impunity. Oil remains the mainstay of transportation while coal dominates electricity generation with natural gas rapidly catching up. New large hydropower sites are limited, and low-head and run-of-river hydro attracts little attention or enthusiasm. Notwithstanding talk of a "nuclear renaissance," that resource seems unlikely to expand very much in the future, especially in light of the Fukushima accident(s) and commitments by several European governments to a fully-renewable electricity grid by 2050. Other non-carbon based energy sources are almost all seen in terms of centralized plants, and none of them has effectively made the jump from demonstration to commercialization, although this could change in the future.

A considerable fraction of current and future growth in renewable energy is in the form of centralized facilities. We have suggested that decentralized renewables, in the form of distributed generation systems, hold numerous advantages over centralized solar and wind plants. We have also proposed – and there is solid evidence to support this second claim – that small-scale renewables hold considerable potential for addressing the energy poverty and health-related problems facing billions of the world's people. This does not mean that such technologies can be transplanted easily – experience over the past four decades illustrates the need for local capacity as well as a reliable supply of spare parts – but microgrids can reduce the need for capital-intensive power plants, and can be installed much more quickly than large-scale transmission corridors.

What of Lovins's argument regarding the authoritarianism of centralized power and the democratic tendencies of decentralized renewables? For better or worse, the history of the past 40 years offers only limited evidence of support for his claim. Centralized power seems largely indifferent to political institutions, although there is ample evidence that the reverse is not the case. Moreover, there is evidence to support the proposition that democracies tend to pay greater attention to the environment than do non-democracies but, once again, this has little to do with scale. But these observations say little or nothing about causality. Many claims have been made for the liberating and democratizing qualities of the Internet, surely one of the most decentralized technological infrastructures to have ever existed. In some instances, social activism through such media seems to have assisted, if not enabled, democracy movements and challenges to authoritarian governments. But the Internet has proven, too, to be easily monitored and managed, especially if a government is not especially concerned about the effects of such control on commerce. To become entirely autonomous, as growing numbers of many defense-related and corporate intranets have done out of the need to protect themselves against hackers, is to be cut off from the world. The parallel with renewable DG is not entirely accurate, but any organized government determined to impose its authority on a population could probably mobilize the force required to garrison and control DG facilities.

We have argued that renewable DG does have the potential to make major impacts on quality of life and human security for billions of the world's inhabitants. Some will argue that the costs of an effort to provide these benefits to the global poor are too great, and that economic growth in the Global North and South will, ultimately, provide more opportunities. Yet, this begs the question. Much of the North's aging electrical infrastructure will have to be replaced over the coming years, while the demand for energy in the South will certainly continue to increase. In other words, literally *trillions* of dollars will be spent on constructing new power plants and distribution systems. The diversion of even a fraction of this investment could go a long way toward penetration of renewable DGs throughout the world, reducing their cost and increasing their reliability, and ameliorating the poverty now experienced by so many. This could also have the additional salutary effects of reducing greenhouse gas emissions, mitigating some degree of future climate change, lessening political and military pressures on supplies of cheap oil and setting the stage for global sustainability. These goals might seem utopian to some, but there is no *technological* or *economic* reason they cannot be achieved over the next 50 years.

Not all roads lead to political utopias, and choosing the path to greater freedom ought not to depend on whether a vehicle's steering tends to the left or the right. What is important is a people's political capacity to make and effect decisions, such as whether to go down the centralized or decentralized path. Technologies are political; technologies have politics; but technologies do not determine what forms political practices will take. That requires activism, action and determination. That, however, is a topic for another chapter in another volume.

## NOTES

1. Of course, electricity generation using fossil fuels can also be highly localized, although this generally requires transport of the fuels from sites of production and processing to end use. "Small" electricity and heat-generating nuclear plants have been proposed and designed at various times over the past 50 years, but none has ever been made commercially available.
2. The current glut of fracked natural gas in the United States has led to a collapse in prices that is undermining the economic viability of renewables, much as happened following the collapse of oil prices in the 1980s.
3. The cost of electricity would depend on a range of factors, so it is impossible in this chapter to compare relative prices from conventional, nuclear and centralized vs. DG generation in 2040.
4. This problem arises because surplus power causes the electric meter to "run backwards." In effect the customer is "storing" electricity in the grid for future use, without paying anything for that function.
5. Which is not to say that growing concentration of assets and control is absent from

global telecommunications and the Internet. But these large corporations are, for the most part, not allotted restricted territories in which they can operate.
6. DSIRE is the "Database of State Incentives for Renewables & Efficiency," maintained by the North Carolina Solar Center at North Carolina State University, and funded by the U.S. Department of Energy.
7. This may under- or over-estimate the actual footprint of this renewable energy capacity; the calculation is based on estimates made at the U.S. National Renewable Energy Laboratory (NREL 2004; Denholm et al. 2009b).

# REFERENCES

AWEA (2012). "Industry Statistics," American Wind Energy Association, August, at: http://www.awea.org/learnabout/industry_stats/index.cfm (accessed September 16, 2012).
Ayres, R. (1975). "Policing Plutonium: The Civil Liberties Fallout," *Harvard Civil Rights-Civil Liberties Law Review* 10: 369–443.
Birol, F. (2004). "Power to the People: The World Outlook for Electricity Investment," *IAEA Bulletin* 46 (1): 9–12, at: http://www.iaea.org/Publications/Magazines/Bulletin/Bull461/46104990912.pdf (accessed September 15, 2012).
BLM (Bureau of Land Management) (2011). "Revised Biological Assessment for the Ivanpah Solar Electric Generating System (Ivanpah SEGS) Project." U.S. Department of Interior.
BrightSource Energy (2012). "Ivanpah and the DOE Loan Guarantee Program," at: http://www.brightsourceenergy.com/ivanpah-and-the-doe-loan-guarantee-program (accessed July 1, 2012).
Cardwell, D. (2012). "Chain Stores Lead Firms in Solar Power Use, Study Finds," *The New York Times*, September 12, at: http://www.nytimes.com/2012/09/12/business/energy-environment/chain-stores-lead-firms-in-solar-power-use-study-finds.html (accessed September 15, 2012).
CEC (California Energy Commission) (2007). "California Rooftop Photovoltaic (PV) Resources Assessment and Growth Potential by County," Navigant Consulting, for the Public Interest Energy Research Program, September, CEC-500-2007-048.
CEC (California Energy Commission) (2011). "Summer 2011 Electricity Supply and Demand Outlook," Staff Report, April, CEC-200-2011-004.
Delucchi, M.A. and M.Z. Jacobson (2011). "Providing all Global Energy with Wind, Water and Solar Power, Part II: Reliability, System and Transmission Costs and Policies," *Energy Policy* 39: 1170–90.
Denholm, P., R.M. Margolis, S. Ong, B. Roberts (2009a). "Break-even cost for residential photovoltaics in the United States: key drivers and sensitivities." National Renewable Energy Laboratory, Golden, CO.
Denholm, P., M. Hand, M. Jackson and S. Ong. (2009b). "Land-Use Requirements of Modern Wind Power Plants in the United States," National Renewable Energy Laboratory, U.S. Department of Energy, NREL/TP-6A2-45834, August, at: http://www.nrel.gov/docs/fy09osti/45834.pdf.
Denholm, P., E. Ela, B. Kirby and M. Milligan (2010). "The Role of Energy Storage with Renewable Electricity Generation," National Renewable Energy Laboratory, Golden, Colo., January, NREL/TP-6A2-47187, at: http://digitalcommons.library.unlv.edu/renew_pubs/5 (accessed April 28, 2012).
Devine-Wright, P. (2005). "Beyond NIMBYism: towards an integrated framework for understanding public perceptions of wind energy," *Wind energy* 8(2): 125–139.
DSIRE (2012). "Financial Incentives for Renewable Energy," Database of State Incentives for Renewables & Efficiency, North Carolina State University/U.S. Dept. of Energy, at: http://www.dsireusa.org/summarytables/finre.cfm (accessed September 16, 2012).

Fears, D. (2012). "Interior Department Sets Aside Millions of Acres for Solar Power," *The Washington Post*, July 24, at: http://www.washingtonpost.com/national/health-science/interior-department-sets-aside-millions-of-acres-for-solar-power/2012/07/24/gJQAS4Co7W_story.html (accessed September 16, 2012).

Fischer, C. (2006). "From Consumers to Operators: the Role of Micro Cogeneration Users," in M. Pehnt, M. Cames, C. Fischer, B. Praetorius, L. Schneider, K. Schumacher and J.-P. Voß (eds), *Micro Cogeneration. Towards a Decentralized Energy Supply*, Heidelberg: Springer, pp. 117–41.

Fthenakis, V., H.C. Kim and E. Alsema (2008). "Emissions from Photovoltaic Life Cycles," *Environmental Science & Technology* 42(2): 2168–2174.

Grueneich, D. (2012). "Energy Challenges Facing the United States and the State of California," talk presented at The Panette Institute for Public Policy, August 22.

Jacobson, M.Z. and M.A. Delucchi (2011). "Providing all global energy with wind, water, and solar power, Part I: Technologies, energy resources, quantities and areas of infrastructure, and materials." *Energy Policy* 39(3): 1154–1169.

Lenton, T., H. Held, E. Kriegler, J. Hall, W. Lucht, S. Rahmstorf and J. Schellnhuber (2008). "Tipping elements in the Earth's climate system," *Proceedings of the National Academy of Sciences* 105(6): 1786–1793.

Lipschutz, R.D. (ed.) (1995). *On Security*, New York: Columbia University Press.

Lipschutz, R.D. and S.T. Romano (2012). "The Cupboard is Full: Public Finance for Public Services in the Global South," Municipal Services Project, Occasional Paper 16, May, at: http://www.municipalservicesproject.org/sites/municipalservicesproject.org/files/publications/Lipschutz-Romano_The_Cupboard_is_Full_May2012_FINAL.pdf (accessed May 20, 2012).

Lovins, A.B. (1976). "Energy strategy: the road not taken," *Foreign Affairs* 55: 186–218.

MacKay, D. (2009). *Sustainable Energy Without the Hot Air*, Cambridge: UIT Cambridge Ltd.

Marnay, C. and O.C. Bailey (2004). "The CERTS micro-grid and the future of the macro-grid," Environmental Energy Technologies Division, Lawrence Berkeley National Laboratory, August, LBNL-55281, at: http://escholarship.org/uc/item/1103m944 (accessed December 29, 2011).

McDonald, N.C. and J.M. Pearce (2010). "Producer responsibility and recycling solar photovoltaic modules", *Energy Policy* 38: 7041–47, at: http://www.sciencedirect.com/science/article/pii/S0301421510005537 (accessed September 15, 2012).

McGregor, R. (2011). "Zhou's Cryptic Caution Lost in Translation," *The Financial Times*, June 10, at: http://www.ft.com/intl/cms/s/0/74916db6-938d-11e0-922e-00144feab49a.html#axzz26Yo8cLd7 (accessed September 15, 2012).

Narula, K., Y. Nagai and S. Pachauri. (2012). "The role of Decentralized Distributed Generation in achieving universal rural electrification in South Asia by 2030," *Energy Policy* 47: 345–357.

Newman, E. (2010). "Critical Human Security Studies," *Review of International Studies* 36: 77–94.

*New York Times* Editors (2012). "Green Civil War: Projects vs. Preservation," January 12, at: http://roomfordebate.blogs.nytimes.com/2010/01/12/green-civil-war-projects-vs-preservation/ (accessed September 16, 2012).

Noble, D.F. (2011). *Forces of Production: A Social History of Industrial Automation*, New Brunswick, N.J.: Transaction, with a new preface by the author.

NREL (2004). "How Much Land will PV Need to Supply Our Electricity," National Renewable Energy Laboratory, U.S. Dept. of Energy, DOE/GO-102004-1835, February, at: http://www.nrel.gov/docs/fy04osti/35097.pdf (accessed September 16, 2012).

Perlin, J. (1999). *From Space to Earth – The Story of Solar Electricity*, Ann Arbor, Mich.: Aatec.

Sahagun, L. (2012). "Canine distemper in kit foxes spreads in Mojave Desert." *Los Angeles Times*, at: http://articles.latimes.com/2012/apr/18/local/la-me-0418-foxes-distemper-20120418 (accessed September 22, 2012).

Scheer, H. (2007). *Energy Autonomy: The Economic, Social and Technological Case For Renewable Energy*, London, Earthscan.
Sen, A. (1999). *Development as Freedom*, New York: Knopf.
SEIA (Solar Energy Industries Association, 2012). "Major Solar Projects List," August 15, at: http://www.seia.org/research-resources/major-solar-projects-list (accessed September 16, 2012).
Shirmohammadi, D. (2010). "Impacts of high penetration of distributed and renewable resources on transmission and distribution systems," *Innovative Smart Grid Technologies*, at: http://ieeexplore.ieee.org/stamp/stamp.jsp?tp=&arnumber=5434736&isnumber=543 4721 (accessed January 27, 2010).
Smil, V. (2010). "Power Density Primer: Understanding the Spatial Dimension of the Unfolding Transition to Renewable Electricity Generation," *Vaclav Smil*, May 8, at: http://www.vaclavsmil.com/wp-content/uploads/docs/smil-article-power-density-primer.pdf (accessed September 15, 2012).
Solar Done Right (2012). at: http://solardoneright.org/ (accessed September 16, 2012).
SVTC (Silicon Valley Toxics Coalition) (2009). "Toward a Just and Sustainable Solar Energy Industry," January 14, http://svtc.org/wp-content/uploads/Silicon_Valley_Toxics_Coalition_-_Toward_a_Just_and_Sust.pdf (accessed September 29, 2012).
Turner, F. (2006). *From counterculture to cyberculture: Stewart Brand, the Whole Earth Network, and the rise of digital utopianism*, Chicago, University Of Chicago Press.
Ullman, R. (1983). "Redefining Security," *International Security* 8(1) (Summer): 129–53.
Williams, N. (2005). *Chasing the Sun: Solar Adventures Around the World*, Gabriola Island, BC, Canada.
Winner, L. (1986). *The Whale and the Reactor: A Search For Limits in an Age of High Technology*, Chicago, University of Chicago Press.
Zweibel, K., J. Mason and V. Fthenakis (2008). "A Solar Grand Plan," *Scientific American*, January: 64–73.

# 23. Human security and energy security: a sustainable energy system as a public good
*Sylvia I. Karlsson-Vinkhuyzen and Nigel Jollands\**

## INTRODUCTION

This chapter is dedicated to the concept of human security and its link to energy and energy governance, particularly global energy governance. Through this focus emerges the need to look at the links between the concept of public goods and energy. Our starting argument is that conventional notions of energy security that are centred on the nation state are insufficient to ensure human security at an individual level (across the globe). Rather, what we refer to as 'deep energy security' is a necessary condition for human security and such security in turn requires a sustainable energy system. We further argue that one approach to strengthen deep energy security is to use the lens of the public goods concept to consider how aspects of a sustainable energy system should be provided.

The chapter is structured as follows. We start by exploring the evolution of the concept of human security and its major components and then analyse the various ways through which energy is linked to this concept. We look at the links between energy and human well-being and security and between energy and human ill-being and insecurity. We then explore the contrast between the concept of human security and the conventional way in which energy security has been framed, contrasting the individual with the collective perspective. We then argue that conventional energy security is not sufficient to deliver human security and propose the notion of deep energy security as a more comprehensive and appropriate concept. This concept is closely linked to the sustainability of energy systems, particularly the global energy system. In the following section we turn our attention to how deep energy security could be provided, with a first step in approaching the sustainability of energy systems as public goods at all levels, particularly the global level. This requires an elaboration of the definition and theory of public goods and how they need to be provided. Acknowledging the need for a multilevel and multilayered provision approach for this public good we examine in more detail what and how much of this good should be provided through global energy governance and then we briefly explore the current practice of global energy governance before drawing some conclusions.

## ENERGY AND HUMAN SECURITY: FROM NATIONAL AND COLLECTIVE TO INDIVIDUAL AND GLOBAL

As we move into the 21st century, there is little doubt that humanity faces daunting energy-related challenges. Access to affordable energy sources is increasingly challenging for countries – a situation that is likely to become worse as fossil fuels prices inevitably increase further. In addition, 2–3 billion people lacking access to modern energy services (AGECC, 2010) and fossil fuels account for 60 per cent of greenhouse gas emissions causing climate change.[1] The need for a transition to sustainable energy production and consumption patterns, not only within national contexts but also across the whole globe, is increasingly acknowledged by a broad spectrum of commentators and decision makers (AGECC, 2010; Bradbrook and Wahnschafft, 2005; ElBaradei, 2008; Jäger and Cornell, 2011; Nilsson et al., 2012; Alabi, Chapter 4 this volume; Peters and Westphal, Chapter 5 this volume). Yet progress is slow. One of the reasons for this, we argue, is the entrapment of energy policymaking within the paradigm of national security. Energy security is considered vital for national security and policies are largely made within that framework even if many countries are paying increasing attention to other concerns (including transnational ones) such as air pollution, climate change and energy poverty and thus various dimensions of sustainability. One implication of this national security paradigm is not only that insufficient attention is given to consequences of energy policies for other countries, their citizens, and the global environment, but also that the potential for cross-border collaboration on energy issues is not realised. Energy choices are focused on affordable energy for the national economy and military needs. Energy security is considered as a national public good with its provision set as a priority for government attention. Collaboration with other countries does not come easily within this paradigm and many win-win opportunities in energy investments, technology cooperation and governance are foregone.

An alternative security paradigm for energy policy is human security. This concept was introduced in 1994 in the UNDP Human Development Report in an effort to make human development challenges more immediate policy concerns. The UNDP identified six types of security beyond security from physical violence: income security, food security, health security, environmental security, personal security, community security and political security (UNDP, 1994). This effort, followed by considerable academic and policy attention particularly at the global level, changes the focus on the object that should be protected and that should be secure – from states and their sovereignty and territorial integrity to individual human beings and their survival, human development, identity and governance (Liotta, 2002). Human security is not only about the physical safety of persons (freedom from fear) but also about their ability to obtain and hold basic goods (freedom from want) (Gasper, 2005). This shift brings a change in attention towards those that are most vulnerable (Owen, 2008). The concept of human security was particularly elaborated by the Ogata-Sen Commission in their report Human Security Now that defined it as 'to protect the vital core of all human lives in ways that enhance human freedoms and human fulfilment' (Commission on Human Security, 2003:4).

Gasper (2005) outlines how the human security concept serves as the container for a discourse and an 'intellectual boundary object' that merges concerns from three other discourses; basic human needs, human development and human rights. The roots in human rights brings 'the unwillingness to sacrifice anyone' and the basic human needs discourse a 'stress on prioritization' (Gasper, 2005:234). Furthermore, human rights tend to highlight not only the rights but the duties of (someone) to ensure those rights and thus a discussion on duties and accountability (Gasper, 2005). The security concept attempts to connect with the national security world where funding and organisational power are at higher levels and indirectly connects to the accountability structure of the UN System and its work on monitoring of humanitarian crises, Millennium Development Goals etc. (Gasper, 2005). The shift from national to human security means that the issue of who can or should provide security for whom changes. Some consider that states should still have the major responsibility to ensure human security but a human security focus raises the concern from being limited to citizens within a particular state to citizens of all states, to global citizens. Can states adopt such a global perspective in their policy-making? We return to this below.[2]

Energy was not mentioned either in the UNDP report of 1994 (UNDP, 1994) or in the Ogata-Sen Commission report (Commission on Human Security, 2003).[3] It is also quite invisible in the academic literature on human security. Nonetheless, the contribution of energy to human security can be explicitly linked to at least four of the six security dimensions that were in the original UNDP definition of human security (UNDP, 1994); economic security, food security, health security, environmental security. Even for the last three categories – personal, community and political security – one can identify links to aspects of energy production or consumption. The links between energy and the various dimensions of human security can be divided into two major categories; the links between energy access/consumption and human well-being/security and between certain types of energy production and consumption and human 'ill-being'/insecurity. The following sections explore these two categories of links in more detail.

**Energy, Human Well-Being and Security**

Whenever anything happens, anywhere, anytime, it is because energy is transformed from one form to another. Not surprisingly, the ability to harness energy for human use has been a vital element of human security for millennia. From learning to use fire for food preparation, heating and protection, to the development of technologies to capture the energy of wind, animals, etc. energy has been used by humans to transform individual and societal life (Smil, 1994). This transformation has often meant increasing individual well-being and security. This development has progressed through the industrial revolution that was very much enabled by the ability to use cheap fossil fuels. All modern technologies that make life easier, healthier and more secure require access to modern energy; from fridges for vaccines to kitchen appliances and cars. While it should be acknowledged that the production and use of energy also has its costs (see discussion below), it is clear that modern energy has had a significant positive impact on many aspects of human well-being and security.

This development of increased well-being and human security has, however, in a global context been very unequal both between and particularly within countries. For example, around 2.7 billion people lack access to modern energy services for cooking and heating and around 1.94 million people in developing countries die each year from respiratory infections as a result of exposure to pollution from solid fuels, see Bhattacharyya (Chapter 19, this volume). The unequal energy use situation is most vividly illustrated by the picture of the Earth at night. Many regions such as urban centres in North America and Europe are bathed in light while others, such as much of Africa, are almost entirely dark.

The links between access to modern energy and human well-being and security are clear and convincing. However, as energy has been so absent in the literature on human security it is also relevant to briefly describe the links to the specific dimensions of security in more detail (the links particularly to economic and health security are discussed in more detail in Chapter 19 of this volume).

- *Economic security* relates to the ability to make livelihoods on a sustainable basis. On a macro scale this is a major reason for governments to pursue secure energy supplies to ensure economic development and indeed national security. Economic development means income security for a larger portion of the population. On a micro scale when the focus is poverty reduction there is clear evidence that access to modern energy services improves livelihoods of the poor in developing countries. Access to cleaner energy options (than traditional fuels for example) can improve working conditions and can provide new job opportunities in sustainable energy sectors (GEA, 2012).
- *Food security* at the household level can also be linked to access to modern energy services. Most staple foods need to be cooked before consumption thus access to heat energy is essential. Liberating the time (and indeed human energy) that women spend collecting fire wood, pump water and mill grains by hand through access to energy services would give more time for work on the productivity of their land (or on other income-generating activities) (GEA, 2012). Modern and affordable energy used for irrigation directly strengthens the food security of the poor (GEA, 2012). Another area where access to modern energy services is linked to improving food security is the reduction of post-harvest losses due to inadequate facilities for harvest, storage and transport (GEA, 2012). Furthermore, on a global scale intensive large scale agriculture increases the availability of food for international trade and this type of agriculture is highly dependent on energy in fertiliser production, irrigation, harvesting and transport (GEA, 2012).
- *Health security* is linked to energy access in several ways. The most prominent link is between the particulates from the combustion of poor-quality cooking fuels and respiratory disorders. The illness and death of millions of primarily women and children every year are directly related to their exposure to the smoke from cooking fuels, see Bhattacharyya (Chapter 19, this volume). There is however also a very strong link between energy and health security in the running of modern health care facilities that require access to electricity and affordable fuels (GEA, 2012).[4]

**Energy, Ill-Being and Insecurity**

The ability to harness an expanding list of energy types through technological innovations is a twin-edged sword. On the one hand it has delivered huge benefits. On the other hand, it has a led to significant insecurity and ill-being for many human beings and indeed states.

- *Environmental security* comes when human beings have a healthy physical environment. Modern energy production and consumption links to human insecurity through environmental degradation and climate change that affects economic, food, health or even personal, community and political security of people. These links can either be manifested in insecurity emerging slowly over time or come as a result of sudden disasters and they can occur within or across state borders.

   The extraction, production and use of many modern forms of energy degrades or even destroys the environment locally or globally. The mining of coal or drilling of oil can create health hazards for workers involved in the process or the population living close by (consider for example oil extraction in the Niger Delta). The use of nuclear power for generating energy can cause disasters with major implications for human security (e.g. Chernobyl and Fukushima). The burning of fossil fuels is a major source of local and regional air pollution with associated health implications in cities around the world.[5] The burning of fossil fuels is also, as mentioned above, the major source of greenhouse gases causing climate change – accounting for around 60 per cent of total global greenhouse emissions.[6] Climate change in turn has a number of implications for human (in)security some of which are related to health such as changed patterns of infectious diseases, others linked to food security (from changed weather patterns and agricultural potential) or physical safety such as extreme weather events (storms, floods and over longer time horizons sea-level rise) which will contribute to migration and associated income insecurity.[7]

- *Personal security* is defined as people's security from physical violence (UNDP, 1994). The most obvious contribution of energy to insecurity in this sense is the enabling of ever more powerful technologies of war including nuclear weapons. Furthermore, after the nationalisation of many countries' fossil fuel reserves in the 1970s some countries in the west started to look at disruptions of energy supply as a national security issue that should be counteracted by military strategy (Peters, 2004). This has obvious implications for human security in relation to physical violence. Energy sources are generally seen as a cause of conflict in international relations (Peters and Westphal, Chapter 5, this volume).[8] Podobnik (2002) argues that the unequal distribution of energy consumption is a risk for future conflicts, and the increasing scarcity of particularly oil may exacerbate this. The scramble for new oil extraction areas for example in the Arctic could be seen as an indication of this.

Another contribution of energy to insecurity is through the so called 'resource curse'. This concept implies that countries with considerable natural resource assets (such as oil), are more prone to internal conflict thus linked to both personal and political security, see for example de Soysa (2002) and Chapter 5 of the GEA (2012).

- *Community security* comes from the fact that '[m]ost people derive security from their membership in a group – a family, a community, an organisation, a racial or ethnic group that can provide a cultural identity and a reassuring set of values' (UNDP, 1994:31). Any impact on communities' ability to indeed live as a community is thus a cause of human insecurity. Among such impacts related to energy production are displacements of whole communities for the building of hydropower dams or drilling of oil on land. The expected forced migrations of even whole countries (low lying islands) from climate changed induced sea-level rise will be a result of consumption of fossil fuels (and other greenhouse gas emitting activities). On the other hand, one can argue that in some communities and cultures excessive consumption of energy is a defining element of their identity, such as communities built on the dependence of car and airplane transport.
- *Political security*, finally, is about people's ability to live in countries where their basic human rights are protected, including their civil and political rights. Here the link to energy production and consumption is tenuous but de Soysa (2002:30) argues that the research community 'needs denser analyses of how resource abundance is associated with conflict through what some observers characterise as the 'spoils politics' of clientelism, corruption, and extrainstitutional "governance," a pervasive feature of politics in resource-abundant countries, particularly in Africa'. On the other hand, the access to electricity is a pre-requisite for the ability to use cell-phones and the internet, and the associated social media that may play a role in supporting more open and democratic societies.

**Deep Energy Security and the Sustainability of Energy Systems**

Clearly, current notions of energy security are not delivering human security in all its dimensions. What is needed is a new approach to energy security that captures the notion that all human beings, including future generations, are entitled to benefit from modern energy services and at the same time be protected from their negative side-effects. In this context, we offer the notion of 'deep energy security' (DES). Deep energy security is energy security that contributes to human security over space (from the local to global) and time (that is, now and for future generations). In this way, deep energy security embraces both a long-term (multi-generational), equitable perspective and holistic approach to the energy system. There are clear parallels between the concept of DES and the sustainability of the energy system. Indeed, we argue in this chapter that DES and the sustainability of the energy system are intertwined. Furthermore, this interconnection could be considered as a nested dependency – where the sustainability of the energy system is a necessary, but not sufficient, condition for DES, and DES is in turn a necessary, but not sufficient, condition for human security. Despite their close relation the concepts of a sustainable energy system and deep energy security are distinct. An energy system with occasional disturbances may still be considered 'sustainable' but not deeply energy secure. The DES concept adds a focus on immediate individual well-being and security while sustainability is a crucial aspect for ensuring DES also for future generations.

At this point, it is important to point out that we take a broad definition of the energy system to include (Karlsson-Vinkhuyzen et al., 2012:13):

- the physical infrastructure needed to extract, transport, transform and use energy;
- the physical impacts on the environment and people of energy extraction, transport, transformation and use;
- the social institutions (such as international agencies, governments and the regulatory frameworks, markets and civil society groups) designed to support the flow of energy services; and
- the individual actors involved in using energy services.

Deep energy security goes beyond the concept of 'comprehensive' energy security. Comprehensive security was a concept developed in Japan in the 1970s and 1980s that was focused still on the security of the state but in a broader sense encompassed both military and economic security (Gasper, 2005).

Deep energy security transcends national boundaries. It is at once global and individual. That is, it accommodates the moral community as a global one but maintains the individual (within a community and nation and dependent on the environment) as the focus of security. An understanding of DES is also useful in that it can help with priority setting. That is, Gasper (2005:241) states, the 'human security discourse is a discourse for getting priority, and priorities, in national and international policy.' Applying this line of logic to DES, it is possible that the proposed concept of DES could serve a similar purpose in the domain of energy policy.

Defining DES is a useful first step in the process of reforming our notion of energy security. In the next section, we take the notion of DES one step further and ask how can deep energy security be achieved – and who, or what, should provide such DES?

## PUBLIC GOODS AND THE SUSTAINABILITY OF THE ENERGY SYSTEM

The section above noted that DES is dependent on sustainability of the energy system. If this is the case, then how can the sustainability of the energy system be provided? We draw on our work in Karlsson-Vinkhuyzen et al. (2012) to argue that a key approach to answering this question is to understand the concepts of private and public goods found in economics literature and to look at the sustainability of the global energy system as a global public good.

### Public and Private Goods

In economics, a good (or service) is considered private or public depending on two dimensions: excludability and rivalry. For a good to be purely 'public' there can be no exclusion of those who refuse to pay for the good or service to enjoy the benefits (non-excludability). A purely public good also requires no rivalry among the beneficiaries of the good or service (non-rivalry). In other words 'if a public good exists . . . anyone can use it regardless of who pays for it' (Daly and Farley, 2004:169).

These concepts deserve attention as they influence opinions as to whether the good should be considered the domain of the public or private sector actors. This is important because wrongly assigning a good or service as private or public can lead to under-provision. And under-provision of an aspect of a sustainable energy system makes DES and thus human insecurity impossible to attain.

A good or service is *excludable* if its 'ownership allows the owner to use it while simultaneously denying others the privilege' (Daly and Farley, 2004:73). It is worth noting that nothing is inherently excludable – policies or social institutions are required to make any good or service excludable (Kaul et al., 2003). For example, governments can decide to privatise what was previously provided freely by public institutions (such as water supply, and exclude people from the service through the application of water-use tariffs) or bring into public ownership what was previously operated privately (such as nationalising oil companies in Iran in 1951 and India in the 1970s for purported public benefits).

On the other hand, some goods or services are inherently non-excludable as a physical characteristic. Examples of such non-excludable goods or services include flood regulation. In the absence of an institution or technology being able to enforce exclusion, these are known as non-excludable goods or services. A good or service is 'rival in consumption' when one person's use reduces the amount available for everyone else. For example, my use of electricity in my home reduces the amount available for my neighbours. Conversely, a non-rival good or service is where its use by one person does not impact on another's use – my reliance on the climate regulation eco-system service to ensure that my London home is not destroyed by a hurricane does not decrease the amount of climate regulation available for my neighbours. In the context of energy, some of the renewable energy sources are virtually non-rival, such as sunlight, and, to a certain extent, wind and wave energy. It is important to note that 'rivaleness is a physical characteristic of a good or service and is not affected by human institutions' (Daly and Farley, 2004:73). Furthermore, non-rivalry is different from abundance. For example, seats at a football arena are rival; if one person occupies a seat, another cannot at the same time. However, if not all seats are filled there is no competition for use and they can be regarded as abundant. We can limit access to a resource in order to keep the resource abundant, but it will always be rival (Karlsson-Vinkhuyzen et al., 2012).

The theories around public goods have been centred in the context of local or national governance, however, the concept is also applicable at a 'global' level. The global attribute lifts the perspective to goods which yield benefits for all countries, people and generations (Kaul et al., 2003:10). Thus, it is possible to identify what are referred to as global public goods – an important concept in the discussion of addressing many global sustainable energy system issues (Karlsson-Vinkhuyzen et al., 2012).

## PUBLIC GOODS, THE PROVISION OF ENERGY SYSTEM SUSTAINABILITY AND HUMAN SECURITY

Economic literature is clear that public goods will not be efficiently (in a Pareto sense) provided by the market (Stiglitz, 2000). In this case, it is argued that there is a clear role for governments in delivering public goods and services. The benefits flowing from the sustainability of the energy system (whether local or global) such as reduced rate of depletion of natural capital or reduced impact on the climate system, can be considered public goods – and in some cases global public goods. This is because it would be difficult to *exclude* anyone from those benefits. Furthermore, my enjoyment of such benefits do not reduce (*rival*) another's ability to take advantage of those benefits. This means that we can look at the sustainability of the energy system at any scale as a public good and particularly at the sustainability of the global energy system as a global public good. It is important to note that it is only possible to look at energy (which in its consumption is both rival and excludable) as a public good by adopting a system perspective. The sustainability of the system (which includes environmental, social and economic sustainability dimensions) is analytically possible to approach as a public good (Karlsson-Vinkhuyzen et al., 2012). This analytical approach makes it very easy to explain why this public good, and DES that relies on it, is most effectively provided by public institutions.

Unfortunately, public goods are usually underprovided as a result of the collective action problem which their publicness generates – particularly in the case of global public goods (Conceição, 2003). This is clearly also the case for a sustainable global energy system. At lower levels of governance it is increasingly common for governments to take the main responsibility for the provision of the public good elements of a sustainable energy system. For example, many local governments are putting in place policies to mitigate the air-quality impact of particulate emissions from heating-fuel combustion (linked to environmental and health security), and most national governments are implementing active energy efficiency and renewable energy policy programmes. There is even increasing collaboration on energy in mostly regional intergovernmental contexts such as IEA, ASEAN, APEC and the EU. But these attempts are still too narrowly focused to be able to address the broad sustainability concerns of the energy system. All together the efforts at local, national and regional levels fall far short of what it takes to build up a sustainable energy system, particularly globally.

In the global governance context the provision of elements of a sustainable energy system is, of course, much more challenging where all joint action is based on voluntary cooperation among close to 200 states with diverse interests. Having established that the provision of the sustainability of the global energy system also requires the contribution of public sector providers, in the next section we consider the governance issues of 'what' and 'how much' should be provided at the global level.

## GLOBAL ENERGY GOVERNANCE AND DEEP ENERGY SECURITY

The switch from state-centric energy security to DES implies a shift in the means and mechanisms that are best suited to secure humanity and make 'people free from fear and want and indignity' (Gasper, 2005:240). Such a switch also raises the key governance issues of what and how much of a sustainable energy system should be provided at the global level if we see such a system as a prerequisite for achieving DES.

**Global Level Provision – What and How Much – In Theory**

Global public goods do not necessarily have to be provided by global (intergovernmental or other) actors. At a general level the provision of a global public good requires a multilayered approach to governance where measures by diverse actors across governance levels aim for a common objective. If actors at local, national and regional level, across the globe, provided substantial contributions to a global sustainable energy system then there may not be much need for action at global level. This is, however, far from the case and we argue there is considerable room for global energy governance. The concept of governance can be seen to imply that there are actors who are taking cooperative action for their common affairs (Commission on Global Governance, 1995). Using this definition, 'global energy governance' would encompass those measures that have as objective to manage energy as a common affair in the international community (Karlsson-Vinkhuyzen, 2010).

Barrett's (2007) categorisation of GPGs linked to their provision needs illustrates why global energy governance is needed. He divides GPGs into four categories:

- single best-effort GPGs where the provision depends on the single best (unilateral or collective) effort (knowledge of cures for diseases);
- weakest link GPGs where provision depends on the weakest individual effort (control of pandemics);
- aggregate effort GPGs where it depends on the combined effort of all countries and actors (for example, the Kyoto Protocol's attempts to address climate change); and
- mutual restraint GPGs where states agree to avoid doing certain things (for example, the Montreal Protocol's efforts to phase out ozone-depleting substances).

The multiple actions required to deliver a global sustainable energy system would fall under the aggregate effort, which as pointed out above could in theory be reached by aggregate efforts at sub global levels. One can of course imagine single countries putting a lot of resources into e.g. fusion research and if that was successful some would argue that the energy provision of the future would be solved. This scenario is unlikely, however, and it would still require an enormous aggregate effort to change all the infrastructure to fit a new type of energy resource. One could also imagine that some would consider nuclear energy as coming with such risks that it should not be part of a sustainable energy system. In that case there would also be need for mutual restraint among countries to avoid building more nuclear power plants.

Having established the need for some level of global energy governance, the question we raise here is *how much* such governance is needed. This can be answered based on various allocation criteria. The authors applied the subsidiarity principle with its associated criteria of effectiveness and necessity used within the European Union to answer this question and concluded that global provision (through collaboration) is (Karlsson-Vinkhuyzen et al., 2012:14):

1) effective
    - when addressing GPG dimensions to policies themselves which are unlikely to be addressed by individual countries or the market such as knowledge and norms promoting sustainable energy;
    - when it aims to strengthen the coherence of the international community's governance (coordinating *ad hoc* efforts and avoiding overlaps).
2) necessary
    - when lower levels of governance do not have the capacity or will to take action to promote sustainable energy; or
    - when global institutions (both norms and organisations) are contributing to preserving a fossil-fuel based unsustainable energy system.

If these two criteria were applied we could imagine a range of initiatives in global energy governance. These could include: collaborative research projects through pooling of research resources, extensive knowledge sharing mechanisms; international norms (including treaties) setting standards for production and consumption parameters related to sustainability; ambitious coordination mechanisms among existing intergovernmental organisations with activities supporting energy for sustainable development or even setting up new organisations for this purpose; large-scale capacity building and awareness raising programmes for countries who are doing very little in the field; and radical reorientation of international trade and financial institutions from implicit and explicit support to fossil fuels towards support for building a sustainable energy system globally and ensuring deep energy security within and across states. This is the picture in theory if the principle of subsidiarity and its effectiveness and necessity criteria were applied. In practice global energy governance looks quite different.

**Global Level Provision – What and How Much – In Practice**

In practice global energy governance, defined as supporting a common objective in the form of a sustainable energy system, is fragmented and dispersed without a strong institutional framework whether in the form of norms or organisations set up for the purpose (Karlsson-Vinkhuyzen, 2010). There are certainly many initiatives and activities by international organisations, by bilateral donor agencies, private-public partnerships and non-governmental organisations that support a sustainable energy system. Descriptions, and to some degree analysis, of these can be found in several reports from the UN system, research reports and papers (Ad Hoc Inter-Agency Task Force on Energy, 2002; Florini and Sovacool, 2011; Gupta and Ivanova, 2009; Karlsson-Vinkhuyzen et al., 2012; Lesage et al., 2009, 2010; Modi et al., 2005; Steiner et al., 2006; UN-Energy, 2006). Compared to the need, however, these activities are still meagre.

Even more meagre, or almost non-existent, are international norms developed explicitly to support a sustainable energy system. The few efforts that have been made to develop soft (non-legally binding) norms have either been very vague and non-committing, see for example the text on energy in the outcome of the United Nations Conference on Environment and Development in 1992 (UNCED, 1993), the World Summit on Sustainable Development in 2002 (United Nations, 2002), and the United Nations Conference on Sustainable Development in 2012 (United Nations General Assembly, 2012).[9] The Commission on Sustainable Development (CSD), a commission under the United Nations Economic and Social Council set up in 1993, has been the only forum where 'energy for sustainable development' has been discussed at any length in an intergovernmental, high-level context. The CSD discussed energy in 2001 when it adopted an outcome text that again was quite general (Commission on Sustainable Development, 2001) and again in 2006/7 when no agreement for an outcome text could be reached due to large disagreements on many of the themes, but particularly on the role of the global level and the UN in following up on any decisions adopted (Karlsson-Vinkhuyzen, 2010). This failure in outcome of the CSD dampened the enthusiasm considerably among countries who had wanted to see more of a role for the UN on sustainable energy. Nonetheless, there were some steps forward after this, particularly with the UN General Secretary's launching of the Sustainable Energy for All initiative and the announcement of 2012 as the UN year for this theme and the UN General Assembly declaring 2014–2024 as the Decade of Sustainable Energy for All.[10]

The Sustainable Energy for All initiative brings together a number of UN organisations as well as international public-private partnerships, business and other organisations. In terms of organisations, however, there is no organisation within the UN system whose exclusive mission is energy, except the International Atomic Energy Agency whose focus is exclusively on nuclear energy. Many UN organisations work with energy but often according to their own mandate and the degree of coordination and cooperation across the UN system is limited (Karlsson-Vinkhuyzen, 2010). The establishment of UN-Energy in 2004 as a permanent inter-agency mechanism was an effort to address this gap in coordination. This non-organisation (it has no staff, no budget etc.) is open to all organisations in the UN system, including the Bretton Woods institutions. Representatives of its members meet regularly, share information, stimulate cooperation and occasionally produce publications, for example with overviews of energy activities in the UN system (UN-Energy, 2006). The future of UN-Energy is, however, uncertain and this mechanism is searching for its role in the governance landscape (UN-Energy, 2010).

Outside the UN system the International Energy Agency has since the late 1990s started to expand on its original mandate to provide security in access to fossil fuels such as oil, and has dedicated efforts to the climate change dimension of energy including energy efficiency and renewable energy (Van de Graaf and Lesage, 2009). In 2009 the first intergovernmental organisation dedicated to renewable energy, the International Renewable Energy Agency (IRENA), was established indicating a strengthened political will for international collaboration in this field. The organisation has in short time attracted 101 states as members and an additional 57 states are signatories pending ratification.[11]

In summary, governments have – with a few exceptions – been reluctant to develop international norms and institutionalise cooperation around energy at the global level, particularly in the only universal multilateral forum, the UN (Karlsson-Vinkhuyzen, 2010). In parallel, the academic and policy literature on energy has only recently started to use the concept of global energy governance, and analyse its possible content, role and main actors (Bradbrook and Wahnschafft, 2005; Florini and Sovacool, 2011; Goldthau and Sovacool, 2012; Goldthau and Witte, 2010; Karlsson-Vinkhuyzen, 2010; Karlsson-Vinkhuyzen et al., 2012; Lesage et al., 2009, 2010; Steiner et al., 2006).

Not even assessments with an explicit global scope such as the World Energy Assessment (United Nations Development Programme et al., 2000), and the energy policy sections in the Fourth Assessment Report of the IPCC (IPCC, 2007a) discussed the role of global energy governance. Neither does the Global Energy Assessment do this systematically, but it does consider the role of international actors and concludes, for example, in its summary for policy makers that there should be a policy focus in the near term on enhancing international cooperation in energy technology research and development and technology standards (GEA, 2012).

## CONCLUSION

The human security concept and discourse long remained silent on energy despite the very strong links between various dimensions of energy production and consumption and human security and insecurity, links which lately are more clearly exposed in debates about for example the Millennium Development Goals. The adoption of the concept of DES as a necessary condition for providing human security provides an opportunity to approach energy as an ethical and even moral issue. The concept moves the perspective of energy security from states and the present to individuals and the future, and it highlights the need to expand the spheres of concern for states from their own citizens to humanity as a whole including future generations. This normative stance is perhaps radical in an energy policy context, but we see this as a fundamental first step in reorienting particularly the governments of the world towards the building of a global sustainable energy system. By approaching a sustainable energy system as a global public good it becomes clear what central role governments have in its provision particularly through global cooperation, a role of which governments are seemingly unaware, or unwilling to admit as indicated by the very humble degree of global energy governance. The current socio-economic-political system has a long way to go to achieve 'Deep energy security' but we argue that a useful first step is some introspective reflections, among individuals and governments, on the values and thus objectives around energy production and consumption that we see as fundamental for sustainable development and human security in the coming years and centuries.

## NOTES

* The contents of the chapter are the views of the authors and not those of the institutions for which they work.
1. See http://epa.gov/climatechange/ghgemissions/global.html
2. The idea of individual security is a liberal thought from the Enlightenment and as it has been treated both as a unique and a collective good it is more difficult to determine

where the responsibility for the ensuring individual security lies (Liotta, 2002) The idea of global citizenship implies the equal value of all human beings and thus concern for the implications of policies far beyond state borders. See e.g. Dower and Williams (2002) for an elaboration of the concept.
3. The exception is the reference to one of the benefits of foreign direct investment through multilateral corporations as being their ability to introduce more and cleaner energy.
4. For an elaboration on the links between health and energy see Bhattacharyya (Chapter 19, this volume) and Chapter 4 of the Global Energy Assessment (GEA, 2012).
5. The GEA (2012) for example, estimates that 2.75 million premature deaths occur annually due to outdoor air pollution from energy systems.
6. See http://epa.gov/climatechange/ghgemissions/global.html.
7. The most comprehensive compilation of climate changes impacts can be found in the Fourth Assessment Report of the Intergovernmental Panel on Climate Change (IPCC, 2007b).
8. Peters and Westphal (Chapter 5, this volume) and Olaniyi (Chapter 4, this volume) discuss the possible conflict areas around energy in more detail, the former focusing on interstate conflicts, the latter on intrastate conflicts.
9. For a discussion and overview of the text and outcome on energy in some of these Summits see Najam and Cleveland (2003) and Spalding-Fecher et al. (2005).
10. See http://www.sustainableenergyforall.org/about-us (accessed 23 February 2013).
11. This figure is as of 5 November 2012, see http://www.irena.org/ (accessed 23 February 2013).

# REFERENCES

Ad Hoc Inter-Agency Task Force on Energy (2002), 'Consolidated Report of Energy Activities', United Nations, available: www.un.org/esa/sustdev/iaenr.htm (Accessed 26 September 2005).
AGECC (2010), 'Energy for a Sustainable Future', The Secretary-General's Advisory Group on Energy and Climate Change (AGECC), Summary Report and Recommendations, New York: United Nations.
Barrett, S. (2007), *Why Cooperate? The Incentive to Supply Global Public Goods*, Oxford: Oxford University Press.
Bhattacharyya, S. C. (forthcoming). 'Energy Poverty, Health and Welfare Issues'. *In*: H. Dyer & J. Trombetta (eds.) *International Handbook of Energy Security* Cheltenham, UK and Northampton, MA, USA: Edward Elgar Publishing, Chapter 19.
Bradbrook, A. J. & R. D. Wahnschafft (2005). 'International Law and Global Sustainable Energy Production and Consumption'. *In*: A. J. Bradbrook, R. Lyster, R. L. Ottinger & W. Xi (eds.) *The Law of Energy for Sustainable Development*, Cambridge: Cambridge University Press.
Commission on Global Governance (1995), *Our Global Neighbourhood: The Report of the Commission on Global Governance*, Oxford: Oxford University Press.
Commission on Human Security (2003), *Human Security Now*, New York.
Commission on Sustainable Development (2001), 'Report on the Ninth Session', 5 May 2000 and 16–27 April 2001, New York: United Nations.
Conceição, P. (2003). 'Assessing the Provision Status of Global Public Goods'. *In*: I. Kaul, P. Conceição, K. Le Goulven & R. U. Mendoza (eds) *Providing Global Public Goods. Managing Globalization*, Oxford: Oxford University Press.
Daly, H. E. & J. Farley (2004), *Ecological Economics: Principles and Applications*, Washington D. C., Island Press.
de Soysa, I. (2002). 'Ecoviolence: Shrinking Pie or Honey Pot?', *Global Environmental Politics*, **2** (4), 2–34.

Dower, N. & J. Williams (eds) (2002), *Global Citizenship. A Critical Introduction*, New York: Routledge.
ElBaradei, M. (2008), 'A Global Agency is Needed for the Energy Crisis', *Financial Times*, 23 July.
Florini, A. & B. K. Sovacool (2011). 'Bridging the Gaps in Global Energy Governance', *Global Governance*, **17** (1), 57–74.
Gasper, D. (2005). 'Securing Humanity: Situating "Human Security" as Concept and Discourse', *Journal of Human Development*, **6** (2), 221–245.
GEA (2012), *Global Energy Assessment – Toward a Sustainable Future*, Cambridge: Cambridge University Press.
Goldthau, A. & J. M. Witte (eds) (2010), *Global Energy Governance. The New Rules of the Game*, Berlin: Global Public Policy Institute and Brookings Institution Press.
Goldthau, A. & B. K. Sovacool (2012), 'The Uniqueness of the Energy Security, Justice, and Governance Problem', *Energy Policy*, **41**, 232–240.
Gupta, J. & A. Ivanova (2009). 'Global Energy Efficiency Governance in the Context of Climate Politics', *Energy Efficiency*, **2**, 339–352.
IPCC (ed.) (2007a), 'Climate Change 2007: Mitigation of Climate Change. Contribution of Working Group III to the Fourth Assessment Report of the Intergovernmental Panel on Climate Change', Cambridge, United Kingdom and New York, NY, USA.
IPCC (2007b), 'Intergovernmental Panel on Climate Change Fourth Assessment Report Climate Change 2007: Synthesis Report. Geneva: Intergovernmental Panel on Climate Change.'
Jäger, J. & S. Cornell (2011), *The Planet in 2050: the Lund Discourse of the Future*, London: Routledge.
Karlsson-Vinkhuyzen, S. I. (2010), 'The United Nations and Global Energy Governance: Past Challenges, Future Choices', *Global Change, Peace and Security*, **22** (2), 175–195.
Karlsson-Vinkhuyzen, S. I., N. Jollands & L. Staudt (2012), 'Global Governance for Sustainable Energy: The contribution of a Global Public Goods Approach', *Ecological Economics*, **83**, 11–18.
Kaul, I., P. Conceição, K. Le Goulven & R. U. Mendoza (2003), 'Why Do Global Public Goods Matter Today?'. *In*: I. Kaul, P. Conceição, K. Le Goulven & R. U. Mendoza (eds) *Providing Global Public Goods. Managing Globalization*, Oxford: Oxford University Press.
Lesage, D., T. Van de Graaf & K. Westphal (2009), 'The G8's Role in Global Energy Governance Since the 2005 Gleneagles Summit', *Global Governance*, **15** (2), 259–277.
Lesage, D., T. Van de Graaf & K. Westphal (2010), *Global Energy Governance in a Multipolar World*, Farnham: Ashgate.
Liotta, P. H. (2002). 'Boomerang Effect: the Convergence of National and Human Security', *Security Dialogue*, **33**, 473–488.
Modi, V., S. McDade, D. Lallement & J. Saghir (2005), 'Energy Services for the Millennium Development Goals', New York: Millennium Project, UNDP, UNEP, the World Bank, ESAMP.
Najam, A. & C. J. Cleveland (2003), 'Energy and Sustainable Development at Global Environmental Summits: An Evolving Agenda', *Environment, Development and Sustainability*, **5**, 117–138.
Nilsson, M., C. Heaps, Å. Persson, M. Carson, S. Pachauri, M. Kok, M. Olsson, I. Rehman, R. Schaeffer, D. Wood, D. v. Vuuren, K. Riahi, B. Americano & Y. Mulugetta (2012), *Energy for a Shared Development Agenda: Global Scenarios and Governance Implications*, Stockholm: Stockholm Environment Institute.
Olaniyi, J. (forthcoming). 'Resource Conflicts: Energy Worth Fighting For'. *In*: H. Dyer & J. Trombetta (eds) *International Handbook of Energy Security* Cheltenham, UK and Northampton, MA, USA: Edward Elgar Publishing, Chapter 4.
Owen, T. (2008), 'The Critique that Doesn't Bite: A Response to David Chandler's "Human Secrity: The Dog That Didn't Bark"', *Security Dialogue*, **39** (4), 445–453.
Peters, S. (2004), 'Coercive Western Energy Security Strategies: 'Resource Wars' as a New Threat to Global Security', *Geopolitics*, **9** (1), 187–212.

Peters, S. & K. Westphal (forthcoming). 'Global Energy Supply Scale, Perception and the Return to Geopolitics'. *In*: H. Dyer & J. Trombetta (eds) *International Handbook of Energy Security* Cheltenham, UK and Northampton, MA, USA: Edward Elgar Publishing, Chapter 5.

Podobnik, B. (2002). 'Global Energy Inequalities: Exploring the Long-Term Implications', *Journal of World-Systems Research*, **VIII** (2), 252–274.

Smil, V. (1994), *Energy in World History*, Boulder, Colorado, Westview Press.

Spalding-Fecher, R., H. Winkler & S. Mwakasonda (2005), 'Energy and the World Summit on Sustainable Development: What Next?', *Energy Policy*, **33**, 99–112.

Steiner, A., T. Wälde, A. Bradbrook & F. Schutyser (2006), 'International Institutional Arrangements in Support of Renewable Energy'. *In*: D. Aßmann, U. Laumanns & D. Uh (eds) *Renewable Energy. A Global Review of Technologies, Policies and Markets*, London: Earthscan.

UNCED (1993), 'Report of the United Nations Conference on Environment and Development', Rio de Janeiro, 3–14 June 1992. New York: United Nations.

UNDP (1994), 'Human Development Report', New York: United Nations Development Programme.

UN-Energy (2006), 'Energy in the United Nations: An Overview of UN-Energy Activities', New York: United Nations.

UN-Energy (2010), 'UN-Energy: Looking to the Future', New York: United Nations.

United Nations (2002), 'Report of the World Summit on Sustainable Development', Johannesburg, South Africa, 26 August–4 September, New York: United Nations.

United Nations Development Programme, United Nations Department of Economic and Social Affairs & World Energy Council (eds) (2000), *World Energy Assessment. Energy and the Challenge of Sustainability*, New York: UNDP.

United Nations General Assembly (2012), 'The Future We Want', Resolution Adopted by the General Assembly, 66/288, New York: United Nations.

Van de Graaf, T. & D. Lesage (2009), 'The International Energy Agency After 35 Years: Reform Needs and Institutional Adaptability', *The Review of International Organizations*, **4**, 293–317.

# PART VII

# CONCLUSIONS

# 24. The political economy of energy security
## Hugh Dyer and Maria Julia Trombetta

The International Energy Agency's (2012) World Energy Outlook devotes an entire section to a focus on energy efficiency 'blueprints', though it struggles to extend this concern to reducing overall demand (and attendant dismal scenarios for energy policy and climate change), as opposed to meeting that growing demand (and more equitable supply) with greater economic efficiency. The conventional assumption seems to be that energy supply should be stabilized, and that shortages endanger everything from individual livelihoods to the global political economy. While these assumptions are reasonable in the context of ongoing business, the overall objective of maintaining current practices through short-term management of supply and price is probably quite unreasonable given the extensive nature of the challenges indicated by the studies in this volume. The arguments and insights provided by the various chapters have underwritten the need to broaden, deepen and transform the understanding of energy security.

This also suggests a tension between a long term perspective on energy security and a short term one. While the former involves political debates, and transformation of the governance of energy, the latter makes evident the implications of the neo-liberal economic approach. The energy security debate is characterized by different voices, which are influenced by the different disciplinary perspectives of the actors involved in the debate. As Cherp and Jewell (Chapter 8) suggest, the sovereign perspective calls for state intervention in assuring, for instance, long-term contracts, such as in the case of gas, and providing stability and order to ensure the smooth functioning of the oil market. National energy champions are protected to ensure the provision of energy services. But this national security perspective, suggesting a zero sum approach, is itself a cause of insecurity (as Karlsson and Jollands have pointed out, Chapter 23) increasing the potential for conflict, not only at global but also at local level (see Alabi, Chapter 4). This suggests the limitations of the national security discourse in providing energy security. On the other hand, arguments for liberalization are often based on the assumption that the market itself will provide solutions, often ignoring the implications of the short term perspective imposed by the market, and the possibility of market failure, including climate change. The approach of the IEA reflects this perspective with its commitment to supporting the market and integrating it with mechanisms to deal with crises and emergencies. In the face of such limited perspectives on energy security, achieving a deeper energy security linked to sustainable human development is a significant challenge. What is required is an integration of the different perspectives, and in this respect reflections on energy security (and, in this context, who deserves to be protected and by what means) may help to address the complexity of a reshaping energy order.

This volume has provided an overview of the existing debates with three aims: to provide an account of the multiplicity of discourses and meanings of energy security and contextualizing them; to use the insights from security studies debates to understand the implications of framing an issue as a security issue; to shed light on the need to integrate different perspectives, overcoming, for instance, the divide between realist/geopolitical perspectives and neoliberal ones, while taking into account environmental and human security considerations.

The initial part on energy security issues outlined the complexity of the perspectives involved, the tensions between them, and the difficulties of reconciling them. Wood examines the meaning of human security in respect of energy issues, where presumed energy imperatives test the quality and integrity of the liberal democracy that defines most post-industrial societies. An end to authoritarianism does not mean support for a green agenda, nor is it guaranteed that newly emancipated electorates would demand what environmental lobbies in established liberal democracies call for. Neither would the emergence of a 'global demos' necessarily result in universal green attitudes and policy: trade-offs will always be required. Political machinations in the international arena confront national self-understandings or self-images. Some liberal democracies have reached a *modus vivendi* with illiberal energy suppliers, and a few have with illiberal customers, which suggests that ready access to cheap energy is the current approach to energy security. On the other side of the energy-climate coin, pressure on human rights and peace-building is set aside in order to get the maximum number of countries signed up to internationally-agreed climate change commitments; some even argue that democratization is a problem.

Brancucci, Pearson and Zeniewski suggest that one's criteria in defining energy security often informs policy outlook. They note that physical interconnections imply both solutions and challenges to security of energy supply. This is particularly true for the European Union, where all three pillars of the EU energy policy (sustainability, competitiveness and security) converge around a common goal of cross-border infrastructure development. Although additional interconnections are often seen as beneficial to security, the accompanying increase in interdependence should be taken into account. Indeed, this chapter has introduced some of the internal challenges and vulnerabilities facing the European Union as it attempts to modernize and inter-connect some of the most complex energy systems on the globe. This entails uncertainties, and special attention has been paid to the problems for security of supply engendered by gas market liberalization, renewable energy deployment and the expansion of ICT systems controlling complex energy networks. These three issues also impact on the external dimension of EU energy security.

As the liberalization of the internal EU gas market challenges long-standing relationships with external suppliers, the problems arising from large-scale deployment of renewables also require external solutions, such as the import of electricity from solar panels in the Sahara desert. Thus, the interdependencies between external and internal dimensions will crucially affect the EU's overall energy security. Alabi noted that because of its significant role in fuelling modern industrial economies and military forces, oil has been the subject of domestic and international conflict, as actors seek to influence the direction of the market. He also noted the interrelationship between the energy security agenda of the Western world and the nature of conflict and underdevelopment in the oil and gas producing countries in the global south. He concluded that most so-called resource conflicts occur where groups or societies have been denied benefits from national resources.

Peters and Westphal emphasized that energy is about geography. Thus, they map hydrocarbon resources as a given underlying reality of producer – consumer and transit relations – and also observe that the landscape of energy is undergoing profound change. They show that it makes a difference how analysts describe and frame development in the markets. 'Energy security' is a question of how this goal is integrated into the global economy, pluralistic structures, good governance in the sector, as well as the perceptions of the tightness of the future supply market and the availability of power projections forces.

The remaining parts of the handbook reviewed the main issues from within different perspectives on energy security.

## SECURITY OF SUPPLY

This part addressed the different strategies to ensure energy of supply. The relevance of the national security perspective is quite apparent, even if many of the tools are shaped by a neo-liberal approach.

Thomson and Boey provide an overview of strategic energy reserves (SERs), a tool that reflects the neoliberal perspective of providing mechanisms to cope with short term crises, and an approach that has been largely endorsed by the IEA. The chapter shows that not all people believe that creating and maintaining SERs is useful. Some believe there is little point in having them because they are costly to build and maintain, and if a supply/price shock is not resolved within a short period of time, the stockpiling is for naught. Moreover, many regard energy supplies as fungible commodities, which challenges the contribution of SERs to price stability vs genuine energy security-related reasons. Hammes provides an update on how the traditional argument of ensuring security through diversification of sources and providers, which has been the mantra since Churchill's arguments that energy security rests on diversity and diversity only, now acquires new meaning as new technological solutions and perspectives emerge. Cherp and Jewell provide a useful account of the issues in measuring energy security, presented in historical context. The interest in measuring energy security results not only from its rising prominence but also from its increasing complexity, which requires relating energy security to a common 'yardstick'. The challenge of measuring energy security is not only to comprehend natural, technological, and economic complexities and uncertainties, but also to acknowledge that it has different meanings for different people. There is no single set of suitable metrics, but despite different choices about the definition of energy security, vital energy systems, key vulnerabilities, indicators and interpretation, all of this can be reflected in principles of an energy systems approach. Schott and Campbell addressed the structure of national energy strategies in the G8 countries and Norway, and evaluated different components of individual strategies. This contributes to the better understanding of the importance and meaning of national energy strategies, differences in the definition of and approach to energy security, and the compatibility of strategies with the emerging world energy order. Energy security issues for major industrialized nations are changing, and it is no longer merely a question of securing adequate supplies of energy to fuel an ever-growing manufacturing base; major industrialized nations are focusing more on becoming major energy exporters of conventional fossil fuels. Gaylord and Hancock indicated that commercial actors depend on reliable energy and are thus hesitant to invest in countries where security of supply is questionable. They focused on energy issues and related strategies of particular relevance to the developing world, which tend to emphasize increasing electricity access for the larger population and long-term economic development for the state. These primary concerns, along with limited financial and military resources for energy security, shape the strategies of developing countries. State strategies at the national level tend to focus on large infrastructure projects while most of the off-grid and alternative energy projects are sponsored by international agencies, NGOs, and community-level organizations with goals and agendas that coexist but can also be at odds with those of the national government. This situation is specific to developing countries, where infrastructure is not fully developed and integrated and the presence of the state throughout a country's territory can be uneven.

## SECURITY OF DEMAND

Romanova attempts to understand the goal of demand security by examining the source of producing countries' political objectives. She observes that energy security is an umbrella term, linking energy, economic growth and political power. Preoccupations of suppliers are not new, but have grown in scope, where demand security is a more recent concept compared to supply security. This represents the other side of the coin and, together with supply security, it forms global energy security. However, despite numerous efforts to establish mutually beneficial cooperation between producers and consumers, energy security remains a combination of liberal efforts to cooperate and realist attempts by both consumers and producers to decrease dependence on each other. It has not become an area of constructive, liberal cooperation, and producers will be increasingly more vulnerable and strategies to minimize their exposure will grow in sophistication. Bahgat offers the International Energy Forum as an example of cooperation between producers and consumers, as a major characteristic of energy is the mismatch between resources and demand. Fluctuation of energy prices plays a key role in the balance of payments, and a major theme of the energy security literature is the importance of diversification of energy mix and energy sources. The less dependent a country is on a single form of energy, the more secure it is. Similarly, the more producing regions there are around the world, the better. Any assessment of OPEC's role in managing global oil prices would be highly controversial, but there is an emerging consensus that stable prices at a reasonable level would serve both the producers' and consumers' interests. Belyi assesses opportunities for energy security governance in the context of the Energy Charter Treaty. A geopolitical, state-centric approach to energy security cannot ignore attempts at international energy governance, since growing interdependence between producing and consuming countries, an increasing number of cross-border investments as well as possible mismatches between transport capacity and supply obligations lead to a demand for predictability. This demonstrates the relevance of the ECT-based energy governance: an overarching framework for protection of investments, including environmental ones; the Most Favoured Nation principle in energy investment field, which could counterbalance the retaliatory reciprocity; a realistic framework of legally binding relations involving the EU, Russia and the transit states. International energy governance requires a certain level of acceptance by both producing and consuming states, while at the same time, the very understanding of energy security remains controversial.

## ENERGY, ENVIRONMENT, SECURITY

Vogler and Stephan consider energy security to be more than just another sectoral security area. This expansive re-definition, they argued, should alert us to the significance of 'that which is to be secured'. Energy and climate are not only materially intertwined, but also interdependent politically. International climate governance will not progress substantially without increased availability of practical and affordable energy technologies to enable climate-friendly economic development. Public goods analysis suggests that the regulatory development should still lead from climate to energy – rather than *vice versa* – because only this arrangement could ensure that long-term strategic foresight prevails over short-term pragmatism. Scheffran tests the links between energy use, climate change and migration. In order to build a secure, equitable and sustainable energy systems that harvests the benefits while minimizing risks and conflicts, principles, criteria and standards for sustainable peace must be implemented in legal norms, certification systems and monitoring processes. Mechanisms for desecuritization, mediation and conflict resolution can support this process, including stakeholder dialogues, participatory decision-making and arbitration. Then geographies of conflict over energy and climate change might be transformed into landscapes of cooperation. Caldés, Lechon and Linares review the environmental implications of a wide range of different energy production technologies showing results for different kinds of pollutant emissions, impacts as well as external costs calculations. In terms of GHG mitigation potential, some renewable technologies seem to attain robust results, while others such as biomass derived electricity and biofuels show a more variable range of results. Of the latter technologies, some aspects of concern are related to the associated indirect effects produced by a large scale deployment; when mitigation of other impacts and pollutants are included in the picture, some renewable technologies show higher potential than others. If policy makers want to promote a sustainable energy system that maximizes social welfare, environmental externalities of all energy technologies must be taken into account. In order to do so, it is necessary to properly identify, quantify and later internalize the external costs in the price of energy. Paskal notes that much energy infrastructure lies in areas that are becoming increasingly physically unstable due to changes in the environment. Thus there are concerns about both older installations not being designed for new conditions and new installations not integrating change into their planning: either of those situations could result in marked decreases in energy output and risks to the installations themselves, in turn affecting energy prices and global security. She suggests it may make sense to focus on building a more decentralized energy structure, preferably based on locally available renewables situated in secure locations. This sort of regional, network-based system might also prove more flexible and adaptive, and therefore more able to cope with the increasing variability and unpredictability caused by environmental change. La Branche observes that climate change is seen by most international organizations as both an amplifier of natural risks and an obstacle to development efforts. He suggests that climate governance is emerging as a 'meta governance' redefining other issues, including energy, but the dynamics are complex, both harmonious and conflictual. In the cases where there is a conflict between both forms of governance, energy takes priority over climate. Thus the challenge is total and global in the sense that it involves all human activities. In this sense, climate change is an epiphenomenon affecting the way we think and perceive the world as a globality.

## ENERGY AND HUMAN SECURITY

Bhattacharyya presents the energy poverty debate and discusses the link between energy poverty and health and welfare. The chapter focuses on the developing country challenges relating to energy poverty highlighting the gravity of the problem and its geographical coverage. The implications of lack of access to energy for health and social welfare are linked with Human Development and economic development indices. The chapter suggests that the climate co-benefits of interventions could be exploited to increase investment in energy poverty reduction. Dyer argues that on the assumption that justice and equity must underwrite the feasibility of any energy strategies, we need an ethical framework for energy which includes as a central concern the lack of human security. A key aspect of the political barriers to ethical practices is the tendency to leave everything to a market economy, allowing current short-term calculations to determine the value of energy sources. It may require some regulatory intervention, or public and energy-consumer action to shift the balance of interests toward sources of modern environmental concern, by identifying attitudes that constrain and support our ecological perspectives. However, even if such approaches make renewable contributions to energy security easier to sell politically, it remains that an ecological ethic more directly reflects the limits we face, and renewable energy carries with it the prospect of ethical as well as ecological sustainability. Okereke and Yusuf argue that energy poverty is obviously a significant issue in Africa, as elsewhere, and that the concept of low carbon development offers plenty of prospects for Africa to grow its economy, achieve energy security while contributing its own quota in the global effort to fight climate change. Given the critical importance of modern energy to well-being and to economic development in general, achieving universal energy access should definitely be the priority of African countries. Emphasis should be on harnessing the abundant renewable natural resources present all across the continent. However, achieving low carbon development will require the massive upscale of climate and clean energy finance, large scale investment in technology and human capacity as well as radical governance reforms. The poor deserve basic energy services like everyone else; moreover, they have made little contribution to climate change. Global justice and equity is therefore at the heart of the debate about climate change, energy security and climate mitigation. Lipschutz and Mulvaney note that socio-technical systems, such as those that make possible electrification of human society, have social tendencies independent of human intention, and that differing technological paths will have different kinds of social impacts and, by extension, effects on human well-being and security. Even a small investment could go a long way towards supporting renewable energy throughout the world, decreasing cost and increasing reliability, and to ameliorating the poverty now experienced by so many. This could also have the additional benefit of reducing greenhouse gas emissions, mitigating some degree of future climate change, lessening political and military pressures on declining supplies of cheap oil, and setting the stage for global sustainability. These goals might seem utopian to some, but there is no *technological* or *economic* reason they cannot be achieved over the next 50 years. Technologies are political, but do not determine what forms political practices will take. Karlsson-Vinkhuyzen and Jollands observe that humanity is facing daunting energy-related challenges, yet progress towards a sustainable energy system is slow. One of the reasons for this is the entrapment of energy policymaking within the paradigm of national security. They argue that coupling the sustainability of the energy system with the human security paradigm can help to break this impasse. However, energy has not received much attention in human security literature, though there are clear links between energy and the multiple dimensions of human security. Karlsson-Vinkhuyzen and Jollands also introduce the notion of 'deep energy security' and an ethical dimension to energy security, making individual humans and humanity as a whole the relevant security holder. This chapter investigates how the sustainability of the energy system (and by implication, human security) can be provided, what should be provided and how much, particularly within global governance.

## CONCLUSIONS

The emergence of 'energy security' reflects an increased sense of urgency around these issues at the heart of state interests and the global political economy. Perhaps the only common ground is acceptance that we currently inhabit a global fossil fuel-based economy, and that in itself has its problems. So far the widely shared logic has been that of neoliberal political economy, if some highly visible failures must bring such assumptions into question. Capitalism may have to reinvent itself in response to the challenge of energy security, and the risks and distributional equity issues arising from it, since these will likely require wide and deep structural reform of high-energy societies that go beyond adjustment within a neoliberal economic framework and question the reasoning of mainstream neoclassical economics. What may present itself as a structural problem also has implications for diverse forms of agency, since changes in the behavior of citizens, civil society organizations, and reorientation of the private sector toward a 'green economy' are all central to long-term energy security, even in a governmental context. Typically, issues of justice and equity are raised in this context to defend the interests of those who benefit less from the petroleum economy and have even more to lose if no preparation is made for a 'low carbon economy', not to mention a 'post-petroleum economy'. Change may bring opportunities in terms of new technologies, new social practices, and new markets, and the possibility of de-materializing economic activity, such that genuine human opportunities and diversity of practices would be welcome and economically beneficial (if they don't need fossil fuels and carbon sinks). Thus energy security is linked with sustainability in the broadest sense, and not least because even in its own terms the current energy system faces challenges of increasing supplies or finding new resources to meet rising demand from a growing and developing population. It is clear that 'energy security' invokes a wide range of long-term security concerns, but not so clear that all of these have been integrated as coherent strategic goals in overall planning and policy-making. In viewing the breadth, depth and transformation of energy security discourses as politically significant developments, we are better able to appreciate the consequences for both structures and agents.

# REFERENCE

International Energy Agency (2012), 'World Energy Outlook 2012', http://www.world energyoutlook.org; http://www.iea.org/media/workshops/2012/energyefficiencyfinance/ 1aBirol.pdf (accessed 12 November 2012).

# Index

Abanda, F. 463, 469, 470
Abbas, M. 416
accountability 64, 74, 75, 87, 448, 509
Aceh region 73, 74
Acheson, R. 333
Ackermann, T. 61
Adejumobi, S. 81
Adelle, C. 313
Adger, W.N. 306, 332
Africa 25, 26, 72, 75–6, 213, 258, 300, 427, 458, 510, 513
   China 28, 29, 95, 122, 220
   climate change 324, 325, 466–9
   electricity 208, 428, 429, 433, 434, 463
   energy consumption 208, 463, 464
   energy poverty 427, 428, 429, 431–2, 433–4, 438, 439–40, 449, 458, 462, 463–6, 499
   climate change and 466–9
   geopolitics 94–5, 99, 101, 102
   low carbon development 462–3, 469–71, 479–80
      benefit of 471–6
      energy security and challenges to 476–9
   northern 46, 59, 99, 101, 102, 142, 213, 324, 464, 470, 473, 499
   nuclear energy 217
   piracy 223
   privatization 213, 215
   renewable energy sources (RES) 34, 218–19, 225, 463, 466, 467–8, 469, 470, 473–4, 476–7, 478, 480, 499
   smuggling 223
   Sub-Saharan *see separate entry*
   US AFRICOM 76
   *see also individual countries*
African Development Bank 218
Agashe, G. 211, 212
Agbemabiese, L. 458
Agnew, J. 95

Ake, C. 75
Akinbami, J.F.K. 465, 470
Alcorta, L. 224
Aleklett, K. 22
Algeria 41, 142, 246, 247, 251, 253, 264, 469
Alhajji, A.F. 155
Alike, E. 84
Amaize, E. 83
Amigun, B. 470
Amin, A. 218
Amin, S.M. 223
Andrews, A. 221
Angola 72, 74, 211, 220, 470
Anozie, A. 466, 470
Anthoff, D. 368
'Arab Spring' 29, 98–9
Arab-Israeli War (1973) 118, 259, 261, 264
Aradau, C. 7
arbitration 277, 278, 279, 280, 282, 289–90
Arctic 105, 186, 305, 307, 394, 395, 396, 410, 512
Argentina 103, 212, 213, 214, 218
Arvizu, D. 354, 355
assessment framework, energy security 146–7, 167–71
   methodological choices 147–9
   stages 150, 167–9
      1: defining energy security 150, 168
      2: vital energy systems 150–4, 168
      3: vulnerabilities 152, 154–7, 168
      4: selecting indicators 157–9, 168–9
      5: making sense of indicators 159–67, 169
Australia 105, 151, 277, 281, 289, 398
   coal 30, 177, 182, 184, 203
   energy strategy 174, 175–6, 177–84, 201, 203
   liberal democracy 22, 23, 30, 33, 34

539

Austria 59
Austvik, O.G. 253
authoritarian oil or gas producers 21, 22, 26–7, 28, 29
Auty, R. 22
Averill, B. 58, 59
Awerbuch, S. 153
Aylon, J.P. 57
Ayres, R. 483
Azcui, M. 212
Azerbaijan 71, 74, 107, 277

Bächler, G. 321
Badea, A. 164
Baer, P. 446
Bahgat, G. 42, 221, 262, 265
Baker, M. 223
Ball, D. 223
Baltic States 54
Bambawale, M.J. 220, 224
Bamberger, R. 128
Bang, G. 310
Bangladesh 219
Baran, Z. 155
Barnett, J. 330, 448
Barrett, S. 518
Barton, B. 260
Bateman, S. 222
Bauer, C. 346, 347, 348
Baumert, K.A. 297
Bayliss, K. 215
Beijing consensus 248–9
Belarus 287–8
Belgium 54, 59
Belyi, A. 273, 279, 283, 287, 288, 290
Benes, J. 97
Bettini, G. 330
Biafra 74
bias 147–8, 168, 313
Bielecki, J. 427
Biermann, F. 331
Bigo, D. 321
biodiversity 219, 322, 323, 324, 335, 358, 360, 364, 374
bioenergy 219, 334–5, 358–61, 449
  biofuels *see separate entry*
  biomass *see separate entry*
biofuels 33, 101, 135, 137, 138, 153, 218, 219, 335, 451, 463, 470, 487

environmental implications 358, 359, 360–361, 370–374, 379, 485
ethics 445, 449, 450–452
food prices 219
oil demand growth patterns 270
United States 197, 310
biogas 136, 138, 141, 219, 358, 478, 490, 492, 499, 500
biomass 152, 169, 188, 208, 225, 337, 443, 450, 474, 477, 478
  coal and 135, 138, 358
  energy poverty 426, 430–431, 438, 439, 440, 462, 463, 472
  environmental implications 358, 359, 360–361, 370, 374, 379
  synthetic natural gas (SNG) 137–8
Birdsall, N. 215
Birol, F. 489
bitumen 188, 189, 190, 191
Blamberger, C. 277, 286
Blank, S. 223
Blatz, W. 23
Blaydes, L. 264
Bodro Irawan, P. 250
Boehlert, G.W. 366
Boëthius, G. 247
Bohi, D.R. 123
Bolivia 103, 212, 213, 214, 221
Bollen, K. 20
Boonstra, J. 28
Booth, K. 11
Bordoff, J. 299
Boro, Issac Adaka 77
Bossel, U. 144
Botswana 472, 473, 474, 475, 476
Boubakri, N. 215
Bowen, A. 462, 467, 469, 472, 477
BP 97, 108, 262, 395, 397
Bradbrook, A.J. 508, 521
Bradshaw, M.J. 101, 301
Brand, Stuart 484
Bratt, D. 216, 217
Brauch, H.G. 297, 321
Brazil 72, 102, 204, 210, 212, 213, 214, 218, 219, 221, 224, 225, 300, 429
Bridge, G. 151, 152
Bringezu, S. 359
Bromley, S. 71
Brooks, A. 198
Bruce, N. 438

Brune, N. 212
Brunei 106, 129
Brzoska, M. 320, 321, 327
Buhaug, H. 325, 330
Burke, M.B. 325
Burkina Faso 472
Burundi 472
Bush, George, II 27
Bush, R. 77
Buyx, A.M. 452
Buzan, B. 7, 297, 320
Byres, E.J. 58
Byrne, J. 11

Calderón, C. 208
Cameroon 70
Campbell, C. 21, 22
Campbell, K.M. 321
Canada 22, 23, 24, 174, 175, 177, 180–181, 188–91, 201, 203
  climate change 324, 394, 395–6
  demand security 246
  exemption: emergency oil reserves 119
  geopolitics 398–9
  national oil companies (NOCs) 210
  oil sands 398–9
  options, energy 31
  provinces 203
  shale gas or oil 102, 202
carbon capture and storage (CCS) 47, 191, 198, 200, 202, 248, 271, 309, 337, 338, 361, 416
carbon lock-in 475
Cardwell, D. 484
Carroll, R. 450
Carter, J. 151
Carter, N.T. 356
Caspian 71, 72, 98, 99, 102, 106–8, 258, 268, 300
Castillo, L. 221
Catley, B. 21
Cayoja, M.R. 212
Central African Republic 472
Central Asia 26, 28, 94–5, 102, 106–8, 221, 223, 251, 278, 279, 286
  *see also individual countries*
Central and Eastern Europe 45, 46, 92, 300
  *see also individual countries*

centralized vs distributed energy strategies 483–503
  costs and benefits 492–9, 500–501
  defining human security 485, 486–8
  roads not taken 489–92
Chad 472
Chalecki, B. 394
Chalk, P. 222, 223
Chang, F.K. 105
Chaturvedi, A. 222
Chavez, Hugo 25, 27–8
Chawkins, S. 391
Chen, M.E. 210
Chen, S. 122
Cherp, A. 3, 4, 5, 146, 147, 148, 153, 155, 222
Chester, L. 4, 40, 41, 146, 147, 202, 207
ChevronTexaco 83, 262
children 438, 439, 466, 472
Chile 122, 213
China 21, 22, 25, 28–9, 72, 92, 155, 204, 210, 213, 225, 258, 299, 301
  Africa 28, 29, 95, 122, 220
  Caspian resources 107
  climate change 325, 395, 398, 412
  climate finance 478
  climate security 304–5, 322
  'synergistic' policy: energy and 309, 311, 314
  coal 126–7, 129, 311, 398, 414
  demand security 251
  diversification 220
  East and South China Seas 105–6, 109
  Energy Charter Treaty (ECT) 277
  energy efficiency 224
  energy poverty 429, 439
  geopolitics 72, 92, 94–5, 102, 105–6, 108, 109
  hydropower 335, 387, 389, 390, 452
  investments in overseas oil assets 151, 211, 220
  land-based imports 223
  national oil companies (NOCs) 211
  nuclear energy 217, 333
  renewable energy sources (RES) 34, 218, 219, 309
  shale gas 100

strategic energy reserves (SERs) 121–2, 125, 126–7, 129, 221
water 398
Chong, A. 215
Chotichanathawewong, Q. 217, 219, 220, 224
Chou En-Lai 484
Chou, Wan-Jung 415
Chum, H. 358
Chung, Chien-peng 223
Churchill, Winston 258, 532
Cinti, G. 143
Ciuta, F. 4
Clark, W.K. 56
climate change 5, 8, 30, 174, 204, 216, 218, 259, 282, 456, 488
  centralized vs distributed energy 485, 489, 492, 495, 496, 501, 503
  conflict, energy and *see* climate change *under* conflicts
  demand security 246
  energy poverty 441, 462, 466–9
  energy-climate governance nexus *see separate entry*
  EU 47, 61, 174, 191, 203, 322, 411
  GHG emissions *see separate entry*
  governance dimensions of climate and energy security *see separate entry*
  human (in)security 512
  infrastructure, energy 386, 387, 388, 389–90, 391, 392, 393, 394–6, 397, 467–8
  Japan 199–201
  liberal democracy 20, 30, 33, 35
  low carbon or clean 489
  OPEC 268, 269–70, 271
  United States 203, 298, 468
  *see also* low carbon development; renewable energy sources
Clinton, Bill 128
coal 19, 71, 151, 258, 299, 409, 410, 443, 484, 489, 501
  Africa 463, 466, 467, 476, 477
  assessment framework 152, 153, 161
  Australia 30, 177, 182, 184, 203
  biomass and 135, 138, 358
  Canada 188
  capital cost 477
  carbon capture and storage 202, 416

China 126–7, 129, 311, 398, 414
Colombia 25
conflict 319
cooking in developing countries 207, 432
development of/increase in use of 412, 414, 416
energy governance 410, 412, 414
environmental change 387, 392, 398
environmental implications 346–7, 370, 374, 379, 416, 488, 512
European Union 191, 196
gasification 141, 302
hydrocarbon energy mix 101
India 309, 311
infrastructure 387, 467, 476, 499
Japan 199
liquefaction 301, 398
miners' strikes 154
source and carrier 133–4
strategic energy reserves (SERs) 117, 126–7, 129, 130
United States 197–8, 414
Cole, B. 223
Collier, J. 280
Collier, P. 22, 73, 74
Colombia 25, 213, 214, 218, 223
combined heat and power 135–6, 138, 139, 144, 356
combined-cycle gas turbines (CCGTs) 44–5, 347–8, 350
Comolli, V. 310
compensation 280
Compston, H. 313
Conceição, P. 517
conciliation 277, 279, 314
conflicts
  climate change 304–6, 319–20
  desecuritization 339–40
  securitization of migration 327–32
  as security and conflict issue 320–327, 412–13
  security and conflict issues of climate policies 332–9
  cyber security/warfare 55–65
  resource *see separate entry*
constructivist approaches 6–7, 304, 447
Correljé, A. 4, 7
corruption 75, 82, 87, 103, 479, 513
Costantini, V. 152

Cramer, C. 214
Creamer, T. 452
Crist, E. 445
Croatia 59
crowding 22–3
curse, resource 22–3, 74, 513
cyber security 55–65, 183, 223, 484
cycles, energy 96–7
Cyprus 100

Dahl, R. 20
Dalby, S. 297, 307
Daly, H.E. 515, 516
Daly, J. 21
Daniel, S. 83
Dannenberg, A. 33
Dannreuther, R. 71, 222
Darbouche, H. 98, 99
Darby, S. 60
Davis, J. 214
De Bruijne, M. 64
De Cara, S. 359
De Changy, F. 32
De Fraiture, C. 360
De Santi, G. 33
de Soysa, I. 513
de Vries, L.J. 65
decentralization 400, 413, 453, 457
  centralized vs distributed energy strategies 483–503
deep energy security 8, 507, 513–15, 517–22, 530
Deepchand, K. 219
definition of energy security 40–41, 176, 178, 180, 192, 194, 197, 201, 202, 207, 299, 448, 532
  ECT 292
  framework 150, 152, 154–5, 168, 169
  producers' security 241, 299, 302
  producing, consuming and transiting countries 243–4, 259–60
deforestation 307, 323, 335, 390, 468, 472, 478, 487
Deichmann, U. 219
Delucchi, M.A. 47
demand, security of 154, 175, 201, 239–40, 245–6, 255, 533–4
  development of concept 240–245
  economic concerns 246–9

governance, energy security *see separate entry*
  instruments 250–255
  oil producers' perspectives 258–61
  OPEC *see separate entry*
  political concerns 249–50
democracy 483, 489, 491, 492, 502, 513
  liberal *see separate entry*
  participatory 413
Democratic Republic of Congo 469, 473
Denholm, P. 490
Denmark 119, 124, 498, 501
Detraz, N. 304
developing countries 101, 152, 298, 304, 322, 445, 446, 469, 477, 510
  carbon space 312–13
  climate aid 306, 477–9, 480
  climate change 323, 332, 467, 468–9
  climate security 304, 311, 314
  national energy strategies 206–24, 224–6, 476
  oil and gas producing 75–6, 87, 271
  OPEC 268, 269, 271
  poverty, energy 425, 430, 432, 433, 434, 438, 439, 441, 468
  renewable energy sources (RES) 33–4, 217–20, 223, 225, 452
  bioenergy 334–5
  *see also individual countries and areas*
Devine-Wright, P. 498
Diaz-Chavez, R. 219
Dickel, R. 249
Dietl, G. 221
Dillon, M. 7, 13
Diog, A. 462, 469, 470, 472, 478
direct carbon fuel cell producing CO 142, 143
Dirmoser, D. 211
disease 207, 306, 322, 363, 364, 390, 407, 438, 439, 466, 497, 512
distributed vs centralized energy strategies 483–503
  costs and benefits 492–9, 500–501
  defining human security 485, 486–8
  roads not taken 489–92
diversification 25, 64, 96, 101, 129, 175, 190, 261, 283, 471
  developing world 216–21

of energy sources 40, 46, 259, 261, 299, 309, 311, 408, 410, 412, 414, 471
  and carriers 133–45
  EU 27, 46, 64–5, 99, 191, 196, 300, 311
  producers 252–3, 271
  of security concept 23
Djebah, O. 77
Djibouti 470
Do, T.M. 221
Dore, J. 276
Dosch, J. 21
Douglas, Oronto 80
Downs, A. 20, 24, 34
Downs, E.S. 121, 129
dual fuel appliances 137, 138
Duffield, M. 75, 76
Duke, S. 26
Dupont, A. 304, 306, 308
Durakoglu, S.M. 215

earthquakes 30, 32, 333, 358, 365, 389, 391, 397
East and South China Seas 105–6, 109
Eastern Europe 3, 59, 100, 109, 174, 283
  see also individual countries
Easton, D. 20, 34
Eberhard, A. 208
Ebinger, C.K. 219, 224
Ebohon, O.J. 466
ecological modernization 8
economic growth 20, 22, 26, 34–5, 242, 270, 313, 444, 448, 503
  Africa 462, 463, 465, 466, 467, 469, 471, 472, 475, 476
Ecuador 103, 220
Edwards, S. 214
efficiency, energy 175, 247, 301, 309, 311, 406, 414, 490, 517, 521, 529
  Canada 190
  China 309, 314
  developing world 209, 223–4, 225, 428, 469, 478
  ECT 278, 281–2
  EU 47, 60, 191, 310, 411
  ICT 60
  India 309, 314
  Japan 199, 200

liberal democracy 25, 30, 34
price of oil and gas 246
rebound effect 415, 452
Russia 186, 187, 245
United States 25, 197, 310
Egenhofer, C. 260
Eggoh, J.C. 478
Egypt 22, 73, 100, 102, 142, 223, 469, 474
Ehrlich, I. 215
Eickhout, B. 360
El-Badri, A.S. 246, 260
ElBaradei, M. 508
El-Hinnawi, E. 352
El-Katiri, M. 246, 253
electricity 133, 183, 190, 218, 455, 456, 457–8, 474–5, 478, 480
  centralized vs distributed energy strategies see separate entry
  cyber security 57, 59–65
  developing world 206, 207, 208, 212, 213–14, 215, 216–17, 224
  energy carrier 134, 144, 152–3
  energy poverty 427, 428–30, 432, 433, 434, 436–8, 440, 449, 462, 463, 464–5
  EU: security of supply 47–55, 57, 59–65, 175, 191, 194–6, 202, 203
  Japan 200
  Norway 185–6
  privatization 213–14
  transmission lines 488, 497
  United States 198
electrolyzers 144
Eleri, E. 464, 465
Ellis, M. 224
emissions trading schemes (ETSs) 30, 33, 191, 314
Endrenyi, J. 62
Energy Charter Treaty (ECT) 107, 244–5, 251, 254, 273, 274, 292, 308
  dispute settlement 277–8, 279, 282, 283, 287, 288, 289–90
  energy governance 274–83
  environmental regime 277
  energy efficiency 278, 281–2
  EU and Russian positions 289–91, 300
  EU-Russian controversies 283–8
  institutional setting 282–3

investment protection 252, 277, 280–281, 284–5, 292
trade and transit governance 277, 278–9, 285–8
energy poverty *see* poverty, energy
energy-climate governance nexus 402–8, 416–17
  energy governance 410–412
    tensions between climate and 412–16
  path dependency 408–10, 411, 415–16
Engbarth, D. 32
Eni 83, 108, 263
environment 512, 534–5
  Canada 190, 191
  China 28
  climate change *see separate entry*
  demand security 241, 242, 245, 259
  ECT 277, 278, 281–2
  energy infrastructure and change in 386–400, 407, 467–8
  implications of energy production for 345–53, 379, 416, 488, 512
    methods to quantify impacts 366–70
    renewable energies 219, 345–6, 354–66, 370–379, 411, 413, 485, 487, 492, 495, 501, 534–5
    results 370–379
  OPEC 268, 269–70, 271
Equatorial Guinea 70
Erickson, W.P. 363
Eritrea 470
Essien, U. 81
Estache, A. 214
Estonia 55
ethical dimensions of renewable energy 443–9, 457–9
  different types of RES 449–57
Ethiopia 469, 470, 472, 475
European Union 79, 151, 155, 164, 225, 255, 302, 517, 530–531
  Canada 190
  China 28, 102
  climate change 47, 61, 174, 191, 203, 322, 411
  climate security 304, 306, 412
    'synergistic' policy: energy and 310–311, 313–14
  ECT 273, 274, 275–7, 292, 308
    Energy Charter Secretariat 277, 282
    EU and Russian positions 289–91, 300
    EU-Russia controversies 283–8
    trade and transit 278–9
  energy strategy 174, 175–6, 191–6, 201, 202, 203
  gas and electricity networks in *see separate entry*
  geopolitics 92, 94, 99, 101, 102, 104, 107–8, 110, 250
  Kyoto Protocol 298
  liberal democracy 24, 26–7, 28, 29, 31, 32–3
  nuclear power 31, 32, 191
  Parliament 274, 291
  renewables *see* European Union *under* renewable energy sources
  Russia *see* European Union *under* Russia
  shale gas 191
  strategic petroleum reserves (SPRs) 119
  taxation 248
  Turkey 31
expropriation 280
  *see also* nationalization
ExxonMobil 83, 108, 209, 262

Faas, H. 9
Fargione, J. 359
Farquharson, E. 213
Farrell, D. 224
Fattouh, B. 239, 241, 242, 246, 266
Favennec, J.-P. 106
Fay, M. 208, 215
Fears, D. 494
Ferguson, C.D. 216
financial crisis (2008/2009) 100, 175, 213, 247, 253
Finnveden, G. 366, 367
Finon, D. 42, 287, 288
Firbank, L.G. 360
Fischer, C. 60, 490
Fleming, J. 338
Fletcher, S. 212
Florini, A. 520, 521
Foell, W. 440

Fogarty, D. 444, 450
Fogarty, E. 21
food 306, 323, 324, 326, 327, 334, 335, 398, 399, 487, 511, 512
  biofuels 450–451, 452
foreign direct investment (FDI), privatize to attract 212–15, 225
Foster, V. 213
Foucault, M. 330, 405
France 33, 59, 218–19, 351, 391, 392, 406, 407
Frankl, P. 374
free riding 47, 312
Friedman, T. 103
Friedrichs, J. 5
Froggatt, A. 309, 313
Frydman, R. 215
Frynas, J.G. 82, 86
Fthenakis, V. 492
fuel blending 135, 137–8, 144, 358
fuel cells 140, 141, 142, 143
fuel-switching 309
future scenarios, energy security in 147, 148–9, 150, 152, 153, 154, 157, 158, 165

G8 175, 187, 203, 204, 243–4, 283, 285, 303, 310
G20 104, 175, 243
G77 304, 322
Gaddafi, M. 21
Gambill, G. 233
García, J.C. 212
Gärtner, S. 358
gas and electricity networks in EU 40–41, 65, 151, 255, 278, 300, 310–311, 314
  cyber security 55–65
  liberalised gas markets 41–7, 242, 248, 251–2, 254, 284–5
  network development 45–7, 52–5
  public funding 46, 47
  security of electricity supply 47–55, 57, 59–65, 175, 191, 194–6, 202, 203
gas-fired power plants 44–5
Gasper, D. 509, 514, 518
GATT (General Agreement on Tariffs and Trade) 276, 278
Gazprom 107, 250, 252, 287, 288

GEA (Global Energy Assessment) 147, 148–9, 150, 151–2, 153, 154, 157, 158, 160, 161, 163–4, 166–7, 170
GECF (Gas Exporting Countries Forum) 100, 245, 254
Geers, K. 56
Geller, H. 443
Gemenne, F. 327
geoengineering 320, 337–9
geopolitics 4–5, 23, 41, 92–5, 109–10, 201, 202, 259, 264, 273–4, 283, 300, 308, 320
  causes of unstable 'energy security' 104–9
  climate change 398–9, 412
  critical 95–6
  definition 95
  demand security 250, 255
  energy cycles and corresponding theoretical views 96–7
  infrastructure and environmental change 398–9
  oil price: 2000s' surge 265
  overview of oil 71–2, 87
  retreat of globalization 101–4
  strategic ellipsis 98
  unprecedented uncertainty 97–101
Georgia 26, 27, 106
geothermal energy 152, 334, 336, 356–8, 406, 410, 443, 445, 454
  Africa 218, 463, 470, 478
Gerbens-Leenes, W. 360
Germany 31, 33, 51, 59, 92, 258, 321–2
  climate change 321–2, 338
  climate security 304
  nuclear power 26, 32, 198, 203, 333, 351, 392, 445
  SPRs 120
Ghana 468, 473
Gheorghe, A.V. 60
GHG emissions 200, 297, 307, 312, 319, 332, 340, 412, 416, 508, 512
  carbon capture and storage *see separate entry*
  development 467, 468–9, 471, 480
  energy poverty 441
  impacts assessment 368, 370–374, 375, 376, 379
  migration 513
  natural gas 201, 470, 489

reduction targets 47, 191, 307, 308, 310, 311
renewables 354, 355, 357, 359, 360, 362, 364–5, 411, 413, 495, 534–5
  DG system 503
  impacts assessment 370–374, 375, 376, 379
Ghilès, F. 246
Gibson, J. 223
Gilbert, N. 451
Gillard, Julia 30
Gilpin, R. 274
Giroux, J. 222, 223
Giuli, M. 27
Gjelten, T. 212
Gleditsch, N.P. 325
Gleick, P.H. 321
Global Environmental Facility (GEF) 477
globalization 95–6, 110, 260, 269, 312
  retreat of 101–4
Goldemberg, J. 216, 217
Goldman, Arnold 495
Goldstein, B. 357
Goldthau, A. 4, 175, 273, 300, 521
Goldwyn, D. 23
Goolsbee, A. 222
Götz, R. 94
governance dimensions of climate and energy security 297–8, 308–9
  climate security 303–5, 403, 412
    preventive 307–8, 313, 314
    reactive 305–6, 312, 313
  dilemma 312–15
  energy security 299–302
    pursuit of global energy security 302–3
  'synergistic' policy 309–11
  see also energy-climate governance nexus
governance, energy security 273–4, 292
  ECT: EU and Russian positions 289–91
  ECT as grounds for 274–83
  EU-Russian controversies 283–8
governance, global energy 3, 518–22
Grave, K. 51
Greece 31
green economy 247, 537
  efficiency, energy see separate entry

renewable energy sources (RES) see separate entry
Greene, D.L. 155
greenfield investment 186, 213, 214
greenhouse gases see GHG emissions
Gross, C. 454
Grubb, M. 153
Grueneich, D. 494
Guatemala 213
Guinée, J.B. 366
Gulf countries 99, 254
  see also individual countries
Gunawardena, U.A.D. Prasanthi 453
Gupta Harsh, K. 389
Gupta, J. 520
Gupta, S. 147, 154, 162, 164, 414

Hadfield, A. 41
Haftendorn, H. 6
Haggett, C. 362
Haghighi, S.S. 278
Haiti 208
Hajer, M.A. 8
Hall, D. 213, 215
Haller, M. 47
Hamel, M. 242, 248
Hampson, Fen Osler 11
Hancock, K.J. 220
Happ, R. 278
Hargrove, E.C. 459
Harper, Stephen 188
Harris, G. 206
Hayward, T. 458
health 407, 477, 478, 486, 487, 492, 502, 510, 511, 512
  disease see separate entry
  energy poverty 423, 427, 438–9, 441, 463
Heinberg, R. 21, 24, 443
Heinrich, A. 41
Helm, D. 155
Hemmes, K. 141, 142
Herberg, M. 210, 211
Hines, P. 222
Hober, K. 280
Höhne, N. 468
Hoicka, C.E. 51
Homer-Dixon, T.F. 321
Hoogeveen, F. 41
Hook, L. 221

Hopkin, J. 20
Hoyos, C. 210
Hsiang, S.M. 325
Hubbert, M.K. 22, 34
Hughes, L. 210, 211, 212
Hughes, R. 56
Hulbert, M. 252
Human Development Index (HDI) 435–8, 439
human rights 21, 35, 275, 509, 513, 530
human security 23, 297, 305, 306, 323, 396, 535–7
   centralized vs distributed energy strategies 483–503
   defining 485, 486–8, 508–13
   ethical dimensions of renewable energy *see separate entry*
   migration 328, 329
   poverty, energy *see separate entry*
   public good: sustainable energy system 507–22
      deep energy security 8, 507, 513–15, 517–22, 530
      public and private goods 515–16
Hungary 124, 335
Huntington, S. 20
Hussain, H. 248
Hussein, Saddam 24
Hussey, H. 21
Huysmans, J. 7
hydrogen 135, 153, 456, 457
   MSMP energy systems 139, 140–142, 143
   as universal energy carrier 143–4, 354, 456
hydropower 33–4, 101, 152, 218, 335–6, 407, 501–2
   Africa 219, 463, 466, 467–8, 469, 473–4, 477, 478
   Bolivia 212
   Brazil 221
   Canada 31, 188
   China 335, 387, 389, 390, 452
   community security 513
   environmental implications 363–5, 374, 387, 487
   ethics 449, 452–3, 456
   European Union 49
   India 214
   infrastructure and environmental change 387, 388–90, 398, 399, 407, 467–8
   Nepal 219
   Norway 185–6
   South Africa 219
   Thailand 219
   United States 198

Ibeanu, O. 85
Iceland 289, 396
ICT 57, 60–1, 457, 493
Igbikiowubo, H. 84
IGCC (integrated gasification combined cycle) plants 200, 346–7, 358
Igure, V.M. 57
Ikelegbe, A. 223
illiberal and liberal democratic regimes 19–21, 23, 25, 28, 35, 530
India 22, 204, 225, 258, 299, 301, 393, 478, 515
   climate security 304, 412
   'synergistic' policy: energy and 309, 311, 314
   demand security 251
   diversification 220–221
   energy poverty 425, 426, 427, 432, 434
   energy strategy 206, 208, 210, 211, 214, 216–17, 219, 220–221, 223, 224
   geopolitics 72, 92, 94, 102, 300
   hydropower 335, 389–90
   national oil company 211
   nuclear energy 216–17, 333, 391
   renewable energy sources (RES) 34, 219
   strategic energy reserves (SERs) 122, 123, 125, 221
Indonesia 73, 74, 223, 264
information 43, 45, 224, 267, 301
   electricity grid 60, 61–2, 500
   exchange of 254, 290
   sharing 64, 521
infrastructure, energy
   environmental change 386–400, 407, 467–8
institutional investors 444, 450
insulation 413, 415

insurance 7, 217, 394, 397, 399, 403
intergenerational equity 448
International Energy Agency (IEA) 97, 102, 109, 151, 152, 175, 212–13, 259, 260, 261, 302, 308, 310, 521, 529–30
 'easy oil' 301–2
 G8 meetings 303
 MOSES *see separate entry*
 strategic energy reserves (SERs) 118, 119, 120, 121, 123, 125, 126, 128, 130
International Energy Forum (IEF) 104, 241, 242, 243, 244, 254, 262, 302
International Energy Programme (IEP) 119
international oil companies (IOCs) 70, 71, 87, 98, 103, 104, 108–9, 501
 1973 Arab-Israeli War 264
 Niger Delta 76, 78, 79, 80, 81, 82, 83, 84, 86
 OPEC 261, 262–3, 264, 265, 266, 267
 Seven Sisters 262–3
International Renewable Energy Agency (IRENA) 253, 521
internationalisation of public policy 76
Internet 57, 63, 64, 502, 513
investment cycle: raw materials 96, 104
Iran 25, 72, 98, 99, 128, 246, 250, 277, 301, 333, 515
 OPEC 263, 264
Iraq 22, 24, 70, 72, 73, 74, 87, 92, 97, 108–9, 121, 262, 333
 hydropower 335
 OPEC 263, 264
 smuggling 222–3
Isihak, S. 439
Israel 29, 100, 118, 259, 261, 264
Italy 59, 124, 142, 333
Ivory Coast 473

Jacobson, M.Z. 334, 494
Jaffe, A.M. 220, 222
Jäger, J. 328, 508
Jakobeit, C. 329–30
Janardhanan, N. 220
Jansen, J.C. 153
Janssen, R. 218

Japan 28, 32, 120, 212, 219, 223, 258, 259, 263, 310, 311
 East and South China Seas 105–6
 Energy Charter Treaty (ECT) 277
 energy strategy 175–6, 194–5, 198–201, 202, 203
 nuclear power 32, 33, 199, 200, 202, 203, 217, 333, 351, 353, 391, 397, 445, 484
 renewable energy sources (RES) 33, 199, 200
Jaureguy-Naudin, M. 52, 54
Jesse, J.-H. 246
Jewell, J. 146, 147, 158
Jiang, J. 211
Johansson, T.B. 147
Jordan 73, 214
Jowit, J. 392
justice 454, 458, 468, 480, 537
 distributional 448, 451

Kalicki, J. 25
Kalitsi, E.A. 469, 474
Kanagawa, M. 463
Kanninen, T. 13
Kanter, J. 392
Kaplan, R. 223
Karl, T.L. 74
Karlsson-Vinkhuyzen, S.I. 303, 514, 515, 516, 517, 518, 519, 520, 521
Kaul, I. 515, 516
Kazakhstan 22, 103, 107, 211, 220, 277
Keating, M. 4–5
Keith, D. 305
Kendell, J.M. 154
Kennedy, A. 211
Kennedy, Robert, Jr 498
Kenya 122, 218, 221, 326, 464, 468, 470
Keohane, R.O. 110, 273
Keppler, J.H. 5, 155
Kessler, G. 333
Khennas, S. 462, 464, 474
Kikeri, S. 214
Kimmins, J.P. 444–5
King, I. 29
Klare, M.T. 7, 24, 70, 73, 82, 94–5, 105, 106, 107, 155
Klawitter, J. 336
Knorr, W. 30
Koch, W. 34

Koike, M. 21
Kojima, M. 221
Koknar, A.M. 222
Köller, J. 363
Komiss, W. 222
Konoplyanik, A. 279, 286, 290
Korea Gas corporation 108
Kotzian, P. 20
Krause, K. 6
Krauss, C. 394
Kumar, A. 365
Kundu, G.K. 215
Kurzman, C. 20
Kuwait 99, 262, 263, 264
Kuzemko, C. 273

La Porta, R. 215
L'Abbate, A. 60
Laborde, D. 359, 374
Ladenburg, J. 362
LaMonica, M. 62
Latin America 25, 103, 206, 208, 210, 211, 212, 213, 217, 218, 300, 335, 464, 468
  *see also individual countries*
Le Billion, P. 71–2, 73
Le Coq, C. 41, 42, 147, 152, 163
least developed countries (LDCs) 432
Lebanon 73, 100
Leboeuf, A. 413
Lechón, Y. 355
Lecointe, C. 374
Lenton, T.M. 324, 489
Lesage, D. 299, 300, 302, 303, 520, 521
Lesotho 474
Letzing, J. 127
Leung, G.C.K. 220
Lewis, A. 365, 366
Lewis, Ian 211
Lewis, Joanna I. 304–5
Li, M. 28
liberal democracy 19–20, 34–5, 530
  external dimension 20–29
  internal dimension 29–34
Libya 120, 128–9, 223, 469
linearization 34–5
Liotta, P.H. 509
Lipschutz, R.D. 23, 485, 487
Lipson, H.F. 63
Lloyd-Roberts, S. 83

LNG (liquefied natural gas) 45, 46, 200, 220, 243, 246, 247, 252, 254, 300
  Australia 177–82, 183
  European Union 196, 198, 252
  geopolitics 99, 100, 102
  strategic gas reserves (SGRs) 123–4
  terminals 45, 46, 187, 198, 201–2
  United States 198, 202
Loftis, J. 212
long-term contracts: natural gas 43–4, 45, 47, 100, 242, 247, 251, 529
Löschel, A. 209
Lovelock, James 445
Lovins, A.B. 483, 485, 489, 492, 502
low carbon development 462–3, 469–71, 479–80
  benefit of 471–6
  energy security and challenges to 476–9
Luciani, G. 128, 251
Ludlow, M. 30
Luft, G. 41, 223
Lujala, P. 73
Lukszo, Z. 222

Ma, W. 123
Maas, A. 337
Mabey, N. 306, 308, 314
Mabro, R. 266
Macalister, T. 210
McDonald, A. 217
McDonald, M. 327
McDonald, N.C. 355, 492
Macfarlane, A.M. 444
MacFarquhar, N. 304
McGregor, R. 484
MacKay, D. 498
McSherry, B. 21
McSweeney, B. 23
Madagascar 470
Madan, T. 211
Maestas, A. 335
Malawi 472
Malaysia 106, 129, 211, 223, 429
Mali 472
Manson, K. 219
Marcel, V. 211
market economy 5, 43, 458–9, 529
Markulec, M. 63

Marnay, C. 490
Marques, A.C. 310
Masera, M. 64
Mattlin, M. 26
Maugeri, L. 97
Mauritania 470
Mazo, J. 306, 307
Meadowcroft, J. 409
mediation 29, 279, 287, 340, 534
Medvedev, Dmitry 244, 251, 252, 254
Megginson, W. 215
Mehlum, H. 22
Melillo, J.M. 359
Mendonca, M. 455, 459
Menichetti, E. 370
meta-governance *see* energy-climate governance nexus
Methmann, C. 321, 330
Mexico 24, 25, 129, 190, 210, 214, 217, 310
MFN principle 280, 284, 292
Middle East 24, 26, 29, 121, 122, 213, 220, 258, 263, 425–6
  climate change 324
  geopolitics 71, 72, 96, 99–100, 101, 102, 283
  *see also individual countries*
migration and climate change 320, 323, 326, 327–32, 513
Miller, A. 250
Miller, D. 391
Minteer, B.A. 458
Mishra, D.K. 390
Mishra, P. 395
Misra, A. 221
Mitee, Ledum 79, 80
Mitrova, T. 278, 288
Mitterrand, Francois 262
Mo, J. 223
Mobjörk, M. 304, 307
Modi, V. 520
Mommer, B. 103
Monagas, Y. 210
morality 447, 522
Moran, D. 4
Morath, E. 393
Morocco 59, 473
MOSES 146, 148–9, 150, 151, 152, 153, 157, 158, 159, 161, 162, 163–4, 165–6, 170

Moyo, D. 109
Mozambique 470, 473–4
Muller, G. 393
Müller-Kraenner, S. 299, 300
multi-fuel appliances 137
multisource multiproduct (MSMP) energy systems 138–43, 144, 145
Mulugetta, Y. 469
Munck, G. 20
Münckler, H. 34
Munro, K. 58, 63
Muradov, N.Z. 143
Myanmar 208, 220, 429
Myers, N. 328
Myrli, S. 55

NAFTA (North American Free Trade Agreement) 190
Namibia 473, 474, 475
Nappert, S. 289
Narula, K. 492
national energy strategies
  developing world 206–24, 224–6, 476
  major industrialized countries 174–6, 203–4
    current energy mix and import dependence 176–7
    explaining differences 201–3
    features of 177–201
national sovereignty over resources 281, 290
nationalism, resource 94, 103, 265, 274, 284, 291, 292, 301
nationalization 103, 104, 210–212, 225, 264, 299, 512, 515
NATO 4
natural gas 31, 41, 144, 258, 278, 284, 299, 309, 410, 501
  Africa 463, 466, 469–70, 473
  assessment framework, energy security 151, 152, 153, 154
  cyber attacks 56–9
  demand security 241–3, 244, 245, 246
    elements of 246–50
    instruments 250, 251–2, 253, 254
  environmental implications of energy production 347–9, 370, 374, 379
  fracking 489, 492

fuel blending 135, 136
gas and electricity networks in EU
  *see separate entry*
geopolitics 92, 93, 94, 96, 98, 99–101, 102, 104, 105, 106–8, 300
hydrocarbon energy mix 101
liberalised gas markets in EU 41–7, 242, 248, 251–2, 254, 284–5
liquefied *see* LNG
long-term contracts (LTCs) 43–4, 45, 47, 100, 242, 247, 251, 529
MSMP systems 141–3
national energy strategies
  Australia 177, 183, 184
  Canada 188
  developing world 210, 213, 214, 220
  European Union 191, 196
  Japan 199, 200–201
  Norway 176, 184, 203
  Russia 187–8, 201–2, 243
  United States 197–8, 203
off-shore/coastal production 392–4
prices 44–5, 100–101, 125, 246–7, 250, 251, 280, 466, 484
resource conflicts 100
  energy security and 72–6, 87
shale gas 93, 100, 110, 175, 187, 188, 189, 191, 198, 202, 300–301, 302, 412
source and carrier 133–4
strategic gas reserves (SGRs) 117, 123–6, 130
Superwind 136, 141–2
synthetic (SNG) 137–8
transit conflicts and ECT 287–8
Nayak, D.R. 362
needs, energy 207–8, 415–16, 424, 427
Nellis, J. 215
neoliberal approaches 4–5, 13, 76, 95, 103, 108, 110, 529, 530, 531–2, 537
Nepal 219, 389
Netherlands 22, 59, 134, 144, 275, 393, 408
network development
  electricity supply in EU 52–5
  liberalised gas market in EU 45–7
Newman, E. 487
Newman, S. 21
Niez, A. 208

Niger 472, 474
Nigeria 22, 25, 70, 72, 220, 464, 466, 469, 470, 473, 475
  Biafra 74
  Niger Delta 71, 72, 74, 76–86, 87, 393, 487, 512
  OPEC 264
  smuggling 222–3
  spending on power sector 476
Nilsson, M. 508
Noble, D.F. 493
Noël, P. 41
non-discrimination 276, 279, 280, 281, 285, 286, 290, 291
non-governmental organizations (NGOs) 23, 75, 226, 329, 336, 467, 476, 499, 533
North, D. 109
North Sea 101, 184, 189, 265, 394
Norway 22, 41, 46, 75, 100, 119, 242, 245, 254, 394
  ECT 277, 281, 289
  energy strategy 174, 175, 176, 178–81, 184–6, 201, 203, 204
nuclear power 152, 169, 202–3, 204, 247, 320, 410, 414, 463, 484, 492, 502, 512
  Canada 31, 203
  climate risks and 333–4, 489
  developing world 216–17, 333
  environmental implications of energy production 351–3, 370, 374, 379, 488
  ethics 445, 449, 456–7
  European Union 31, 32, 191
  France 33, 351, 391, 392, 407
  Germany 26, 32, 198, 203, 333, 351, 392, 445
  IAEA 217, 333, 489, 521
  infrastructure and environmental change 387, 391–2, 398, 407
  Japan 32, 33, 199, 200, 202, 203, 217, 333, 351, 353, 391, 397, 445, 484
  liberal democracy 19, 20, 26, 31–2, 33, 35
  MSMP systems 143
  Russia 186, 290
  United States 197–8, 203, 353, 391, 483, 488

waste 32, 217, 333, 351, 352, 353, 374, 456, 488
weapons 333–4, 512
Nussbaumer, P. 426
Nussbaumer, T. 361
Nye, J.S., Jr 3

Obama, Barack 105
Obayuwana, O. 83, 84
Occidental 108, 263
ocean acidification 306
ocean energy 188, 218, 336, 365–6, 397, 455, 516
Oduniyi, M. 83
OECD countries 96, 98, 103, 104, 308, 322, 468
   SERs 119–20, 123–4, 128, 130
Oels, A. 327, 330
O'Hara, S. 71
oil 21, 133, 258, 278, 284, 309, 398, 416, 463, 466, 501
   assessment framework, energy security 151, 152, 153, 154, 155, 161, 162, 166–7
   British Navy: switch from coal to 151, 258
   China: investments in overseas oil assets 151, 211
   cold climate infrastructure 394–6
   community security 513
   curse, resource 22–3, 74
   cyber attacks on SCADA systems 58
   demand security 239, 240–241, 242, 245, 246
      elements of 246–50
      instruments 250, 252, 253, 254
   embargos 79, 118, 152, 155, 221, 259, 261, 264, 299
   environmental implications of energy production 349–50, 370, 374, 379
   geopolitics 71–2, 87, 92, 94, 96–9, 101–4, 105, 106–10, 398–9
   growth in demand 310
   hydrocarbon energy mix 101
   Icelandic ash cloud 396
   national energy strategies
      Australia 177, 182, 183, 184
      Canada 31, 188–90, 203
      developing world 210, 211, 212, 213, 214, 220, 221–3, 225–6
      European Union 194
      Japan 199
      Norway 176, 184–5, 203
      Russia 203, 243
      United States *see below*
   off-shore/coastal production 392–4
   peak 22, 96, 97, 266, 302, 402, 409–10, 448
   pipeline explosions 488
   prices 3, 72, 73, 74, 76, 81, 83, 87, 96, 101–4, 109, 110
      1973–1974 oil embargo 118, 259, 261, 264
      2000s: demand-driven surge 265
      Africa 466, 468, 471–2, 473, 474, 475
      Asian markets 252
      broken pipeline 395
      Canada 189–90
      development 242
      economic cycles 281
      energy efficiency 411
      formula 265
      GDP 472
      high status of energy security 320
      hurricanes 393
      market psychology 267
      mid-1980s 265
      non-conventional oils 301, 302
      OPEC 261, 262–8, 269, 271–2
      SERs 117, 118–19, 120, 121, 122, 123, 127, 128, 130
      Seven Sisters 262–3
      stability 241, 246, 262, 263, 271–2
      storage and idle capacity 250
      upstream investments 280
      vicious circle 239
   producers' perspectives 258–61
      OPEC *see separate entry*
   resource conflicts *see separate entry*
   source and carrier 133–4
   strategic petroleum reserves (SPRs) 117–23, 128, 129–30, 221–2, 225–6, 259
   United States 21, 22, 23, 24–5, 94, 102, 108–9, 197, 198, 301
      Canada 188–9, 398–9
      Carter doctrine 151

conflicts 70, 71, 72, 73, 76, 79, 87, 92
domestic production 197, 310
economic prosperity 263
mid-1980s: low prices 265
national oil companies (NOCs) 210
off-shore production 393
oil embargo 118, 259, 261, 264
SPRs 120, 128, 130
Trans-Alaska pipeline 394–5
unconventional 197
Ojakorotu, V. 81
Okafor, E. 465
Okereke, C. 467, 468, 471, 479
Okoro, O.I. 464, 465, 470
Oldag, A. 104
Oliveira, D. 223
Oliver, C. 125
Olivier, L. 417
O'Loughlin, J. 325
Omoweh, D. 77
Onosode, G. 85
Onuoha, F. 223
OPEC (Organization of Petroleum Exporting Countries) 3, 98, 119, 189, 240, 242, 245, 246, 254–5, 261–7, 299, 302
  Energy Charter Treaty (ECT) 281
  long-term strategy 270–272
  objectives and perspectives 267–70
Orbán, A. 41
O'Rourke, D. 223
Osaghae, E.E. 78, 86
OSCE 243, 249
Oseni, M. 464
Oswald-Spring, Ú. 332
Ouedraogo, N. 467
Owen, N.A. 301, 472
Owen, T. 509

Pachauri, S. 219, 423–4, 426
Page, S.E. 359
Paik, I. 119, 121
Pakistan 220, 277, 306
Palacios, L. 214
Palazuelos, E. 105
Palestinian Authorities 100
Papua New Guinea 129
Parfomak, P. 223

Parker, D. 215
Pascual, C. 41
Patel, A. 390
path dependency 408–10, 411, 415–16
Paust, J. 29, 155
peak oil 22, 96, 97, 266, 302, 402, 409–10, 448
Peelen, W.H.A. 142
Pemberton, M. 395
Perez, Carlos 262
Perlin, J. 499
permafrost 394–6
Peru 213, 220, 221
Peters, S. 24, 72, 508, 512
Philippines 105, 106, 122, 123, 214, 221, 429
Piebalgs, A. 312
Pierson, P. 408
pig-cycles 96, 104
piracy 222, 223
Pirani, S. 288
Plecash, C. 190
Podobnik, B. 512
Poland 124, 214
Polsky, D. 438
Ponseti, M. 335
Popvici, V. 100
Portugal 59, 124
poverty, energy 175, 207, 208, 209, 441, 449, 458, 462, 463–6, 508
  climate change and 441, 462, 466–9
  concept 423–6
  decentralized renewables 499, 502
  energy security linkage and 427
  incidence of 428–35
  options for remedial interventions 439–41
  subsidies 301, 439
  welfare issues and 435–9
Praiwan, Y. 122
Prange-Gsthol, H. 291
precautionary approach 7, 340
Preiss, P. 368
prices 258, 259, 260, 301, 303, 311, 312, 313, 399, 468
  food 452, 468
  low carbon development 471
  natural gas 44–5, 100–101, 125, 246–7, 250, 251, 280, 466, 484

oil *see* prices *under* oil
rebound effect 415, 452
privatization 210, 212–15, 225, 284, 515
Profant, T. 93
prosumers 49, 411, 413
Przeworski, A. 20
public goods 47, 250, 312, 315, 339
  storage and idle capacity 250
  strategic energy reserves (SERs) 118
  sustainable energy system 507–22
    deep energy security 8, 507, 513–15, 517–22, 530
    public and private goods 515–16
Putin, Vladimir 27–8, 243, 255

Qatar 22, 75, 98, 99, 102, 252, 264
al-Quaida 57, 87

Radetzki, M. 301
Rahman, A. 321
Raphael, S. 73, 76
Rasmussen, M.V. 7
Ratner, M. 41
Ravindranath, N.H. 359
raw materials investment cycle 96, 104
Rayner, S. 339
rebound effect 415, 452
Redford, Alison 190
Reed, T. 58
refugees 73, 306, 324, 328, 329–31
renationalization/nationalization 103, 104, 210–12, 225, 264, 299, 512, 515
renewable energy sources (RES) 31, 143, 144, 169, 188, 189, 309, 313, 397
  biofuels *see separate entry*
  biomass *see separate entry*
  centralized and decentralized power production 486, 489–503
  costs 32–4, 54
  demand security 247, 248, 249
  developing world 33–4, 217–20, 223, 225
  environmental implications 219, 345–6, 354–66, 370–379, 411, 413, 485, 487, 492, 495, 501, 534–5
  ethical dimensions of 443–59

European Union 31, 32–3, 47–55, 61, 62, 65, 191, 196, 203, 310, 411, 413
  direct carbon fuel cell 142
  transport 137, 451
geopolitics 101, 105, 320
geothermal energy *see separate entry*
hydropower *see separate entry*
infrastructure and environmental change 397, 400
Japan 33, 199, 200
land use conflicts 334–7
Norway 186
producers of oil and gas 252–3
social un/acceptability 413–14, 415
solar energy *see separate entry*
taxation 248
uncompetitive options 467
United States 197, 413, 450, 451, 455, 494–9
wind power *see separate entry*
Republic of Benin 474
research agenda 169–70, 226
resilience perspective 156, 157
resource conflicts 70–76, 86–7, 92, 93, 100, 155, 324, 412, 512, 513
  Niger Delta 71, 72, 74, 76–86, 87
resource curse 22–3, 74, 513
resource nationalism 94, 103, 265, 274, 284, 291, 292, 301
Reuveny, R. 328, 330
Riahi, K. 147
Rickels, W. 337, 339
Rigobón, R. 210, 212
Ripley, C. 215
risks 4, 260, 271, 319–20, 323, 486, 491–2, 501
  assessment framework 154–69
  meta-risk 404–5, 407–8
  risk assessment 397–8
  *see also* diversification
Roberts, J. 99
Rockefeller, J.D. 71
Rogers, P. 306, 308
Romania 214, 217
Rosenberg, D. 223
Ross, M. 22–3, 73, 74
Rothschild, E. 6
Rowell, A. 78, 79, 80

Roxas, F. 215
Rubin, J. 301
Rudd, Kevin 30
Russia 3, 7, 22, 25, 27–8, 45, 72, 155, 258, 268
　China 211, 220
　climate change 304, 322, 324, 394, 395, 398
　demand security 243–6, 247, 248, 249–50, 251–2, 253, 254, 292
　Energy Charter Treaty (ECT) 107, 244–5, 251, 252, 254, 273, 274, 276–7, 292
　　break-down of Soviet Union 278–9
　　conciliation 279
　　Energy Charter Secretariat 277, 282
　　EU and Russian positions 289–91, 300
　　EU-Russian controversies 283–8
　energy strategy 174, 175–6, 177, 180–181, 186–8, 201–2, 203, 252, 253
　European Union 26, 27, 28, 41, 46, 92, 94, 99, 106–8, 273, 274
　　demand security 248, 250, 251–2, 254, 255
　　Energy Charter Treaty *see above*
　geopolitics 72, 92, 94, 98, 99, 100, 106–8, 109, 300, 398
　infrastructure and environmental change 394, 395, 398
　nuclear power 186, 290
　privatization 214
　resource nationalism 103
　strategic petroleum reserves (SPRs) 122, 129
　*see also* Soviet Union
Ruus, K. 56
Rwanda 122, 221, 472, 474, 476, 478–9
Ryu, D.H. 57

Sachs, J.D. 22, 214, 444
Sahagun, L. 497
Sala, O.E. 360
Sanchez, T. 467
Saro Wiwa, Ken 78–9, 80
Sathaye, J. 410

Saudi Arabia 21, 22, 29, 99, 250, 253, 301, 393
　Energy Charter Treaty (ECT) 277
　OPEC 263, 264–5
Savonis, M.J. 391
SCADA systems 56–9, 62, 64
Scafetta, N. 30
Scandinavia 247
　*see also* individual countries
Scheepers, M. 162, 164
Scheer, H. 443
Scheffran, J. 320, 321, 324, 325, 326, 329, 330, 332, 334, 335, 337
Schienke, E.W. 458
Schlesinger, J. 20
Schultz, C.L. 3
Schwartz, P. 304, 412
Scott, S.V. 304
Seliverstov, S. 274
Sen, A. 487
Senegal 306
Sennes, R. 212
Shadrina, E. 187
Sharma, A. 206
Shaw, T. 57, 58
Shea, J. 445
Shehadi, K.S. 214
Sheives, K. 223
Sheldon, F. 223
Shell 78–80, 82, 83–4, 85, 86, 108, 223, 262, 394
Shirmohammadi, D. 490
Shrader-Frechette, K.S. 456
Shtilkind, T. 244, 251, 252, 254
Shukman, D. 396
Shultz, J. 74
Sieg, L. 32
Silvestre, B. 215
Simmons, M. 99
Simpson, T.W. 360
Sinai, J. 222
Singapore 105, 129–30, 223, 393
Singer, C. 319
Slovakia 335
small-island developing states (SIDS) 304, 322
Smil, V. 11, 485, 510
Smith, Ben 72
smuggling 222–3
Sokolov, Y.A. 217

Sokona, Y. 462, 467, 470, 477
Solana, Javier 305
solar energy 33, 34, 49–51, 52, 60, 61, 65, 143, 145, 225, 336, 410
   Africa 219, 253, 463, 470, 472, 477
   Bangladesh 219
   Canada 188
   capital cost 477
   centralized vs DG system 483–4, 486, 490, 491, 492, 494–7, 498–9, 500, 501, 502
   China 218
   environmental implications 354–6, 370, 374, 379, 485, 487
   ethics 447, 449, 455–6, 457
   India 219, 309
   infrastructure and environmental change 397
   MSMP systems 138, 141–3
   non-rival 516
   Russia 186
   social acceptability 415
   United States 198, 202
solar radiation management (SRM) 337, 338
Sonatrach 253
Sorenson, H.C. 365
Sorrell, S. 415
South Africa 72, 211, 217, 219, 467, 470, 473, 474, 475–6
South America 101, 329
   *see also individual countries*
South Sudan 73–4
Sovacool, B.K. 11, 147, 150, 209, 427, 438
sovereignty over resources 281, 290
sovereignty perspective 155, 156, 157, 529
Soviet Union 265
   former (FSU) 276, 277, 278, 287, 288
Soyinka, W. 78, 79–80
Spain 59, 124, 333, 392, 473
Spero, J. 70, 73
stakeholders 64–5, 147–8, 149
Stanculescu, A. 216
Standard Oil 71, 262
Statoil 108
Stefanini, A. 59
Steinberg, F. 210

Steiner, A. 520, 521
Stern, N. 389
Stevens, P. 210
Stiglitz, J.E. 73, 516
Stirling, A. 153, 154, 220
Stokes, D. 25, 108
Strait, A.L. 117
strategic energy reserves (SERs) 117, 130
   coal 117, 126–7, 130
   controversies around release of 120, 127–9
   countries not holding 129–30
   gas: SGRs 117, 123–4, 125–6, 130
   history and evolution of 118–19
   petroleum: SPRs 117–23, 128, 129–30, 221–2, 225–6, 259
   rationale for and operation of 117–18
strategic protection reserves (SPRs) 23–4
Styles, G. 128
Sub-Saharan Africa 75–6, 329, 462, 470
   energy poverty 428, 429, 431–2, 433–4, 438, 439–40, 449, 463, 464–5, 466
   national energy strategies 206, 208, 213, 215, 218, 219
   *see also individual countries*
subsidiarity principle 46, 500, 519
Sudan 29, 73, 211, 220
Superwind 136, 140–142
supply chains 134–6, 144–5, 260, 287, 447, 448
supply, security of 531–3
   assessment framework, energy security *see separate entry*
   diversification of energy sources and carriers 133–45
   electricity supply in EU 47–55
   liberalised gas markets in EU 44–5, 46
   national energy strategies *see separate entry*
   Niger Delta 79
   SERs *see* strategic energy reserves
   timescales 136, 137, 143
sustainable energy system as public good 507–22

Swart, R. 321
Swaziland 474
Sweden 155
Switzerland 32, 286, 333
Syed, A. 33
Syria 73, 211, 223

Taiwan 31, 32, 106, 122
Tandon, S. 311
Tanzania 129, 464, 470, 474
Tänzler, D. 332
Tavits, M. 19
taxes 32, 33, 189, 191, 212, 219, 221, 226, 241, 471, 500
   import duties and 248, 249, 268–9, 270
   *Yukos* case 289
Taylor, I. 21
Taylor, J. 129
Tebekaemi, T. 77
technology transfer 271, 314, 334
terrorism 59, 120, 222, 223, 260, 297, 306, 329, 339
Thailand 122, 211, 217, 429
Thakkar, H. 390
Thavasi, V. 220
Thomson, E. 126
tidal power 188, 218, 336, 365–6, 397, 455, 516
Tilman, D. 359, 360
Togo 473
Tol, R. 325, 368
Toon, O.B. 334
Tordo, S. 210, 211
transparency 167, 201, 202, 242, 244, 255, 268, 290, 300, 302
   G8 175
   liberalised gas markets in EU 45
   SERs 121, 123
Treat, R. 57
Trebing, H.M. 215
tri-generation systems 136, 138, 139, 140–2, 144
Trombetta, M.J. 8, 24, 72, 321
Trudeau, N. 224
Trudeau, Pierre Elliot 189
Tuathail, G.Ó. 95
Tunisia 211
Turkey 31, 100, 107, 214, 335
Turkmenistan 22, 28, 98, 107, 279

Turner, F. 484
Turton, H. 152

Udo de Haes, H.A. 367
Uganda 122, 129, 221, 464, 470, 472
Ukraine 41, 45, 92, 106, 244, 278, 279, 285, 287–8, 290, 300, 310, 311
Ullman, R. 485
uncertainty 24, 30, 54, 96, 157, 302, 388, 488
   business: legal 253
   climate change 5, 307, 309, 323, 330, 368, 393, 408
   energy system: unprecedented 97–101
   liberalised gas markets in EU 43, 44
   OPEC: long-term strategy 270–1
UNESCO 445, 452, 458
United Arab Emirates 22, 75, 99, 264
United Kingdom 22, 27, 119, 121, 155, 184, 242, 322, 392, 396
   climate security 304
   energy poverty 424
   energy strategy 151
   Navy: from coal to oil 151, 258
   nuclear power 391
United Nations 75, 243, 303, 310, 509, 520–521
   Charter 277
   Development Program 207, 225, 403, 411, 412, 413, 508–9
   Environment Program 218, 311, 404
   Framework Convention on Climate Change 268, 298, 303, 304, 307, 309, 339
   Human Rights Commission 79
   Kyoto Protocol 268, 298, 313, 518
   Millennium Development Goals 207, 403, 466, 509, 522
   Security Council 282, 304, 309, 322
United States 105, 121, 212, 218, 225, 241, 258, 324, 414, 468, 487
   China 28, 94
   climate security 304, 403, 412
   'synergistic' policy: energy and 309–10, 311, 313–14
   DG system 491
   Energy Charter Treaty (ECT) 276, 290, 291

energy strategy 25, 174, 175–6, 192–3, 196–8, 201, 202, 203
Federal Energy Regulators Commission (FERC) 198
geopolitics 23, 71, 72, 92, 94–5, 102, 105–6, 107, 108–9, 398–9
infrastructure and environmental change 386–7, 393, 396, 397, 398–9
Kyoto Protocol 298
liberal democracy 21, 22, 23, 24–5, 27, 28, 29, 34
monopsonistic power 188–9
nuclear power 197–8, 203, 353, 391, 483, 488
oil *see* United States *under* oil
al-Quaida 57, 87
renewable energy sources (RES) 197, 413, 450, 451, 455, 494–9
resource conflicts 70, 71, 72, 73, 76, 79, 87, 92, 105–6, 155
Russia 27, 94
shale gas 93, 100, 110, 175, 198, 202, 300–301
shale oil 102, 110
strategic energy reserves (SERs) 120, 124, 128, 130
Upadhyaya, H. 390
USSR 58, 103
Uzbekistan 28

Valdes, C. 218
Van de Graaf, T. 244, 521
Van der Linde, C. 249, 252
van der Vleuten, E. 59, 64
van Vliet, M.T.H. 392
Vaughan-Williams, N. 6
Vazquez, I. 214
Venezuela 25, 27–8, 58, 103, 211, 212, 214, 220, 263, 277, 301
Victor, D.G. 175
Vietnam 106, 122, 221, 429
Vivoda, V. 210, 211
Voser, P. 175

Waelde, T. 273, 277, 281, 289
Wæver, O. 7, 23, 24, 320
Wahab, B. 223
Wake, H. 350
Walt, S. 6

Waltz, K. 273
Wanneau, K. 403
Warner, K. 327
Washington consensus 249
waste 339, 347, 349, 350, 457, 488
  computing technologies 486
  nuclear 32, 217, 333, 351, 352, 353, 374, 456, 488
  renewables 152, 219–20, 225, 357–8, 359, 360, 361, 365, 485
water 306, 323, 324, 327, 363, 386, 398, 404, 407, 412
  biofuels 449, 451
  biomass 360, 361
  desalination 354
  DG system 492
  geopolitics 398–9
  geothermal energy 357
  hydropower *see separate entry*
  nuclear energy 351, 352, 391–2, 407
  ocean acidification 306
  ocean energy 188, 218, 336, 365–6, 397, 455, 516
  off-shore/coastal production 392–4
  pollution 339, 347, 348–9, 350, 351, 357, 360, 361, 493, 500
  solar energy 356
  uncompetitive options 467
Watson 219
Watts, M.J. 73, 78
Wei, Y.M. 221
Weitzman, M.L. 308
Wertheim, P. 212
West, J. Robinson 29
Westphal, K. 3, 100, 107
White, G. 82
Wicks, M. 155
Widen, J. 51
Willems, A.R. 284
Williams, A. 33
Williams, J. 21
Williams, N. 499
Williams, R.I. 57
Willrich, M. 240
wind power 33, 49–51, 52, 60, 61, 101, 143, 145, 219, 225, 336, 410, 516
  Africa 219, 470, 477
  Canada 188
  capital cost 477

centralized vs DG system 490, 492, 494, 497–8, 500, 501, 502
China 218
  environmental implications 361–3, 485
  ethics 449, 453–4, 456, 457
  India 219
  infrastructure and environmental change 397
  MSMP systems 136, 138, 140–2
  Norway 186
  Russia 186
  social un/acceptability 313, 413, 415, 454–5, 497–8
  South Africa 219
  Thailand 219
Winner, L. 483, 491
Winzer, C. 13, 40, 147, 150, 154, 209
Wiser, R. 362, 363
Wolde-Rufael, Y. 470
Wolman, D. 395
women 427, 438, 439, 466, 472, 511
Wood, S. 26, 31, 32
World Bank 85, 218, 218–19, 225, 403, 408–9, 412, 465, 468, 475, 480
  coal fired power plant 467, 476
Wright, S. 329, 330
WTO 251
Wüstenhagen, R. 454

Xie, J. 219
Xu, Y. 207, 219

Yafimava, K. 273, 279, 288
Yan, P. 121
Yar'Adua, President 80
Yemen 122, 223
Yépez-García, R. 208
Yergin, D. 96, 118, 120, 151, 155, 222, 260, 319
Yi, L. 51
Youngs, R. 21, 24, 26, 29, 35, 300
Yu, T. 126
Yu, W. 222
Yueh, L. 303

Zambia 122, 129, 221, 470, 473, 476
Zangger, C. 32
Zanoyan, V. 260
Zhang, D.D. 325
Zhang, Y. 215
Zhang, Z. 151
Zhou, W. 220
Zimbabwe 473
Zimmermann, H. 26
Zissis, C. 212
Zoll, R. 336
Zweibel, K. 494